中国植物病理学会
2017年学术年会论文集

◎ 彭友良　李向东　主编

Proceedings of the Annual Meeting of Chinese Society for Plant Pathology (2017)

中国农业科学技术出版社

图书在版编目（CIP）数据

中国植物病理学会2017年学术年会论文集 / 彭友良，李向东主编 .—北京：中国农业科学技术出版社，2017. 7

ISBN 978-7-5116-3184-8

Ⅰ. ①中… Ⅱ. ①彭…②李… Ⅲ. ①植物病理学-学术会议-文集 Ⅳ. ①S432. 1-53

中国版本图书馆CIP数据核字（2017）第166220号

责任编辑 姚 欢 邹菊华
责任校对 李向荣

出 版 者 中国农业科学技术出版社
北京市中关村南大街12号 邮编：100081
电　　话 (010)82106636(发行部) (010)82106631(编辑室)
(010)82109703(读者服务部)
传　　真 (010) 82106631
网　　址 http://www.castp.cn
经 销 者 各地新华书店
印 刷 者 北京富泰印刷有限责任公司
开　　本 889 mm×1 194 mm 1/16
印　　张 33. 25
字　　数 1 000千字
版　　次 2017年7月第1版 2017年7月第1次印刷
定　　价 100. 00元

《中国植物病理学会2017年学术年会论文集》

编辑委员会

前　言

由中国植物病理学会主办、山东农业大学植物保护学院等单位承办的中国植物病理学会2017年学术年会将于2017年7月25—29日在山东省泰安市举行。会议主题为“植物病理学与农产品质量安全”，主要交流我国植物病理学科基础理论与实践应用方面的研究进展。

自2017年1月18日正式发布第一轮会议通知，国内外同仁踊跃报名与投稿，截至2017年6月10日，共收到论文和摘要460余篇。大会论文编辑组对所收到论文和摘要进行了分类整理和格式编辑，从论文和摘要的题目看，涵盖了植物病理学科的各个方面，尤其是真菌与真菌病害、病毒、病害防治及抗病性方面的研究成果颇丰，基本反映了我国植物病理学科的发展现状和研究水平。

本论文集收录的论文和摘要数量较多，多数论文和摘要集中在5月31日至6月10日间提交，编辑工作量大，时间紧迫，我们抽调了7名骨干教师参与了论文的分类整理与格式编辑工作，召开了4次论文编辑会议，本着尊重作者原意和文责自负的原则，对论文主体内容未做修改，只是按照会议论文及摘要的格式要求进行了统一整理与规范，保留了作者原有的写作风貌；在最后一次编辑会议通稿时对明显的文字拼写错误进行了修改。尽管力求严谨，由于时间仓促和经验不足，对于论文集的分类与格式的不妥之处，恳请作者和读者谅解。

大会筹办过程中，得到了中国植物病理学会、山东农业大学、青岛农业大学、山东省农业科学院植物保护研究所、山东省植物保护总站、山东植物病理学会、山东省科学技术协会等多家单位及众多专家的鼎力支持，在此表示由衷感谢！

最后，谨以此论文集祝贺中国植物病理学会2017年学术年会召开，预祝大会圆满成功！

编　者

2017年6月

目　录

第一部分　真菌

第二部分　卵菌

第三部分　病毒

第四部分　细菌

第五部分　线虫

第六部分　抗病性

第七部分 病害防治

第八部分　其他

第一部分　真　菌

黄芪根腐病菌侵染过程中几种细胞壁降解酶的变化*

岳换弟[1**]，秦雪梅[1]，王梦亮[2]，高　芬[2***]
（1. 山西大学中医药现代研究中心，太原　030006；
2. 山西大学应用化学研究所，太原　030006）

摘　要：细胞壁降解酶（CWDEs）是镰刀菌（*Fusarium*）侵染寄主使其致病的主要手段之一。评价病原菌 CWDEs 在致病过程中的作用，主要依据病原菌在体外产生降解酶的能力和降解酶破坏植物组织结构的能力。镰刀菌属真菌是黄芪根腐病的优势致病菌群。目前，对于其生化致病机制的研究还很薄弱，特别是 CWDEs 在致病过程中的作用还未见报道。

本研究以山西黄芪根腐病优势病原菌 *F. acuminatum*、*F. solani* 为研究对象，对其在侵染黄芪过程中产生的多聚半乳糖醛酸酶（PG）、果胶甲基半乳糖醛酸酶（PMG）、羧甲基纤维素酶（Cx）和 β-葡萄糖苷酶（βG）变化规律进行研究。试验采用离体回接法将 2 种致病菌菌块接种于健康黄芪根部，不同时间段（0d、4d、8d、12d、19d）采集发病组织，经典方法提取 CWDEs，DNS 比色法测定各酶活性。结果表明：随着接种天数的增加，黄芪的发病程度逐渐加重。*F. acuminatum* 接种后，4d 时黄芪根部即开始出现浅褐色病斑，病情指数为 41.67；随时间延长，病斑颜色逐渐加深，在 12~19d 病斑组织开始呈现纤维状腐烂，19d 时病情指数达 75.00。接种 *F. solani* 后，黄芪的发病情况与 *F. acuminatum* 侵染相似，病情指数在 4d 时为 47.50；19d 时为 73.61。在这一过程中，两种病菌侵染黄芪后均能产生 PG、PMG、Cx 和 βG，但不同病菌侵染导致的各酶活性变化规律和活性大小并不相同。*F. acuminatum* 侵染黄芪后，发病组织中的 CWDEs 活性顺序依次为 PG、PMG、Cx 和 βG，各酶的活性均随时间延长增强，PG 和 PMG 在 12d 达高峰后开始逐渐降低，而 Cx 和 βG 在第 19d 才达到最大酶活。*F. solani* 侵染后 CWDEs 活性的顺序与 *F. acuminatum* 侵染后存在差异，依次为 PMG、PG、Cx 和 βG，但活性变化趋势基本一致。由此可以看出，无论是哪种菌侵染，在致病过程中果胶酶总是先于纤维素酶作用于细胞壁。从各酶的活性大小来看，*F. solani* 侵染黄芪后产生的 PMG 和 βG 活性高于 *F. acuminatum* 侵染时的活性，分别为其的 1.30 倍和 1.13 倍；而 *F. acuminatum* 侵染产生的 PG 和 Cx 活性高于 *F. solani* 侵染时的活性，约为 1.18 倍和 1.39 倍。可见，不同根腐病菌在致病中过程中降解细胞壁成分的能力并不相同。*F. acuminatum* 通过 PG 水解果胶中的多聚半乳糖醛酸和通过 Cx 水解纤维素中的半纤维素基质多糖的能力强于 *F. solani*；而 *F. solani* 通过 PMG 水解果胶酸脂和通过 βG 降解纤维二糖的能力较强。上述各种 CWDEs 协同作用共同降解黄芪的细胞壁，从而使其致病。

本研究初步明确了这 2 种病原真菌致病过程中主要 CWDEs 的作用特点，但由于病原真菌对寄主细胞壁的降解机制非常复杂，各种 CWDEs 如何协调作用，还有待进一步深入研究。

关键词：黄芪根腐病；镰刀菌；细胞壁降解酶；DNS 分光光度法

* 基金项目：山西省中药现代化关键技术研究振东专项（No. 2014ZD0501-2）

** 第一作者：岳换弟，硕士研究生，主要从事根腐病菌与黄芪互作研究；E-mail：476848042@qq.com

*** 通信作者：高芬，副教授，主要从事药用植物病害和病原菌与植物互作研究；E-mail：gaofen@sxu.edu.cn

马铃薯黄萎病菌营养亲和型及致病力分化的研究

赵晓军，东保住，张 贵，张 键，张园园，周洪友，赵 君
（内蒙古农业大学农学院，呼和浩特 010018）

摘 要：马铃薯黄萎病是由大丽轮枝孢菌侵染引起的系统侵染的土传病害，严重的为害马铃薯的产量和品质。为了明确马铃薯黄萎病菌的营养亲和型和致病力分化，笔者对从不同地点的马铃薯地块中采集了 29 株大丽轮枝菌进行了营养亲和性，生理小种，交配型以及致病力分化的研究，结果表明供试的 29 株菌株被划分为 3 个营养亲和群即 VCG2B（11 份菌株）、VCG4B（3 份菌株）和 VCG4A（15 份菌株）。生理小种的鉴定结果表明所有供试的马铃薯菌株均为 2 号生理小种；交配型测定中发现所有供试的菌株均为 MAT1-2-1 交配型。采用灌根法接种进行致病力测定的结果表明不同的马铃薯黄萎病菌菌株间的致病力存在差异，且 VCG4B 型菌株的整体致病力表现为最高，其次为 VCB2B 型菌株，而 VCG4A 型菌株的致病力最弱。

关键词：马铃薯黄萎病；VCG；生理小种；交配型；致病力分化

同一地块向日葵核盘菌和小核盘菌生物学特性和致病力的比较

贾瑞芳[1]，李　敏[2]，张　键[1]，孟庆林[3]，周洪友[1]，赵　君[1]
（1. 内蒙古农业大学农学院，呼和浩特　010018；2. 内蒙古农牧业科学院植保所，呼和浩特　010031；3. 黑龙江省农业科学院植物保护研究所，哈尔滨　150086）

摘　要： 向日葵菌核病是世界性范围的真菌病害，在我国各向日葵产区均有发生。由于轮作倒茬的困难，使得向日葵菌核病在我国向日葵主产区呈现逐年加重的趋势，个别年份会造成部分地区向日葵的绝收。引起菌核病的核盘菌［*Sclerotinia sclerotiorum*（Lib.）de Bary］属于子囊菌亚门、盘菌纲、柔膜菌目、核盘菌科、核盘菌属，是具有非常广泛寄主范围的一种腐生营养型致病菌。目前，世界上至少400种植物包括许多重要的经济作物如大豆、油菜、向日葵等都能受到其不同程度的侵染，导致菌核病的发生。本研究以同一地块中采集的向日葵核盘菌和小核盘菌各10株菌株作为供试材料，对其生物学特性及致病力进行了比较研究，结果表明，培养3d后核盘菌平均直径达到8.34cm，而小核盘菌的平均直径仅为5.38cm。草酸含量和PG酶活力的测定结果表明核盘菌草酸含量的平均值为27.67μg/mg，PG酶活力平均值为26.87U/mg，而小核盘菌上述两个指标的平均值分别为18.91μg/mg和21.58U/mg。在菌核形成方面，小核盘菌不仅菌核形成速度明显快于核盘菌，而且其形成的菌核颗粒的数量明显多于核盘菌。而致病力的结果表明，接种2d后核盘菌在离体叶片上形成病斑的平均直径为3.11cm，显著高于接种相同时间点小核盘菌形成的病斑直径的平均值1.71cm。因此，除了菌核的形成时间和数量外，同一地块中向日葵核盘菌菌株生物学特性各项指标以及致病力均高于小核盘菌菌株。

关键词： 核盘菌；小核盘菌；同一地块；生物学特性；致病力；菌核的形成

内蒙古中部地区马铃薯根系及根际土 AMF 种群多样性分析

田永伟[1]，严志纯[2]，张　建[1]，赵晓军[1]，张之为[1]，郑红丽[1]，Ton Bisseling[2]，赵　君[1]
（1. 内蒙古农业大学农学院，呼和浩特　010018；
2. 瓦赫宁根大学植物科学系，瓦赫宁根，荷兰　6700AP）

摘　要：马铃薯是世界第四大粮食作物之一，同时又是内蒙古地区的主要粮食作物。本研究以内蒙古自治区包头市达茂旗大井村、乌兰察布市凉城县徐麻夭村和察哈尔右翼后旗红格尔图村采集的马铃薯根系及根际土壤样本，以及包头市达茂旗大井村马铃薯不同发育期和连作地块的根系及根际土壤样本为研究材料，构建 AMF 群体的基因文库。通过测序分析建立马铃薯根际和根系的 AMF 群落的系统发育树，从而明确不同供试地点、不同发育阶段以及连作对马铃薯根系丛枝菌的组成和分布的影响。3 个不同地点马铃薯土样中 AMF 群落聚类分析的结果表明乌兰察布市察哈尔右翼后旗红格尔图村马铃薯根系及根际土壤样本中 AMF 聚类的多样性最高，共有 9 个 AMF 类群；其次为达茂旗大井村样本，共有 8 个 AMF 类群；凉城县徐麻夭村样本的 AMF 丰富度最低，只有 6 个 AMF 类群。同时，不同地点马铃薯根际土样本中 AMF 的多样性均都高于根系群体。Rhizophagus group 只在根系中出现，并且在所有的 AMF 种类中所占比例最高；而 Glomus group B 和 Glomus group C 在 3 个不同地点根际土和根系样本中都有一定比例的分布。马铃薯不同发育阶段根系及根际土壤样本中 AMF 多样性分析结果表明：马铃薯苗期 AMF 群体种类最多，聚类结果共包括了 9 个 AMF 类群；其次为盛花期，共有 7 个 AMF 类群；块茎膨大期 AMF 丰富度最低，有 6 个 AMF 类群。和不同地点 AMF 群落的分析结果相似，同一地点马铃薯不同生育期根际土样中 AMF 多样性都高于根系，而且 Rhizophagus group 只在根系样本中分布，所占比例也最高。而 Glomus group B 和 Glomus group C 在根际土和根系样本中均具分布。通过研究连作对其马铃薯根系及根际土样本 AMF 多样性的影响，发现连作地块中根际土中 AMF 种群数量呈现下降的趋势，但对根系中 AMF 群落的组成并无明显影响。

关键词：马铃薯；丛植菌根；多样性

纹枯病菌致病力相关 miRNA 的鉴定及其对玉米靶基因的调控研究*

汪少丽[1]，储昭辉[2]，丁新华[1]，李晓明[1**]
（1. 山东省农业微生物重点实验室，山东农业大学植物保护学院，泰安　271018；
2. 山东农业大学农学院，泰安　271018）

摘　要：立枯丝核菌（*Rhizoctonia solani*）引起的纹枯病是影响玉米产量的主要真菌病害之一，对其分子致病机理的研究急需加强。miRNA 作为近年来发现的调控因子，在动植物、微生物中均具有重要的调控作用。本研究构建了玉米、正常培养的纹枯病菌和侵染玉米的纹枯病菌混合样本 3 个小 RNA 文库，通过小 RNA 高通量测序技术全面预测了纹枯病菌的 miRNA，共得到 2 790条保守 miRNA 和 126 条新预测的 miRNA 序列，利用荧光定量 PCR 方法验证了测序数据的可靠性。通过对不同组织材料中 miRNA 的表达量差异分析，得到了 2 023个与纹枯病菌致病力相关的 miRNA，其中 905 个 miRNA 是纹枯病菌在侵染玉米后诱导表达的。对这些 miRNA 的启动子顺式元件分析表明，这些 miRNA 的表达受到 ABA，GA，MeJA，MYB，真菌启动子、胁迫反应等因素的调控。应用生物信息学方法分析了纹枯病菌致病力相关 miRNA 在玉米中的靶基因，并进行功能注释，它们参与的具体生物学通路主要包括信号转导途径、抗氧化胁迫机制、转录调控途径、抗病相关代谢途径以及物质转运等途径，这与纹枯病菌侵染玉米过程中影响的相关路径是一致的。最后，分析了部分 miRNA 在纹枯病菌生长过程及侵染玉米过程中的表达量，选取纹枯病菌 Rs-miR263 和 Rs-miR421 的玉米靶基因突变体进行了功能验证，发现玉米突变体在接种纹枯病菌后病斑长度明显长于亲本材料，表明纹枯病菌可能通过 miRNA 调控玉米基因的表达增强了自身致病力。

关键词：纹枯病菌；miRNA；玉米靶基因；调控

* 基金项目：山东省自然科学基金青年项目（ZR2014CQ044）
** 通信作者：李晓明，讲师，主要从事分子植物病理学研究；E-mail：lxmlmh@163.com

Pathogenicity Differentiation of *Cochliobolus heterostrophus* in Fujian province*

Dai Yuli**, Gan Lin, Shi Niuniu, Ruan Hongchun,
Du Yixin, Chen Furu, Yang Xiujuan***

(*Fujian Key Laboratory for Monitoring and Integrated Management of Crop Pests, Institute of Plant Protection, Fujian Academy of Agricultural Sciences, Fuzhou, Fujian* 350013, *China*)

Abstract: In order to provide evidences for selecting resistant corn varieties to southern corn leaf blight, forecastingand controlling it in Fujian province, pathogenicity level of *Cochliobolus heterostrophus* isolates were determined. Two hundred and ninety-eight isolates of *C. heterostrophus* were isolated from 21 counties of 7 main corn production regions in Fujian province. The pathogenicity of these isolates on Shangpin, a susceptible corn variety, was assessed by spraying conidia suspension on corn leaves at the stage of seedling. The results indicated that all tested isolates had pathogenicity on corn variety Shangpin, with the disease index ranging from 13.50 to 80.95, and the average value of 49.49. According to the pathogenicity test results, *C. heterostrophus* isolates were roughly divided into weak, moderate and strong pathotypes, and the frequency of isolates in each pathotype was 29.87%, 63.42% and 6.71%, respevtively. The moderate pathotype was dominant. There were significant differences in pathogenicity of *C. heterostrophus* isolates from different regions. The isolates from fields of continuous corn cropping was significant higher than that of normal rotation cropping fields in Fuzhou region. The pathogenicity of different isolates exhibited abundantly diversity on different corn varieties, when eleven corn varieties were inoculated with 15 strong virulence isolates of *C. heterostrophus*. Among the tested 298 isolates, MH141 and JY142 had stronger pathogenicity; on the contrary, 'Rongtian 1' and 'Yongzhen 7' showed better resistance to *C. heterostrophus* among the tested 11 corn varieties. The pathogenicity has an obvious differentiation in isolates of *C. heterostrophus* collected from different regions in Fujian province. The differences in the pathogenicity of different isolates on the samevariety were observed. The pathogenicity of the same isolate on eleven different corn varieties was also obviously different.

Key words: Southern corn leaf blight; *Cochliobolus heterostrophus*; Pathogenicity differentiation; Fujian province

---

* 基金项目：福建省自然科学基金项目（2016J05073）；福建省农业科学院青年科技英才百人计划项目（YC2016-4）；福建省农业科学院博士启动基金项目（2015BS-4）；福建省属公益类科研院所专项（2014R1024-5）

** 第一作者：代玉立，安徽霍邱人，助理研究员，博士，研究方向：真菌学及植物真菌病害；E-mail：dai841225@126.com

*** 通信作者：杨秀娟，福建建瓯人，研究员，研究方向：植物病理学；E-mail：yxjzb@126.com

Development of a Rapid PCR-Based Method for Detection of Mating Types of *Cochliobolus heterostrophusand* Its Adaptability to Infected Corn Leaf Samples*

Dai Yuli**, Lin Gan, Ruan Hongchun, Shi Niuniu,
Du Yixin, Chen Furu, Yang Xiujuan***

(*Fujian Key Laboratory for Monitoring and Integrated Management of Crop Pests, Institute of Plant Protection, Fujian Academy of Agricultural Sciences, Fuzhou* 350013, *China*)

Abstract: A simple, rapid and sensitive method for detection of mating types of *Cochliobolus heterostrophus* in infected corn leaves using amolecular technique of PCR was established in this study. Two specific primers Ch01 − 1 and Ch02 − 1 were designed according to the nucleotide sequences of two different matingtypes of *C. heterostrophus*. This PCR-based assay can accurately differentiate two different matingtypes of *C. heterostrophus*. This sensitive PCR technique can detect 100 pg for MAT1−1 or 100 fg for MAT1−2 from total genomic DNA, five diseased spots of *C. heterostrophus* on infected corn leaf tissue or ten conidia of *C. heterostrophus* on corn leaf, indeed in the present of 20 ng total genomic DNA of *Zea mays*. This PCR assay was also successfully implemented for detection of mating types of *C. heterostrophus* in open field infections.

Key words: Detection; PCR; *Cochliobolus heterostrophus*; Matingtype; *Zea mays* L.

---

* 基金项目：福建省自然科学基金项目（2016J05073）；福建省农业科学院青年科技英才百人计划项目（YC2016-4）；福建省农业科学院博士启动基金项目（2015BS-4）；福建省属公益类科研院所专项（2014R1024-5）

** 第一作者：代玉立，安徽霍邱人，助理研究员，博士，研究方向：真菌学及植物真菌病害；E-mail：dai841225@126.com

*** 通信作者：杨秀娟，福建建瓯人，研究员，研究方向：植物病理学；E-mail：yxjzb@126.com

Baseline Sensitivity and Cross-Resistance of *Cochliobolus heterostrophus* to Fluazinamin Fujian Province[*]

Dai Yuli[**], Gan Lin, Ruan Hongchun, Shi Niuniu,
Du Yixin, Chen Furu, Yang Xiujuan[***]

(*Fujian Key Laboratory for Monitoring and Integrated Management of Crop Pests, Institute of Plant Protection, Fujian Academy of Agricultural Sciences, Fuzhou* 350013, *Fujian Province, China*)

Abstract: To determine the sensitivity of *Cochliobolus heterostrophus* to fluazinam and evaluate the-cross-resistance between fluazinam and other fungicides with different modes of action, a total of 73 *C. heterostrophus* isolates collected from seven regions in Fujian Province were testedby the method of measuring the mycelial growth on the fungicide-amended media. The results indicated that the tested isolates were very sensitivity to fluazinam, and the EC_{50} of these isolates for fluazinam was in the range of 0.002 3~1.286 3μg/mL, with the mean value of 0.257 3±0.254 0μg/mL. The frequency distribution was a continuous and unimodal curve, and followed the normal distribution. Hence, the mean EC_{50} valueof 0.257 3±0.254 0μg/mL could be severed as baseline sensitivity of *C. heterostrophus* to fluazinam in Fujian province. Cross-resistance analysis results showed that a strong positive relation was observed between fluazinam and iprodione (r=0.478 0, P<0.001). Nevertheless, there was no notable evidence of positive cross-resistance between the fluazinam and carbendazol (r=−0.241 0, P<0.061 3), fluazinam and chlorothalonil (r=−0.078 9, P<0.545 8), and fluazinam and propiconazole (r=0.045 8, P< 0.725 7). This research is expected to be instructive for guidance on reasonable selection of fungicides for effective management of southern corn leaf blight.

Key words: *Cochliobolus heterostrophus*; Fluazinam; Baseline sensitivity; Cross-resistance

---

* 基金项目：福建省自然科学基金项目（2016J05073）；福建省农业科学院青年科技英才百人计划项目（YC2016-4）；福建省农业科学院博士启动基金项目（2015BS-4）；福建省属公益类科研院所专项（2014R1024-5）

** 第一作者：代玉立，安徽霍邱人，助理研究员，博士，研究方向：真菌学及植物真菌病害；E-mail：dai841225@126.com

*** 通信作者：杨秀娟，福建建瓯人，研究员，研究方向：植物病理学；E-mail：yxjzb@126.com

玉米大斑病间时空模式初步分析

柳 慧，郭芳芳，王世维，吴波明

摘 要：玉米大斑病是影响玉米产量的主要叶部病害之一。过去几十年玉米大斑病在我国东北、华北北部和南方冷凉山区曾几度流行，一般年份减产20%左右，严重流行年份减产可达50%以上。对玉米大斑病的时空分析可以更好地理解病害的时空动态和病害发展的机制，对制定病害治理策略有重要作用。本研究在中国农业大学上庄实验站种植玉米大斑病易感品种先玉335，从中选取42行×84株的区域（行60cm，列30cm）定期调查每株玉米的大斑病发病情况，记录位置及病斑数。结果表明病情随时间增长的曲线符合逻辑斯蒂模型（$\ln(y/(1-y)) = 0.1645x-41.878$，其中d是时间，即一年中的天数，$R^2=0.9528$）。移动窗口平均方差法分析结果显示窗口大小是2~9时，各时期空间相关性指数都接近0，表明玉米大斑病的分布是随机分布，大田中没有明显的发病中心。Joint-counts方法检验的结果也表明实际发病与随机分布的假设相差不大，故玉米大斑病的分布是随机的。结果表明，2016年试验地中的玉米大斑病流行可能主要是由外来菌源引起的。这一结论还需要多年多点来验证，从而进一步了解玉米大斑病在大田中的发病历程。

关键词：玉米大斑病；时空；发病

稻瘟病菌生理小种与水稻抗病育种的研究进展*

张源明**，彭丹丹，舒灿伟，周而勋***
（华南农业大学农学院植物病理学系，广州 510642）

摘 要：由稻巨座壳［*Magnaporthe oryzae*（Hebert）Barr］［无性态：稻梨孢（*Pyricularia oryzae* Cav.）］引起的水稻稻瘟病是水稻三大病害之一。近年来，该病在中国、韩国、日本、越南和美国等水稻主产区发生与为害严重。迄今为止，全球许多实验室都致力于稻瘟病菌生理小种的鉴定及其在抗病育种上的应用研究。本文从以下几方面对稻瘟病菌生理小种鉴定及其在水稻抗病育种上的应用研究概况进行较为全面的综述，旨在为稻瘟病的相关研究提供一个较为全面的概貌：

（1）稻瘟病菌的分布及其为害：简述了这种由稻巨座壳引起的稻瘟病是世界性的、是对水稻造成危害最为严重的真菌性病害之一，每个水稻种植区几乎都会受到不同程度为害。

（2）稻瘟病菌的变异机制：叙述了影响稻瘟病菌产生变异的各种因素，普遍被接受的是以下 8 种原因：突变、菌丝融合、准性生殖、有性生殖、转座子、异核现象、寄主的定向选择、迁移。其中最主要的原因是：①突变；②准性生殖；③异核现象；④寄主的定向选择。

（3）稻瘟病菌的侵染机理：目前普遍接受的机理是稻瘟病菌附着孢形成假说，附着孢产生侵染栓穿透寄主组织的角质层和表皮细胞壁，继而在寄主细胞内生长并侵染临近表皮细胞及深入叶肉细胞，此假说主要适用于解释稻瘟病菌对水稻地上部分侵染。

（4）稻瘟病菌生理小种的鉴定及其命名：1976 年，我国自主确定了由特特普、珍龙 13、四丰 43、东农 363、关东 51、合江 18、丽江新团黑谷 7 个水稻品种作为中国稻瘟病菌生理小种鉴别品种，并根据全国稻瘟病菌生理小种联合试验组的命名方法对其生理小种进行命名。

（5）稻瘟病菌生理小种鉴定在育种方面的应用研究：根据基因对基因假说，水稻品种上有抗病基因，那么或迟或早都会出现一个新的小种或基因，使抗病基因失效。因此，及时并全面地了解稻瘟病菌生理小种对水稻育种的作用具有深远的意义。

（6）分子生物技术在抗病育种上的应用：只具有单一抗源的品种，容易因为稻瘟病菌的变异而导致失效。因此，通过分子技术将多个抗病基因聚合到一个品种上，培育出多抗品种，普遍认为是可行的。稻瘟病在全球大规模发生且造成严重的危害，本文较为深入地阐述了稻瘟病菌生理小种的变异及其侵染机理以及分子技术的应用等方面的研究进展，旨在为稻瘟病相关方面的研究提供一些有意义的参考资料。

关键词：稻瘟病；稻瘟病菌；生理小种；分子生物技术；抗病育种

---

* 基金项目：国家公益性行业（农业）科研专项资助项目（201403075）

** 第一作者：张源明，硕士生，研究方向：植物病原真菌及真菌病害；E-mail：740331752@ qq. com
*** 通信作者：周而勋，教授，博导，研究方向：植物病原真菌及真菌病害；E-mail：exzhou@ scau. edu. cn

海藻糖酶在水稻纹枯病菌菌核发育过程中的功能分析*

王陈骄子**，江绍锋，舒灿伟，周而勋***
（华南农业大学农学院植物病理学系，广州　510642）

摘　要：水稻纹枯病是我国发病面积最大的水稻真菌病害，引起巨大的经济损失。水稻纹枯病菌（*Rhizoctonia solani* AG1-IA）菌核在该病的病害循环中起着重要作用。据报道，在真菌中，活性氧可诱导菌核分化，而海藻糖作为一种活性氧清除剂，在生物体响应氧化胁迫过程中起着关键作用。在真核细胞中，海藻糖酶可以调控内源海藻糖水平，是真菌体内降解海藻糖的关键酶。然而，海藻糖代谢途径调控水稻纹枯病菌菌核发育的分子机制尚不清楚。本实验室在前期研究中采用 RNA-seq 技术，挖掘出在菌核发育过程中表达水平显著差异的海藻糖酶基因，本研究分别以水稻纹枯病菌菌核发育过程中 7 个阶段（培养 36h，48h，60h，72h，5d，7d，14d）的菌体为研究材料，测定了水稻纹枯病菌 GD-118 菌株菌核发育 7 个阶段海藻糖酶的活性与海藻糖含量的变化。为了进一步研究海藻糖酶的功能，笔者克隆了海藻糖酶基因并通过荧光定量 PCR（qRT-PCR）测定了海藻糖酶基因在菌核发育 7 个阶段的表达量。结果表明：海藻糖酶在水稻纹枯病菌培养 5d 后（即成熟菌核形成）时的活性最高，海藻糖含量在 5d 时最少；与此同时，qRT-PCR 结果显示，病菌培养 5d 时海藻糖酶基因的表达量也处于峰值，该结果表明海藻糖酶响应了 GD-118 菌株菌核的发育。另外，本研究发现在培养基中添加外源海藻糖后菌核量降低，DCHF-DA 荧光探针检测活性氧量明显减少。因此，本研究结果初步表明，海藻糖代谢途径参与调控水稻纹枯病菌菌核发育。本研究将为水稻纹枯病新型杀菌剂的研发提供新靶标，对菌核发育调控和水稻纹枯病的有效防治都将具有重要的理论和实践意义。

关键词：水稻纹枯病菌；海藻糖酶；菌核发育；荧光定量 PCR

* 基金项目：国家自然科学基金项目（31271994）
** 第一作者：王陈骄子，博士生，研究方向：植物病原真菌及真菌病害；E-mail：513425099@ qq. com
*** 通信作者：周而勋，教授，博导，研究方向：植物病原真菌及真菌病害；E-mail：exzhou@ scau. edu. cn

水稻纹枯病菌黑色素形成的转录组分析*

江绍锋**，王陈骄子，舒灿伟，周而勋***
（华南农业大学农学院植物病理学系，广州 510642）

摘　要：由立枯丝核菌（*Rhizoctonia solani* Kühn）引起的水稻纹枯病是世界范围内分布最为广泛的水稻三大病害之一。目前国内外的研究表明，黑色素的多少与立枯丝核菌抗逆性的大小和致病力的强弱成正相关关系，但人们对立枯丝核菌黑色素形成的分子机制所知甚少。为了从分子水平上揭示水稻纹枯病菌生长发育过程中黑色素形成的机制，本研究分别以不同处理条件下（空白处理 Rs＝褐色，50μg/mL 儿茶酚处理 RsC＝深褐色或黑色，300μg/mL 莨菪碱处理 RsH＝白色）培养 4d 的水稻纹枯病菌的菌体为研究材料，提取其总 RNA，以 Illumina Hi-seq 2000 平台技术进行高通量转录组测序，并进行生物信息学分析，旨在筛选黑色素形成的差异表达基因。研究结果表明：经过过滤低质量 Reads 后，对 Rs、RsC 和 RsH 进行了 Tophat2 分析，分别保留了 67 777 960对、66 965 558对和 57 937 624对 Reads，其中 64.21%、61.59%和 61.26%的 Reads 能准确比对到参考基因组上，说明 RNA-Seq 结果和参考基因组可靠。通过 DESeq 分析差异表达基因，结果发现当 Rs 与 RsC 相比较时，有 385 个基因显著上调，而显著下调的基因则有 208 个；当 Rs 与 RsH 相比较时，显著上调的基因 69 个，而显著下调的基因有 98 个；而 RsC 与 RsH 相比较时，上调的基因有 62 个，而下调的基因则为 50 个。GO 富集分析表明，DEGs 描述基因的分子功能有 14 种、所处的细胞组分 16 种和参与的生物过程有 73 种。KEGG 注释富集分析表明，有 3 769个基因富集在 42 条信号通路上，这些差异表达基因主要参与糖代谢、氨基酸代谢、蛋白质翻译、信号转导和疾病传播等生物代谢过程，而 KEGG 分析筛选得到的色素合成基因主要分布于酪氨酸代谢通路和黑色素合成代谢通路。将水稻纹枯病菌转录组拼接后建立本地数据库，根据色素合成基因序列比对出水稻纹枯病菌生长发育的 24 种同源基因序列，设计特异引物，通过荧光定量 PCR（qRT-PCR）方法分别对这 24 种基因的转录水平进行检测，qRT-PCR 分析结果表明，24 个色素相关基因的表达模式与 RNA-Seq 分析结果变化趋势一致。这些研究结果为进一步研究水稻纹枯病菌黑色素形成的分子机制奠定了理论基础，并为水稻纹枯病菌生长发育相关基因的结构及功能研究奠定了基础。

关键词：水稻纹枯病菌；黑色素；转录组；差异表达基因；荧光定量 PCR

---

* 基金项目：国家自然科学基金项目（31271994）

** 第一作者：江绍锋，博士生，研究方向：植物病原真菌及真菌病害；E-mail：jiangshaofeng98@foxmail.com

*** 通信作者：周而勋，教授，博导，研究方向：植物病原真菌及真菌病害；E-mail：exzhou@scau.edu.cn

水稻纹枯病菌与水稻互作基因的筛选和表达分析*

赵　美**，周而勋，舒灿伟***
（华南农业大学农学院植物病理学系，广州　510642）

摘　要：水稻纹枯病（rice sheath blight）是水稻的重要病害之一，对水稻的生产造成严重的影响。据最新的病害预测，2016 年我国水稻纹枯病菌发病面积达 2.6 亿亩，发病面积远高于稻瘟病。其病原菌立枯丝核菌（*Rhizoctonia solani*）是重要的土传真菌，除了为害水稻病害外，还可以侵染多达 260 种植物。由于缺乏可靠稳定的遗传转化系统，目前对该菌致病机理的研究还比较少。因此，本研究拟通过建立寄主诱导的基因沉默（host-induced gene silencing，HIGS）技术体系研究水稻纹枯病菌关键致病基因的功能。首先利用水稻纹枯病菌基因组数据库 RSIADB 中的基因序列进行生物信息学分析，包括基因筛选、功能富集（GO 和 KOG）和表达趋势分析（short time-series expression miner，STEM），筛选出水稻纹枯病菌与水稻互作的候选基因；再利用水稻纹枯病菌强致病力菌株 GD-118（其全基因组序列已经公布）活体接种水稻 9311 品种，分别在 6 个时间段（10h，18h，24h，32h，48h 和 72h）收集水稻叶片，利用荧光定量 PCR 技术验证、筛选基因的表达模式。最后选择在侵染初期高度上调表达的基因，构建 HIGS 表达载体，再通过根癌农杆菌介导的遗传转化（*Agrobacterium tumefaciens*-mediated transformation，ATMT）方法转化水稻，获得了转基因的水稻植株。根据 RSIADB 中的分泌蛋白、病原菌与寄主互作数据库（PHI）、碳水化合物活性酶数据库（CAZy）信息和转录组测序结果，利用 Venn 分析最后筛选出 15 个高 FPKM 值的候选基因，GO 分析发现这些候选基因主要参与水解酶活性、碳氧裂解酶活性和转移酶活性等生物学功能，KOG 分析发现这些基因与细胞壁生物合成、细胞骨架和碳水化合物转运和代谢有关，STEM 分析把这 15 个基因归类为 5 个不同的表达趋势。荧光定量 PCR 结果发现 5 个候选基因（AG1IA_ 00256，AG1IA_ 09356，AG1IA_ 06529，AG1IA_ 06159 和 AG1IA_ 03930）的表达量在侵染早期（10h 和 18h）高度上调，与转录组测序的结果基本一致，因此将其选为 HIGS 体系的候选基因。根据这 5 个候选基因的序列，利用 pBDL03 构建沉默载体后通过 ATMT 方法转化水稻，获得了稳定的转基因后代植株。本研究通过生物信息学和荧光定量 PCR 分析，获得了 5 个在侵染水稻早期上调表达的基因，建立了 HIGS 技术体系，并获得了转基因水稻植株，为进一步阐明这些基因的致病过程和机理奠定基础。

关键词：水稻纹枯病菌；互作基因；寄主诱导基因沉默；分泌蛋白；荧光定量 PCR

---

* 基金项目：广东省自然科学基金（博士启动）项目（1614050001896）；国家自然科学基金项目（31271994）
** 第一作者：赵美，硕士生，研究方向：植物病原真菌及真菌病害；E-mail：429605899@ qq. com
*** 通信作者：舒灿伟，副教授，研究方向：植物病原真菌及真菌病害；E-mail：shucanwei@ scau. edu. cn

Phylogenetic Study of *Valsa* Species from *Malus pumila* in Shanxi Province[*]

Yin Hui[**], Zhou Jianbo, Lv Hong, Chang Fangjuan,
Guo Wei, Zhang Zhibin, Qin Nan, Zhao Xiaojun[***]

(*Institute of Plant Protection, Shanxi Academy of Agricultural Sciences; Shanxi Key Laboratory of Integrated Pest Management in Agriculture, Taiyuan* 030031, *China*)

Abstract: To determine the pathogens caused apple *Valsa* canker from the apple planting areas in Shanxi province. Morphology, phylogeny and pathogenicity were integrated to evaluate the 78 isolates from 8 apple planting areas of Shanxi Province. The results showed that *Valsa* spp. from Shanxi Province belongs to *V. mali* and *V. pyri*. is the dominant species, located each apple planting areas in Shanxi Province, accounting for 58. 97% of the total number of strains. The isolation rate of *V. pyri* was 100% in Jinbei apple planting area (Xinzhou, Shuozhou) of Shanxi province. Morphological study indicated that colony color of *V. mali* divided into yellow-brown colony and off white colony on PDA. Off white colony produced pycnidium on PDA after 4 day, with the size of 358. 17 ~ 735. 68μm. Yellow-brown colony don't produced pycnidiumon PDA. The colony color of *V. pyri* was milk-white colonyon PDA, produced pycnidiumon PDA after 3 day, with the size of 817. 5 ~ 1 159. 7 μm. Pathogenicity test showed that *V. mali*and *V. pyri* could cause apple *Valsa* canker. The disease spots of *V. mali* expanded twigs of apple trees at the temperatures of 4℃ and relative humidity 45% for 7 day, with the size of 5. 8 cm. *V. mali* produced black pycnidium on twigs of apple trees after 15 day, with yellow spore horns. The disease spots of *V. pyri* slowly expanded twigs of apple trees, with the size of 2. 8 cm. *V. pyri* produced off white pycnidium on twigs of apple trees after 10 day.

Key words: Apple tree; *Valsa* species; Phylogenetic systematics; Morphology; Pathogenicity

---

* 基金项目：山西省重点研发计划项目（农业方面，201603D221013 - 3）；山西省应用基础研究计划青年基金项目（201601D202073）；山西省农业科学院优势课题组自选项目（YYS1716）

** 第一作者：殷辉，硕士，助理研究员，主要从事果蔬病害病原菌多样性研究；E-mail：yinhui0806@ 163. com

*** 通信作者：赵晓军，博士，研究员，主要从事果蔬病害病原学、病原菌抗药性及综合治理研究；E-mail：zhaoxiaojun0218@ 163. com

西瓜枯萎病菌 *Argonaute* 基因的克隆与功能分析*

Cloning and Functional Analysis of *Argonaute* Genes in *Fusarium oxysporum* f. sp. *niveum*

曾凡云**，丁兆建，漆艳香，彭　军，谢艺贤，张　欣***
（中国热带农业科学院环境与植物保护研究所，海口　571101）

摘　要：西瓜枯萎病（*Fusarium oxysporum* f. sp. *niveum*，Fon）是西瓜生产中危害最为严重的一种世界性土传病害。海南是我国主要冬春季商品西瓜生产基地之一，西瓜枯萎病菌的为害严重阻碍了海南西瓜种植产业的发展。*Argonaute* 基因在真核生物发生机制及干扰中发挥重要的不可缺的作用，动植物中该基因的研究相对成熟，而真菌中研究的较少，有相关报道表明 *Argonaute* 基因对尖孢镰刀菌的致病力并无太大影响，但是均能抑制其早期分生孢子的形成。为明确 *Argonaute* 基因与西瓜枯萎病菌分生孢子形成及致病力之间是否存在相关性，本实验以西瓜枯萎病菌基因组 DNA 为模板，分别克隆出 2 个 *Argonaute* 基因，分析其序列特征、预测蛋白的保守结构域，利用 Split-PCR 技术对 2 个 *Argonaute* 基因进行敲除，进而探讨西瓜枯萎病菌中 *Argonaute* 基因的功能。结果表明：*Fon-ago*1 基因 cds 全长序列为 3 201bp，编码 1 067个氨基酸；*Fon-ago*2 基因 cds 全长序列为 2 964bp，编码 988 个氨基酸；两个 Argonaute 蛋白具有典型的 DUF1785、PAZ、Piwi 保守结构域。目前已获得部分敲除突变体，后续的相关的基因功能研究正在进行之中。

关键词：西瓜枯萎病；*Argonaute* 基因功能；RNA 干扰

---

* 基金项目：国家自然科学基金项目（No. 31401685）

** 第一作者：曾凡云，助理研究员，从事热带果树病害防控技术研究；E-mail：fanyunzeng83@163. com

*** 通信作者：张欣，研究员，从事热带果树病害防控技术研究；E-mail：zhxppi@163. com

香蕉枯萎病菌三个同源 *CYP*51 基因的克隆与初步功能分析*

Cloning and Preliminary Functional Analysis of Three Paralogous *CYP*51 Genes in *Fusarium oxysporum* f. sp. *cubense*

刘远征**，漆艳香，曾凡云，丁兆建，彭　军***，谢艺贤，张　欣
（中国热带农业科学院环境与植物保护研究所，海口　571101）

摘　要：香蕉枯萎病（*Fusarium oxysporum* f. sp. *cubense*，Foc）是破坏香蕉维管束导致植株死亡的毁灭性土传病害，其病菌腐生能力很强，在土壤中可长期存活。唑类药剂是目前用于防治该病害的主要杀菌剂之一，其靶标为细胞色素 P450 甾醇 14α 脱甲基酶（CYP51），是甾醇生物合成中的关键酶。在镰刀菌属真菌中含有 3 个 *CYP*51 基因，根据基因进化聚类分析分为 3 类，分别为 *CYP51A*，*CYP51B* 和 *CYP51C*。为明确 3 个 *CYP*51 基因与唑类杀菌剂敏感性的关系，以及与香蕉枯萎病菌致病力之间是否存在相关性，本实验以香蕉枯萎病菌基因组 DNA 为模板，分别克隆出 3 个基因，分析其序列特征、推测蛋白的保守结构域，并构建了 pSilent-1 沉默载体，利用 RNA 干扰技术探讨香蕉枯萎病菌 3 个 *CYP*51 基因的功能。结果表明：*CYP51A* 基因 DNA 和 cDNA 全长序列分别为 1 574bp、1 521bp，编码 506 个氨基酸；*CYP51B* 基因 DNA 和 cDNA 全长序列分别为 1 761bp、1 584bp，编码 527 个氨基酸；*CYP51C* 基因 DNA 和 cDNA 全长序列分别为 1 659 bp、1 512bp，编码 503 个氨基酸。3 个 *CYP*51 基因在蛋白质水平存在 59. 60%的同源性，具有典型的 p450 保守结构域。该沉默载体已构建完成，为下一步获得沉默突变体以及分析其基因功能奠定良好基础。

关键词：香蕉枯萎病；*CYP*51 基因功能；RNA 干扰

---

* 基金项目：现代农业产业技术体系专项（No. CARS-32-04）；国家自然科学基金项目（No. 31471738，No. 31571957）
** 第一作者：刘远征，在读硕士研究生，研究方向为农业微生物学；E-mail：liuyuanzheng1234@ 126. com
*** 通信作者：彭军，副研究员，从事热带果树病害防控技术研究；E-mail：swaupj2@ 126. com

香蕉枯萎病菌厚垣孢子形成体系的建立及优化*

Establishment and Optimization of Chlamydospores Formation in *Fusarium oxysporum* f. sp. *cubense*

丁兆建**，漆艳香，曾凡云，彭　军，谢艺贤，张　欣***

（中国热带农业科学院环境与植物保护研究所，海口　571101）

摘　要：香蕉枯萎病菌（*Fusarium oxysporum* f. sp. *cubense*）是威胁我国乃至世界香蕉（*Musa* spp.）生产的重要植物土传病原真菌，其厚垣孢子可在土壤中存活至 30 年，是轮作防控香蕉枯萎病的主要障碍，故所引起的香蕉枯萎病的防控难度极大。迄今，香蕉枯萎病菌厚垣孢子形成机制的研究尚未见报道，为解析该病原菌厚垣孢子的形成机制，最终为香蕉枯萎病的综合防控提供理论指导和技术支持，笔者建立及优化了厚垣孢子的诱导形成体系，研究发现该病原菌菌丝在含有七水硫酸镁（$7H_2O \cdot MgSO_4$）和磷酸二氢钾（KH_2PO_4）两种盐溶液体系中，添加作为碳源的葡萄糖（Glucose）或碳酸镁（$MgCO_3$）浓度为 1 mg/L 时，在 28℃ 黑暗静止条件下培养 4d，诱导几乎所有的菌丝顶端和部分菌丝中间产生了圆形和椭圆形的厚垣孢子，葡萄糖或碳酸镁的浓度增大或降低均导致厚垣孢子产量的减少，这说明碳源在该病原菌厚垣孢子的形成过程中起着开关作用；此外，碳酸镁作为碳源优于葡萄糖作为碳源时产生厚垣孢子的数量。该诱导体系的建立为香蕉枯萎病菌厚垣孢子形成机制的研究提供了原材料，相关后续工作正在进行之中。

关键词：香蕉枯萎病菌；厚垣孢子；形成条件

---

* 基金项目：现代农业产业技术体系专项（No. CARS－32－04）；中央级公益性科研院所基本科研业务费专项（No. 2015hzs1J013）；海南省自然科学基金面上项目（No. 20163103）

** 第一作者：丁兆建，助理研究员，从事香蕉真菌病害的致病机理研究；E-mail：dingzhaojian@163. com

*** 通信作者：张欣，研究员，从事热带果树病害防控技术研究；E-mail：zhxppi@163. com

Rab GTPase 家族蛋白 Sec 4 调控大丽轮枝菌致病性功能分析*

黄彩敏**，宋爽爽，刘 燕，孙维霞，田 李***
（曲阜师范大学生命科学学院，曲阜 273165）

摘 要：大丽轮枝菌是引起多种农作物黄萎病的土传性病原真菌，严重威胁农业生产安全。胞外分泌蛋白调控在病原侵染寄主过程中扮演重要角色，其中 Rab GTPase 家族蛋白 Sec 4 是蛋白分泌晚期的重要调控因子，参与了蛋白的膜泡运输，但在大丽轮枝菌中一直未见报道。本研究克隆了大丽轮枝菌 Rab GTPase 家族蛋白 Sec 4（*VdSec* 4），基因全长 850bp；利用同源重组方法构建了 *VdSec*4 基因敲除突变体，分别以细胞壁组分（纤维素、果胶）为单一碳源鉴定表明，突变体菌株较野生型的生长能力延缓，暗示突变体果胶酶、纤维素酶等胞外蛋白的分泌能力受限；致病力检测表明，突变体菌株对寄主棉花的致病力显著下降，菌体生物繁殖量仅为野生型菌株43%；进一步利用蛋白定量技术（iTRAQ）比较了突变体菌株和野生型菌株在细胞壁组分诱导培养基下胞外蛋白谱，发现 332 个蛋白因 VdSec 4 缺失无法分泌。上述结果说明，大丽轮枝菌 VdSec 4 介导了胞外蛋白的分析，进而调控了植物细胞壁降解能力和对寄主植物的致病功能。

关键词：大丽轮枝菌；Sec 4 蛋白；致病性；蛋白定量

---

* 基金项目：国家自然科学基金（31501588）

** 第一作者：黄彩敏，硕士生，主要从事植物病原真菌的致病机理研究；E-mail：874362206@ qq. com

*** 通信作者：田李，副教授，主要从事植物病原真菌的致病机理研究；E-mail：tianlister@ 163. com

油菜抗感品种接种根肿菌防御酶活性及转录组分析*

郭　珍**，陈国康***，马冠华，张　炜，张苏芸
（西南大学植物保护学院，重庆　400716）

摘　要：由芸薹根肿菌（*Plasmodiophora brassicae* Woron.）引起的十字花科根肿病是世界上重要的土传病害之一，严重影响十字花科油料和蔬菜作物的产量及品质。我国十字花科根肿病最早在台湾发现，现已蔓延到华南、西南、华北及东北等地。根肿病作为典型的土传病害，其休眠孢子甚至可在土壤中存活20年以上，导致传统的病害防治手段收效甚微，而选育和利用抗病品种，则是目前十字花科根肿病防控较为有效的途径。深入探索十字花科寄主的抗病机制，既是抗病品种筛选的基础，又可为寄主品种的抗性评价提供依据。

为了探究抗病和感病两种油菜品种抗病机理及基因表达差异，首先通过水培法对感病品种（中双11号）和抗病品种（6M80）接种根肿菌，然后利用紫外分光光度法和Seq-RNA技术分别测定根部防御酶的活性及根部所有的转录序列。结果表明：接种根肿菌12d后抗病品种中CAT（Catalase）、SOD（Dismutases）、POD（Peroxidase）的活性达到峰值并且显著高于感病品种，分别为328.5U/（g·min），2 044.44U/（g·h）和55.70U/（g·min）。中双11号和6M80分别接种根肿菌0d、3d、6d、9d、12d后相比较，在比对到参考基因组唯一位置的229 408 692个基因中，有10 080个差异表达基因（Different expression genes，DEGs），其中有4 494个DEGs表达上调，5 090个表达下调。将DEGs进行功能注释，表明大多数DEGs参与信号转导、生物代谢、转录及防御机制。在抗病和感病品种之间存在121与抗性相关的DEGs，这些DEGs与病菌相关分子模式（Pathogen-associated molecular patterns，PAMPs）、钙离子流编码基因、转录因子、激素信号、防御酶及细胞壁修饰相关。在抗病品种中尽管POD相关的基因下调，但是CAT和SOD相关的基因上调，并且转录因子、激素信号传导以及细胞壁修饰相关基因表达与感病品种之间存在差异。

本研究探索油菜抗感品种接种根肿菌后相关生化指标表型上的差异及了解转录水平上油菜抗病品种与感病品种抗根肿病之间的差异，为进一步了解根肿菌侵染油菜后抗性基因的分子代谢提供了依据。

关键词：根肿菌；油菜；抗性；防御酶；转录组

* 基金项目：重庆市“十三五”社会事业与民生保障科技创新专项资金资助（cstc2015shms-ztzx0129；cstc2015shms-ztzx80009）；重庆市农委“榨菜现代产业技术体系2015—2020”

** 第一作者：郭珍，硕士研究生，植物病理学；E-mail：guozhenguorui@163.com

*** 通信作者：陈国康，副教授，植物病理学；E-mail：chenguokang@.swu.edu.cn

海南菠萝一种叶腐病的调查及病原初步鉴定*

罗志文[1**]，范鸿雁[1]，何　凡[1]，陈业光[1]，郭利军[1]，胡福初[1]，
余乃通[2]，韩　冰[1]，刘志昕[2]，胡加谊[1,3]，李向宏[***]

（1. 海南省农业科学院热带果树研究所/农业部海口热带果树科学观测实验站/海南省热带果树生物学重点实验室/海南省热带果树育种工程技术研究中心，海口　571100；
2. 中国热带农业科学院热带生物技术研究所/农业部热带作物生物学与遗传资源利用重点实验室，海口　571101；3. 黄埔出入境检验检疫局，广州　510730）

摘　要：笔者在菠萝病虫害调研中，于 2014 年在海南澄迈的果园内发现了一种叶腐病。该病多于多雨季节排水不畅的园地发病，以近地面叶片或接触土壤的叶片受害较多，带伤口叶片易受害，多从叶尖侵入，发病速度快，害病后 2d 即造成全叶腐烂，叶背常产生灰白色霉层。该病虽在田间偶然发生，但致死率高，危害较大。笔者对病样进行了组织分离，通过柯赫氏法则验证，确定了病原菌株，其形态学特征主要为：在 PDA 上菌落呈絮状，粉白色、浅粉色或肉色，菌落背面略显紫色，菌丝白色质密。病菌常产生小型和大型分生孢子，小型分生孢子着生于单生瓶梗上，无色，单胞，卵圆形或肾形，大小为（4.8~11.8）μm×（2.0~3.7）μm；大型分生孢子镰刀形，稍弯曲，多为 3 隔 4 胞，两端细胞稍尖，大小为（19.6~39.4）μm×（3.5~5.0）μm。根据病原菌培养性状及形态学特征，结合相关资料，将该病原菌初步鉴定为半知菌类（Imperfecti fungi）、从梗孢目（Moniliales）、瘤座孢科（Tuberculariaceae）、镰刀菌属（*Fusarium*）的尖孢镰刀菌（*Fusarium oxysporum* Schl.）。

关键词：海南菠萝；叶腐病；病原鉴定

---

* 基金项目：公益性行业（农业）科研专项经费项目（201203021）；海南省重点研发计划项目（ZDYF2016035）；海南省农业科学院农业科技创新专项经费项目（CXZX201410）

** 第一作者：罗志文，男，硕士，助理研究员，研究方向：果树植物病理学；E-mail：zhiwenluo@163.com

*** 通信作者：李向宏；E-mail：lxh.0898@163.com

重庆市大棚番茄移栽后茎枯病病原鉴定与病害田间调查*

叶思涵[1]**，董　鹏[2]，孙现超[1]***

（1. 西南大学植物保护学院，重庆　400715；2. 重庆市农业技术推广总站，重庆　400020）

摘　要：番茄是茄科番茄属一年生或多年生草本植物，由于产量高、供应季节长，用途非常广泛，因而在我国南北方广泛栽培。由于目前番茄大多采用温室设施种植，而温室内的小气候环境一般湿度较高，且冬春季节平均温度也高于室外，很容易滋生病害，其中灰葡萄孢菌（*Botrytis cinerea*）引发的灰霉病就是一种较为常见的病害。

重庆市地处四川盆地东南部，属亚热带季风气候，立体性强，雨量充沛，适宜各种蔬菜的周年生产栽培。但该地区温室种植的番茄常年都有灰霉病发生，对番茄茎、叶、花、果均可为害，但主要为害果实，通常以青果发病较重。极大影响了番茄产量，严重时减产达20%~30%。

重庆市潼南农业科技园区番茄嫁接苗于2月底进行移栽，3月中旬大棚内出现番茄嫁接部上边茎枯死。在发病初期，茎部和叶部呈水渍状且有不规则病斑，但病部尚未萌发灰霉菌子实体。笔者从棚内采集了疑似感染灰霉病的番茄幼苗，在防虫网室内进行了保湿培养，7d后幼苗茎部感病处生出灰褐色霉层，用挑针挑取菌丝制片进行镜检，经查对有关资料，根据田间症状及形态学初步鉴定该病害的病原为灰葡萄孢菌（*Botrytis cinerea*）。通过显微镜观察，该病原为分生孢子梗丛生，有隔，褐色，上端有分枝，梗顶膨大呈棒头状，其上密生小柄，柄上着生分生孢子。通过显微测量测得，分生孢子梗长短不一，大致为（1 429.3~3 207.0）μm×（15.5~25.0）μm；分生孢子呈圆形、椭圆形，单细胞，无色，大小为（4.85~12.50）μm×10.5μm。随后笔者将该菌置于PDA培养基进行菌株的分离纯化，将扩增纯化后的单克隆菌落进行ITS序列的DNA测序，测序结果在NCBI数据库中进行BLAST同源性比较，通过比对发现该菌株与GenBank中已知的灰葡萄孢菌（*Botrytis cinerea*，登录号KX766413.1）ITS序列高度同源，同源性为100%。

采用对角线五点取样法进行取样，每个方位选取一定株数，每点20株，共100株，对发病株数和病害级数进行记录，最后将记录的数据进行统计分析，计算发病率、病情指数。调查结果显示，3月中旬番茄灰霉病平均发病率为56%，病情指数为25。在随后几周，通过适当延长下午放风时间，避免阴雨天浇水、控制浇水量，及时摘除病叶病果病枝，该病得到了有效控制。

关键词：茎枯病；病原鉴定；田间调查

---

* 基金项目：重庆市社会事业与民生保障创新专项（cstc2015shms-ztzx80011，cstc2015shms-ztzx80012）；中央高校基本科研业务费专项资金资助项目（XDJK2016A009，XDJK2017C015）

** 第一作者：叶思涵，女，硕士研究生，从事植物病理学研究

*** 通信作者：孙现超，博士，研究员，博士生导师，主要从事植物病毒学及植物病害生物防治研究；E-mail：sunxianchao@163.com

网斑病菌感染花生叶片组织的转录组分析*

张　霞**，许曼琳，吴菊香，陈殿绪，迟玉成***
（山东省花生研究所，青岛　266100）

摘　要：花生网斑病是我国北方花生产区最重要的一种叶部病害，目前主要从形态结构和生理生化方面探讨花生对网斑病的防御作用，而分子水平的研究则鲜有报道。笔者课题组在完成花生网斑病菌全基因组测序的基础上，应用高通量 RNA-Seq 测序技术，从转录组学水平上，完成网斑病菌侵染花生过程中差异表达基因的分析，筛选出响应网斑病菌的候选基因。笔者研究发现，无论是以野生型花生 *Arachis duranensis* 的基因组为参考基因组，还是以野生型花生 *A. ipaensis* 的基因组为参考基因组，接种 7d 后即可出现大量差异表达基因；同不接种网斑病菌的花生叶片相比，差异表达基因的数量呈现出先增加后降低的趋势。后续还会开展候选基因的功能研究，揭示花生响应网斑病菌胁迫的分子机制，为花生病害的综合防治提供一定的技术支撑。

关键词：花生网斑病；RNA-Seq；差异表达基因；功能研究

---

* 基金项目：科技重大专项（2060901）——花生主要土传病害病原鉴定及绿色防控技术研究
** 第一作者：张霞，博士，助理研究员，植物病理学；E-mail：zhangxia2259@126.com
*** 通信作者：迟玉成，博士，副研究员，植物病理学；E-mail：87626681@163.com

花生 *HyPGIP* 原核表达载体的构建及蛋白表达活性鉴定*

王麒然[1**]，王　琰[2]，王志奎[3]，张茹琴[1]，迟玉成[4]，夏淑春[1]，鄢洪海[1***]
（1. 青岛农业大学农学与植物保护学院，青岛　266109；2. 山东省乳山市农业局，青岛　264500；3. 青岛市即墨市农业局，青岛　266200；4. 山东省花生研究所，青岛　266101）

摘　要：依据已知多聚半乳糖醛酸酶抑制蛋白 *PGIP* 基因序列设计 *HyPGIP* 的扩增引物，以 *HyPGIP* 质粒为模板，扩增 *HyPGIP* 基因片段，通过 10g/L 琼脂糖凝胶电泳检测 PCR 扩增产物。回收 PCR 扩增产物，与 pGEM-T Easy 载体连接，转化大肠杆菌感受态细胞 DH5α。经 PCR 检测后送测序，提取质粒，利用 *BamH* I 和 *Sac* I 进行酶切，回收目的片段，同时酶切 pET28 质粒，利用 T4-DNA 连接酶连接目的片段和载体，转化 BL21 大肠杆菌感受态细胞，进行原核表达分析，通过上述方法成功构建了表达载体 HyPGIP-pET28，转化 BL21 表达菌株，阳性菌株质粒 PCR 获得了 951bp 大小的去除信号肽的目的蛋白的基因序列。

挑取单菌斑接种于含有相应抗生素的 LB 液体培养基中，于 37℃振荡培养过夜。次日，将菌液按 1∶100 转接到含相应抗生素液体培养基中，37℃震荡培养至 OD_{600} 为 0.3，加入 IPTG 至终浓度 0.3 mmol/L，诱导目标蛋白表达。继续于摇床培养 3h。10 000r/min 离心 5min，收集已诱导的细胞，取上清液和沉淀物分别加入等体积的 2×SDS 上样缓冲液，充分悬浮菌体，沸水中煮沸变性 5min，冰浴 2min，SDS-PAGE 电泳检测。结果发现含有重组质粒 HyPGIP-pET28 的 BL21 菌株，经过 0.1mmol/L IPTG，18℃，180r/min 12h 诱导之后，在 23ku、40ku 中间处出现一条明显诱导带，与预期的 37ku 去除信号肽后的大小一致。

将菌液按 1∶100 的比例转接于培养基中，37℃摇菌至 OD_{600} 约为 0.3 h，加入终浓度为 0.3mmol/L 的 IPTG，28℃ 200r/min 培养 3h，诱导目的蛋白表达。4℃，4 000r/min 离心 20min，去上清，沉淀重悬于 20mL 裂解缓冲液（20mmol/L Tris-HCl，150mmol/L NaCl，pH8.0），然后进行超声破碎，4℃，10 000r/min 离心 20min，取上清加入 1mL 50%的 Ni-NTA 树脂，4℃轻微振荡 30min。将结合有目的蛋白的 Ni-NTA 树脂装柱。用不同浓度的咪唑漂洗液（20mmol/L、25mmol/L、30mmol/L）漂洗 3 次。用 250mmol/L 咪唑溶液洗脱目的蛋白。His 标签蛋白纯化-亲和层析处理后，只有经 IPTG 诱导和沉淀中出现明显条带，说明目标蛋白以不可溶的方式的包涵体存在。

关键词：花生多聚半乳糖醛酸酶抑制蛋白（*HyPGIP*）；原核表达载体构建；IPTG 诱导；蛋白鉴定

---

* 基金项目：山东省自然科学基金项目（ZR2011CL005）；山东省“泰山学者”建设工程专项经费资助（BS2009NY040）

** 第一作者：王麒然，男，山东潍坊人，硕士研究生，主要从事分子植物病理学研究；E-mail：814532720@qq.com

*** 通信作者：鄢洪海，男，吉林长春人，教授，博士，主要从事植物病理学与玉米病害研究；E-mail：hhyan@qau.edu.cn

我国冬小麦主产区茎基腐亚洲镰孢菌群体遗传多样性研究*

赵 芹**，邓渊钰，李 伟，孙海燕，陈怀谷***
（江苏省农业科学院植物保护研究所，南京 210014）

摘 要：亚洲镰孢菌（*Fusarium asiaticum*）是我国冬小麦主产区茎基腐的主要病原菌，本研究应用筛选的 15 条简单序列重复区间（Inter-simple sequence repeat，ISSR）分子标记对我国冬小麦主产区 6 个省、直辖市的 128 份亚洲镰孢菌标样进行遗传多样性研究，利用 NTSYSpc version 2.11 进行菌株遗传相似性与聚类分析，基于 POPGENE version l.32 软件计算各项遗传多样性参数与群体聚类。本研究共获得 229 个稳定 DNA 片段，其中 205 个呈多态性，多态性比例为 89.50%，平均每个引物扩增 15.27 个条带，扩增产物集中在 150~2 000bp，菌株相似系数变化范围 0.66~0.93，相似系数为 0.75 时，将供试菌株分为 4 大类，各地区菌株遗传相似性较高，河北省菌株遗传多样性较为丰富。6 个地理群体多态性位点数在 40~186，平均为 122，多态位点百分率在 17.47%~81.22%，平均为 53.35%，有效等位基因数（Ne）在 1.123 5~1.521 9，平均为 1.345 0，Shannon's 信息指数（I）在 0.105 6~0.433 9，平均为 0.292 5；在地理种群水平上的 Nei's 基因多样性指数（H）为 0.072 4~0.295 5，平均为 0.198 1，表明不同地理群体间存在一定遗传差异。河北与四川地理种群的 Shannon's 信息指数和 Nei's 基因多样性指数分别为最高与最低，说明这两个地区菌株间遗传多样性最丰富与最低。遗传相似性分析证明，江苏省亚洲镰孢种群和河北省种群最近，四川省种群和重庆市种群最远。6 个地理群体间的遗传分化系数 Gst 为 0.184 4，群体内为 0.874 9，群体内多样性明显高于群体间多样性。6 个地理群体间的居群每代迁移数（Nm）为 2.211 9，说明不同地理群体的亚洲镰孢菌株间存在较大基因流动。6 个地理群体总基因多样度 Ht 为 0.242 9，各地理群体内基因多样度 Hs 为 0.188 4，地理群体间基因多样度 Dst 为 0.198 1，说明相同地理来源的菌株具有较近亲缘关系。聚类分析将 6 个地理群体划分为 3 个不同类群，第Ⅰ类群包括江苏、河北、安徽与河南，重庆种群属于第Ⅱ类群，四川省种群为第Ⅲ类群。本研究为小麦茎基腐病抗病育种以及综合防治提供了重要理论参考，并为亚洲镰孢菌分类鉴定及系统发育研究提供依据。

关键词：小麦茎基腐；亚洲镰孢菌；简单序列重复区间；遗传多样性

* 基金项目：国家公益性行业（农业）项目（201503112）
** 第一作者：赵芹；E-mail：zhaoqin0802@126.com
*** 通信作者：陈怀谷；E-mail：huaigu@jaas.cn

Cloning and Expression Analysis of *CqWRKY*1 Gene Induced by Fusaric Acid in *chieh-qua* [*Benincasin hispida* (Thunb.) Cogn. var. *chieh-qua*][*]

Zhao Qin[1**], Zhang Jinping[1,2**], Ma Sanmei[2], He Xiaoming[1], Xie Dasen[1]

(1. *Vegetable Research Institute*, *Guangdong Academy of Agricultural Sciences*, *Guangzhou* 510640, *China*; 2. *Department of Biotechnology*, *Jinan University*, *Guangzhou* 510632, *China*)

Abstract: Based on ESTs of SSH library from interaction analysis between chieh-qua and *Fusarium oxysporum* f. sp. *chieh-qua*, full-length cDNA sequence of a WRKY transcription factor (*CqWRKY*1) was cloned from ‘A39FA’ cultivar treated with fusaric acid by RACE. The GenBank accession number was KX702278. Sequence analysis showed that cDNA sequence length of *CqWRKY*1 was 2 088bp with one open reading frame of 1 746 bp, encoding 581 amino acids. The deduced amino acids sequence contained two typical conserved domains of WRKY transcription factors and one Zinc finger motif of C_2H_2 ($C\text{-}X_4\text{-}C\text{-}X_{22-23}\text{-}H\text{-}X_1\text{-}H$), which was classified into Class I WRKY transcription factor. Tertiary structure of itsencoding protein included four antiparallel β sheets, three random coils and a zinc-binding pocket at theend of β sheet. CqWRKY1 transcription factor was predicted to be localized in nucleus with two nuclear localization signals of PTKKKVE and PEAKRWR, and possessed transcription, transcription regulation functions. Amino acids sequence alignment and phylogenetic analysis indicated that CqWRKY1 shared high similarities with WRKY transcription factors from other plants, in which the closest evolutionary relationship was shared with WRKYP2 from *Cucumis sativus* with homology of 92%. The expression patterns of *CqWRKY*1 in different tissues and treatments were analyzed by real-time quantitative PCR. The results showed that *CqWRKY*1 expression in chieh-qua exhibited tissue specificity, and the highest level was in roots, followed by stem and leaves. Expression level of *CqWRKY*1was strongly induced under fusaric acid stress, signal molecules salicylic acid (SA) and methyl jasmonate (MeJA), suggesting that *CqWRKY*1 may involved in chieh-qua defense response of fusarium wilt disease, and its regulation may depend on signal pathways mediated by SA or jasmonic acid (JA).

Key words: Chieh-qua; WRKY; Gene cloning; Expression analysis; Real-time quantitative PCR

---

* Funding: Natural Science Foundation of Guangdong Province (1045106400100606)

** First authors: Zhao Qin and Zhang Jinping; E-mail: zhaoqin0802@126.com

棉花枯萎病菌新生理型菌株的分子检测*

郭庆港**，王培培，鹿秀云，董丽红，赵卫松，张晓云，李社增，马　平***
（河北省农林科学院植物保护研究所，河北省农业有害生物综合防治工程技术研究中心，
农业部华北北部作物有害生物综合治理重点实验室，保定　071000）

摘　要：由尖孢镰刀菌萎蔫专化型（*Fusarium oxysporum* f. sp. *vasinfectum*）引起的棉花枯萎病是棉花生产上的主要病害之一，在各地棉花主产区均有发生。种植抗病品种是防治棉花枯萎病最为经济有效的措施之一，但由于病原菌的遗传变异，品种丧失抗性的现象时有报道。棉花枯萎病菌存在高度的遗传分化，通过传统的鉴别寄主法进行小种划分，棉花枯萎病菌被划分为8个不同的生理小种和一个澳大利亚生理型，我国存在3号、7号、8号小种。通过扩增片段长度多态性（AFLP）技术对采自我国主要植棉区的80株棉花枯萎菌进行了遗传多样性分析，发现5株与3号、7号、8号小种具有较大遗传的菌株。通过对这5株枯萎菌进行elongation factor（EF-1α）、β-tubulin（BT）和phosphate permase（PHO）基因序列分析，证明这5株枯萎菌与澳大利亚棉花枯萎菌生理型菌株具有较近的亲缘关系。利用4个不同抗性的棉花品种测试了这5株新生理型菌株的致病性，结果表明这5株枯萎菌均能引起棉花枯萎病，表现中等致病力。进一步通过对3号、7号、8号小种和新生理型菌株的EF-1α基因序列进行比对，发现新生理型菌株存在一些特异性碱基。根据这些特异性碱基设计了一对针对新生理型菌株的特异性引物，证明该引物可以特异性检测新生理型菌株。利用该引物对采自我国主要植棉区的90株棉花枯萎菌进行筛选，获得2株潜在的新生理型菌株。从这2株棉花枯萎菌中克隆出EF-1α和BT基因序列，通过构建系统发育树，证明新筛选出的这2株棉花枯萎菌为新生理型菌株，并且这类新生理型菌株与澳大利亚生理型菌株具有较近的亲缘关系。

关键词：尖孢镰刀菌；特异性；分子检测；多样性

* 基金项目：国家现代棉花产业技术体系（CARS-18-15），河北省财政专项（F17C10007）

** 第一作者：郭庆港，博士，副研究员，主要从事植物病害生物防治研究；E-mail：gqg77@163.com
*** 通信作者：马平，博士，研究员，主要从事植物病害生物防治研究；E-mail：pingma88@126.com

山东省玉米南方锈病菌遗传多样性及初侵染来源分析*

隋鹏飞[1**]，王　琰[2]，张茹琴[1]，夏淑春[1]，王志奎[3]，鄢洪海[1***]

（1. 青岛农业大学农学与植物保护学院，青岛　266109；2. 山东省乳山市农业局，威海　264500；3. 即墨市农业局，青岛　266201）

摘　要：从20世纪90年代以来，由多堆柄锈菌（*Puccinia polysora* Underw.）引起玉米南方锈病近年来在我国发生呈上升趋势，在东部沿海和黄淮海玉米种植区时常暴发，对玉米生产的威胁明显增大，已成为这一地区的主要病害。由于该病害在中国的发生与流行规律迄今尚不清楚，这给控制这一病害带来了很大难度，病害严重发生时，病菌完全覆盖感病玉米的叶片，导致叶片干枯，植株早衰，造成减产。

由于国际上没有玉米南方锈病病菌生理小种的统一鉴别寄主，以及该病菌孢子寿命短，且不能长时间脱离寄主的原因，导致较少有关于对该病菌遗传变异和致病性分化方面的研究报道。采用分子生物学技术研究我国玉米南方锈病菌的遗传多样性，并以此来解析病害发生的一些规律和特点，将极可能为该病害的协同防控提供重要帮助。本研究利用11对多态性ISSR引物对2014年、2015年和2016年采集自山东及国内其他4个省市的60个菌株进行扩增，分析种群的遗传多态性及亲缘关系。结果表明：玉米南方锈病菌山东省群体具有丰富遗传多样性，在群体水平上，多态位点百分率（*P*）为96.34%，观察等位基因数（*Na*）为5.33，有效等位基因数目（*Ne*）为3.66，Nei's基因多样性指数（*H*）为0.76，Shannon信息指数（*I*）为1.33。遗传多样性研究结果还表明，山东多堆柄锈菌遗传多样性与其他地区种群之间有一定程度的差异，但江苏亲缘关系最近，其次是浙江，与海南的群体亲缘关系最远。

玉米南方锈病病菌是典型的专性寄生菌，已有的研究表明：在我国冬孢子很少发生，病菌的夏孢子生命力较短，新鲜孢子一般不足30d，并且不耐低温，无法在北方越冬存活。另外，迄今也未发现多堆柄锈菌存在转主寄主，因此推断该菌只能通过不断地产生夏孢子完成病害侵染循环和扩散流行以及病菌自身的生活史。美国发生的玉米南方锈病能证明这一点，病菌就是从南向北逐渐扩展的，根本原因是南部地区玉米为1年两作，锈病菌在南方冬季玉米种植区越冬，早期发病后的病菌通过气流逐渐向北方玉米区扩散并引起北部地区夏末初秋的玉米植株发病。另外，非洲西部地区1949年暴发的玉米南方锈病也被证明病菌就是从位于大西洋西部的加勒比海地区经过风的作用传入的，造成当地玉米受到玉米南方锈病的严重危害。

本研究证明，冬季在海南省三亚发生的玉米南方锈病，病原菌与其他地区的比较，遗传相似性最低，遗传距离最远。因此，笔者推断，发生在海南省的玉米南方锈病不是山东省夏玉米南方锈病的初侵染菌源，山东省的初侵染源更可能来自周年可以种植玉米的菲律宾或中国台湾省等地，通过台风传递到江浙沿海地区种植的玉米上，然后依次向西北扩展，到达山东及内陆其他地区。因为从笔者的研究结果看，山东菌株群体与江苏、浙江菌株群体间相似性较高，山东菌株群体与江苏、浙江、河南菌株间遗传距离较近，而与海南菌株群体都很远。对不同地域间种群基因流的分析也表明，山东等北方地区玉米南方锈病的初侵染源可能与海南病害无关。

关键词：玉米南方锈病菌；山东群体；遗传多样性；初侵染来源分析

---

* 基金项目：山东省自然科学基金项目（ZR2011CL005）；山东省“泰山学者”建设工程专项经费资助（BS2009NY040）

** 第一作者：隋鹏飞，男，山东潍坊人，硕士研究生，主要从事植物病理学与玉米病害研究；E-mail：suipf@163.com

*** 通信作者：鄢洪海，男，吉林长春人，教授，博士，主要从事植物病理学与玉米病害研究；E-mail：hhyan@qau.edu.cn

Pathogenic Fungi Associated with *Camellia sinensis* and Allied Species*

Liu Fang, Cai Lei**

(*Chinese Academy of Sciences*, *Institute of Microbiology*, *State Key Laboratory of Mycology*, *Beijing*, 100101, *China*)

Abstract: *Camellia* Linnaeus (Theaceae) species are of important commercial and ornamental value, especially *C. sinensis* (tea plant). *Camellia* production is heavily affected by a number diseases caused by fungal pathogens, such as anthracnose, gray blight and ring spot. During the past five years, we carried out an investigation of fungi associated with diseases of *C. sinensis* and allied *Camellia* spp. in 42 locations in the main cultivation provinces in China, and 1 042 fungal strains were isolated using single spore isolation and tissue isolation methods. ITS analyses of these strains suggested that they belonged to at least 162 taxa in 68 genera, 38 families, 20 orders, and the top five dominant genera were *Colletotrichum* (25.7%), *Diaporthe* (10.5%), *Pestalotiopsis* (10.7%), *Setophoma* (8.3%) and *Cladosporium* (6.6%). Further morphological examinations and multi-locus phylogenetic analyses were employed to identify fungi in the above five genera to species level, which revealed 11 *Colletotrichum* spp., 9 *Diaporthe* spp., 15 *Pestalotiopsis* spp., 7 *Setophoma* spp. and 9 *Cladosporium* spp. associated with plant diseases of *Camellia* spp., including 20 novel species. This study revealed a high proportion of unrecorded and undescribed species on *C. sinensis*, which increased the known species associated with tea plant from 280 to 384 species, a significant improvement of our understanding on the diversity of fungal pathogens on *Camellia* species.

Key words: Phylogeny; Plant pathogen; Species delimitation; Taxonomy

---

* Funding: This study was financially supported by the NSFC (31400017)

** Corresponding author: Cai Lei; E-mail: cail@ im. ac. cn

一种丹参新病害的病原鉴定及致病力测定*

鹿秀云[1**]，李社增[1]，年冠臻[1]，温春秀[2]，郭庆港[1]，张晓云[1]，马　平[1***]

（1. 河北省农林科学院植物保护研究所，农业部华北北部作物有害生物综合治理重点实验室，河北省农业有害生物综合防治工程技术研究中心，保定　071000；

2. 河北省农林科学院经济作物所，石家庄　050051）

摘　要：丹参（*Salvia miltiorrhiza* Bge.）为唇形科（Labiatae juss）鼠尾草属多年生草本药用植物，以根入药，广泛应用于心脑血管疾病的治疗。随着规模化种植年限延长，丹参病害为害逐年显现，生产上为害严重的丹参病害主要包括枯萎病、根腐病、根结线虫病3种，次要病害包括白绢病和紫纹羽病等。2015年笔者在河北省安国市丹参种植基地发现了一种新的病害，罹病植株根部不腐烂；地上部从植株下部叶片开始发病，逐渐向上扩展，发病叶片边缘变软，上卷，叶片变黄；剖开植株发现维管束组织变褐；严重时整株枯死。该病害发病率高，危害严重。为了有效防治该病害，开展了丹参新病害病原菌分离鉴定及致病力测定实验。采用常规组织分离法，在丹参病株根部维管束组织和叶柄维管束组织分离并单孢纯化获得15个真菌菌株。显微镜检查发现15个菌株的菌丝均能观察到清晰的轮状分支，结合菌落能够产生黑色放射状微菌核这一特征，初步将罹病丹参分离物确定为轮枝菌属真菌（*Verticillium* spp.）。采用分子技术利用真菌通用引物ITS1/ITS4和大丽轮枝菌特异引物DB19/DB22鉴定，分离的15株真菌均为大丽轮枝菌（*V. dahliae*）。柯赫法则回接试验表明，接种丹参分离物大丽轮枝菌孢子悬浮液14d开始丹参植株出现轻微的叶片变黄、萎蔫；21d时出现了较为明显的田间发病症状，经叶柄维管束组织分离重新获得了大丽轮枝菌菌株。实验结果充分证明安国市丹参种植基地发现的丹参新病害为大丽轮枝菌引起的丹参黄萎病。采用切根蘸孢子法评价了10株丹参大丽轮枝菌在棉花上的致病力。实验获得2株强致病力菌株，4株中等致病力菌株，4株弱等致病菌株，菌株致病力之间存在差异，均低于棉花黄萎菌强致病力菌株WX-1的致病力。以上结果将为防治丹参新病害提供重要依据。

关键词：丹参；大丽轮枝菌；病原鉴定；致病力

---

* 基金项目：公益性行业科研专项（201503109）；国家现代棉花产业技术体系（CARS-18-15）

** 第一作者：鹿秀云，副研究员，主要从事植物病害生物防治研究；E-mail：luxiuyun03@163.com

*** 通信作者：马平，博士，研究员，主要从事植物病害生物防治研究；E-mail：pingma88@126.com

稻瘟菌胞外蛋白 MoGAS1/MoGAS2 的重组表达、纯化及晶体生长

席玉轩*，王　超，李国瑞，彭友良，刘俊峰**
（中国农业大学植物保护学院植物病理学系，北京　100193）

摘　要：由稻瘟菌引起的稻瘟病是一种世界性的水稻病害，严重影响着我国水稻的生产和粮食安全，如何防治稻瘟病，减少稻瘟病对水稻产量和质量的损害，提高水稻产量已成为全世界共同追求的目标。多种类农药的发明和使用为此提供了可能，而针对药物靶标进行合理化分子设计成为农药研发的重要手段。已有的研究表明稻瘟菌中 MoGAS1 和 MoGAS2 参与了蛋白激酶 PMK1 途径，在调控附着胞渗透能力和病原菌致病力中扮演着重要的角色，缺失编码这些蛋白的基因会导致附着胞渗透能力下降和病原菌致病力减弱，是潜在的药物设计靶点。获得 MoGAS1 和 MoGAS2 稳定均一的适合晶体生长的蛋白，解析其晶体结构，将为深入阐释其生物学功能和绿色农药设计提供重要的结构基础。

本研究利用大肠杆菌原核表达系统，首先将 MoGAS1 和 MoGAS2 全长基因分别构建到相应的表达载体中，摸索适宜的表达体系进行蛋白表达纯化，发现全长的重组蛋白在不同载体中多数不可溶或不表达，只有 MoGAS2 在 pETMBP-XE 载体中获得稳定表达且均一的蛋白。通过晶体条件筛选在 15% PEG3350 为主沉淀剂的条件下获得晶体，晶体衍射分辨率为 9Å，正在尝试重复和优化晶体生长条件以获得更高分辨率的数据。除此之外，作者拟通过 MoGAS1 和 MoGAS2 二级结构和结构域的预测结果，综合考虑进行相应的截短，将不同截短体构建到相应表达载体中，筛选能稳定表达可溶且均一的蛋白样品，摸索合适的晶体生长条件，为解析这两个胞外蛋白的结构以及基于靶标的合理化绿色农药设计奠定坚实的基础。

关键词：稻瘟菌；MoGAS1/MoGAS2；蛋白纯化；晶体生长

* 第一作者：席玉轩，在读硕士研究生，主要从事分子植病研究；E-mail：victorxyx@ foxmail. com
** 通信作者：刘俊峰，教授，主要从事植物与病原菌的分子互作的结构基础；E-mail：jliu@ cau. edu. cn

云南、四川、贵州杧果炭疽病病原鉴定*

卜俊燕[1**]，莫贱友[2,3]，郭堂勋[2,3]，唐利华[2,3]，黄穗萍[2,3]，李其利[2,3***]，余知和[1***]
（1. 长江大学生命科学学院，荆州　434025；2. 广西农业科学院植物保护研究所，
南宁　530007；3. 广西作物病虫害生物学重点实验室，南宁　530007）

摘　要：杧果是世界五大水果之一，其广泛种植于热带、亚热带地区，目前在我国主要种植于广西、海南、云南、四川、贵州、广东、福建等省区。杧果炭疽病是杧果生长期和采后贮藏期的主要病害之一，为害较为严重。全面了解杧果炭疽菌的病原种类及优势种群，明确其致病力分化情况，有助于掌握该病的发生流行规律，也是研究杧果炭疽病绿色防控新策略、新方法和新药剂的基础。长期以来，我国杧果炭疽病的病原一直未获得详细而系统的研究，一般都是对部分地区的病原进行鉴定。目前认为杧果炭疽病主要由胶胞炭疽复合群（*C. gloeosporioides* complex）和尖孢炭疽复合群（*C. acutatum* complex）引起，其中胶胞炭疽菌为主要病原。胶胞炭疽复合群内可分为22个种，而尖孢炭疽复合群可分为31个种，世界各地报道的杧果炭疽菌种类存在显著差异，我国各杧果产区生态环境差别大、产期不一致，病原种类也可能存在显著差异。

本研究分别从云南的华坪县、红河县、西双版纳、元阳县、元江县；四川的攀枝花仁和区、攀枝花东区、米易县、盐边县；贵州的望谟县、册亨县、兴义市等杧果产区采集病害样本，通过常规组织分离法共分离到215个菌株。通过初步形态学鉴定和ITS序列分析，从各地分别选择出杧果炭疽菌代表菌株72株进行进一步的形态学鉴定和ITS、ACT、CHS-1、GAPDH、TUB2的多基因分子系统学分析。形态学鉴定结果表明：供试菌株在PDA上的菌落形态初期灰白色绒毛状，后期呈现白色、灰黑色或灰褐色，部分菌株产生扇形结构，且边缘不整齐；有的菌株产生墨绿色色素，部分菌株产生灰色或黑色色素；在培养到5~7d后，部分菌株可产生大量粉红色的分生孢子团，孢子无色透明，椭圆形至圆柱形，大小为（7.58~18.62）μm（平均16.26μm）×（3.26~8.68）μm（平均4.54μm）。多基因分子系统学分析结果表明：采集于云南、四川和贵州的杧果炭疽菌分别与NCBI上的*C. asianum*、*C. cliviae*、*C. fructicola*、*C. siamense*、*C. brevisporum*、*C. endophyticum*、*C. karstii*、*C. kahawae*、*C. citri*、*C. musae*等10个种的基因序列同源性为99%~100%。根据形态学和分子系统学鉴定结果，云南、四川和贵州三省的杧果炭疽菌可分为10种，其中云南发现7种杧果炭疽菌：*C. asianum*、*C. cliviae*、*C. fructicola*、*C. siamense*、*C. brevisporum*、*C. endophyticum*、*C. karstii*，且*C. asianum*为云南的优势种群；贵州发现7种杧果炭疽菌：*C. asianum*、*C. fructicola*、*C. citri*、*C. karstii*，*C. kahawae*，*C. musae*，*C. siamense*，且*C. fructicola*是贵州的优势种群，四川发现2种杧果炭疽菌，分别为*C. asianum*和*C. siamense*，且*C. siamense*是四川的优势种群。本研究发现的杧果炭疽菌*C. citri*、*C. musae*、*C. brevisporum*、*C. endophyticum*为世界杧果上首次报道，*C. cliviae*、*C. fructicola*、*C. siamense*、*C. karstii*为中国杧果上首次发现。

关键词：杧果；炭疽菌；形态学；多基因分子系统学

---

* 基金项目：国家自然科学基金（31560526，31600029）；国家现代农业产业技术体系广西杧果创新团队建设专项；广西留学回国重点基金（2016GXNSFCB380004）；广西农业科学院科技发展基金（桂农科2016JZ24，2017JZ01）

** 第一作者：卜俊燕，硕士，微生物学；E-mail：571437597@qq.com

*** 通信作者：李其利，副研究员，主要研究方向为植物真菌病害及植物病害生物防治；E-mail：liqili@gxaas.net
余知和，教授，主要研究方向为真菌学；E-mail：zhiheyu@hotmail.com

Identification and Characterization of Brown Spot Disease on Fruit of Chinese Torreya

Zhang Shuya*, Li Ling*, Zhang Chuanqing

(*School of Agricultural and Food Science*, *Zhejiang Agriculture and Forest University*, *Lin'an* 316300, *Zhejiang Province*, *China*)

Abstract: Chinese torreya (*Torreya grandis*) is one of the important economic foresee crops in the hilly hills and mountain slopes naturally, particularly in Zhejiang province, south China. A new disease, brown spots, appeared on fruits of *Torreya grandis* recently. Initially, small dark brown spots and necrotic center lesion surrounded by a chlorotic halo appeared on fruit, and then the fruit shrank and fell off gradually with the development of disease. In order to clarify the causal agent of brown spot, the pathogenic fungus was identified based on the analysis of morphologic characteristics, rDNA-ITS sequences and pathogenicity tests. Colonies of the pathogen on PDA were initially gray white and dark gray or green brown gradually. Conidiophores of the fungus were solitary or clustered, without branching, and conidiophores were brown, obpyriform or oval. The average size of spores 19.258μm×9.048μm, with 1~7 transverse septum, 0~3 longitudinal septum, columnar beaks or no. These characteristics wereconsistent with *Alternaria alternata*. Sequence of rDNA-ITS (Accession no. KU525533) analysis showed 100% similarity with *A. alternata* from GenBank database. Taken together, both morphological and molecular proofs indicated that brown spot disease was caused by *A. alternata*.

Key words: *Torreya grandis*; Brown spot; *Alternaria*; Biological characteristics

---

* Frist authors: Zhang Shuya, Li Ling; E-mail: cqzhang@zafu.edu.cn

效应因子 AVR-Pia 与其受体结合结构域复合物的结构解析

郭力维，彭友良，刘俊峰
（中国农业大学植物保护学院植物病理学系，北京　100193）

摘　要：植物病害严重影响其产量和品质，为抵御病原物入侵，植物 R 蛋白作为免疫识别受体，通过 NB-LRRs 直接或间接相互作用来识别病原物效应因子，引发过敏性坏死反应提高寄主抗性。水稻与稻瘟病菌是研究植物与病原物互作的模式系统之一，二者相互作用符合“基因对基因”假说。目前，已克隆水稻抗病基因 23 个和稻瘟病菌无毒基因 9 个。因此，研究稻瘟病菌效应因子与其水稻受体的相互作用机理具有理论和实践意义。

在已经确定的水稻 R 蛋白中，RGA5-A 作为 Pia（RGA4/RGA5-A）的一个重要组分，通过其 C 末端 LRR 区域下游的 RATX1 结构域与 AVR-Pia 直接相互作用，但其结合机制尚未明确。

本研究利用原核表达系统将 RGA5-A_S（RATX1）和 AVR-Pia 进行重组表达，通过筛选不同的重组表达载体和菌株组合，借助层析技术纯化重组表达蛋白，已建立了简单、有效的目标蛋白纯化流程。在酵母双杂实验体内验证二者相互识别的同时，利用凝胶排阻层析和等温量热滴定等技术体外验证重组表达的 RGA5-A_S（RATX1）和 AVR-Pia 存在直接的相互作用，二者间结合的解离常数约为 1.52μmol/L。采用座滴气相扩散法进行复合物结晶条件的筛选，以期利用结构生物学等手段解析水稻抗病蛋白 Pia RATX1 结构域识别效应因子的结构基础，这将为深入分析植物抗病基因识别病原物的分子机理和水稻抗病基因的改良等方面奠定基础。

关键词：稻瘟病菌；效应蛋白；水稻抗病蛋白；复合物结构

Screening and Analysis of Effector Proteins from *Puccinia triticina*

Li Jianyuan, Wei Jie, An Zhe, Zhang Yue, Yang Wenxiang, Liu Daqun
(*Department of Plant Pathology, Agricultural University of Hebei/Biological Control Center of Plant Diseases and Plant Pests of Hebei Province/National Engineering Research Center for Agriculture in Northern Mountainous Areas, Baoding* 071001)

Abstract: Leaf rust caused by the fungus *Puccinia triticina* (*Pt*) is a major constraint to wheat production worldwide. The molecular events that underlie *Pt* pathogenicity are largely unknown. Like all rusts, *Pt* forms a specialized infection structure called haustoria to take nutrients and water from host plant tissue, and to secrete pathogenicity factors called effector proteins. In this study, we extracted the total RNA of urediniospores, germinated urediniospores and infected wheat leaves harvest at 6d from three different virulence strains of *Pt* and used Illumina Hiseq sequencing platforms to assemble the transcriptomes and then we got 46008 Unigenes. Candidate effectors were screened through Signal P 4.1, Target P 1.1, TMHMM 2.0 and Effector P, and 325 genes which constitute candidate effectors were identified. qRT-PCR analysis confirmed the expression patterns of 96 candidate effectors and 42 genes showed aspecial increase in various stages which may be involved in the pathogenic process. 33 candidate effectors were cloned successfully and 17 were identified to be able to suppress the cell death caused by BAX in tobacco cells. This analysis of candidate effectors is an initial step towards functional research and established the foundation forclarify the pathogenic mechanism of leaf rust pathogens.

Key words: *Puccinia triticina*; Effector proteins; Screening

Botryosphaeria spp. 侵染诱导桃乙烯应答因子表达载体的构建及转化*

刘　勇[1]，何华平[1**]，龚林忠[1]，王富荣[1]，王会良[1]，艾小艳[1]，王泽琼[2]
（1. 湖北省农业科学院果树茶叶研究所，武汉　430064；
2. 武汉生物工程学院，武汉　430415）

摘　要：乙烯应答因子是介导植物防卫反应的重要基因资源，在植物与病原菌互作过程中起着重要的调控作用，迄今已有大量的 ERF 基因在其他植物中被克隆和研究，而桃中的 ERF 基因还没有被充分发掘。前期的研究中，笔者采用 RNA-seq 测序技术从被 *Botryosphaeria* spp. 侵染的“丰光”油桃中筛选到一个高丰度表达的乙烯应答因子（PNGERF-1）。该基因在感病桃树和健康桃树中表达差异倍数显著，表明其在桃与 *Botryosphaeria* spp. 的互作过程中可能发挥着重要作用。鉴于目前桃遗传转化技术还不成熟，笔者通过拟南芥遗传转化技术对其生物学功能进行了初步探讨。

根据已获得的基因序列信息，通过 RT-PCR 方法从一年生“丰光”油桃枝条中克隆出受 *Botryosphaeria* spp. 侵染诱导的 PNGERF-1 基因，构建到植物表达载体 pCAMBIA1300-35S 上，并命名为 pCAMBIA1300-35S-PNGERF-1。通过农杆菌介导法转化野生型拟南芥，经 PCR 证明 PNGERF-1 基因在拟南芥中整合并表达。获得的转基因阳性植株一共表现 4 种不同的生物学表型：①不抽薹，只停留在营养生长阶段；②能够抽薹开花，但不结实；③能够开花结实，但种子产量低；④正常开花结实，与对照组无明显差异。这些研究结果的获得为进一步探讨 PNGERF-1 在桃与 *Botryosphaeria* spp. 的互作过程中的作用机制提供了有益的启示。

关键词：*Botryosphaeria* spp.；桃；乙烯应答因子；转化

---

* 基金项目：湖北省自然科学青年基金（2015CFB224）；农作物重大病虫草害防控湖北省重点实验室开放基金（2015ZTSJJ1）

** 通信作者：何华平，研究员，主要从事桃种质资源评价与创新利用；E-mail：hbgchhp@163.com

Genome Sequence of an isolate of *Fusarium oxysporum* f. sp. *melongenae*, the Causal Agent of the Fusarium Wilt of Eggplant*

Dong Zhangyong[1], Hsiang Tom[2], Luo Mei[1], Xiang Meimei[1]**
(1. *Department of Plant Protection*, *Zhongkai University of Agriculture and Engineering*, *Guangzhou*, 510225, *China*; 2. *School of Environmental Sciences*, *University of Guelph*, *Canada*, *N1G 2W1*)

Abstract: *Fusarium* wilt of eggplant caused by *Fusarium oxysporum* f. sp. *melongenae* (Fomg) is an economically important soil-borne disease limiting eggplant production worldwide. This pathogen was initially reported in Japan in 1958 and the first reportin China was in 2005. Here, we report a draft assembly for Fomg isolate 14004, a field strain collected in Guangdong Province, China, where *Fusarium* wilt of eggplant was observed over ten years ago and is currently endemic.

Onemicro gram of genomic DNA was sent for sequencing in Genome Quebec (Montreal, Canada) specifying 100 bp paired-end reads with 300 bp insert. Over 43 million paired-end reads totaling 10. 7 Gb were received. The genome was assembled using the programs Velvet v. 1. 2. 10, ABySS v. 2. 0. 1 and SOAPdenovo v. 2. 04 with odd-numbered kmers between 21 and 91. Assembly quality was assessed by examining the N50 value and by examining the total number of scaffolds produced by the programs. The initial assembly was 54 488 475bp in length with an N50 value of 568 281bp, resulting from 4 617 scaffolds. After removing scaffolds that were smaller than 200bp by using a PERL script, the final assembly consisted of 1 631 scaffolds with a genome size of 53 986 354bp (G + C content, 46. 4%). A total of 16 485 protein-coding genes were predicted from the assembly with the highest N50 (kmer = 77, ABySS) using AUGUSTUS v. 3. 2. 2 based on gene models from *Fusarium oxysporum.*

* Funding: This work was supported by National Natural Science Foundation of China (31301627) and Natural Science Foundation of Guangdong Province (S2012040006912)

** Corresponding author: Xiang Meimei; E-mail: mm_xiang@163. com

Study on Diverse *Colletotrichum* Species Associated with Pear Anthracnose in China*

Fu Min[1,2], Bai Qing[3], Zhang Pengfei[2], Hong Ni[1,2], Wang Guoping[1,2**]

(1. *National Key Laboratory of Agromicrobiology*, *Huazhong Agriculture University Wuhan*, *Hubei* 430070, *China*; 2. *The Key Laboratory of Plant Pathology of Hubei*, *Huazhong Agricultural University*, *Wuhan*, *Hubei* 430070, *China*; 3. *Department of Plant Pathology*, *Washington State University*, *Pullman*, *WA* 99164-6430, *USA*)

Abstract: *Colletotrichum* spp. are highly variable in species and corresponding disease symptoms toward pear fruits and leaves in China. Since 2007, pear anthracnose was increasingly occurred in many cultivars with a widespread in distribution, resulting in bad quality of pear fruits and substantial yield loss for famers. To determine the diversity of the causing agents in genus *Colletotrichum* and related disease symptoms on pear fruits and leaves, we collected 319 samples with three types of anthracnose symptoms (normal anthracnose necrosis, off-white necrosis and black spot with small diameter less than 1mm) from *Pyrus pyrifolia*, *P. bretschneideri* and *P. communis* from 2013 to 2017 in 7 provinces of China. Through isolation and purification, we finally gained 505 strains. The morphological characters of all strains were observed and compared, and multi-locus phylogenetic analyses (ITS, ACT, CAL, CHS-1, GAPDH, and TUB2) were performed on 82 representative strains. Twelve *Colletotrichum* species were identified, including nine well-characterized species (*C. aenigma*, *C. conoides*, *C. fructicola*, *C. gloeosporioides*, *C. siamense*, *C. fioriniae*, *C. citricola*, *C. karstii*, and *C. cliviae*) and three novel species. *C. conoides*, *C. siamense*, *C. citricola*, *C. karstii*, and *C. cliviae* were reported for the first time from pear. Pathogenicity testing showed that all species can infect pear leaves. Notobly, *C. fructicola* was the only pathogen to cause black spot with diameter less than 1mm.

* Funding: Pear Modern Agro-industry Technology Research System (CARS-29-10) and Chinese Ministry of Agriculture, Industry Technology Research Project (grant no: 201203034-02)

** Corresponding author: G. P. Wang; E-mail: gpwang@ mail. hzau. edu. cn

Transcriptome Profile of *Trichoderma afroharzianum* LTR-2 Induced by Oxalic Acid*

Wu Xiaoqing[1**], Ren He[2], Lv Yuping[2], Zhou Hongzi[1],
Zhao Xiaoyan[1], Zhang Guangzhi[1], Zhang Xinjian[1***]
(1. *Shandong Provincial Key Laboratory of Applied Microbiology*, *Ecology Institute*, *Shandong Academy of Sciences*, *Jinan* 250014, *China*; 2. *College of Life Science*, *Shandong University of Technology*, *Zibo* 255049, *China*)

Abstract: Oxalic acid (OA) has been proved to be a pathogenic factor of grey mold fungus *Botrytis cinerea*, which causes pre-and postharvest crop losses worldwide. *Trichoderma afroharzianum* LTR-2 is an effective bio-control strain against *Botrytis cinerea*, and able to eliminate OA at the same time. Although biocontrol mechanisms of *Trichoderma*, such as mycoparasitism, antibiosis and competition are well studied, the biocontrol meaning of OA elimination effect is still vague. In this article, we analyzed the transcriptome changes of strain LTR-2 induced by 0 and 20 mmol/L OA using RNA-seq technology, each treatment set 3 biological replicates. Totally 653 up-regulated genes and 402 down-regulated genes were found among differentially expressed genes (DEGs), of which the gene functions were mainly related to metabolic process and catalytic activity according to GO terms. Based on pathway enrichment analysis, the DEGs were primarily involved in metabolic pathways, biosynthesis of secondary metabolites and biosynthesis of antibiotics. We selected 4 genes of interest and found oxalate decarboxylase (OXDC) gene (g2863), chitinase gene (g7837), glucanase gene (g5151) and protease gene (g5140) were all significantly up-regulated, and the results were confirmed by RT-PCR analysis. The transcriptome analyses preliminarily showed that OXDC gene was involved in OA elimination, and OA might be a signal which could regulate mycoparasitism in *Trichoderma*.

Key words: *Trichoderma*; Oxalic acid; RNA-seq; Bio-control

---

* Funding: the Research Award Fund for Excellent Young and Middle-aged Scientists of Shandong Province (BS2015SW029), by the Natural Science Foundation of China (31572044)

** First author: Wu xiaoqing; E-mail: xq_ wu2008@163. com

*** Corresponding author: Zhou Hongzi; E-mail: zhangxj@sdas. org

真菌病毒对梨轮纹病菌差异基因表达的影响*

王利华[1,2]，罗　慧[1,2]，洪　霓[1,2]，王国平[1,2]，王利平[1,2]**
（1. 华中农业大学植物科学技术学院，湖北省作物病害监测与安全控制重点实验室；
2. 华中农业大学，农业微生物学国家重点实验室，武汉　430070）

摘　要：梨轮纹病是由葡萄座腔菌［*Botryosphaeria dothidea*（Moug：Fr）Ces & De Not］引起的真菌病害，能够造成梨树枝条枯死、枝干粗皮和溃疡，果实腐烂。轮纹病在我国南北梨产区的为害呈逐年加重趋势，弱毒相关病毒导致植物病原真菌致病力衰退，可作为梨轮纹病生物防治的另一条重要途径。真菌病毒导致植物病原真菌的弱毒相关特性是病毒与寄主在长期进化过程中相互作用的结果，被病毒侵染后寄主基因表达受到影响。研究介导梨轮纹病菌毒力衰退相关真菌病毒与寄主互作，从转录水平上分析病毒与寄主互作关系，旨在明确真菌病毒引起梨轮纹病菌致病力衰退的关键基因及梨轮纹病菌应答病毒侵染的分子机制，挖掘防治梨轮纹病潜在的基因资源，为梨轮纹病害防治提供新途径。

本研究以复合感染产黄青霉病毒（BdCV1）和双分体病毒（BdPV1）的梨轮纹菌株 LW-1、单独感染 BdCV1 的 LW-C、单独感染 BdPV1 的 LW-P 以及无病毒感染的 Mock 菌株作为四组材料，构建了 4 个文库（分别命名为 LW-CP，LW-C，LW-P 和 Mock），进行 *de novo* 测序，共获得 30 058个 *Unigene* 基因，平均长度为 2 128bp，在 7 大基因数据库进行了注释，明确了功能分类。分析真菌病毒感染条件下梨轮纹病菌 3 组对比文库，获得了 LW-CP/Mock 上调表达基因 5 605个，下调基因 3 301个；LW-C/Mock 上调表达基因 4 478个，下调基因 4 311个；LW-P/Mock 上调表达基因 2 596个，下调表达基因 2 328个。以上结果表明：受 BdCV1 病毒感染的差异表达基因数量最多，推测 BdCV1 对梨轮纹病菌寄主基因表达影响明显。采用 RT-qPCR 对病毒感染条件下候选的 9 个差异基因表达量进行分析，检测结果与转录组测序得到的表达趋势一致，验证了 RNA-seq 测序结果的可靠性，明确了这些差异基因在病毒感染条件下的表达模式。另外，在梨轮纹病菌中检测到参与基因沉默途径的几个关键基因（*Ago*、*Dicer* 和 *RdRp*）且在病毒感染条件下为上调表达。对病毒感染条件下 3 组对比文库的差异表达基因进行 GO 功能注释和 KEGG 通路分析，结果显示，这些基因主要在代谢途径以及生物合成二级代谢途径中显著富集，参与细胞复制、信号转导、抗病防御等途径；相关关键基因功能研究还在进一步试验中。

关键词：沙梨；梨轮纹病；葡萄座腔菌；*De novo* 测序；真菌病毒；产黄青霉病毒

---

* 基金项目：国家自然科学基金（31471862）；国家梨产业技术体系（CARS-29-10）

** 通信作者：王利平，副教授，研究方向为果树病理学；E-mail：wlp09@ mail. hzau. edu. cn

TcLr19PR1、*TcLr19PR2* 和 *TaLr19TLP1* 的酵母双杂交诱饵载体的构建及鉴定

王 菲，张艳俊，梁 芳，张家瑞，王海燕，刘大群
（河北农业大学植物保护学院，河北省农作物植物病虫害生物防治工程技术研究中心，
国家北方山区农业工程技术研究中心，保定 071000）

摘 要：在前期研究的基础上，已成功克隆得到 3 个病程相关蛋白 PR1、PR2 和 PR5 的序列，分别命名为 *TcLr19PR1*、*TcLr19PR2* 和 *TaLr19TLP1*。将这 3 个基因分别连接到酵母双杂交诱饵载体 pGBKT-7 上，并转化到感受态菌株 Y2HGold 中，检测其毒性和自激活性。结果显示，成功构建了包含目的基因的诱饵重组载体 pGBKT-7-TcLr19PR1、pGBKT-7-TcLr19PR2 和 pGBKT-7-TaLr19TLP1；将转化产物涂布于 SD/-Trp/X 平板上，生长良好，并出现阳性克隆；毒性检测中，将 3 个诱饵重组载体与空载体在 SD/-Trp 液体培养基中的生长情况进行对比，发现诱饵无毒性；自激活检测实验中，重组载体无法在二缺、三缺和四缺培养基中正常生长，证明诱饵无自激活性。因此，本研究成功构建的 3 个诱饵重组载体可用于 3 个 PR 蛋白互作蛋白的筛选，为进一步研究其在小麦与叶锈菌互作中的分子机理奠定基础。

关键词：病程相关蛋白；酵母双杂交；诱饵载体；自激活检测；毒性检测

Aquaporin1 Regulates development, Secondary Metabolism and Stress Responses in *Fusarium graminearum*[*]

Ding Mingyu, Li Jing, Fan Xinyue, Yu Xiaoyang, He Fang,
Xu Houjuan, Liang Yuancun, Yu Jinfeng
(*College of Plant Protection*, *Shandong Agricultural University*, *Taian* 271018, *China*)

Abstract: The Ascomycete fungus *Fusarium graminearum* is the causative agent of Fusarium head blight, and has become the predominant model organism for study of fungal phytopathogens. Aquaporins (AQPs) have been implicated in the transport of water, glycerol, and a variety of other small molecules in plants and animals. However, the role of these proteins is poorly understood in plant pathogenic fungi. Here, we identified and attempted to elucidate the function of five aquaporin genes in the *F. graminearum* genome. A phylogenetic analysis of aquaporins revealed that AQPs are divided into two clades, AQP1 belongs to the first clade. Deletion of *AQP*1 is responsible for the regulation of development, stress response, sexual and asexual reproduction. The *ΔAQP*1 mutant was fully pathogenic similar as the wild type, whilethe mutant was increased in DON production. In addition, we attempt to characterise the cellular localisation of AQP1 using GFP fusions, and we found that AQP1 appears to be located to the nuclear membranes and vacuole membranes in conidia. Notably, except to localize to the nuclear membranes and vacuole membranes, we also found that AQP1 is localized to the endoplasmic reticulum inmycelial growth stage. Taken together, these results suggest that AQP1 is involved in development, stress response, sexual and asexual reproduction.

Key words: Aquaporin; *Fusarium graminearum*; Development; Secondary metabolism; Stress responses

* Funding This work was supported by the Nature Science Foundation of China (31171806 and 31520103911) and the Wheat Innovation Team of Shandong Province Modern Agricultural Industry Technology System (SDAIT-01-09)

多隔镰刀菌的生物学特性研究*

马　迪**，王桂清***
（聊城大学农学院，聊城　252000）

摘　要：国槐溃疡病是一种日益严重的枝干病害，多隔镰刀菌（*Fusarium decemcellulare*）是其致病菌之一。文章旨在研究该菌的生物学特性，探索温度、pH 值、光照、培养基、碳源及氮源等条件对其菌丝生长速率和产孢量的影响。研究结果表明：该菌在 PDA 培养基中生长最快，产孢量最大；25℃、pH 值 8.0~9.0、全黑暗是菌丝生长的最适条件，光暗交替更利于产孢；该菌能利用多种碳源和氮源，以麦芽糖、牛肉浸膏最适。该研究为进一步系统研究多隔镰刀菌引起的病害发病规律和防控措施提供了理论依据。

关键词：国槐；溃疡病；多隔镰刀菌；生物学特性

Biological Characteristics of *Fusarium decemcellulare*

Ma Di，Wang Guiqing
（*College of Agronomy*，*Liaocheng University*，*Liaocheng* 252059，*Shandong*，*China*）

Abstract：*Fusarium decemcellulare* is the pathogenic fungus of *Sophora japonica* which can cause a seriously canker disease. The effects of temperature，pH，illumination，medium and the sources of nitrogen and carbon on the mycelium growth and the production of conidiospore of *F. decemcellulare* were studied. Results showed that the fungus had a better growth in PDA medium. The optimum temperature for growth is 25℃，and optimum pH value is 8.0~9.0. Darkness is favorable for the mycelium growth while alternate with illumination is favorable for production of conidiospore. Data shows that maltose and beef extractare the best C and N sources for the culture of the pathogen. The biological characteristics studied above of *F. decemcellulare* are beneficial for the better underst and and control of canker diseasein *S. japonica*.

Key words：*Sophora japonica*；Canker；*Fusarium decemcellulare*；Biological characteristics

国槐（*Sophora japonica*）为我国特有树种，易栽培，寿命长，抗性及适应能力强，常被用作行道树，也为聊城市园林绿化的基调树种。但近年来，国槐病害的发生愈加严重，尤以溃疡病发病最重，因此，对国槐溃疡病病因的调查和防治迫在眉睫。笔者从病害发生地分离鉴定其致病菌，其中之一为多隔镰刀菌（*Fusarium decemcellulare*），本文探究了该菌的生物学特性，为进一

* 基金项目：山东省自然科学基金（ZR2012CL17）；聊城市科技发展计划项目（2014GJH10）
** 第一作者：马迪，女，硕士研究生，研究方向为种植资源利用和有害生物防治
*** 通信作者：王桂清，女，河北泊头人，博士，教授，主要从事植物保护的教学与科研工作

步深入系统研究该菌引起的病害发病规律和防治措施提供了理论依据。

1　材料与方法

1.1　供试菌种与培养

供试菌种为国槐溃疡病致病菌多隔镰刀菌（*F. decemcellulare*），由聊城大学植物病理研究室提供。采用 PDA 培养基（25±1）℃的恒温培养箱内培养 5d，用打孔器制取 0.7cm 的菌饼接种备用。

1.2　温度对菌丝生长和产孢量的影响

将菌饼接种于 PDA 培养基平板中央，1 皿 1 饼，分别置于 5℃、10℃、15℃、20℃、25℃、30℃、35℃、40℃恒温光照培养箱中培养，L：D=12h：12h，每处理重复 3 次，5d 后采用十字交叉法测量菌落直径；5d 后每皿用 5mL 无菌水洗脱孢子，血球计数器计产孢量。以下培养天数及菌丝生长、产孢量测量方法相同。

1.3　光照对菌丝生长和产孢量的影响

分别在全光照、光暗交替（L：D=12h：12h）、全黑暗条件下，利用 PDA 培养基于（25±1）℃恒温培养箱中培养。

1.4　pH 值对菌丝生长和产孢量的影响

用 1.0mol/L HCl 和 1.0mol/L NaOH 调节 PDA 培养基的 pH 值分别为 3～12，接种后置于（25±1）℃恒温光照培养箱中培养，L：D=12h：12h。

1.5　培养基对菌丝生长和产孢量的影响

供试培养基为 PDA（马铃薯葡萄糖琼脂：马铃薯 200g、葡萄糖 20g、琼脂 20g、蒸馏水 1 000mL，pH 值自然）；mSDA（改良沙氏葡萄糖蛋白胨琼脂基：葡萄糖 20g、蛋白胨 10g、琼脂 20g、蒸馏水 1 000mL，pH 值自然）；CzA（查氏培养基：KH_2PO_4 0.5g、$MgSO_4 \cdot 7H_2O$ 0.5g、NaCl 0.1g、天门冬酰胺 5g、K_2HPO_4 0.6g、$CaCl_2$ 0.1g、葡萄糖 20g、琼脂 20g、蒸馏水 1 000mL，pH 值自然）；BCM（基础固体培养基：牛肉浸膏 3g、蛋白胨 10g、NaCl 5g、K_2HPO_4 1g、琼脂 20g、蒸馏水 1 000mL，pH 值自然），分别接种菌饼，1 皿 1 饼，3 次重复，于（25±1）℃恒温光照培养箱中培养，L：D=12h：12h。

1.6　碳源对菌丝生长和产孢量的影响

以 CzA 为基础培养基，以供试碳源替换其中等量的葡萄糖制成不同碳源培养基，以无碳及无氮、碳培养基为空白对照。供试碳源为蔗糖、麦芽糖、乳糖、果糖、海藻糖、阿拉伯糖、可溶性淀粉、微晶纤维素。接种后置于（25±1）℃恒温光照培养箱中培养，L：D=12h：12h。

1.7　氮源对菌丝生长和产孢量的影响

以 CzA 为基础培养基，以供试氮源替换其中等量的天门冬酰胺制成不同氮源培养基，以无氮及无氮、碳培养基为空白对照。供试氮源为硫酸铵、氯化铵、牛肉膏、酵母浸膏、尿素、硝酸钠、蛋白胨、甘氨酸。接种后置于（25±1）℃恒温光照培养箱中培养，L：D=12h：12h。

2　结果与分析

2.1　温度对菌丝生长和产孢量的影响

试验结果表明，温度对多隔镰刀菌菌丝生长及分生孢子数量影响基本一致。该菌菌丝生长温度范围 10～35℃，适宜温度为 20～30℃，其中 25℃菌落直径最大，为 3.55cm；该菌可产孢温度范围 15～35℃，最适温度 20～25℃，25℃产孢量高达 2.675×10^7个。说明 25℃为该菌的最适生长及产孢温度。

2.2 光照对菌丝生长和产孢量的影响

在各光照强度下，该菌均能良好生长。全黑暗菌落直径最大，为 3.87cm，光暗交替下直径最小，为 3.55cm；光暗交替最利于产孢，产孢量达 2.675×10^{7} 个，黑暗时产孢量相对较小。表明该菌对光照条件不敏感或适应性较强，全黑暗条件相对更利于该菌生长，光暗交替更利于产孢。

2.3 pH 值对菌丝生长和产孢量的影响

多隔镰刀菌对酸碱度适应范围较广。该菌在 pH 值 5～12 均能生长，pH 值 8～9 和酸性条件更适合菌丝生长，pH 值<4 时，菌丝几乎未见生长；该菌在 pH 值 5～11 均可产孢，pH 值=9 时产孢量最大，达 3.725×10^{7} 个，pH 值<4 及 pH 值>11 时，几乎未见产孢。表明 pH 值对菌丝生长及产孢影响基本相同，该菌适合在酸性至偏碱环境中生长。

2.4 培养基对菌丝生长和产孢量的影响

该菌在受试的 4 种培养基中均可生长。在 PDA 和 mSDA 上菌落直径较大，分别为 3.55cm 和 3.32cm；在 PDA 和 CzA 中产孢量较大，分别为 2.675×10^{7} 个和 2.205×10^{7} 个。表明 PDA 为该菌最适培养基。

2.5 碳源对菌丝生长和产孢量的影响

碳源对菌丝生长和产孢量的影响，各种碳源中，该菌在以麦芽糖及乳糖为碳源下生长较快，直径分别为 3.43cm 和 3.2cm，在可溶性淀粉中生长最慢，为 1.81cm；在阿拉伯糖为碳源下最易产孢，产孢量达 3.975×10^{7} 个，其次是蔗糖和麦芽糖，空白对照直径虽大，但菌落稀疏，产孢量最少。因此，麦芽糖和乳糖最利于该菌菌丝生长，阿拉伯糖最利于该菌产孢。

2.6 氮源对菌丝生长和产孢量的影响

实验结果表明，在各供试氮源中，该菌在以酵母浸膏、蛋白胨和牛肉膏为氮源的培养基上均良好生长，菌落生长致密且较快。以酵母浸膏直径最大，为 3.53cm，在硫酸铵中生长最慢，为 0.76cm；在酵母浸膏中产孢最强，高达 4.625×10^{7} 个；其次为硝酸钠和甘氨酸，产孢量分别是 2.3×10^{7} 个和 2.075×10^{7} 个，在硫酸铵和氯化铵中不产孢。空白对照直径虽大，但菌落稀疏，产孢量低。因此，酵母浸膏为该菌生长和产孢最适氮源。

3 结论与讨论

本文从营养特性和生态特性两个方面研究了多隔镰刀菌的生物学特性，这些特性的深入系统研究对了解该病原菌的适应能力、明确它引起的病害特点以及防控均具有重要意义。

研究结果表明：多隔镰刀菌对温度、pH 值和营养条件均具有较强的适应性，其中菌丝生长温度范围为 10～35℃，适宜温度 20～30℃，产孢温度 15～35℃，最适范围 20～25℃，说明该菌属于中温型微生物，与单隔镰刀菌（*F. dimerum*）、内生镰刀菌（*Fusarium*GI024）、茄病镰刀菌（*F. solani*）的适应温度基本相吻合。在 pH 值 5～12 范围内，该菌均能生长，且最适生长及产孢 pH 值为 5～9；不同镰刀菌对酸碱度的适应性存在差别，拟枝孢镰刀菌（*F. sporotrichioides*）适应偏碱环境，而单隔镰刀菌的最适 pH 值为 7，尖孢镰刀菌（*F. oxysporum*）最适 pH 值为 6.5。光照条件对该菌生长发育的影响不显著，黑暗对菌丝生长相对更有利，光暗交替更利于产孢，光照影响对内生镰刀菌同样不敏感，而光暗交替更适合尖孢镰刀菌生长。该菌能利用蔗糖、麦芽糖、乳糖、果糖、微晶纤维素和葡萄糖等多种碳源，以及牛肉膏、酵母浸膏和蛋白胨等多种氮源，可见，该菌具有较广的营养适应性，这与其他多种镰刀菌的报道基本一致。

国槐溃疡病是国槐苗期较为普遍发生的一种枝干病害，危害严重，死亡率高。目前对该病害的了解仍局限于一般的病害调查、发生规律及简单防治等方面，而病害的侵染循环、致病机理及预测预报等均不清楚。对其致病病原菌生物学特性的系统研究将能够为这些研究提供基础，加快其研究进度，有效防控该病害的发生，确保园林绿化的质量。

参考文献

［1］　桑娟萍，席忠诚．国槐枝枯病的综合防控［J］．中国林业，2012，5：78-80.

［2］　朱迎迎，高兆银，李敏，等．火龙果镰刀菌果腐病病原菌鉴定及生物学特性研究［J］．热带作物学报，2016，37（1）：164-171.

［3］　蒋继宏，陈凤美，朱红梅，等．银杏内生镰刀菌 GI024 生物学特性［J］．浙江林学院学报，2004，21（3）：299-302.

［4］　杨静美，陈健，冯岩，等．番木瓜茄病镰刀菌的生物学特性研究［J］．中国热带农业，2011，38（1）：56-58.

［5］　潘龙其，张丽，杨成德，等．紫花苜蓿根腐病原菌——拟枝孢镰刀菌的鉴定及其生物学特性研究［J］．草业学报，2015，24（10）：88-98.

［6］　孔琼，王云月，朱有勇，等．香荚兰尖孢镰刀菌生物学特性［J］．西南农业学报，2015，18（1）：47-49.

［7］　徐法燕．山东聊城市城区国槐衰弱死亡原因的初步研究［J］．中国园艺文摘，2016，9：82-84.

分离于澳大利亚的核盘菌菌株病毒多样性分析*

穆　凡[1,2**]，成淑芬[1,2]，谢甲涛[1,2]，程家森[1,2]，付艳苹[1]，陈　桃[1]，姜道宏[1,2***]
（1. 湖北省作物病害监测和安全控制重点实验室（华中农业大学），武汉　430070；
2. 农业微生物学国家重点实验室（华中农业大学），武汉　430070）

摘　要：核盘菌（*Sclerotinia sclerotiorum*）是一种典型的死体营养性病原真菌，可以侵染多种重要的经济作物以及油料作物等，引起作物菌核病。真菌病毒是侵染真菌的病毒，核盘菌中也蕴含着丰富的真菌病毒。本研究对来自于澳大利亚多种寄主上的核盘菌菌核进行纯培养，获得了84株纯培养菌株，对这些菌株在菌落形态、致病力和生长速度等方面呈现多样性。为了分析分离菌株中含有的病毒的种类，进行高通量宏病毒组测序分析。发现了59种不同的核盘菌病毒信息，这些病毒序列中62.7%是未曾报道新病毒序列。根据核酸类型，59个病毒中包含dsRNA病毒、+ssRNA病毒、-ssRNA病毒和ssDNA病毒；基于病毒复制酶序列，这些病毒主要隶属于以下10个病毒科（目）的病毒，即内源RNA病毒科（5个新的病毒）、类双生病毒科（核盘菌弱毒相关DNA病毒1的不同株系）、低毒病毒科（2个新病毒）、真菌单股负链RNA病毒（4个新的病毒）、裸露病毒科（11个新的病毒）、双分病毒科（2个新的病毒）、欧密尔病毒属（1个新的病毒）、番茄丛矮病毒科（2个新的病毒）、全病毒科（首次在核盘菌中发现全病毒科的病毒）和芜菁花叶病毒目（7个新的病毒）；此外，还发现了两个未分类的病毒，一个与病毒Botrytis porri RNA virus 1有同源性，另一个与新报道的dsRNA病毒Aspergillus fumigatus tetramycovirus-1有同源性。本研究首次报道来源于澳大利亚的核盘菌菌株中蕴含丰富的病毒多样性，研究结果极大的丰富了真菌病毒的种类，对研究病毒进化、病毒性生防资源提供了丰富的生物材料。

关键词：核盘菌；菌株；多样性

* 资助基金：油菜现代产业技术体系岗位科学家科研专项资金（CARS-13）
** 第一作者：穆凡，女，硕士研究生，主要从事植物病害生物防治（真菌病毒）相关研究；E-mail：1031662808@qq.com
*** 通信作者：姜道宏，教授，主要从事分子植物病理学及生物防治的相关研究；E-mail：daohongjiang@mail.hzau.edu.cn

根癌农杆菌介导尖孢镰刀菌的遗传转化及致病性缺陷突变体的筛选[*]

董燕红[**]，刘丽媛，刘力伟，王树桐[***]，曹克强[***]
（河北农业大学植物保护学院，保定 071001）

摘 要：苹果再植病害（Replant disease of apple，ARD）在世界苹果主产区广泛发生。该病害的病原较为复杂。本研究组前期研究中发现尖孢镰刀菌（*Fusarium oxysporum*，Fo）是河北省苹果再植病害的重要致病菌之一。当前关于引起苹果根部病害的致病尖孢镰刀菌致病机理的研究还较少。为了解尖孢镰刀菌的致病机理，从基因层面揭示病原菌生长发育及致病相关基因，探索病害防治的新靶标，本研究对根癌杆菌介导的尖孢镰刀菌的遗传转化的体系进行优化并构建其突变体库，获得了致病力缺陷的单拷贝插入突变菌株。对影响 Fo-HS2 菌株农杆菌介导转化效率的主要因子进行了单因子条件测验，得到其遗传转化体系为：预诱导 4h，分生孢子浓度 10^6 个/mL，农杆菌浓度 $OD_{600}=0.3$，乙酰丁香酮浓度 200μmol/mL，19℃，共培养时间 48h。利用这一体系进行 Fo-HS2 菌株突变体库的构建，以对潮霉素抗性基因的 PCR 检测呈阳性的转化子作为插入突变体。经过在不含潮霉素的 PDA 培养基上 5 代培养，验证 T-DNA 插入突变体的遗传稳定性。对稳定遗传的转化子进行产孢和致病力试验分析得到的致病性缺陷的突变体中，用Southern blot 技术验证致病力变弱的突变菌株 T-DNA 插入拷贝的拷贝数，选取 8 株突变体进行Southern杂交分析，结果显示其中 6 株突变体为单拷贝插入，1 株为双拷贝插入。用 TAIL-PCR 技术对产孢量和致病力下降且为单拷贝插入的突变体分别进行了侧翼序列扩增，共获得了 3 条侧翼序列，经过分析，其中突变菌株 HS2-100 的右翼序列与尖孢镰刀菌假定蛋白的 mRNA 序列同源，HS2-100 的左翼序列与层出镰刀菌 D02945 的 *des* 基因、*P*450-4 基因和 *P*450-1 基因为同源序列，HS2-1107 左翼序列与构巢曲酶 FGSCA4 c 的 VIII 染色体的 2 848 590~2 848 704bp 区域序列同源。

关键词：尖孢镰刀菌；农杆菌介导遗传转化；TAIL-PCR；苹果再植病害

---

* 基金项目：国家苹果产业技术体系（CARS-28）和河北省自然科学基金（c2016204140）
** 作者简介：董燕红，在读硕士研究生，研究方向：植物病害流行与综合防治；E-mail：2550558743@qq.com
*** 通信作者：王树桐，博士，教授，从事植物病害流行与综合防治研究；E-mail：bdstwang@163.com
曹克强，博士，教授，从事植物病害流行与综合防治研究；E-mail：ckq@hebau.edu.cn

根肿菌 MAPK 级联在病原菌致病过程中生物学功能初步研究*

赵艳丽[1,2**]，陈　桃[1,2]，谢甲涛[1]，程家森[1]，付艳苹[1]，姜道宏[1,2***]
（1. 湖北省作物病害监测和安全控制重点实验室（华中农业大学），武汉　430070；
2. 农业微生物学国家重点实验室（华中农业大学），武汉　430070）

摘　要：有丝分裂原活化蛋白激酶（MAPK）在细胞生长、繁殖和生存中扮演着重要的角色。在模式植物、哺乳动物、酵母和真菌中，MAPK 级联相关基因已经被广泛的研究，但是在根肿菌（*Plasmodiophora brassicae*）与寄主互作中，该类基因的功能还未曾被报道。由根肿菌引起的十字花科植物根肿病是一种重要病害，对油菜等作物的危害日渐严重。前期笔者课题组完成了根肿菌基因组测序工作，分析预测到根肿菌基因组中含有 7 个 *MAPK*、3 个 *MAPKK* 和 9 个 *MAPKKK* 基因。转录组分析表明，其中的 6 个 *MAPK*，3 个 *MAPKK* 和 2 个 *MAPKKK* 基因在根肿菌的皮层侵染阶段上调表达。通过酵母双杂交分析，在酵母中证实了 PbMAKKK7 与 PbMAKK3 存在相互作用，PbMAKK3 和 PbMAK1，PbMAKK3 和 PbMAK3 存在相互作用。因此，推测在根肿菌侵染寄主植物过程中可能存在 PbMAKKK7-PbMAKK3-PbMAK1/PbMAK3 级联反应。U0126 是 MAPKK 的一种特异性的抑制剂，它可以抑制根肿菌休眠孢子的萌发。用 U0126 抑制剂处理根肿菌的休眠孢子后接种油菜，液体培养 35d 后，统计结果表明 U0126 抑制剂可以减轻根肿病的病情指数，降低发病根部根肿菌的含量，并且推迟了根肿菌的发育状态，阳性对照植株根肿菌处于休眠孢子状态，而抑制剂处理后根肿菌仅处于次级游动孢子囊的状态。综合以上研究结果，MAPK 信号通路可能在根肿菌的生长、发育和致病过程中起着重要的作用。本研究结果将为根肿菌的致病机理提供分子生物学证据，也为研究根肿菌的基因功能的方法提供参考。

关键词：根肿菌；致病；分子生物学

---

* 资助基金：油菜现代产业技术体系岗位科学家科研专项资金（CARS-13）

** 第一作者：赵艳丽，女，硕士研究生，主要从事分子植物病理学相关研究；E-mail：2548774230@ qq. com
*** 通信作者：姜道宏，教授，主要从事分子植物病理学及生物防治的相关研究；E-mail：daohongjiang@ mail. hzau. edu. cn

广西珍珠李真菌性溃疡病研究初报*

杨 迪，杜婵娟[1]，付 岗[1**]，潘连富[1]，张 晋[1]，叶云峰[2]，陈伯伦[2]
（1. 广西农业科学院微生物研究所，南宁 530007；
2. 广西农业科学院园艺研究所，南宁 530007）

摘 要：珍珠李是从广西天峨县重要的特色水果之一，该品种是从当地野生李中选育出的一个特晚熟李品种，2009 年获得品种审定。近年来珍珠李产业已发展成为当地的重要扶贫产业。随着种植面积的逐年扩大，珍珠李病害的发生也日益严重，溃疡病就是其中为害严重的病害之一。该病主要为害李树枝干，主干、主枝和各级侧枝均可受害。果树修剪时形成的剪锯口是病菌最佳侵入途径，其他机械伤口或皮孔也是重要侵染点。横切发病枝干，可见横截面髓部及木质部变褐色、红褐色或黑褐色。变色部位以枝干分叉处面积最大，沿分叉处向两端延伸，褐变面积逐渐变小。受害枝干进入生长季后，常发生流胶症状。感病枝条上的叶片稀少，部分叶片沿边缘向内变枯黄，后期落叶。

采用组织分离法，从染病枝干处分离获得病原菌 Bd1 菌株。该菌在 PDA 培养基上菌落呈灰黑色，分生孢子长梭型或长椭圆形，顶端钝圆，基部平截，单胞无色，大小为（4.7~8.3）μm×（13.6~31.5）μm。分生孢子器黑色，近球型，丛生或单生。病菌经单孢纯化后，进行 rDNA 的 ITS 序列测定，获得 558 bp 的序列（GenBank 登录号：MF040305）。Blast 同源比对结果表明：菌株 Bd1 与葡萄座腔菌 *Botryosphaeria dothidea* 的同源性最高，为 99%。结合形态学观测结果，将该菌鉴定为葡萄座腔菌。

关键词：珍珠李；溃疡病；*Botryosphaeria dothidea*；鉴定

---

* 基金项目：2016 年中央引导地方科技发展专项资金
** 通信作者：付岗，博士，副研究员，主要从事植物病害及其生物防治研究；E-mail：fug110@ gxaas. net

The Identification of Tea Gray Blight Disease in Huishui County, Guizhou Province, China*

Li Dongxue[1**], Zhao Xiaozhen[1], Bao Xingtao[1],
Ren Yafeng[1], Wang Yong[2], Li Xiangyang[1],
Yu Haiyou[3], Tan Xiaofeng[4], Chen Zhuo[1***], Song Baoan[1]
(1. *Key Laboratory of Green Pesticide and Agricultural Bioengineering*, *Ministry of Education*, *Guizhou University*, *Guiyang*, 550025, *China*; 2. *Agricultural College*, *Guizhou University*, *Guiyang*, 550025, *China*; 3. *The Joint Office of Tea industry of Guizhou Province*, *Guiyang* 550001, *China*; 4. *Plant Protection Station of Guizhou Province*, *Guiyang*, 550025, *China*)

Abstract: Tea gray blight disease is an important disease of the tea tree, which can harm tea seedlings, as well as the shoots, fresh leaves and mature leaves of trees for various growth periods. The disease can cause the tea plants weakened and severethe production loss of tea leaves. Tea gray blight disease is one of the most important diseases in Guizhou tea region. In this work, the pathogenic fungus was isolated, purified from the diseased tea leaves, and confirmed the pathogenicity of the pathogen according to Koch's rule. The isolated strain of DX - 1 was further identified by the morphology and molecular biology. The study indicated that DX-1 represents the features as following: on PDA medium, the colonyis obviously wheel round like, margin regular of the colonies; Reverse represents light yellow colony with white or colorless mycelium. The observation using the optical microscope and SEM indicated that theconidia represent fusi form with 4 diaphragms and 2 thin ends, and 2 - 3appendages and 1appendage exist onthe top and the tail end of the cell, respectively. Conidia with (16.07±1.24) μm×

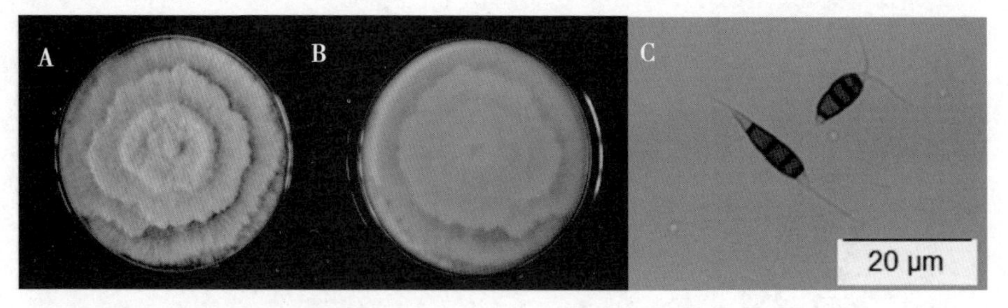

Figure. The morphology of DX-1 on PDA medium (A) and (B, reverse), and the conidia observed by optical microscope (C)

* Funding: Major Science and Technology Projects in Guizhou Province (No. 2012-6012), Program for New Century Excellent Talents in University (NCET-13-0748), Projects for Guizhou Provincial Department of Science and Technology-The People's Government of Qiannan Cooperation on Agricultural Science and Technology (No. 2013-01). We also thank Prof. Song Baoan for designing the research work

** First author: Li Dongxue; E-mail: 1973266776@ qq. com

*** Corresponding author: Chen Zhuo; E-mail: gychenzhuo@ aliyun. com

(4. 23±0. 20) μm is colorless in the early stage, and the 3 septum near the center change brown in the late stage. The genes of *EF*－1α, β-*tubulin* and *LSU* were further amplified, purified and sequenced, and then the phylogenetic trees were constructed using MEGA6. 0 software according to Neighbor-joining method. The results indicated that DX-1strain was closely related to *Pestalotiopsis* sp. AAHE1 (KJ623232. 1) and YN1B4 (KU252565. 1) against two phylogenetic treesof *EF*-1α and β-*tubulin*; and then clustered together with *P. theae* strain 3. 919 2 (JN940840. 1) in a branch, and closely related to *P. adusta* stain 3. 909 8 (JN940825. 1) for the phylogenetic tree of *LSU*.

Key words: Tea gray blight disease; Pathogenicity; *Pestalotiopsis theae*

A New Pathogen of Tea Diseases was Found in Huishui County, Guizhou Province, China*

Li Dongxue[1**], Zhao Xiaozhen[1], Bao Xingtao[1],
Ren Yafeng[1], Li Xiangyang[1], Tan Xiaofeng[2],
Yang Song[1], Chen Zhuo[1***], Song Baoan[1]
(1. *Key Laboratory of Green Pesticide and Agricultural Bioengineering*, *Ministry of Education*, *Guizhou University*, *Guiyang*, 550025, *China*; 2. *Plant Protection Station of Guizhou Province*, *Guiyang*, 550025, *China*)

Abstract: The tea diseases are frequently occurred in Guizhou tea region because of being always cloudy, rainy, and low temperature. The tea diseases cause greatly loss of the tea qualityand production for being an amount of disease kinds and occurring in large area of Guizhou. It will contribute to the prevention and control of the tea disease for the classification and identification of the pathogen of tea tree diseases. In order to explore the pathogen of tea diseases in Qiannan area of Guizhou Province, we isolated, purified the pathogen from diseased tree leaves, and then determinedits pathogenicity according to Koch's rule for the sample of Qilichong tea garden in Huishui County. The isolated strains of HS-1 ~ HS-4 were identified by using the optical microscope, SEM and gene sequence. The results indicated that the colonies on PDA medium represent white, sparse aerial mycelium. Reverse represent black brown near the center in the early stage; After 3 days of inoculation, the colonies gradually change to gray, aerial mycelium woolly. Reverse represent black brown, with being gradually expanded in the area. The growth rate of the colonies on OA medium was significantly slower than that of PDA medium. Conidia can produced on OA medium, single-spore, ovoid [(12.79 ± 1.41) μm× (6.68 ± 0.24) μm], with one slightly pointed end and/oranother blunt end, smooth cell wall in the early stage. Conidia display a shallow stripes, a septum on the center, and the colorof the cell with inhomogeneityin the late stage. Along with the incubation time being prolonged, the color of cell changes from colorless to black brown. The genes of *ITS*, *β-tubulin*, *EF*-1α and *LSU* were amplified, purified and sequenced, and then identified 4 strains were *Lasiodiplodia theobromae* by Blast software. This work is firstly found *L. theobromae* can cause tea leaves diseases, and the pathogenic mechanism is worthy of being studied in future.

Key words: Tea diseases; *Lasiodiplodia theobromae*; Identification; Pathogenicity

* Funding: This work was supported by the Major Science and Technology Projects in Guizhou Province (No. 2012-6012), Projects for Guizhou Provincial Department of Science and Technology-The People's Government of Qiannan Cooperation on Agricultural Science and Technology (No. 2013-01). We also thank Prof. Song Baoan for designing the research work

** First author: Li Dongxue; E-mail: 1973266776@ qq. com

*** Corresponding author: Chen Zhuo; E-mail: gychenzhuo@ aliyun. com

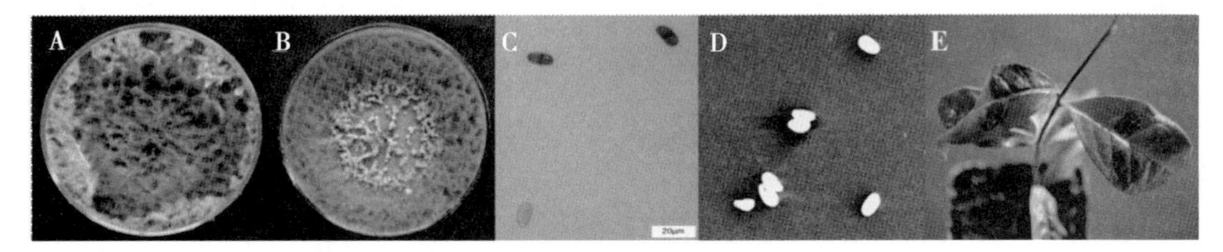

Figure. The morphology of HS-2 on PDA medium（A）, OA medium（B）, theconidia observed by optical microscope（C）and SEM（D）, and the tea leaves is inoculated by the strain（E）for 6 dpi

The Identification of a Pathogenic Fungus from *Gloeosporium theae-sinensis* Miyake in Yuqing County, Guizhou Province, China*

Li Dongxue[1**], Zhao Xiaozhen[1], Bao Xingtao[1],
Ren Yafeng[1], Li Xiangyang[1], Yu Haiyou[2],
Tan Xiaofeng[3], Chen Zhuo[1***], Song Baoan[1]
(1. *Key Laboratory of Green Pesticide and Agricultural Bioengineering, Ministry of Education, Guizhou University, Guiyang*, 550025, *China*; 2. *The Joint Office of Tea Industry of Guizhou Province, Guiyang* 550001, *China*; 3. *Plant Protection Station of Guizhou Province, Guiyang*, 550025, *China*)

Abstract: *Gloeosporium Theae-Sinensis* Miyake is an important tea disease, which mainly damage the mature leaf. The disease can cause an amount of fallen leaves when it occurs seriously. The disease mainly occurs in the season of high humidity and rainy day, and then become a main disease of tea in Guizhou Province. This work isolated the pathogen from the diseased leaves from the tea, and determined the pathogenicity of the pathogen according to Koch's rule. Based on this, we obtained 3 strains of GZYQ-1, GZYQ-2, and GZYQ-3. GZYQ-2 strain was further studied and the results indicated that margin regular of the colonies with white and villiform, aerial mycelium developed near the center, along with being upheaved; reverse faint yellow near the center, with the white margin on PDA medium in the early stage; The colonies with light color near the center, and deep color along the margin. Reverse the area of the colonies with faint yellow being gradually increased, and deep color near the center, light color along the margin, with few black dots. The conidium did not form on PDA medium. The genes of *ITS* and *β-tubulin* were amplified and sequenced, and the sequence alignment was conducted using BLAST software. GZYQ-2 strain and the strains with high identity were used to construct the phylogenetic trees using MEGA 6.0 software according to Neighbor-Joining method. The results indicated that GZYQ-2 is closely related to *Colletotrichum ignotum* isolate Col1 (JX515292.1) and *C. kahawae* (GU174550.1) for ITS gene. GZYQ-2 and *C. kahawa* (AY245019.1) is converged to a branch, closely related to *C. tropicale* CMM4243 (KU213604.1).

Key words: *Camellia sinensis*; *Gloeosporium theae-sinensis* Miyake; Identification

---

* Funding: This work was supported by the Major Science and Technology Projects in Guizhou Province (No. 2012-6012). We also thank Prof. Song Baoan for designing the research work

** First author: Li Dongxue; E-mail: 1973266776@qq.com

*** Corresponding author: Chen Zhuo; E-mail: gychenzhuo@aliyun.com

A Pathogenic Fungus *Phoma* sp. was Identified from Tea Leaf Spot Disease in Shiqian County, Guizhou Province, China*

Zhao Xiaozhen[1**], Li Dongxue[1], Bao Xingtao[1], Duan Changliu[2], Ren Yafeng[1], Li Xiangyang[1], Yu Haiyou[3], Yang Song[1], Chen Zhuo[1***], Song Baoan[1]

(1. *Key Laboratory of Green Pesticide and Agricultural Bioengineering, Ministry of Education, Guizhou University, Guiyang*, 550025, *China*; 2. *Tea Administration of Shiqian County, Shiqian Guizhou* 558000, *China*; 3. *The Joint Office of Tea industry of Guizhou Province, Guiyang* 550001, *China*)

Abstract: The tea leaf spot disease is always taken place in the region with the characteristics of the low-temperature and high-humidity containing with Shiqian County, Guizhou Province. The disease always infects the tender leaves and shoots, which lead to the lowest tea quality and the significant production loss. To identify the pathogen of the tea spot disease, this work isolated and identified the pathogen of this disease from the tea leaves, and determined the pathogenicity of the pathogen according to Koch's rule. The study indicated that the isolated strain of GZSQ-4 represent the features as following: on PDA medium, margin regular of the colonies, aerial mycelium closely growing, white mycelium in the early stage; then gradually change to grey brown, black brown in the center, with white margin in the latter stage; reverse brown to black brown near the center, with the white margin. On OA medium, margin regular of the colonies, aerial mycelium woolly, pale brown to brown near the center, and produce a lot of black pycnidia. Using the optical microscope and SEM indicated that the septate mycelium contains the inclusion granule; Pycnidia represent subglobose, glabrous, thin-walled, solitary or confluent, 50~190μm, 1 or 2 ostioes with a neck; Conidia represent ellipsoidal or ovoid, thin-walled, smooth, hyaline, aseptate, (2~4) μm× (1~2.5) μm, with 0~3 guttules in two polar ends. The genes of *ITS*, *β-tubulin* and *LSU* were further amplified, purified and sequenced, then the phylogenetic trees were constructed using MEGA6.0 software and Neighbor-joining method. The results indicated that the strainof GZSQ-4 was closely related to the strains of *Phoma* sp. QC-2014b (LC 1633) and Didymella bellidis (CBS 714.85). Integrated with the above results, the pathogen isolated by our research group are preliminarily identified to *Phoma* sp..

Key words: *Phoma* sp.; Tea leaf spot disease; Pathogenic fungus

* Funding: This work was supported by the Major Science and Technology Projects in Guizhou Province (No. 2012-6012). We also thank Prof. Song Baoan for designing the research work

** First author: Zhao Xiaozhen; E-mail: xzzhao@aliyun.com

*** Corresponding author: Chen Zhuo; E-mail: gychenzhuo@aliyun.com

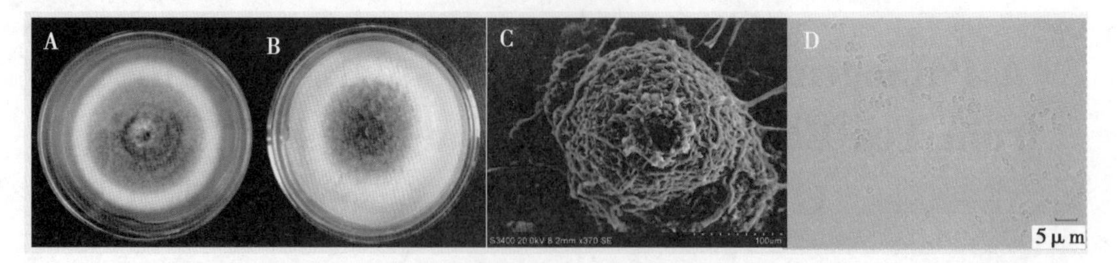

Figure. The morphology of *Phoma* sp. about the colony on PDA medium（A）, colony on OA medium（B）, the pycnidium observed by SEM（C）and the conodia observed byoptical microscope（D）

The Correlative Study between the Four Tea Diseases and Endophytes in Tea Tree*

Bao Xingtao[1**], Zhao Xiaozhen[1], Li Dongxue[1], Ren Yafeng[1], Li Xiangyang[1], Wang Yong[1,2], Yu Haiyou[3], Yang Song[1], Chen Zhuo[1**], Song Baoan[1***]

(1. *Key Laboratory of Green Pesticide and Agricultural Bioengineering*, *Ministry of Education*, *Guizhou University*, *Guiyang*, 550025, *China*; 2. *Agricultural College*, *Guizhou University*, *Guiyang*, 550025, *China*; 3. *The Joint Office of Tea industry of Guizhou Province*, *Guiyang* 550001, *China*)

Abstract: The endophytes of tea tree play an important role during the occurrence and development of the tea diseases. At present, there has no research report about the correlations study between the tea diseases and endophytes. In this study, the samples of tea blister blight, tea white scab, tea round red spot and tea anthracnose were collected in tea garden from Meitan, Wuchuan, Dejiang, Shiqian, Duyun, Weng'an County in Guizhou Province and Anxi County in Fujian Province in 2016. *ITS* gene for the pathogens for four tea diseases and specific *ITS* gene for three endophytes was amplified and sequenced. The results indicated that *Trichoderma* sp. exists in tea blister blight, tea white scab and tea round red spot, *Guignardia mangiferae* exists in tea blister blight, and *Trichoderma longibrachiatum* exists in tea blister blight, tea white scab and tea anthracnose. In addition, two fungus of *T. longibrachiatum* combined with *T.* sp. or *G. mangiferae* was simultaneously detected from tea blister blight, two fungus of *T. longibrachiatum* combined with *T.* sp. was simultaneously detected from tea white scab, and only endophyte was detected from tea anthracnose. This study provided apreliminary implication for the interaction relation between tea diseases and endophytes, and was worthy to further be studied in future.

Key words: Tea diseases; Endophytes; Identification; Correlative study

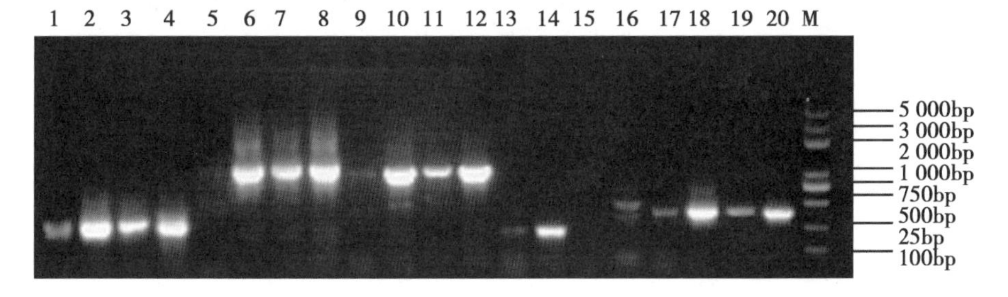

Figure. The detection of *ITS* gene for three endophytes by agarose gel electrophoresis. Lane 1-4 represents*T. longibrachiatum* from Wuchuan, Anxi, Dejiang, Meitan County, Lane 6-8 represents *T.* sp. from Anxi, Dejiang, Meitan and lane 17-20 represents *G. mangiferae*from Wuchuan, Anxi, Dejiang, Meitan.

---

* Acknowledgments: This work was supported bythe Major Science and Technology Projects in Guizhou Province (No. 2012-6012). We also thank Prof. Song Baoan for designing the research work

** First author: Bao Xingtao; E-mail: xingtaobao@ outlook. com

*** Corresponding author: Chen Zhuo; E-mail: gychenzhuo@ aliyun. com

禾谷镰刀菌甾醇 14α-脱甲基酶（CYP51B）与三唑类杀菌剂互作研究*

钱恒伟[1**]，迟梦宇[1]，赵 颖[1]，黄金光[1,2***]
（1. 青岛农业大学农学与植物保护学院，青岛 266109；
2. 山东省植物病虫害综合防控重点实验室，青岛 266109）

摘 要：由禾谷镰刀菌引起的小麦赤霉病是全球性的麦类真菌性病害，不仅降低产量造成巨大的经济损失。另外，染病麦粒中所含有的真菌毒素对人畜健康安全产生了严重的威胁。目前，生产上对于小麦赤霉病的防治主要依靠化学药剂防治。近年来在生产上，三唑类药剂（戊唑醇，三唑酮）逐渐成为防治小麦赤霉病的主要杀菌剂。该类杀菌剂主要作用于甾醇生物合成中的14α-脱甲基酶（CYP51B），CYP51B 蛋白氨基酸的突变是病原菌对三唑类产生抗药性的一个重要机制。为了探究禾谷镰刀菌 CYP51B 蛋白与三唑类杀菌剂精细互作，明确禾谷镰刀菌对三唑类杀菌剂潜在的抗性风险与机制。作者试图通过结构生物学的方法来解析禾谷镰刀菌 CYP51B 蛋白与三唑类杀菌剂之间的互作机制以及进一步发现在互作过程中发挥作用的氨基酸残基。通过以 AfCYP51B 蛋白晶体结构为模板对 FgCYP51B 蛋白进行了同源建模，并在此基础上与三唑类杀菌剂戊唑醇、三唑酮、丙环唑进行了分子对接，并找到了 FgCYP51B 蛋白与三唑类杀菌剂互作中发挥重要作用的位点，包括 Phe511，Val136，le374，Ala308，Ser312，Try137，并结合不同物种 CYP51 蛋白的保守氨基酸序列分析，利用基因定点突变技术将禾谷镰刀菌 FgCYP51B 蛋白的 123 137位酪氨酸改变为组氨酸，并获得转化子，测定了突变体对戊唑醇、丙环唑和三唑酮的敏感性。其中，Y123H 突变体对戊唑醇、丙环唑和三唑酮的敏感性均无变化，而 Y137H 突变体对戊唑醇杀菌剂敏感性降低，出现抗性现象。说明 FgCYP51B 与戊唑醇互作过程中 137 位酪氨酸起着一定的作用，但对丙环唑和三唑酮的敏感性均无变化。

关键词：禾谷镰刀菌；CYP51B；三唑类杀菌剂；原核表达；同源建模；定点突变；敏感性

---

* 基金项目：国家自然基金项目（31471735）；青岛市现代农业产业技术体系（6622316106）
** 第一作者：钱恒伟，硕士研究生，植物病理专业
*** 通信作者：黄金光，教授，研究方向：蛋白质结构生物学，分子植物病理学；E-mail：jghuang@ qau. edu. cn

The Occurrence and Pathogen Identification of Powdery Mildew on *Avena sativa* in China*

Xue Longhai**, Li Chunjie***

(*State Key Laboratory of Grassland Agro-Ecosystems, College of Pastoral Agriculture Science and Technology, Lanzhou University, P. O. Box* 61, *Lanzhou* 730020, *China*)

Abstract: Oat (*Avena sativa*) is extensively planted as a fodder crop in China. Most of the oats are used for livestock feed. In March 2017, powdery mildew was observed on *A. sativa* in the Forage Germplasm Nursery in Xinjin county, Sichuan Province, China. Approximately 40% of plants (about 0. 2hm^2 in total) were affected. The purified isolates were determined for pathogenicity with Koch's postulates. In order to clear this pathogen and it's taxonomic status, the pathogen was identified by morphology and molecular biology identification. Based on morphological characteristics of the anamorph and the teleomorph, the fungus was identified as *Blumeria graminis* (DC.) Speer. Conidiophores were unbranched and cylindrical with swollen bases, measuring (6. 3~13. 2) μm× (23. 4~67. 5) μm, and borne vertically on hyphae. Each conidiophore produced 3~7 conidia (mostly 6 conidia) in a chain. The conidia were oval, one-celled, and colorless, measuring (11. 2~17. 5) μm× (20. 2~43. 6) μm. Cleistothecia and asci were not observed. Sequence analysis of rDNA-ITS region showed that the identification had 100% homology with this of *B. graminis* on *A. sativa* in Swizerland (AF011284) in GenBank. To our knowledge, this is the first report of powdery mildew of oat caused by *Blumeria graminis*in China. Its confirmation is a significant step toward management recommendations for growers.

Key words: *Avena sativa*; Identification; *Blumeria graminis*

* 基金项目：国家公益性行业农业科技专项——草地病害防治技术研究与示范（201303057）

** 第一作者：薛龙海，男，河南开封人，研究生，主要从事植物病理学方向的研究；E-mail：xuelh16lzu. edu. cn

*** 通信作者：李春杰，男，二级教授，主要从事植物病理学和禾草内生真菌方向的研究；E-mail：chunjie@ lzu. edu. cn

Identification of a new root rot disease on *Aconitum carmichaelii* in Sichuan [*]

Xue Longhai[**], Liu Yong[***]

(*Plant Protection Institute*, *Sichuan Academy of Agricultural Science*, *Chengdu* 610066, *China*)

Abstract: A new root rot disease was found on *Aconitum carmichaelii* and caused severity loss in Jiangyou district of Sichuan Province. The purified isolates were determined for pathogenicity with Koch's postulates and sequenced based on ribosomal DNA-ITS. The results showed that the isolates were the new pathogen of *A. carmichaelii*. The pathogen formed white colony on clarified PDA agar, often forming mycelial fans. Pure cultures were prepared by transferring single hyphal tips to PDA. The radial mycelial growth was 16 mm/day, numerous globoid sclerotia were formed on PDA after 14 days of growth. Sclerotia were initially white becoming dark brown with age and were 0.85 to 1.16 mm (average = 1.02 mm, n = 50) in diameter at maturity. These are typical features of *Sclerotium rolfsii*. DNA sequences of rDNA-ITS shared the highest similarity and had 99% homology with known *Athelia rolfsii*. These results confirmed that the pathogen caused root rot disease on *A. carmichaelii* was *A. rolfsii*. According to our knownledge, this is the first report of new host of *A. rolfsii* and new disease of *A. carmichaelii*.

Key words: *Aconitum carmichaelii*; Identification; *Athelia rolfsii*

* 基金项目：国家公益性行业农业科技专项——草地病害防治技术研究与示范（201303057）

** 第一作者：薛龙海，男，河南开封人，研究生，主要从事植物病理学方向的研究；E-mail：xuelonghai55@126.com

*** 通信作者：刘勇，男，研究员，主要从事植物病理学方向的研究；E-mail：liuyongdr@163.com

两个不同致病类型小麦叶锈菌株差异表达分析及其分泌蛋白的筛选

韦　杰，王逍东，杜东东，张　娜，孟庆芳，杨文香，刘大群
（河北农业大学植物病理学系/河北省农作物病虫害生物防治工程技术研究中心/国家北方山区农业工程技术研究中心，保定　071001）

摘　要：小麦叶锈菌是小麦的重要病原物，防治由该菌引起的小麦叶锈病最有效、经济、安全的方法是培育抗病品种。开展对小麦叶锈菌致病性差异研究对于揭示小麦叶锈菌致病分子机制及揭示毒性易变区域，有效利用抗病基因具有重要意义。本试验采用两种不同致病类型叶锈菌 08-5-361-1（THTT）和 09-12-284-1（THTS），分别接种到感病小麦品种 Thatcher 上，在吸器大量出现时提取接种小麦叶片总 RNA，使用 Illumina Solexa（深圳华大基因公司）平台对两个 RNA 样品（样品 ID 分别为 Tc284 和 Tc361）进行数字表达谱测序，分别命名为数据库 Tc284 和 Tc361，以已有的数据库 Tc15_ 2 为参考数据库。以样品 Tc361 为处理组，以 Tc284 为对照组进行基因差异表达分析，一共得到了 21 172个表达量不同的基因，其中有 1 367个基因在Tc361-1 中特异表达，有 1 222个基因在 Tc284-2 中特异表达。经过差异表达基因筛选（FDR≤0.001 且倍数差异不低于 2 倍），得到了 2 784个差异基因，其中 1 708个上调，1 076个下调。在 Tc361_ 1VSTc284_ 2 上调的基因中，45 个基因在 Tc361-1 特异表达；在下调的基因中，有 26 个基因在 Tc284-2 中特异表达。笔者选取了 11 个基因进行 RT-qPCR 分析，这些结果为今后分泌蛋白的研究及阐明小麦叶锈菌致病机制奠定基础。

关键词：小麦叶锈菌；RNA-seq 分析；差异表达分析；RT-qPCR

尖孢镰刀菌细胞壁降解酶活性研究*

艾聪聪，惠金聚，王桂清**，马 迪
（聊城大学农学院，聊城 252000）

摘 要：为了明确尖孢镰刀菌引起百合鳞茎腐烂病的致病机理，采用比色法测定了其不同细胞壁降解酶的活性。研究结果表明：尖孢镰刀菌可以产生纤维素酶、半纤维素酶、果胶酶、木质素酶和蛋白酶，说明细胞壁降解酶是尖孢镰刀菌主要的致病因子之一；该菌产生的 Cx 酶、果胶酶、木聚糖酶和蛋白酶活性较高，均大于 0.2 U/mL，说明这些酶系是尖孢镰刀菌主要的致病酶系；该菌可以产生木质素酶，但仅局限与胞内（活力为 0.437 4~28.493 8U/mL），而不分泌到胞外，胞外活力均为负值，说明木质素酶不是其主要的致病酶系。

关键词：尖孢镰刀菌；细胞壁降解酶；百合鳞茎腐烂病

Study on cell wall degrading enzymeactivity of *Fusarium oxysporum*

Ai Congcong, Hui Jinju, Wang Guiqing, Ma Di
(*College of Agronomy*, *Liaocheng University*, *Liaocheng* 252059, *Shandong*, *China*)

Abstract: In order to clarify the pathogenic mechanism of lily bulb rot caused by *Fusarium oxysporum*, the activity of different cell wall degrading enzymes was determined by colorimetry. The results showed that *F. oxysporum* could produce cellulase, hemicellulase, pectinase, ligninase and protease, which indicated that cell wall degrading enzymesare one of the major pathogenic factors of *F. oxysporum*. The activities of Cx, pectinase, xylanase and protease were higher than 0.2U/mL, indicating that these enzymes are the main pathogenic enzymes of *F. oxysporum*. It could produce ligninase, but only limited tointracellular (0.437 4~28.493 8U/mL), but not secreted to extracellular, extracellular activity is negative, indicating that ligninase is not its main pathogenic enzyme system.

Key words: *Fusarium oxysporum*; Cell wall degrading enzymes; Lily bulb rot

镰刀菌属（*Fusarium*）是农作物最重要的病原真菌之一，是一种最为常见的寄生病原菌，严重影响植物的生长及生产，可侵染多种植物，造成萎蔫、根腐、穗腐等各种类型的腐烂病，导致严重的减产，造成重大的经济损失。尖孢镰刀菌（*F. oxysporum*）是引起百合鳞茎腐烂病的主要致病菌。目前学者对镰刀菌的致病机理研究较多，但未见关于百合鳞茎腐烂病致病机理的研究报道。本文将从纤维素酶、果胶酶等方面，测定细胞壁降解酶活性，以期探讨尖孢镰刀菌侵染百合

* 项目来源：山东省自然科学基金（ZR2012CL17）；聊城大学大学生科技文化创新基金项目（26312158803）
** 通信作者：王桂清，女，河北泊头人，博士，教授，主要从事植物保护的教学与科研工作

引起鳞茎腐烂病的致病机理。

1　材料与方法

1.1　供试菌株与培养

供试菌株为百合鳞茎腐烂病致病菌尖孢镰刀菌（*F. oxysporum*），保存于聊城大学植物病理学实验室。PDA 培养基（25±1）℃恒温光照培养箱培养 5d，备用。

1.2　细胞壁降解酶的提取

1.2.1　胞内酶的提取

用直径 7mm 的打孔器打取菌片，接种于 PD 培养液中，每瓶 20 个菌片，于（25±1）℃ 150r/min 震荡培养箱中培养 7d。真空抽滤，菌丝体自然干燥后，取 1g 放入预冷的研钵中，加入 5mL 0. 1M pH 值 7. 0 的磷酸缓冲液，冰浴研磨成匀浆，转入离心管，4℃下静置 12h，10 000r/min 4℃下离心 20min，取上清液，即为酶粗提液，-20℃冰箱中保存备用。

1.2.2　胞外酶的提取

打孔器打取菌片，纤维素酶和果胶酶接种于 Maravs 等（1986）方法配制的培养液中，木质素酶和蛋白酶接种于 PD 液体培养液，每瓶 20 个菌片，于（25±1）℃ 150r/min 震荡培养 15d（纤维素酶、半纤维素酶）或 10d（果胶酶）或 7d（木质素酶、蛋白酶）。真空抽滤，滤液转入离心管，4℃ 1 0000r/min 下离心 20min，上清液保存于-20℃冰箱中备用。

1.3　酶活力的测定

参照李宝聚等[1]的方法，测定了纤维素酶（羧甲基纤维素酶（Cx）、C1 酶、滤纸酶（FPA）和 β-葡萄糖苷酶）和多聚半乳糖醛酸酶（PG）、果胶甲基半乳糖醛酸酶（PMG）的活性；参照梁勤[2]的方法，测定了半纤维素酶（木聚糖酶和木糖苷酶）和木质素酶（木质素过氧化物酶（LiP）、锰过氧化物酶（MnP）和漆酶（Laccase））的活性；参照李忠福等[3]的方法测定了果胶总酶（PP）的活性，果胶裂解酶（PL）的活力测定参照冯文宇[4]的方法，果胶甲基酯酶（PE）的活力测定参照高洪敏等[5]的方法；蛋白酶活力测定参照郭圣茂[6]的方法，蛋白含量测定采用考马斯亮蓝 G250 方法。

1.4　数据分析方法

试验结果采用 Microsoft Excel 软件进行数据整理和数据线性回归，差异显著性分析依据 Duncar 新复极差法，由 DPS 9. 50 专业版完成。

2　结果与分析

本研究采用比色法测定了尖孢镰刀菌五类细胞壁降解酶的活性，结果见表 1 和图 1～图 3。

Table 1　The activity of cell walldegrading enzymes

Cell wall degrading enzymes		Activity（U/mL）	
	Intracellular	Extracellular	
Cellulase	β-Glucosidase	0. 175 6±0. 007 0	0. 004 3±0. 000 3
	Cx	0. 158 6±0. 010 6	0. 017 4±0. 001 1
	C1	0. 069 5±0. 002 9	0. 009 8±0. 000 1
	FPA	0. 070 3±0. 002 2	0. 004 4±0. 000 4
Hemicellulose	Xylanase	0. 917 8±0. 019 9	1. 201 8±0. 017 3
	Xylosidase	0. 001 5±0. 000 1	0. 000 3±0. 000 0

（续表）

Cell wall degrading enzymes	Activity（U/mL）		
	Intracellular		Extracellular
Pectinase	PP	0. 623 0±0. 003 2	0. 390 6±0. 004 6
	PMG	0. 221 0±0. 004 4	0. 219 5±0. 004 8
	PG	0. 261 8±0. 003 6	0. 257 9±0. 008 0
	PE	3. 880 0±0. 072 1	3. 953 3±0. 050 3
	PL	0. 164 3±0. 011 7	0. 079 7±0. 003 5
Ligninase	Laccase	21. 728 4±0. 372 8	−14. 653 9±0. 523 4
	MnP	28. 493 8±0. 085 5	−1. 087 8±0. 071 2
	LiP	0. 437 4±0. 012 1	−30. 703 4±0. 799 4
Protease	27. 250 0±2. 046 3	24. 916 7±1. 464 9	

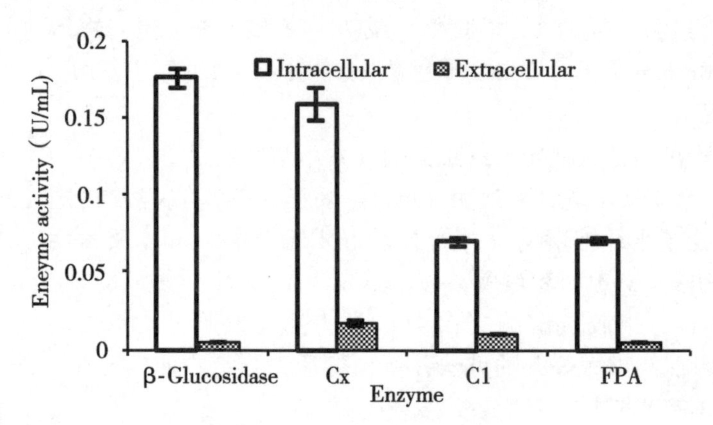

Fig. 1　The activity of cellulase

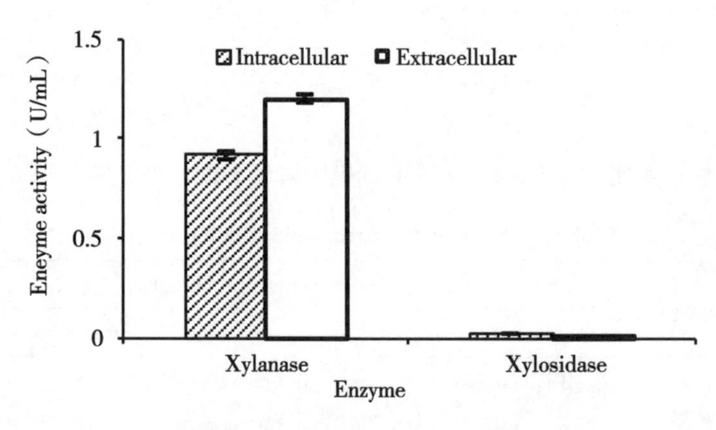

Fig. 2　The activity of hemicellulase

由表 1 和图 1 可以看出，尖孢镰刀菌的胞内与胞外产生的纤维素酶活性存在差异，胞内活性明显高于胞外。胞内纤维素酶活性为 0. 069 5～0. 175 6U/mL，β－葡萄糖苷酶最高，活性为 0. 175 6U/mL，其次 Cx 酶活性为 0. 158 6U/mL；胞外纤维素酶活性范围为 0. 004 3～0. 017 4U/mL，Cx 酶活性最高，为 0. 017 4U/mL。总体来看，尖孢镰刀菌胞内和胞外 Cx 酶活性均较高，

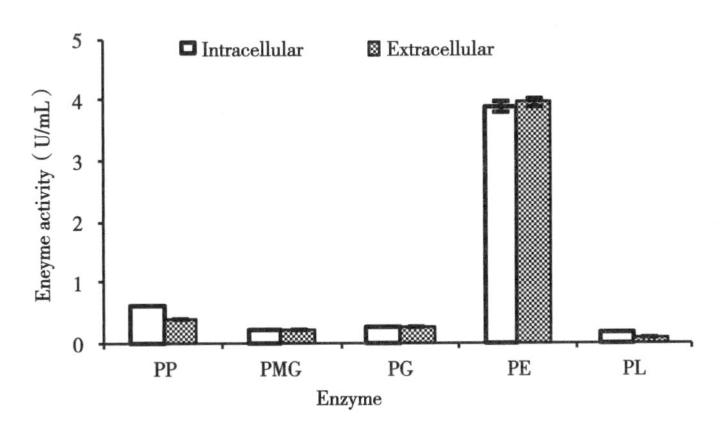

Fig. 3　The activity of pectinase

说明 Cx 酶为该菌分泌的主要纤维素酶，在侵染过程中起主要作用。

尖孢镰刀菌的胞外半纤维素酶活性高于胞内（表 1、图 2）。其胞内半纤维素酶活性为 0.001 5~0.917 8U/mL，胞外活性为 0.000 3~1.201 8U/mL，胞内和胞外的木聚糖酶的活性都明显高于木糖苷酶，是木糖苷酶的 337~9 006倍，由此可以看出，木聚糖酶是该菌分泌的主要半纤维素酶。

对于果胶酶而言，胞内酶活性与胞外酶活性相差不大（表 1、图 3）。果胶甲酯酶活性最高，胞内为 3.88U/mL，胞外为 3.953 3U/mL，均明显高于其他 4 种果胶酶系，说明该酶是尖孢镰刀菌在降解果胶质时分泌的主要细胞壁降解酶，为优势致病酶系，在致病过程中发挥主导作用。

由表 1 可知，尖孢镰刀菌虽然可以产生木质素酶，但其只存在于胞内，而分泌不到胞外。其胞内锰过氧化物酶和漆酶的活性均较高，分别为 28.493 8U/mL 和 21.728 4U/mL，而木质素过氧化物酶活性较低，仅为 0.437 4U/mL。其木质素酶未分泌到胞外的原因，可能是因为诱导培养基不适合，也有可能是因为该镰刀菌在侵染过程中没有向胞外分泌木质素酶。

尖孢镰刀菌分泌蛋白酶能力较强，其胞内胞外酶活性均较高（表 1）。其胞内蛋白酶活性为 27.250 0U/mL，胞外酶活性为 24.916 7U/mL。该菌产生并分泌蛋白酶以破坏寄主植物的防御系统并产生致病症状。

3　结论与讨论

本研究采用比色法测定了尖孢镰刀菌分泌产生的细胞壁降解酶活性，比较了胞内胞外不同纤维素酶、半纤维素酶、果胶酶、木质素酶和蛋白酶活力的差异，表明尖孢镰刀菌可以产生活性较高的 Cx 酶、木聚糖酶、果胶甲酯酶和蛋白酶等，说明细胞壁降解酶在尖孢镰刀菌侵染寄主过程中起非常重要的作用，是该菌重要的致病因子之一。高增贵[7]等研究发现玉米茎腐病菌（*Pythium aphanidermatum*、*F. graminearum*）在活体内和活体外都能分泌细胞壁降解酶，其产生的果胶酶、纤维素酶等都对玉米胚根有明显的浸解作用，表明细胞壁降解酶是玉米茎腐病菌的重要致病因子。高晓敏等[8]介绍了尖孢镰刀菌可以分泌针对植物细胞壁成分的细胞壁降解酶来降解植物的细胞壁，降解物阻塞寄主植物的导管，阻碍寄主植物吸收水分，导致植物萎蔫致死。本研究结果与前人的研究结果一致。

果胶是组成植物细胞壁主要成分之一，通常存在于植物初生壁和细胞之间，百合鳞茎中果胶成分占总体的 5%，为其主要成分[9]。本研究测定的果胶酶系活性都较高，在 0.2~5U/mL，说明果胶酶是尖孢镰刀菌分泌的主要细胞壁降解酶。Kang Z 等[10]的研究证明了引起小麦根腐的禾谷镰孢菌（*F. graminearum*）分泌果胶酶活性高的菌株致病性相对强，且果胶甲基半乳糖醛酸酶

（PME）和多聚半乳糖醛酸酶（PG）不仅比其他酶活性高，而且它们间还存在协同作用，突显果胶酶是该病菌的主要致病因子。本研究测定的果胶甲基半乳糖醛酸酶（PMG）、多聚半乳糖醛酸酶（PG）、果胶甲酯酶（PE）等的活性普遍高于纤维素酶的活性，这与 Kang Z 等的研究结果一致。

木聚糖是植物半纤维素的重要组分，在植物细胞壁中的含量仅次于纤维素，约占细胞干重的35%[11]。陈晓林等[12]证明了苹果树腐烂病菌（*Valsa mali*）在培养基诱导和病组织侵染后均能产生木聚糖酶等系列细胞壁降解酶，且木聚糖酶活性最高，是苹果腐烂病的主要致病因子之一。本研究结果表明：尖孢镰刀菌产生分泌的木聚糖酶活性高，为 0.9~1.2U/mL，该菌通过木聚糖酶降解细胞壁的重要组分—半纤维素木聚糖，破坏植物细胞壁，导致寄主产生坏死症状或加速病程发展，说明木聚糖酶是尖孢镰刀菌侵染百合产生的主要致病酶系，与陈晓林等的研究结果一致。

百合鳞茎富含蛋白质，含量高达 4%。本研究测定了尖孢镰刀菌的蛋白酶活性，胞内与胞外活性都在 20U/mL 以上，说明蛋白酶在尖孢镰刀菌致病过程中扮演着十分重要的角色，既能提供该菌蛋白合成所需要的氨基酸，也能降解百合中与抗性有关的蛋白质。杨钟灵[13]研究发现胡萝卜软腐果胶杆菌（*Pectobacterium carotovorum*）通过产生蛋白酶等细胞壁降解酶破坏寄主植物的防御系统并产生致病症状，与笔者的研究结果一致。

参考文献

[1] 李宝聚，周长力，赵奎华，等．黄瓜黑星病菌致病机理的研究Ⅱ［J］．植物病理学报，2001，30（1）：17-18.

[2] 梁勤．我国银耳种质资源遗传多样性及木质素、纤维素酶活测定［D］．成都：四川农业大学，2014：33-35.

[3] 李忠福，徐建国．分光光度法测定果胶酶活性方法的研究［J］．黑龙江医药，2002，15（6）：428-430.

[4] 冯文宇．果胶裂解酶特性研究的课程设计［D］．齐齐哈尔：齐齐哈尔大学，2012.

[5] 高洪敏，陈捷，何晶，等．禾谷镰刀菌产生的细胞壁降解酶的特性研究［J］．玉米科学，2005，13（3）：112-113.

[6] 郭圣茂，杜天真，邱业先，等．亚热带植物叶蛋自酶活性研究［J］．精细化工，2006，23（9）：887-890，899.

[7] 高增贵，陈捷，高洪敏，等．玉米茎腐病菌产生的细胞壁降解酶种类及其活性分析［J］．植物病理学报，2000，30（2）：149-152.

[8] 高晓敏，王琚刚，马立国，等．尖孢镰刀菌致病机理和化感作用研究进展［J］．微生物学通报，2014，41（10）：2143-2148.

[9] 周全军．百合种植技术［M］．南京：江苏科学技术出版社，1984.

[10] Kang Z，Buchenauer H. Ultrastructural and cytochemical studies on cellulose，xylan and pectin degradation in wheat spikes infected by *Fusarium culmorum*［J］．J Phytopathol，2000，148（5）：263-275.

[11] 邵蔚蓝，薛业敏．以基因重组技术开发木聚糖类半纤维素资源［J］．食品与生物技术，2002，21：88-93.

[12] 陈晓林，牛程旺，李保华，等．苹果树腐烂病菌产生细胞壁降解酶的种类及其活性分析［J］．河北农学报，2012，27（2）：207-212.

[13] 杨钟灵．胡萝卜软腐果胶杆菌与黄花马蹄莲不同方式互作的蛋白比较分析及相应致病基因的研究［D］．南京：南京农业大学，2012：34-36.

江西省十字花科蔬菜根肿病影响因子研究*

黄瑞荣[1**]，黄 蓉[1***]，胡建坤[1]，华菊玲[1]，李信申[1]，丁云花[2]
（1. 江西省农业科学院植物保护研究所，南昌 330200；
2. 北京市农林科学院蔬菜研究中心，北京 100097）

摘 要：采用威廉士鉴别系统鉴定病原菌生理小种，研究病菌致病性分化与蔬菜品种抗病性关系。调节土壤中根肿病菌休眠孢子数量，测定休眠孢子含量与寄主病情的关系。以硫酸亚铁和氧化钙为土壤酸碱度调节剂，研究土壤 pH 值对根肿病情的影响。在根肿病田间病圃分期播种小白菜，分析气象因素对根肿病的影响。设计水旱轮作和旱旱轮作田间试验并结合田间调查，研究轮作对根肿病的防控效果。采用 9 号小种和 4 号小种苗期接种，鉴定小白菜、油菜单播及其两者混播各自的根肿病情变化。并在 9 号小种和 4 号小种菌源地块试验小白菜、油菜单播及其两者混播各自的根肿病情变化。透过种稻数十年的稻田改种小白菜诱发根肿病现象，分析病原菌侵染来源。

研究结果，江西省蔬菜根肿病菌致病性分化明显，鉴定出 1 号、3 号、4 号、5 号、8 号、9 号、12 号、16 号共 8 个生理小种。种群多样，结构复杂，鉴定出强致病力 4 号小种和弱致病力 5 号小种，以 9 号小种为优势小种。不同种类的十字花科蔬菜对根肿病的抗性差异显著，甘蓝、油菜等抗 9 号小种，大白菜、小白菜、紫菜薹等感 9 号小种，多数十字花科作物对 4 号小种抗性较弱，但萝卜对 4 号小种的抗性相对较强。土壤中病菌休眠孢子含量与根肿病关系密切，每克土壤中休眠孢子数低于 10^2 个根肿病不发生或偶发生；介于 10^3～10^4 个病株率和病指上升；高于 10^5 个病害可重度发生。土壤 pH 值影响根肿病病情严重度，适宜根肿病发生的土壤 pH 值 4.0～6.5，最适发病土壤 pH 值 4.5～5.5。高温影响病情严重度，低温不利于根肿病发生，以月平均气温 20～30℃ 病害发生较重，25～28℃ 为病害发生最适温度。降雨量对根肿病未见明显影响，适温条件下强降雨可加重病情，月平均降雨量 250mm 以下病害发展无规律性变化。十字花科作物与玉米、西瓜、水稻短期轮作根肿病情未见减轻。在特定环境条件下种间作物混播似存化感现象，小白菜与抗 9 号小种的油菜混播，根肿病病情轻于单播，但与感 4 号小种的油菜混播这一现象不复存在；人工接种也再现这一现象。油菜单播及与小白菜混播，根肿病病情未发生显著变化。田间调查种稻数年甚至数十年的稻田改种小白菜有诱发根肿病现象，究其菌源传播途径，推测应与灌溉水有关。

结果表明：江西省根肿病菌致病力分化明显，存在多个小种。品种的抗病性因小种不同而异。根肿病与土壤中病菌休眠孢子含量密切相关，孢子数含量低于 10^3 个根肿病不发生或偶发生，高于此值病害发生风险加大。根肿病好发生于酸性土壤，适宜根肿病发生的土壤 pH 值4.0～6.5。气象因素以气温对根肿病情影响较大，适宜根肿病发生的月平均气温 20～30℃，最适气温 23～25℃。轮作无明显控病效果，但在特定条件下十字花科属种间抗性不同的作物混播似有化感现象。灌溉水助推根肿病扩散。

关键词：江西省；十字花科蔬菜；根肿病；影响因子

---

* 基金项目：国家公益性行业（农业）科研专项（201003029）；江西省科技支撑计划项目（20151BBF60068）；江西省科研院所基础设施配套项目（20151BBA13034）；江西现代农业科研协同创新专项（JXXTCX2015005-004）；江西省现代农业蔬菜产业体系病虫害防治岗位专项（JXARS）；江西省重点研究计划项目（20161ACF60015）

** 第一作者（通信作者）：黄瑞荣，男，研究员，硕士，从事植物病害防治研究；E-mail：huangruirong@sohu.com

*** 第一作者：黄蓉，女，副研究员，博士，从事植物病害防治研究；E-mail：huangrong8229@163.com

胶孢炭疽菌（*Colletotrichum gloeosporioides*）黑色素合成关键蛋白 PKS 的生物信息学分析*

刘朝茂**，魏玉倩，陈　摇，韩长志***
（西南林业大学林学院，云南省森林灾害预警与控制重点实验室，昆明　650224）

摘　要：胶孢炭疽菌（*Colletotrichum gloeosporioides*）是引起植物炭疽病的主要病原菌，可以侵染诸如桃、板栗、杧果、枇杷、番木瓜、麦冬、橡胶等众多植物，给人们的生产生活造成较大的经济损失，严重威胁着各国林业的健康可持续发展。目前国内学者对该菌的研究主要集中在生物学特性以及对其防治的化学药剂筛选等方面，而关于其遗传多样性、致病相关基因等报道较少，这在一定程度上限制了对其侵染过程和致病机制的深入认识，从而制约着用于防治该病菌的新作用靶点的药剂研发。黑色素作为植物病原丝状真菌分生孢子入侵植物必要的合成物质之一，学术界对诸如稻瘟菌、玉米大斑病菌等植物病原菌黑色素合成基因进行了较为深入的研究，然而尚未见有关胶孢炭疽菌黑色素合成关键基因的生物信息学分析的报道。本研究对其黑色素合成关键蛋白 PKS 进行蛋白质二级结构、保守结构域、跨膜结构域、细胞信号肽、亚细胞定位及疏水性等生物信息学分析，明确①蛋白质二级结构含有 α 螺旋和无规则卷曲，没有 β 转角，无序结构组成由高到低；②蛋白为跨膜蛋白，跨膜区域的氨基酸都具有强疏水性；③具有分泌信号肽，属于分泌蛋白；④亚细胞定位于胞外上，最高值为 28. 26。

为进一步深入研究胶孢炭疽菌 PKS 蛋白在致病过程中所发挥的作用打下坚实的理论基础，也为深入解析重要经济林木炭疽病的病原侵染作用机制提供重要的理论参考。

关键词：胶孢炭疽菌（*Colletotrichum gloeosporioides*）；黑色素；PKS；生物信息学分析

---

* 基金项目：国家级大学生创新创业训练计划项目（项目编号：201610677001）；云南省大学生创新创业训练计划项目（核桃炭疽病菌黑色素合成关键基因的生物信息学分析与克隆）

** 第一作者：刘朝茂，硕士，助理实验师，研究方向：林木病理学；E-mail：lcm1987swkx@ 163. com

*** 通信作者：韩长志，博士，副教授，研究方向：经济林木病害生物防治与真菌分子生物学；E-mail：hanchangzhi2010@ 163. com

抗连作障碍草莓新品种根系分泌物对尖孢镰刀菌的影响及其组分分析*

齐永志[1**]，贾　薇[1]，苏　媛[1]，甄文超[2***]
（1. 河北农业大学植物保护学院，保定　071001；2. 河北农业大学农学院，保定　071001）

摘　要：草莓枯萎病的发生与自身根系分泌物密切相关。本研究以 2 个抗连作障碍草莓新品种（连达和连童）及各自亲本（达赛莱克特和童子 1 号）为试材，通过沙培试验收集根系分泌物，测定根系分泌物对尖孢镰刀菌草莓专化型孢子萌发、产孢量和菌丝生长的影响，分析抗连作障碍新品种及其亲本根系分泌物中主要自毒物质、氨基酸和可溶性糖含量差异。结果表明：童子 1 号和达赛莱克特各浓度根系分泌物均明显促进了尖孢镰刀菌菌丝生长，提高了产孢量和孢子萌发率；连童根系分泌物对菌丝生长和孢子萌发均未产生明显影响，而较高浓度连达根系分泌物明显抑制了菌丝生长。连童、连达及其亲本根系分泌物均未明显影响菌丝干重。连童、连达根系分泌物中对羟基苯甲酸（培养第 100d 时除外）、苯甲酸和阿魏酸含量均分别明显低于各自亲本童子 1 号和达赛莱克特，邻苯二甲酸与各自亲本未达差异显著水平。与达赛莱克特相比，检测到连达根系分泌物中阿魏酸、丁香酸和香草酸的时间约晚 40d；之后，连达根系分泌物中阿魏酸、丁香酸与亲本差异不明显。连童和连达根系分泌物中对尖孢镰刀菌有抑制作用的天冬氨酸和谷氨酸含量均明显高于各自亲本，天冬氨酸增幅分别约为 20. 56% 和 9. 10%，谷氨酸增幅分别约为 33. 91% 和 27. 01%。而对病原菌有促进作用的丝氨酸含量却低于亲本，降幅分别为 33. 92% 和 29. 79%；连童和连达根系分泌物中苏氨酸和精氨酸出现时间较晚。童子 1 号和达赛莱克特根系分泌物中葡萄糖和蔗糖含量均分别明显高于连童和连达。本研究发现在连作条件下草莓根系分泌物中的酚酸、氨基酸和可溶性糖是影响根际土传病原菌活性的重要因子。

关键词：草莓；连作障碍；根系分泌物；尖孢镰刀菌

---

* 基金项目：国家自然科学基金（31140064）
** 第一作者：齐永志，博士，讲师，主要从事植物生态病理学研究；E-mail：qiyongzhi1981@163. com
*** 通信作者：甄文超，博士，教授，博士生导师，主要从事农业生态学与植物生态病理学研究；E-mail：wenchao@hebau. edu. cn

胶孢炭疽菌 G 蛋白 α 亚基的生物学功能*

柯智健**，柳志强，张月凤，李晓宇***
（海南大学热带农林学院，海口　570228）

摘　要：胶孢炭疽菌（*Colletotrichum gloeosporioides*）是引起炭疽病的一种重要病原菌，可以侵染多种作物，造成严重的经济损失。胶孢炭疽菌不仅寄主范围广泛，其侵染策略也多种多样，目前对于该病菌的防治也较为困难。G 蛋白是能与鸟嘌呤核苷酸结合，具有 GTP 酶活性的蛋白。异源三聚体 G 蛋白是由 α，β 和 γ 3 种亚基组成的，目前还未有关于胶孢炭疽菌 G 蛋白 α 亚基生物学功能的研究。本研究从胶孢炭疽菌中鉴定了 3 种 Gα 亚基，命名为 CgGa1、CgGa2 和 CgGa3。利用同源重组技术获得了 3 种 Gα 亚基的敲除突变体，表型分析发现 *CgGa*1 敲除突变体表现为营养生长缓慢，产孢量减少，孢子萌发率降低，无法形成附着胞，对 NaCl 和 SDS 更加敏感，致病力减弱；*CgGa*2 敲除突变体表现为营养生长缓慢，产孢量减少，对 NaCl 和 H_2O_2 更加敏感，对 SDS 的耐受性增强；*CgGa*3 敲除突变体表现为产孢量减少，对 H_2O_2 更加敏感。结果表明 G 蛋白 α 亚基参与调控胶孢炭疽菌营养生长，分生孢子产生及萌发，附着胞形成，氧化应激反应，渗透压响应和致病性等多个过程。

关键词：胶孢炭疽菌；G 蛋白信号途径；Gα 亚基；分生孢子；致病性

---

* 基金项目：国家自然科学基金（31560045）；海南省重点研发计划（ZDYF2016155）
** 作者介绍：柯智健，硕士研究生；E-mail：mike1136@ qq. com
*** 通信作者：李晓宇，博士，副教授，主要从事植物病原真菌学研究；E-mail：hdlixiaoyu@ 126. com

辣椒病原真菌多样性及其影响因素研究

刁永朝*，蔡 磊**
（中国科学院微生物研究所真菌学国家重点实验室，北京 100101）

摘 要：辣椒是一种重要蔬菜，全世界大多数国家均有种植，但其生产受到真菌病害的严重影响，辣椒上的真菌病害超过 15 种。虽然生产中杀菌剂被大量应用在辣椒病害防治中，但每年因真菌病害造成的减产仍达 20%~50%，除了病原菌产生抗药性导致药效下降以外，对病害和病原真菌不够了解，尤其对一些多种病原菌复合侵染引起的病害，严重影响了杀菌剂的正确选择和病害的科学防治，同时有些次要病害受到生态因素影响会变为主要病害，也会影响到病害的有效防控，因此揭示辣椒上病原真菌多样性对其病害的有效科学防控具有重要意义。

本研究从 19 个国家的辣椒病害样品上收集了 2 000株菌株，通过 ITS 测序，病症及形态观察进行初步鉴定，结果发现链格孢属、刺盘孢属、镰刀菌属是其中最重要的 3 个病原菌类群。对 2015 年和 2016 年连续两年中国辣椒上病原真菌多样性进行分析发现，主要病害是黑斑病、炭疽病、腐烂病、叶霉病，但不同地区的主要病害存在一定差异。对其他国家的病害情况统计发现，主要病害与中国相似，黑斑病和炭疽病是分布较为广泛的病害。结合地理信息、辣椒品种以及辣椒素含量，分析各个国家辣椒上病原真菌多样性与其相关性，结果发现辣椒素含量越高的样品，其真菌多样性越低，并具有显著相关性。地理距离更近的国家，真菌菌群相似度更高。相比辣椒品种，地理距离和辣椒素含量对辣椒真菌多样性的影响更大。

关键词：辣椒病害；真菌多样性；生态因素

* 第一作者：刁永朝，男，助理研究员，主要研究方向为真菌多样性；E-mail：diaoyongz@163.com
** 通信作者：蔡磊，研究员；E-mail：cail@im.ac.cn

我国蓝莓主要真菌病害发生情况及防治建议*

祝友朋**，韩长志***
（西南林业大学林学院云南省森林灾害预警与控制重点实验室，昆明 650224）

摘 要：蓝莓属杜鹃花科越橘属植物，为多年生落叶灌木或小灌木，果实富含花青素，被认为是当前世界极具发展潜力的新兴果树。该植物具有生长适应性强，分布范围广和经济价值高等明显优势。目前，国内外关于蓝莓的研究主要集中在产业发展、丰产栽培技术、病虫害防治等方面。中国蓝莓从 2000 年开始种植，在全国 27 个省（区）市均有种植，自 2009 年起进入高速发展期，2014 年全国总栽培面积已经飞速增长到了 26 068hm^2。目前蓝莓加工的产品共有五大类 20 种，包括果酒、果汁饮料、乳制品、糖果、干果、烘焙食品、保健品和化妆品以及果酱。随着蓝莓种植面积的逐年扩大，其病害的发生发展情况日益严重，极大地阻碍着该产业的健康、有序和快速发展。

我国蓝莓栽种面积广，可将蓝莓种植区分为 5 个区：吉林、黑龙江地区，分布在长白山和大兴安岭的山区；辽东半岛地区，分布在丹东、庄河和大连等地；胶东半岛地区，分布在青岛、威海、烟台到连云港等地；长江中下游地区，长江流域的华东、江浙一带，以广东、广西、福建沿海为主；云贵高原地区，分布在云南、贵州等地。就云南省而言，在曲靖、昆明等地均有蓝莓的栽植。

蓝莓品种主要有矮丛蓝莓、半高从蓝莓、兔眼蓝莓、高丛蓝莓四类。我国栽种面积广、品种多样，其中矮丛蓝莓极适于东北高寒山区大面积商业化栽培；半高丛蓝莓是由高丛蓝莓和矮丛蓝莓杂交获得的品种类型，适应北方寒冷地区栽培；兔眼蓝莓适应于我国长江流域以南、华南等地区的丘陵地带栽培；高丛蓝莓包括南高丛蓝莓和北高丛蓝莓两大类，南高丛蓝莓喜湿润、温暖气候条件，适于我国黄河以南地区如东北、华南地区发展，北高丛蓝莓喜冷凉气候，抗寒力较强，有些品种可抵抗-30℃低温，适于我国北方沿海湿润地区及寒地发展。

目前，对于蓝莓上病害的研究报道较多，常见病害有灰霉病、炭疽病、僵果病、枝枯病、锈病、白粉病、叶斑病、根腐病等，而对于蓝莓上的病害尚缺乏较为系统性的研究。因此，为了更好地开展蓝莓上主要病害的发生和防治研究，本研究基于前人研究结果，从蓝莓的主栽品种以及分布范围入手，通过对主要病害的病原、为害症状、发病规律以及防治措施进行总结，以期为开展云南省蓝莓种植过程中病害防治提供重要的理论基础。

关键词：蓝莓；病害；发生情况；防治措施

---

* 基金项目：西南林业大学大学生科技创新项目（C16094）

** 第一作者：祝友朋，云南曲靖市人，2014 级本科生；E-mail：3420204485@ qq. com

*** 通信作者：韩长志，河北石家庄市人，博士，副教授，研究方向：经济林木病害生物防治与真菌分子生物学；E-mail：hanchangzhi2010@ 163. com

重组酶聚合酶等温扩增技术快速检测油菜茎基溃疡病菌

雷　荣，邵　思，陈乃中，吴品珊
（中国检验检疫科学研究院，北京　100176）

摘　要：油菜黑胫病又称油菜黑腿病，其病原菌主要有两种：油菜茎基溃疡病菌（*Leptosphaeria maculans*）和油菜黑胫病菌（*L. biglobosa*），油菜茎基溃疡病菌具有极强的致病力，是引起世界各地油菜黑胫病广泛传播及油菜产量损失的主要原因。该病菌引起的产量损失一般约为10%～20%，严重时可达30%～50%或更高。在加拿大、美国、澳大利亚和欧洲等油菜产区同时存在这两种致病菌。我国油菜产区气候与世界其他产油菜国家相似，且目前的油菜栽培品种高度感病，因该病菌对我国油菜产业的潜在巨大风险，2007年油菜茎基溃疡病菌列入我国检疫性有害生物名录。目前，用于油菜茎基溃疡病菌的检测方法主要有形态观察和传统PCR检测。等温扩增方法，环介导等温扩增（LAMP）也用于检测油菜茎基溃疡病菌，但LAMP方法普遍存在假阳性的问题。重组酶聚合酶扩增（Recombinase Polymerase Amplification，RPA）技术被认为是可以取代PCR的核酸检测技术，捉操作简单，不需要复杂的仪器即可进行检测实验，已在转基因水稻，病原微生物检测领域有所应用。

本研究采用RPA方法开发了油菜茎基溃疡病菌 *Leptosphaeria maculans* 的快速检测。实时荧光RPA曲线表明在39℃恒温条件下，在6min就能检测到明显的荧光信号，在20min以内就可完成检测（图）。笔者设计的引物和探针可检测到0.021 4ng *L. maculans* 基因组DNA，而不扩增 *L. biglobosa* 和其他黑胫病真菌。RPA快速扩增的优点可用于开发口岸大量样品检测的快速筛查方法。

关键词：油菜茎基溃疡病；重组酶聚合酶扩增；实时荧光；快速检测

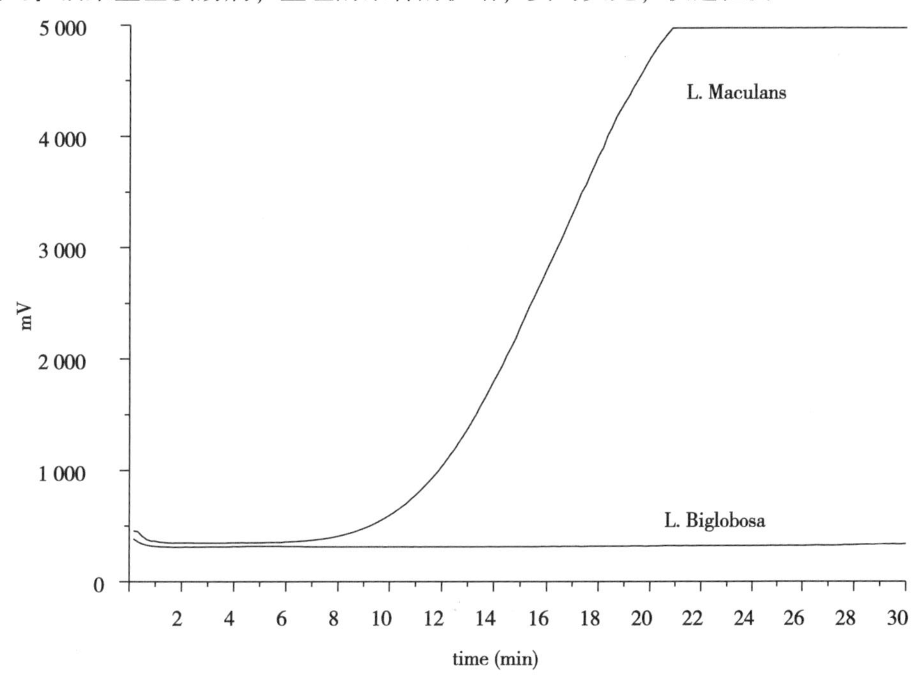

图1　油菜茎基溃疡病菌（1）和油菜黑胫病菌（2）的实时荧光RPA扩增曲线

Hydrogen Peroxide Induces Apoptosis Mediated by Mitochondria and Metacaspase in *Fusarium graminearum*[*]

Li Jing, Ding Mingyu, Fan Xinyue, Yu Xiaoyang, He Fang,
Xu Houjuan, Liang Yuancun
(*College of Plant Protection*, *Shandong Agricultural University*, *Taian* 271018, *China*)

Abstract: *Fusarium graminearum* is a fungal pathogen that infects many plant species, particularly cereals and corn. It is the causal agents of Fusarium Head Blight, leading to reduced yields and subsequent economic losses. Apoptosis is a form of programmed cell death with the importance for development and homeostasis of multicellular organisms and is controlled by a complex regulatory network which can be activated by external signal and internal processes. Hydrogen peroxide (H_2O_2) is a kind of strong oxidant which can be used as an inducer of apoptosis. In this study, treatment with H_2O_2 resulted in many apoptotic hallmarks, including nuclei deformation stained with DAPI, plasma membrane translocation of phosphatidylserine stained with annexin V-FITC and propidium iodine, loss of the mitochondrial membrane potential proved by JC-1, and DNA fragmentation observed with TUNEL assay. Furthermore, expression levels of metacaspase genes and activation were significantly increased after H_2O_2 treatment. Importantly, pretreatment with caspase inhibitor and oligomycin A dramatically reduced apoptotic ratetreated with hydrogen peroxide, indicating that H_2O_2-induced apoptosis was dependent mitochondria and metacaspase. These results indicated that hydrogen peroxide induces apoptosis mediated by mitochondria and metacaspase in *Fusarium graminearum*.

Key words: Apoptosis; Hydrogen peroxide; Metacaspase; *Fusarium graminearum*

* Funding: This work was supported by the Nature Science Foundation of China (31171806 and 31520103911) and the Wheat Innovation Team of Shandong Province Modern Agricultural Industry Technology System (SDAIT-01-09)

橡胶树胶孢炭疽菌转录因子 CgZF1 的克隆及功能分析*

李晓宇**，柯智健，张月凤，柳志强***
（海南大学热带农林学院，海口　570228）

摘　要：胶孢炭疽菌是引起橡胶炭疽病的一种重要病原菌，也是威胁我国天然橡胶产业发展的重要生物因素之一。分生孢子在胶孢炭疽菌侵染过程中起着重要的作用，目前关于该病菌分生孢子发育及侵染的分子机制研究还不够深入。本研究从橡胶树胶孢炭疽菌中鉴定了一个新的转录因子基因 *CgZF*1，该基因编码一个 491 个氨基酸的锌指蛋白，含有 4 个相邻的 C_2H_2 型锌指结构域。利用同源重组的方法获得 *CgZF*1 基因的敲除突变体，并在突变体的基础上获得了互补株。与野生型相比，*CgZF*1 敲除突变体分生孢子产量显著减少，萌发率降低，附着胞形成滞后且形成率显著降低、黑色素产量减少，对 H_2O_2 和 SDS 抗性增加，无法在健康橡胶叶片上致病，而互补株可以恢复以上表型缺陷。由此可见，转录因子 CgZF1 参与调控橡胶树胶孢炭疽菌分生孢子产生、萌发，附着胞的形成，黑色素产生及致病性等过程，本研究为深入认识胶孢炭疽菌分生孢子发育及侵染的分子机制奠定了一定基础。

关键词：橡胶炭疽病；胶孢炭疽菌；转录因子；分生孢子；致病性

---

* 基金项目：国家自然科学基金（31560045）；海南省重点研发计划（ZDYF2016155）
** 作者介绍：李晓宇，博士，副教授，主要从事植物病原真菌学研究；E-mail：hdlixiaoyu@ 126. com
*** 通信作者：柳志强，博士，副教授，主要从事植物真菌病害及生物防治的研究；E-mail：liuzhiqiang80@ 126. com

新疆发生梨锈病

李金霞[1*]，王杰花[2]，高　霞[2]，陈卫民[2**]
（1. 伊犁州农科所，伊宁　83500；2. 新疆伊犁职业技术学院，伊宁　835000）

摘　要：梨生产中，锈病是为害梨树的重要病害之一。但至今新疆未见梨锈病发生的报道。作者于2016年在新疆特克斯县梨园中进行林业有害生物普查时发现此病，发病田块病株率在30%~70%。为害梨树的叶片、叶柄、嫩梢和幼果，叶片发病初期为橙红色近圆形病斑，病斑中部生出黑色小点（性孢子器），后病斑处变肥厚，叶正面凹陷，背面隆起，并生出淡黄色后期为黄褐色毛状物（锈子器）。一个病斑上可生出十几个到几十个毛状物，破裂散出黄褐色粉状锈孢子，此后病斑变黑，病斑多时引起叶片枯死或脱落。幼果上发病，病部稍凹陷，中间密生性子器，而在病斑周围产生锈子腔，病斑由橙红色变黑，病果停止生长、畸形早落。依据症状和病原形态特征，鉴定该病为梨锈病，病原为梨胶锈菌 *Gymnosporangium asiaticum*，属担子菌亚门胶锈菌属。田间调查表明，特克斯县梨园旁为苗圃，苗圃中栽有圆柏。圆柏从关内调入，作为绿化树种，其上有大量冬孢子角。另外，2016年特克斯县春季多雨，温度在13~20℃，这样的气候条件满足了梨锈病担孢子萌发的最适温度和相对湿度，导致梨锈病在特克斯县较大范围内发生为害。

* 第一作者：李金霞，农艺师，从事果树病虫害防治
** 通信作者：陈卫民，教授，硕士生导师，从事果树病害的教学和研究防治工作

杧果炭疽菌铁通透酶基因 *CgFTR*1 生物学功能初步研究*

刘文波[2**]，何其光[2]，孙茜茜[2]，靳鹏飞[2]，范鸿雁[1***]，秦春秀[2***]
（1. 海南省热带果树生物学重点实验室，海口　571100；
2. 海南大学热带农林学院，海口　570228）

摘　要：克隆杧果炭疽菌铁通透酶基因 *CgFTR*1（Fe transporter permease），分析其生物信息学特征、亚细胞定位情况，并敲除该基因，验证其在致病性中的作用，为研究 *CgFTR*1 对胞内铁代谢功能的影响。通过 RT-PCR 及基因克隆方法，从杧果炭疽菌总 RNA 中克隆得到基因 *CgFTR*1；应用生物信息学方法分析其 DNA 序列及其编码的蛋白序列特性，使用 MEGA5.0 对 *CgFTR*1 氨基酸序列及其同源序列进行多序列比对，构建同源物种间的系统进化树，将不同浓度的 Fe^{2+} 置于相同的培养基同一温度下观察其生长速度及菌丝的生长状态进行观察，并进行 RT-qPCR 表达分析；利用 PCR 介导的同源重组方法构建基因缺失菌株；再利用原子吸收光谱方法测定基因缺失菌株胞内铁含量的变化，并对基因缺失菌株在缺铁条件和菌丝诱导条件下的生长状况进行研究；通过代谢转换实验，研究 *CgFTR*1 对细胞液泡功能的影响。结果表明：利用 RT-PCR 方法从杧果炭疽菌（*Colletotrichum gloeosporioides*）材料中克隆得到全长为 1 038 bp 的 *CgFTR*1 基因，序列比对及生物信息学分析表明，其基因预测编码 345 个氨基酸，分子量为 37.66ku，理论等电点为 6.32，具有三价铁结合位点 Glu-Xaa-Xaa-Glu 的结构单元，具有 7 个可能的跨膜螺旋区；和兰生炭疽菌（*C. chlorophyti*）和曲霉菌（*Aspergillus lentulus*）的 *FTR*1 相似性为 92%和 55%，推测 *CgFTR*1 基因可能与其他蛋白结合参与杧果炭疽菌的铁转运；亚细胞定位显示定位在细胞膜上与 PSORT Prediction 预测的亚细胞定位一致；RT-qPCR 结果分析表明该基因，铁匮乏条件会诱导 *CgFTR*1 的表达，而富铁条件则会抑制其表达；*CgFTR*1 是一种低铁应答基因，维持杧果炭疽菌胞内铁离子吸收功能方面具有重要作用。下一步将开展 *CgFTR*1 缺失菌株致病力、胁迫应答及铁代谢相关基因分析。

关键词：杧果炭疽菌；铁通透酶；*CgFTR*1；调控

---

* 基金项目：海南省自然科学基金（20153131、20163040）；农业部 2013 年热作农技推广与体系建设项目（13RZNJ-20）；农业部 2012 年热作农技推广与体系建设项目（12RZNJ-19）资助

** 第一作者：刘文波，男，硕士，主要从事热带植物病理研究；E-mail：saucher@163.com
*** 通信作者：范鸿雁，女，硕士，副研究，从事果树物病害等研究；E-mail：157172189@qq.com
秦春秀，女，助理研究员，从事植物病理学研究；E-mail：qin_401@126.com

砂梨、白梨和西洋梨干枯病病原菌种类的比较鉴定*

郭雅双，白　晴，洪　霓，王国平**

（华中农业大学植物科学技术学院，农业微生物学国家重点实验室，武汉　430070）

摘　要：我国的栽培梨，南方主要为砂梨（*Pyrus pyrifolia*）、北方主要为白梨（*P. bretschneider*），局部地区也有西洋梨（*P. communis*）。近年在这3种梨上新发生一种由拟茎点霉属（*Phomopsis*）真菌引起的枝干病害，导致枝干枯死，严重时整株死亡。因该病在东方梨（砂梨、白梨）和西洋梨上的症状有异，分别称为“梨干枯病”和“洋梨干枯病”。

本研究对东方梨和西洋梨干枯病的病原菌种类进行了比较鉴定，2013—2017年从我国12个省（市）的砂梨、白梨和西洋梨上采集干枯病病样，共分离纯化获得304个拟茎点霉菌株，其中277个菌株来源于东方梨、27个菌株来源于西洋梨。根据培养特性、分生孢子形态和rDNA-ITS序列分析，来源于东方梨的菌株鉴定出5种拟茎点霉菌（*Phomopsis fukushii*、*Diaporthe eres*、*P. amygdail*、*P. longicolla*和*D. neotheicola*），而来源于西洋梨的菌株仅鉴定到2种拟茎点霉菌（*P. fukushii*和*D. eres*）。来源于东方梨和西洋梨的*P. fukushii*或*D. eres*的不同菌株在系统进化上没有表现出遗传距离，其菌落形态、菌丝生长速率和分生孢子形态，以及在相同的培养基上诱导产生的分生孢子器内部构造和泌出的分生孢子角均无明显差异。但交互接种试验结果显示，两类不同来源的菌株存在一定的寄主选择性，东方梨菌株的致病力明显强于西洋梨菌株。研究结果为进一步研究梨干枯病的病原种类多样性奠定了基础，也为加深对拟茎点霉菌的认识提供了新的信息。

关键词：鉴定；病原；梨干枯病

---

* 基金项目：梨现代农业技术产业体系（CARS-29-10）；国家农业部公益性行业计划（201203034-02）

** 通信作者：王国平；E-mail：gpwang@ mail. hzau. edu. cn

An updated investigation of the proteins in sclerotial exudates of *Sclerotinia sclerotiorum*[*]

Chen Caixia[**], Sun Henan, Liang Yue[***]
(*College of Plant Protection*, *Shenyang Agricultural University*, *Shenyang* 110866)

Abstract: *Sclerotinia sclerotiorum* (Lib.) de Baryis one of the most devastating necrotrophic phytopathogens capable of infecting more than 400 plant species. This fungal pathogen can produce sclerotia, which consist of aggregated mycelia, as long-term survival structures. Sclerotia are critical for the *S. sclerotiorum* life and disease cycles. Sclerotial production involves three stages including initiation, development, and maturation. The formation of liquid droplets on the sclerotial surface (i. e., sclerotial exudate) is a characteristic of the development stage. However, the exudate and its potential functions have not been comprehensively studied in *S. sclerotiorum*. In this study, wild-type *S. sclerotiorum*was routinely cultured on potato dextrose agar, and the liquid exudates collected from the surface of developing sclerotia were concentrated by lyophilization. The protein composition of the sclerotial exudate was analyzed by liquid chromatography-tandem mass spectrometry. More than 400 proteins were preliminarily identified as being associated with several functional groups, including carbohydrate and amino acid metabolism, secondary metabolism, energy, and unknown functions. Furthermore, the interaction between four enzymes (i. e., mannanase, amylase, β-galactosidase, and cellulase) and acid phosphatase to degrade plant cell wallswas verified in a bioactivity assay. The mannanase, amylase, and cellulase in sclerotial exudates were determined to hydrolyze their corresponding substrates. Therefore, these enzymesmay be functional during sclerotial development and in fluence pathogenicity. In summary, the protein composition of the sclerotial exudate was comprehensively investigated. Some of the identified hydrolyticproteinslikely help mediatefungal morphogenesis and pathogenesis. The presented data may provide novel insights into *S. sclerotiorum* and other sclerotium-forming fungi.

* 基金项目：沈阳农业大学引进人才科研启动费项目（20153040）
** 第一作者：陈彩霞，硕士研究生，研究方向为植物病理学；E-mail：chencaixia9@ 163. com
*** 通信作者：梁月，教授，博士生导师；E-mail：liangyuet@ 126. com

Biological characterization of a *Colletotrichum* sp. responsible for sunflower anthracnose*

Sun Huiying**, Guan Gege, Liang Yue***

(*College of Plant Protection*, *Shenyang Agricultural University*, *Shenyang* 110866)

Abstract: Sunflower (*Helianthus annuus* L.) is an important oilseed crop. Anthracnose caused by-*Colletotrichum* spp. is a serious disease affecting some Asteraceae plants. In this study, sunflower stems with dark brown lesionswere collected and the putative pathogenic fungi were isolated and identified according to Koch's postulates. An analysis of the colony morphology of a fungal isolate revealed that the cottony aerial mycelium wasinitiallywhite, but eventually turned darkgray. Dark-brown andround acervuli with brown and straight setaeand conidiophores were also scattered among the colonies. Moreover, conidia were hyaline and cylindrical, with rounded ends (13.16 ~ 20.21μm× 3.05 ~ 5.2μm). Hence, the cultural and morphological characteristics were similar to those described for *Colletotrichum* spp. Meanwhile, the rDNA internal transcribed spacer region was amplified and sequenced using the universal primer pair ITS1 and ITS4. This sequence was 98%identical to a *Colletotrichum* sp. sequence. The biological characteristics of the putative pathogenic fungus were investigated under different growth conditions (i. e., various media, carbon and nitrogen sources, ambient pHs, and temperatures). The fungus grew best onpotato dextrose agar, with solublestarch and ammonium sulfateas idealnutrient sources. Optimal growth was observed at pH 8.0and 28℃. In summary, a *Colletotrichum*sp. isolated from infected sunflower stems was preliminarily identified and morphologically and biologically characterized for a more complete disease diagnosis.

---

* 基金项目：沈阳农业大学引进人才科研启动费项目（20153040）

** 第一作者：孙慧颖，博士研究生，研究方向为植物病理学；E-mail：sunhuiying0715@126.com

*** 通信作者：梁月，教授，博士生导师；E-mail：liangyuet@126.com

Isolation and Identification of Endophytic Fungi from Arecanut (*Areca catectu* L.) *

Song Weiwei**, Niu Xiaoqing, Tang Qinghua, Qin Weiquan***

(*Coconut Research Institute*, *Chinese Academy of Tropical Agricultural Science*, *Wenchang*, *China*, 571339)

Abstract: Endophytes with many varieties are widely distributed and almost exist in all aquatic and terrestrial plants. Endophytes are rich in secondary metabolites and play an important role in biological control of plant diseases. The host plants infected by endophytes usually grow fast, have a strong resistance to adversity and diseases, and are immune to animal attack compared to uninfected plants. *Areca catechu* L. is the main economic crop in Hainan, which is one of the most important southern herbal medicine resources. The diversity of endophytic fungi in arecanut (*Areca catechu* L.) was discussed in this study. A total of 47 endophytic fungi were isolated from the arecanut by tissue separation method. Then rDNA-ITS sequence analysis was used for these isolates, identification and the phylogenetic tree was constructed. The results showed that 47 isolates belonged to 8 genus, including *Penicillium*、*Fusarium*、*Phyllosticta*、*Curvularia*、*Nigrospora*、*Colletotrichum*、*Acremonium* and *Exserohilum*. The dominant endophytic genus in arecanut were *Penicillium* and *Fusarium*, which accounted for 31.91% and 27.66% respectively among the identified fungi. The research provided theoretical basis for subsequent development and utilization of endophytes in arecanut.

Key words: Arecanut (*Areca catechu* L.); Endophytic fungi; Identification; rDNA-ITS sequence; Phylogenetic analysis

* Funding: The Key Project of Hainan Province (ZDYF2016058)

** First author: Song Weiwei, Doctor; E-mail: songweiwei426@aliyun.com

*** Corresponding author: Qin Weiquan, Professor; E-mail: QWQ268@163.com

茄科作物土壤镰刀菌的分离与鉴定*

唐　琳**，李　睿，黄龙梅，吴瑞雪，赵　冉
（洛阳师范学院生命科学学院，洛阳　471934）

摘　要：镰刀菌是引起土传枯萎病的重要病原菌，近年来，在茄科作物的栽培生产过程中发生严重，造成不同程度的减产，经济损失较为严重。因此，明确茄科土壤镰刀菌的系统发育关系和种群分布特点，为进一步监测和防控茄科作物枯萎病奠定理论基础。本实验采用土壤稀释分离法，从洛阳以及三门峡等 62 个不同的地方采集到发生枯萎病的茄科作物周围土壤，并进行镰刀菌的分离纯化。本实验总共分离出 209 株镰刀菌，采用形态学鉴定和 EF-1α 序列分析对供试菌株进行鉴定。结果表明：分离出的镰刀菌分属于 7 个种，分别为尖孢镰刀菌（*Fusarium oxysporum*），三线镰刀菌（*Fusarium tricinctum*）、腐皮镰刀菌（*Fusarium solani*）、轮枝镰刀菌（*Fusarium verticillioides*），水稻恶苗病菌（*Fusarium fujikuroi*）、*Fusarium* sp. DI、*Fusarium* sp. 4. 6。其中尖孢镰刀菌（*F. oxysporum*）在洛阳和三门峡的所有土壤中均有分布，其分离频率均最高，分别为 66. 19%和 58. 83%，而水稻恶苗病菌（*F. fujikuroi*）的分离频率最低，仅为 2. 56%，在洛阳白马寺佃庄的茄科辣椒土壤中存在。本研究表明不同地区茄科作物土壤的镰刀菌种群存在差异，EF-1α 序列分析能体现镰刀菌种间和种内的亲缘关系及遗传差异性，各个自然种群之间的遗传亲缘关系与地理分布之间无相关性。因此，调查茄科作物土壤镰刀菌种类及种群分布，对防控茄科作物土传枯萎病、改善栽培制度及抗病品种选育具有重要的意义。

关键词：茄科作物；镰刀菌；菌落形态；EF-1α

---

* 基金项目：河南省高等学校重点科研项目计划（17A210025）

** 通信作者：唐琳，博士，讲师，主要从事生物防治研究；E-mail：tanglin869@163.com

甘蔗褐锈病菌巢式 PCR 分子检测方法的建立*

汪　涵[1,3**]，吴伟怀[2**]，杨先锋[1]，李　锐[2]，郑金龙[2]，黄　兴[2]，
梁艳琼[2]，习金根[2]，贺春萍[2***]，易克贤[2***]
（1. 中国热带农业科学院环境与植物保护研究所/农业部热带农林有害生物入侵检测与控制重点开放实验室/海南省热带农业有害生物检测监控重点实验室，海口　571101；2. 农业部橡胶树生物学与遗传资源利用重点实验室/省部共建国家重点实验室培育基地—海南省热带作物栽培生理学重点实验室/农业部儋州热带作物科学观测实验站，儋州　571737；
3. 海南大学热带农林学院，海口　570228）

摘　要：由黑顶柄锈菌（*P. melanocephala* Sydow）引起的甘蔗褐锈病是我国甘蔗种植区普遍发生的一种病害。本研究利用真菌 DNA 内源转录间隔区通用引物 ITS1/ITS4 扩增甘蔗黑顶柄锈菌 DNA，并对其扩增产物进行克隆测序，获得序列于 NCBI 网站进行比对分析，并于该序列多态性丰富区域设计了 2 对引物 PM2F/R 与 PM3F/R。通过对同寄主不同病原菌、以及不同属的锈菌 DNA 进行 PCR 检测以验证引物的特异性。结果表明：在优化的单一 PCR 反应体系与程序条件下，2 对引物均仅能从甘蔗黑顶柄锈菌中扩增出约 474bp 和 363bp 的特异条带，而从其他真菌 DNA 中均扩增不出任何条带。进一步将引物 PM2F/R 作为第一轮扩增引物，PM3F/R 作为第二轮扩增引物进行巢式 PCR 扩增后，其检测灵敏度在 DNA 水平上可达 0.001ng/μL，较常规 PCR 提高 100 倍。由此表明，通过本研究所设计的 2 对引物而建立的一种快速、灵敏、准确的甘蔗黑顶柄锈菌的检测技术，对病原菌的早期诊断、快速检测及病害流行学研究具有重要意义。

关键词：甘蔗褐锈病；黑顶柄锈菌；巢式 PCR；分子检测

---

* 基金项目：中国热带农业科学院橡胶研究所省部重点实验室/科学观测实验站开放课题（RRI-KLOF201506）
** 第一作者：汪涵，硕士研究生，主要从事热作锈菌的分子检测；E-mail：18738310063@163.com
吴伟怀，男，博士，副研究员，研究方向：植物病理；E-mail：weihuaiwu2002@163.com
*** 通信作者：贺春萍，女，硕士，副研究员，研究方向，植物病理；E-mail：hechunppp@163.com
易克贤，男，博士，研究员，研究方向：分子抗性育种；E-mail：yikexian@126.com

PCR-Mediated Molecular Detection of *Hemileia vastatrix* in Coffee*

Wu Weihuai[1**], Wang Han[1,2**], Li Le[1,2], He Chunping[1], Liang Yanqiong[1], Zheng Jinlong[1], Li Rui[1], Huang Xin[1], Yi Kexian[1***]

(1. *Environment and Plant Protection Institute, CATAS/Ministry of Agriculture Key Laboratory for Monitoring and Control of Tropical Agricultural and Forest Invasive Alien Pests/Hainan Key Laboratory for Detection and Control of Tropical Agricultural Pests, Haikou, Hainan* 571101, *China*; 2. *Institute of Tropical Agriculture and Forestry, Hainan University, Haikou* 570228, *China*)

Abstract: Coffee leaf rust is the most important fungal disease affecting coffee production, which worldwide caused by the biotrophic pathogen Hemileia vastatrix (Hv). Thus, the rapid, accurate and identification of the pathogen is urgently required to manage and prevent the spreading of coffee leaf rust. In order to establish a molecular biological way for rapid test that has the characteristics of high specificity, high sensitivity and simple operation. This study were analysised by multiple comparisons among coffee rust rDNA-ITS and its proximal sequences, then gets the specific rDNA-ITS regions of Hv that was designed a pair of primers, Hv-ITS-F/R. The specificity of PCR assay were evaluated on the genome DNA from Hv, coffee brown spot pathogen, hyperparasite of Hv, coffee Colletotrichum and other fungi belonging to different Genus. The result showed that they only amplified 396 bp specific bands from Hv, not from other close or similar pathogenic fungi. The sensitivity of PCR assay was also evaluated. The detection sensitivity of the nested PCR was approximately10 pg/μL at the DNA level. Via a further test of healthy and diseased coffee plants, the examining method has a great application prospect. These results illustrated that the primers Hv-ITS-F/R can be used to detect Hv in coffee tissue. Therefore, this method will be very useful in forecast of coffee rust early disease, growth and disease epidemiology.

Key words: Coffee; *Hemileia vastatrix*; Molecular detection

---

* 基金项目：中央级公益性科研院所基本科研业务费专项资金（中国热带农业科学院环境与植物保护研究所，NO. 2015hzs1J009）；FAO/IAEA（No. 20380）；海南省农业厅外来入侵有害生物防治项目；中国热带农业科学院基本科研业务费专项资金（1630042017021）；农业部国际交流与合作项目“热带农业对外合作试验站建设和农业走出去企业外籍管理人员培训”

** 第一作者：吴伟怀，男，博士，副研究员，研究方向：植物病理；E-mail：weihuaiwu2002@163. com
汪涵，硕士研究生，主要从事热作锈菌的分子检测；E-mail：18738310063@163. com

*** 通信作者：易克贤，男，博士，研究员，研究方向：分子抗性育种；E-mail：yikexian@126. com

香榧裂皮病病原菌鉴定及其生物学特性

张书亚[1*]，李　玲[1]，韩国柱[2]，朱家卫[2]，张传清[**]

（1. 浙江农林大学 农学与食品科学学院，临安　311300；2. 嵊州市森防站，嵊州　312400）

摘　要：香榧（*Torreya grandis* Fort. ex Rindl.）为种子植物门裸子植物亚门裸子植物松杉目红豆杉科，是中国特有树种，其香榧果实，具有很高的食用和药用价值。近年来，香榧上出现一种新病害——裂皮病，造成主干和枝条树皮和韧皮部大量开裂、龟裂、变褐色，随后大块脱落，裸露的木质部也变为褐色。该病冬春季新梢发病少，夏季病害严重，叶片萎蔫，果实脱落，甚至整个枝条枯死。分别采集变褐木质部和韧皮部的香榧样品，通过病原菌分离纯化、形态特征观察、核糖体DNA内转录间隔区（rDNA-ITS）、翻译延伸因子（TEF-1α）和微管蛋白（β-tubulin）基因序列测定及致病性测定，将该病原菌鉴定为可可毛色二孢（*Lasiodiplodia theobromae*）。该病菌在PDA上菌落边缘不整齐，初期灰白色，后转为墨绿色或黑色；气生菌丝发达，绒毛状，有分隔，不规则分支；分生孢子椭圆形或卵形，初期单胞无色，老熟后变褐色，近中部有一横隔；分生孢子大小为（19.6～27.0）μm×（6.3～12.2）μm；该病原菌菌丝生长的适温为28～30℃，致死温度为50℃，在pH值5～7的范围内生长良好，碳源和氮源分别以葡萄糖和氯化铵为宜。本研究为香榧裂皮病的首次报道。

关键词：香榧；裂皮病；可可毛色二孢；生物学特性

* 第一作者：张书亚，硕士研究生，主要从事植物病理学研究；E-mail：yesadam@ sina. com

** 通信作者：张传清，博士，教授，主要从事植物病害治理及杀菌剂药理学与抗药性研究；E-mail：qzhang@ zafu. edu. cn

香蕉枯萎病颉颃木霉 gz-2 菌株的 GFP 标记研究*

杜婵娟[1,2]，吴礼和[3]，付岗[2**]，杨迪[2]，张晋[2]，潘连富[2]，王维[2]
（1. 广西作物病虫害生物学重点实验室，南宁 530007；
2. 广西农业科学院微生物研究所，南宁 530007；
3. 贺州学院食品与生物工程学院，贺州 542899）

摘　要：木霉是广泛存在于自然界中的一种真菌，常被用于植物病害的防治。gz-2 菌株是从土壤中筛选出的一株对香蕉枯萎病菌有较强抑制效果的哈茨木霉。为了进一步开展 gz-2 菌株的定殖规律及环境适应性等方面的研究，就需要对该菌进行相应的分子标记。

绿色荧光蛋白（GFP）标记是相对成熟的标记技术，为提高哈茨木霉 gz-2 菌株原生质体的形成率和转化率，从而快速、高效地获得绿色荧光标记菌株。本研究采用 PEG-$CaCl_2$介导的原生质体转化体系，分别从菌株的菌龄及酶解时间、转化体系中质粒的含量及再生培养基中潮霉素 B 的浓度等方面对哈茨木霉 gz-2 菌株绿色荧光标记的条件进行了优化。明确 gz-2 菌株 GFP 标记的最佳条件为：木霉培养 36h，菌丝酶解 3.5h，每 240μL 转化体系中质粒 DNA 含量为 40μL，转化子再生培养基中含潮霉素 B 浓度为 600μg/mL。最终筛选获得 1 株表达稳定的转化子 GFP-gz-2 菌株。该菌株在菌落形态、菌丝生长速率及产孢量上与出发菌株 gz-2 无显著差异，且具有较好的遗传稳定性。

关键词：香蕉枯萎病；哈茨木霉；原生质体；GFP 标记

---

* 基金项目：广西自然科学基金（2016GXNSFBA380173）；广西作物病虫害生物学重点实验室开放基金（2015-KF-1）；广西农业科学院基本科研业务专项（桂农科 2016JZ15，2015YT79，2015JZ61）

** 通信作者：付岗，博士，副研究员，主要从事植物病害及其生物防治研究；E-mail：fug110@ gxaas. net

Characterization of VdASP F2 secretory factor from *Verticillium dahliae* by a fast and easy gene knockout system

Xie Chengjian[1,2], Yang Xingyong[1,2 *]

(1. *School of Life Sciences*, *Chongqing Normal University*, *Chongqing* 401331, *China*;
2. *The Chongqing Key Laboratory of Molecular Biology of PlantEnvironmental Adaptations*,
Chongqing Normal University, *Chongqing* 401331, *China*)

Abstract: The vascular wilt fungus *Verticillium dahliae* produces persistent resting structures known as microsclerotia, which enable long-term survival of this plant pathogen in soil. The completed genome sequence of *V. dahliae* has facilitated large-scale investigations of individual gene functions using gene-disruption strategiesbased on *Agrobacterium tumefaciens*-mediated transformation (ATMT). However, the construction of gene-deletion vectors and screening of deletion mutants have remained challenging in *V. dahliae*. In this study, we developed a fast and easy gene knockout system for *V. dahliae* using ligation-independent cloning and fluorescent screening. We identified secretory factor VdASP F2 in a T-DNA insertion library of *V. dahliae* and deleted the VdASP F2 gene using the developed knockout system. Phenotypic analysis suggests that VdASP F2 is not necessary for *V. dahliae* growth on potato dextrose agar under various stress conditions. However, on semisynthetic medium or under limited nutrient conditions at lower temperatures, the VdASP F2 deletion mutant exhibited vigorous mycelium growth, less branching, and a significant delay in melanized microsclerotial formation. Further assessment revealed that VdASP F2 was required for the expression of *VDH*1 and *VMK*1, two genes involved in microsclerotial formation. Cotton inoculated with the VdASP F2 deletion mutant wilted, demonstrating that VdASP F2 is not associated with pathogenicity under normal conditions. However, after inducing microsclerotial formation and incubation at low temperatures, cotton infected with the VdASP F2 deletion mutant did not exhibit wilt symptoms. In conclusion, our results show that VdASP F2 plays an important role in the response of *V. dahliae* to adverse environmental conditions and is involved in a transition to a dormant form for prolonged survival.

* Corresponding author: Yan Xingyong; E-mail: yangxy94@ swu. edu. cn

苹果树腐烂病菌 qPCR 检测方法的建立*

祁兴华**，郭永斌，王树桐***，曹克强***
（河北农业大学植物保护学院，保定 071001）

摘 要：苹果树腐烂病（*apple tree Valsa canker*）是我国各主要苹果产区普遍发生且危害严重的一种真菌病害，该病害由苹果黑腐皮壳菌（*Valsa mali* Miyabe et Yamada）侵染所致。潜伏侵染是腐烂病侵染过程的一个重要特征。为了对该病害进行早期诊断，实时检测，本研究建立了一套实时定量 PCR 菌检测方法。利用通用引物获得苹果腐烂病菌菌株 Vmm45 的 ITS、β-微管蛋白和 EF-1α 的基因序列，与 GenBank 中的序列对比，在设计的 16 对引物中筛选出基于 EF-1α 基因序列的引物对 VE-F/VE-R。20μL 反应体系中，引物 VE-F 和 VE-R 各 0.5μL，模板 DNA 1μL，含荧光染料的 Mix 10μL，ddH_2O 8μL；反应条件为：95℃预变性 300s，95℃变性 12s，58.0℃退火 12s，72℃延伸 14s，40 个循环；罗氏 Light Cycler 96 程序生成的溶解曲线程序是 95℃ 10s，65℃ 60s（2.20℃/s），97℃ 1s（0.20℃/s）。用该引物对梯度稀释的 DNA 样品进行定量扩增，结果表明：该体系可精确定量到 2.4 fg/μL 的病菌基因组 DNA，其灵敏度在 fg/μL级别。该引物特异性强，可从苹果轮纹病菌、链格孢、层出镰刀菌、木霉、青霉等苹果树上的常见真菌中特异性的检测出苹果树腐烂病菌。通过对苹果树树皮组织和叶片的检测，证明本体系可以检测出未发病部位带菌。利用这一 qPCR 检测体系与已有的检测体系对比发现，本检测体系灵敏度高，特异性强，检测结果可靠。可用于苹果树腐烂病菌的快速检测，为苹果树腐烂病的早期诊断提供新方法。

关键词：苹果腐烂病菌；实时定量 PCR；检测方法

---

* 基金项目：国家苹果产业技术体系（CARS-28）和河北省自然科学基金（c2016204140）

** 作者简介：祁兴华，在读硕士研究生，研究方向：植物病害流行与综合防治；E-mail：990587038@qq.com

*** 通信作者：王树桐，博士，教授，从事植物病害流行与综合防治研究；E-mail：bdstwang@163.com
曹克强，博士，教授，从事植物病害流行与综合防治研究；E-mail：ckq@hebau.edu.cn

Analysis of the community compositions of soil fungifrom different cropping pattern fields of soybean and maize

Yan Li[1], Zhou Huanhuan[1], Wang Wei[1], Chang Xiaoli[1 *]
(1. *Institute of Ecological Agriculture*, *Sichuan Agricultural University*)

Abstract: In order to find out the difference of pathogenic fungi in soil from different cropping pattern fields and pick out the dominant pathogens, we used four different soils sampled from the fields of Maize-soybean strip intercropping planting pattern, Maize-soybean intercropping planting pattern, Maize monoculture and Soybean monoculture for two years continuous cropping to analysis the taxonomic community of fungi high-incidence of maize and soybean diseases. A total of 934 973 modified internal transcribed spacer (ITS) sequences were obtained from four different soil samples using Illumina Miseq PE250/PE300 high-throughput sequencing. The dominant eumycote fungal were identified to be Ascomycota and Basidiomycota in the four soil samples. Different cropping pattern of soybean and maize affected the diversity of fungi in soil and the dominant pathogens are *Fusarium*, *Alternaria* and *Ceratobasidium*. *Fusarium*, the most pathogenic fungi for those two crops we detected, got a much larger proportion in monoculture fields than in intercropping fields. As for the sexual status of *Fusarium*, the situation of *Gibberella* is broadly the same as *Fusarium*, but we detected much more *Gibberella* in Maize-soybean intercropping field than Maize-soybean strip intercropping field, which we speculate to be the longer distance from maize to soybean in Maize-soybean strip intercropping pattern make it more difficult for different *Fusarium* from different plants to get enough plant disability to over-winter by the sexual status.

Key words: Maize; Soybean; High-throughput sequencing; *Fusarium*

* Corresponding author: Chang Xiaoli; E-mail: 510835763@qq.com

Functional identification of toxin ecoding gene (*Cas*) of *Corynespora cassiicola*-the pathogen of brown leaf spot disease on kiwi in Sichuan

Xu Jing[1], Qi Xiaobo[1], Chang Xiaoli [1], Cui Yongliang[2], Gong Guoshu[1 *]

(1. *College of Agronomy & Key Laboratory for Major Crop Diseases, Sichuan Agricultural University*, *Chengdu* 611130, *P. R. China*;

2. *Sichuan Academy of Natural Resource and Sciences*, *Chengdu* 610041, *Sichuan*, *China*)

Abstract: The brown leaf spot caused by *Corynespora cassiicola* is a main disease on kiwifruit, which affected fruit setting and burgeoned. The toxin was purified from the culture solution and named "cassiicolin", cassiicolin is a small secreted protein, it was found that the toxin protein is a host-selective toxin (HST), which is closely related to the pathogenicity of *C. cassiicola* and is an important causative effect factor of *C. cassiicola*.

PCR was performedby specific primers to obtain the *Cas*2 toxin ecoding gene sequence from *C. cassiicola*, construction of expression vector with *Cas*2 toxin gene, the subcellular localization observation on the cigarette showed that the storage protein could be located on the cell membrane, it is suggested that the toxin protein may combine or degrade the target protein on the host cell membrane, and the pathways to the host disease should be further studied. Construction of *Cas* toxin gene knockout vector, using the principle of homologous substitution by PEG-$CaCl_2$-mediated protoplast transformation method to obtain *Cas* gene mutant strain, phenotypic analysis showed that there were significant differences in the development of aerial hyphae and the color of colony surface compared with the wild type strain, but no significant difference in growth rate, pigment production and sporulation. Inoculated kiwi leaves with wild type strain and mutant strain, experiments showed that there was a significant difference in the rate of lesion extension, lesions formed on leaves with mutant strains extended slowly and were not typical. It was proved that *Cas* gene played an important role in the pathogenesis of the host, and there may be other genes that work together, need further study.

Key words: Kiwi; *Coyrnespora cassiicola*; *Cas* gene; Functional identification

* Corresponding author: Gong Guoshu; E-mail: guoshugong@ 126. com

RNA-Seq Reveals *Fusarium proliferatum* Transcriptome and Candidate Effectors during the Interaction with Tomato Plants*

Gao Meiling[1**], Xu Pinsan[1,4], Luan Yushi[1], Yu Haining[1], Cui Jun[1], Sun Wenhui[1], Pu Zhongji[1], Liu Huiquan[3***], Bao Yongming[1,2***]

(1. *School of Life Science and Biotechnology*, *Dalian University of Technology*, *Dalian*, 116024, *China*; 2. *School of Food and Environmental Science and Technology*, *Dalian University of Technology*, *Panjin* 124221, *P. R. China*; 3. *State Key Laboratory of Crop Stress Biology for Arid Areas and College of Plant Protection*, *Northwest A&F University*, *Yangling*, *Shannxi* 712100; 4. *School of Life Science and Medicine*, *Dalian University of Technology*, *Panjin* 124221, *China*)

Abstract: Leaf spot disease caused by *Fusarium proliferatum* (Matsushima) Nirenberg is a destructive disease of tomato plants in China. Typical symptoms of infected tomato leaves and stems are soften and wilt, with eventual death of the whole plant. However, the molecular pathogenicity mechanisms of *F. proliferatum* is not well understood. Effectors are known to act as pathogenicity or virulence determinants to manipulate the interactions between the pathogen and specific host plants. Here, we provided transcriptional profile analysis to gain insight into the repertoire of effectors during *F. proliferatum*-tomato interaction. Mycelia collected from 7-day-old potato dextrose agar culturesand *F. proliferatum*-infected tomato leaves at 96h post-inoculation were subjected to RNA-Seq analysis. A total of 61 544 598 clean reads were *de novo* assembled to a *F. proliferatum* reference transcriptome that was generated with 89 716 transcripts expressed from 75 044 unigenes. After comparisons with public databases, 95.78% of the unigenes were annotated and 14 607 unigenes were assigned into 276 KEGG pathways; 16 734 unigenes (22.3%) had high homology with the genes from *F. fujikuroi*. 18 075 differentially expressed genes (DEGs) were discovered and performed GO and KEGG enrichment analysis. Analysis showed that 184 candidate effectors were identified from the putative secreted protein genes from the DEGs and 79.89% of them were up-regulated expression. Moreover, 17 randomly selected *F. proliferatum* DEGs (contain 10 candidate effector genes) were validated by quantitative real-time PCR (qRT-PCR). In addition, we also further measured the expression levels of 10 candidate effector genes in tomato leaves at different time points. The study demonstrated that transcriptome analysis was an effective method to identify the repertoire of candidate effectors and provided an invaluable resource for future functional analysis of *F. proliferatum* pathogenicity in the evolution of *F. proliferatum* and tomato plants host interactions.

---

* Funding: This work was supported by grants from Panjin Campus for Food Science and Technology Research Initiative, Dalian University of Technology and the National Natural Science Foundation of China (No. 31471880)

** First author: Gao Meiling, Ph. D candidate; E-mail: gaomeiling@mail.dlut.edu.cn

*** Corresponding authors: Bao Yongming; E-mail: biosci@dlut.edu.cn

Liu Huiquan; E-mail: liuhuiquan@nwsuaf.edu.cn

玉米弯孢叶斑病菌 *ClNPS4* 基因克隆与功能分析*

高艺搏**，路媛媛#，肖淑芹，薛春生***
（沈阳农业大学植物保护学院，沈阳　110866）

摘　要：微生物的非核糖体肽合成酶（Nonribosomal Peptide Synthetase，NPS）负责合成低分子量次生代谢产物，其中 NPS6 和 NPS2 在多个丝状真菌中结构保守，功能稳定，分别与胞外铁载体生物合成和胞内铁离子运输密切相关，但对 NPS4 作用研究较少。据报道 *Cochliobolus heterostrophus* 的 *ChNPS*4 和 *Fusarium graminearum* 的 *FgNPS*4 缺失后，病原菌表面疏水能力下降，致病力下降。从 NCBI 和 JGI 数据库中新月弯孢菌株 m118 和 CX-3 全基因组中获得 *NPS*4 的同源序列，该基因全长为 21606bp，包含一个开放阅读框，编码了 7201 个氨基酸。设计分别带 *Sal* Ⅰ、*Xba* Ⅰ和 *Kpn* Ⅰ、*EcoR* Ⅰ 酶切位点的特异性引物，利用 PCR 方法克隆得到 567bp、795bp 的 *ClNPS*4 侧翼片段，连接至双元载体 pPZP100，同时将来自 pPCT74 的潮霉素和绿色荧光蛋白基因也连接至 pPZP100，最终获得 pPZP100ClNPS4，电击法将 pPZP100ClNPS4 转入农杆菌菌株 AGL-1 中，ATMT 技术获得了玉米弯孢叶斑病菌 *NPS*4 基因缺失突变体。Δ*Clnps*4 的菌落形态、生长速率以及分生孢子形态、产孢数量和孢子萌发率均与野生型无显著差异，而 Δ*Clnps*4 在液体培养后的菌丝干重与野生型相比差异显著；同时将 Δ*Clnps*4 接种感病自交系黄早 4，与野生型相比其致病力显著下降。新月弯孢菌 *ClNPS*4 生物合成的次生代谢产物及其作用机理正在研究过程中。

关键词：玉米弯孢叶斑病菌；非核糖体肽合成酶；NPS4

---

* 基金项目：国家自然科学基金（31271992）；辽宁省自然科学基金（20170540811）
** 第一作者：高艺搏，本科生
*** 通信作者：薛春生，教授，博士生导师；E-mail：cshxue@ sina. com

调节稻瘟菌附着胞形成多肽的原核表达及纯化

郭娇*，赵彦翔，郭力维，刘强，彭友良，刘俊峰**
（中国农业大学植物保护学院植物病理学系，北京 100193）

摘　要：水稻作为世界三大经济作物之一，是全球50%以上人口的主食，由稻瘟菌（*Magnaporthe oryzae*）引发的稻瘟病病害是水稻产量损失最为严重的病害之一。目前，化学防治仍是防治稻瘟病的主要措施之一，但长期使用化学药物导致的病菌抗药性问题日益突出，同时农药残留，生态失衡的问题也日益严重。抗菌肽由于其高效、环境友好且不易产生抗药性等特点，是理想的防治稻瘟菌杀菌剂。由于抗菌肽的天然资源有限，且利用化学合成成本较高。本研究尝试通过操作简单、成本低的大肠杆菌（*Escherichia coli*）表达系统探索获取大量抗菌肽的途径，同时进一步利用结构生物学方法研究小分子抗菌肽的三维结构并探究其生物学功能，为抗菌肽的作用机制提供理论基础。

前人研究发现，抗菌肽 MgAPI16 能够通过改变稻瘟菌附着胞的形状从而降低稻瘟病菌的致病力。本研究中，笔者分别采用将抗菌肽 MgAPI16 与促溶标签 YFP 融合表达；MgAPI16 基因多拷贝同向串联融合表达和抗菌肽 MgAPI16 与标准 α-螺旋相互融合成杂合抗菌肽进行表达。将编码上述 3 种抗菌肽基因插入到携带 YFP 大标签的 pRSET-Y 载体中，之后将构建好的载体分别转化到 BL21（DE3）、Rosetta（DE3）、Shuffle T7 等表达菌株中进行表达，已筛选出较优的表达菌株及表达诱导条件。在此基础上，通过亲和层析、离子交换层析、分子筛层析等方法进一步纯化出单一性较好的目标蛋白。目前，笔者正在利用气相液滴法对纯化好的蛋白展开大量晶体条件筛选。此研究有助于进一步获取相关抗菌肽晶体，了解其作用方式，为开发新型防治稻瘟病的绿色高效的杀菌剂提供重要基础。

关键词：稻瘟菌；抗菌肽；蛋白纯化

---

* 第一作者：郭娇，硕士研究生，植物病理学专业；E-mail：15600910650@163.com
** 通信作者：刘俊峰，教授；E-mail：jliu@cau.edu.cn

交配型基因 *ClMAT* 对新月弯孢（*Cochliobolus lunatus*）有性生殖及生长发育的影响*

刘克心**，孙玉鑫，肖淑芹，薛春生***

（沈阳农业大学植物保护学院，沈阳　110866）

摘　要：玉米弯孢叶斑病菌属异宗配合，有性生殖主要是由交配型位点（MAT）控制。该基因控制雌雄配子亲和和两性亲和细胞的激素调节机制，控制性别分化和有性世代，以及子囊孢子的产生。同时，对弯孢菌无性态的生长发育也起到一定的调控作用。

利用农杆菌介导遗传转化（ATMT）技术分别将 *C. lunatus* 的 CX-3 和 ZD958 菌株的 *MAT*1-2-1 和 *MAT*1-1-1 基因敲除，获得突变体 Δ*MAT*1-2-1 和 Δ*MAT*1-1-1。通过有性态诱导可知，菌株 CX-3×ZD958 可以产生成熟的假囊壳，而菌株 CX-3×Δ*MAT*1-1-1、ZD958×Δ*MAT*1-2-1 及 Δ*MAT*1-2-1×Δ*MAT*1-1-1 的杂交均无法产生假囊壳，这一结果证明*MAT*1-2-1 和 *MAT*1-1-1 基因是有性生殖过程中必不可少的。通过比较野生型和突变体的生物学测定可知，当 *MAT*1-2-1 和 *MAT*1-1-1 缺失，两个突变体与野生型 CX-3 和 ZD958 菌株在 PDA 培养基和 CM 培养基上生长速度差异不明显，但菌丝干重存在明显差异，两个突变体的菌丝干重明显低于野生型菌株。同时，*MAT*1-2-1 基因还调控分生孢子的萌发率。

关键词：新月弯孢菌（*Cochliobolus lunatus*）；交配型；*MAT*1-2-1；*MAT*1-1-1

---

* 基金项目：国家自然科学基金（31271992）；辽宁省自然科学基金（20170540811）

** 第一作者：刘克心，硕士研究生，从事玉米病害研究

*** 通信作者：薛春生，教授，博士生导师；E-mail：cshxue@ sina. com

NADPH 氧化酶基因对玉米弯孢叶斑病菌生长发育与致病力的影响*

毛秀文**，刘雨佳，肖淑芹，薛春生***
（沈阳农业大学 植物保护学院，沈阳 110866）

摘 要：为研究 NADPH 氧化酶的主要组分对玉米弯孢叶斑病菌致病力的影响，利用 ATMT 技术获得了 NADPH 氧化酶基因突变体 Δ*Clnox*1、Δ*Clnox*2 和 Δ*Clnox*1*nox*2，又采用 over-lapping PCR 方法获得了互补转化子 Com-Clnox1、Com-Clnox2。突变体生物学特性检测果表明：Δ*Clnox*1、Δ*Clnox*2 和 Δ*Clnox*1*nox*2 与野生型 CX-3 相比附着胞形成率分别下降 73.03%、22.03%、77.37%；Com-Clnox1、Com-Clnox2 则与对照菌株之间无差异，突变体在菌落形态、产孢量、附着胞形态和分生孢子形态方面与野生型不存在差异。在 PDA 和 CM 两种培养基上，Δ*Clnox*1*nox*2 生长速率较野生型有下降的趋势（*P*>95%），其他突变体与野生型无明显差异。Δ*Clnox*2 4h 的分生孢子萌发率为 93.33%，低于野生型 CX-3 和 Δ*Clnox*1 的 97%和 96.67%，6h 时后 CX-3 和 Δ*Clnox*1 无差异，表明 ClNox2 使分生孢子萌发滞后；Δ*Clnox*1*nox*2 在 4h、6h 分生孢子萌发率分别为 82.33%、92.33%，与 CX-3 存在差异（*P*>95%）；Δ*Clnox*1、Com-Clnox1 和 Com-Clnox2 与野生型 CX-3 相比无明显差异。接种 8 叶期玉米幼苗的致病力观察试验结果表明：Δ*Clnox*1、Δ*Clnox*2 和 Δ*Clnox*1*nox*2 致病力显著下降，其平均病斑面积分别为 3.82 mm^2、4.76 mm^2、3.43 mm^2，与 CX-3 病斑面积 7.05 mm^2 相比存在明显差异（*P*>95%）；Com-Clnox1、Com-Clnox2 病斑面积与对照相比无明显差异。

关键词：玉米弯孢叶斑病菌；NADPH 氧化酶；致病力

---

* 基金项目：国家自然科学基金（31271992）；辽宁省自然科学基金（20170540811）
** 第一作者：毛秀文，硕士研究生，从事玉米病害研究
*** 通信作者：薛春生，教授，博士生导师；E-mail：cshxue@ sina. com

四川省大豆不同部位病害致病镰孢菌的群体多样性研究

周欢欢[1]，严雳[1]，杨文钰[1]，常小丽[1*]
（1. 四川农业大学农学院，农业部西南作物植物病理重点实验室，成都 611130）

摘　要：镰孢菌（*Fusarium*）是引起大豆根腐病的重要致病菌，其中，以尖孢镰孢菌（*F. oxysporum*）和腐皮镰孢菌（*F. solani*）报道最多、危害最重。近年来，镰孢菌引起的大豆病害在生产中有加重趋势。为明确四川省大豆不同部位病害致病镰孢菌的群体多样性，于 2015—2016 年先后在四川省内不同大豆栽种区随机采集大豆茎腐病、根腐病和荚腐病的病害样品，通过形态学鉴定和分子鉴定，初步了解引起大豆不同部位病害致病镰孢菌的状况，并采用大豆黄化苗刺伤接种法来评价确定不同镰孢菌的致病力情况。

结果表明：从 170 株分离物中获得 127 株镰孢菌分为 9 个种，包括木贼镰孢菌（*F. equiseti*），尖孢镰孢菌（*F. oxysporum*），腐皮镰孢菌（*F. solani*），层生镰孢菌（*F. proliferatum*），禾谷镰孢菌（*F. graminearum*），轮枝镰孢菌（*F. verticillioides*），*F. commune*，燕麦镰孢菌（*F. avenaceum*），藤仓镰孢菌（*F. fujikuroi*）。分离频率依次为 32. 28%、23. 62%、17. 32%、7. 87%、7. 09%、5. 51%、3. 15%、2. 36%和 0. 79%。大豆不同部位病害的镰孢菌种类和分离频率也存在差异：禾谷镰孢菌、腐皮镰孢菌和尖孢镰孢菌在茎部、根部和豆荚部分离频率都高，木贼镰孢菌、燕麦镰孢菌和轮枝镰孢菌主要分离自根部和豆荚部位，藤仓镰孢菌和 *F. commune* 只在大豆荚腐病中分离得到。致病性测定表明，以上 9 个种的镰孢菌均能引起幼苗发病，其中尖孢镰孢菌、木贼镰孢菌和禾谷镰孢菌致病性最强；同时，各镰孢菌的致病力与大豆生长指标具有相关性，与株高及鲜重成负相关。

综上所述，四川省大豆不同部位病害优势致病镰孢菌为腐皮镰孢菌、尖孢镰孢菌和木贼镰孢菌，禾谷镰孢菌分离频率较低但致病性最强，对于四川地区大豆病害发生具有潜在威胁。本研究为四川地区大豆生产提供了有价值的参考。

关键词：大豆；茎腐；根腐；荚腐；镰孢菌；鉴定

* 通信作者：常小丽；E-mail：510835763@ qq. com

玉米大斑病菌（*Setosphaeria turcica*）效应因子筛选和功能验证*

王芬**，闫丽斌**，肖淑芹，薛春生***

（沈阳农业大学植物保护学院，沈阳　110866）

摘　要：大斑病是一种世界性玉米叶部病害，多发生于全球冷凉地区的玉米种植区，常造成严重经济损失。致病菌为大斑刚毛球腔菌［*Setosphaeria turcica*（Luttrell）Leonard & Suggs］，营养类型为半活养寄生的病原真菌，侵染寄主初期为活体营养阶段，待成功侵入之后转入死体营养阶段，这一侵染过程中必然存在着复杂的分子机制。研究玉米大斑菌侵染寄主过程中效应因子的作用机制对于进一步揭示在与寄主互作过程中的病原菌的侵染机制非常重要。本实验室前期根据玉米大斑病菌株 Et28A 全基因组信息，获得 60 个大部分功能未知的候选效应蛋白。

本试验对 60 个候选效应因子进行信号肽预测，并对候选效应因子进行内含子分析，发现 60 个候选效应因子中有 24 个不含内含子，36 个含有一个或多个内含子。为保证效应因子能够在烟草细胞质内稳定地表达并行使功能，依据去掉信号肽后的效应因子 cDNA 序列设计上游引物，从而保证后续研究中基因表达产物不被分泌到烟草细胞外。分别以玉米大斑病菌 Et28A 基因组 DNA 和 cDNA 为模板，扩增出了 24 个不含内含子，7 个含内含子的效应因子，将所有经高保真酶扩增得到的回收产物经过酶切后连接到相应酶切位点的载体 pGR106 中，转化大肠杆菌，待长出菌落之后，利用菌落 PCR 方法验证，验证正确后导入农杆菌。下一步本试验以能够引起烟草过敏性坏死反应的细胞凋亡诱导因子 Bax、卵菌中疫霉菌特征分子（PAMP）INF1 为阳性对照，以 GFP 和空白载体为阴性对照，并利用农杆菌介导的病毒表达系统在本氏烟上瞬时表达效应因子，筛选出可直接引起植物反应的效应因子，检测候选效应因子功能。

关键词：玉米大斑病菌（*Setosphaeria turcica*）；效应因子；瞬时表达

---

* 基金项目：国家自然科学基金（31271992）；辽宁省自然科学基金（20170540811）

** 第一作者：王芬，博士研究生，从事玉米病害研究；闫丽斌

*** 通信作者：薛春生，教授，博士生导师；E-mail：cshxue@ sina. com

东北地区玉米镰孢菌穗腐病病原菌种类鉴定与分布*

许佳宁**，肖淑芹，孙佳莹，王芬，路媛媛，薛春生***
（沈阳农业大学植物保护学院　沈阳　110866）

摘　要：玉米穗腐病是为害我国玉米产区的重要穗部病害，其主要致病菌为镰孢菌属（*Fusarium*）。为了明确东北地区玉米穗腐病病原菌的种类，2016 年从辽宁、吉林等 8 个市县、18 个不同地区调查取样，并对采集回来的病样进行病原菌分离和单孢纯化，共获得镰孢菌标样 101 株。采用形态学和镰孢菌种的特异引物扩增的方式进行鉴定。结果表明：从 101 株镰孢菌标样中共鉴定出 5 个种，主要致病菌为拟轮枝镰孢菌（*Fusarium verticillioides*），其分离频率为 71.29%，其次为禾谷镰孢菌（*Fusarium graminearum*），约占 10.89%，其他为层出镰孢菌（*Fusarium proliferatum*）和亚黏团镰孢菌（*Fusarium subgutinans*），分离频率分别为 9.9%和 7.92%。禾谷镰孢菌均在吉林省分离得到，表明玉米穗腐病镰孢菌种类分布存在地区间差异。

关键词：玉米穗腐病；镰孢菌；特异性引物

---

* 基金项目：国家重点研发计划（SQ2017ZY060067）；现代农业玉米产业技术体系（CARS-02）

** 第一作者：许佳宁，硕士研究生，主要从事玉米病害研究

*** 通信作者：薛春生，教授，博士生导师；E-mail：cshxue@ sina. com

新月弯孢（*Curvularia lunata*）CX-3的*ClNPS2*基因敲除载体构建*

苑明月**，路媛媛#，肖淑芹，薛春生***
（沈阳农业大学植物保护学院，沈阳　110866）

摘　要：玉米弯孢菌叶斑病（Corn Curvularia Leaf Spot）是我国东北、华北和西南等地区常见的一种玉米叶部病害，由弯孢属的多个种引起，其中新月弯孢（*Curvularia lunata*）为优势种。非核糖体肽合成酶2（Nonribosomal Peptide Synthetase 2，*NPS2*）是非核糖体肽合成酶家族中结构保守、功能稳定的一类酶。

本研究以玉米弯孢叶斑病菌菌株CX-3为研究对象，利用生物信息学方法，在NCBI（https：//www. ncbi. nlm. nih. gov/）网站上公布的数据库中，搜索并得到了新月弯孢CX-3菌株的*NPS2*基因（GenBank：KY471558. 1），全长15 333bp，一个开放阅读框。采用DNAMAN软件分析并设计带有SalⅠ和EcoRⅠ酶切位点的特异引物，以新月弯孢菌CX-3菌株基因组DNA为模板，扩增出1 139bp片段；将该片段纯化后，T载体连接，得到pMD19-T-NPS2，用SalⅠ和EcoRⅠ双酶切骨架载体pPPZP100和pMD19-T-NPS2，连接*NPS2*片段，获得pPPZP100NPS2，最后用XbaⅠ和KpnⅠ双酶切载体pPPZP100NPS2和PCT74，在侧翼片段中间插入绿色荧光蛋白（GFP）和潮霉素抗性基因（HYG B），最终成功构建含有绿色荧光蛋白（GFP）和潮霉素抗性基因（HYG B）的*NPS2*基因的敲除载体pPPZP100HGNPS2。并采用电击法成功转入农杆菌AGL-1菌株中。

拟通过农杆菌介导的遗传转化方法，获得*NPS2*基因敲除突变体，并比较突变体与野生型菌株的表型及致病力的差异，以明确*NPS2*的功能及其调控机制，为玉米弯孢病菌致病分子机制的研究提供理论基础。

关键词：新月弯孢（*Curvularia lunata*）；*ClNPS2*；敲除载体构建

---

* 基金项目：国家自然科学基金（31271992）；辽宁省自然科学基金（20170540811）
** 第一作者：苑明月，硕士研究生，从事玉米病害研究
*** 通信作者：薛春生，教授，博士生导师；E-mail：cshxue@ sina. com

玉米弯孢叶斑病菌致病力分化及遗传多样性分析*

张　丹**，王　芬**，肖淑芹，薛春生***
（沈阳农业大学植物保护学院，沈阳　110866）

摘　要：弯孢叶斑病是由新月弯孢［*Curvularia lunata*（Wakker）Boed.］侵染引起的玉米生产上常发性叶部病害之一，曾对我国玉米安全生产产生了严重的影响。2010-2016 年，本研究从黑龙江、吉林、辽宁、河北、河南、山东、安徽、湖北、四川、贵州和云南等 11 个我国玉米主产区共获得 208 个标样，依据传统的形态学分类结合分子生物学方法，共鉴定出 189 个 *C. lunata*，采用沈 135、CN165、78599、Mo17、E28、B73、掖 478、鲁原 92 和黄早四等 9 个鉴别寄主对上述菌株的致病类型进行了划分，按照病级标准和病情指数，将菌株划分为 6 种类型。利用改进的 CTAB 法提取供试菌株基因组 DNA，进行双酶切及预扩增，再用 16 对引物进行选择性扩增，优化 AFLP 试验条件，拟通过 AFLP 技术对其中 105 株玉米弯孢叶斑病菌的 DNA 扩增片段长度多态性进行分析。在此基础上进行基于 AFLP 标记划分的 AFLP 类群与基于鉴别寄主病情指数划分的致病力类群相关性分析。

关键词：玉米弯孢叶斑病菌；致病力分化；AFLP；遗传多样性

* 基金项目：国家自然科学基金（31271992）；辽宁省自然科学基金（20170540811）
** 第一作者：张丹，硕士研究生，从事玉米病害研究；王芬
*** 通信作者：薛春生，教授，博士生导师；E-mail：cshxue@ sina. com

浙江省3种中药材常见叶片病害的初步研究*

张佳星[1**]，李玲[1]，戴德江[2]，张传清[***]
（1. 浙江农林大学农学与食品科学学院，临安 311300；
2. 浙江省农药检定管理所，杭州 310004）

摘 要：浙江省作为传统中药生产大省，中药材产业已成为省内十大农业主导产业之一，磐安白术、桐乡杭白菊、乐清铁皮石斛等已是当地农民重要的经济收入来源。为鉴定引起白术、杭白菊和铁皮石斛常见叶部病害的病原菌，本文分别从浙江省磐安县、桐乡市和临安市采集白术、杭白菊和铁皮石斛叶片病样，采用组织分离法对病原菌进行分离纯化，综合生物学特征及rDNA-ITS序列对病原菌进行鉴定，并采用离体叶片回接法对3种中药材叶片进行回接。结果初步确定人参叶点霉（*Phyllosticta panax*）是引起白术叶片斑枯病的主要病原菌，细极链格孢（*Alternaria tenurissima*）和茶树炭疽菌（*Colletotrichum fructicola*）是铁皮石斛叶片黑斑病和炭疽病的主要病原菌，杭白菊叶枯病病原菌为豆类刺盘孢菌（*Colletotrichum truncatum*）。

关键词：白术；杭白菊；铁皮石斛；病原菌；分离鉴定

* 基金项目：浙江省公益项目（2016C32002）
** 第一作者：张佳星，硕士研究生，主要从事植物病理学研究；E-mail：275893077@qq.com
*** 通信作者：张传清，博士，教授，主要从事植物病害治理及杀菌剂药理学与抗药性研究；E-mail：cqzhang@zafu.edu.cn

稻瘟菌促分裂原活化蛋白激酶MoOsm1的重组表达、纯化与晶体生长

张圆圆*，周锋，彭友良，刘俊峰**，张国珍**
（中国农业大学植物保护学院植物病理学系，北京 100193）

摘 要：水稻是世界上主要的粮食作物之一，由稻瘟菌（*Magnaporthe oryzae*）引起的水稻稻瘟病是为害我国水稻安全生产的三大病害之一。稻瘟菌在生长发育以及侵染寄主的过程中需要多条信号通路共同参与，MAPK（Mitogen-activated Protein Kinase）通路是其中非常重要的通路之一。稻瘟菌中的Hog1-MAPK级联通路是一条单独控制菌丝生长和孢子形态的信号通路，主要包括MoSsk2、MoPbs2和MoOsm1三个级联蛋白激酶，MoOsm1介导的渗透调节通路在稻瘟菌的胁迫响应、有性和无性发育等方面发挥着重要的作用，能直接与下游的C_2H_2型锌指转录因子MoMsn2相互作用来调控稻瘟菌对逆境的响应，具有重要的生物学功能。本研究尝试重组表达和纯化该激酶和筛选到其晶体生长的条件，以解析MoOsm1的三维结构和为其在信号通路中的调节机制及其与下游转录因子的调控机制提供直接的结构基础。

本研究利用原核表达体系，构建了MoOsm1全长及截短体的不同表达载体，分别热激转化到*Escherichia coli*BL21（DE3）和*E. coli*Rosetta（DE3）菌株中进行重组表达，通过亲和层析技术初步筛选到合适的表达载体和表达菌株的组合，进而采用离子交换层析及凝胶过滤层析技术进一步分离纯化目的蛋白，探索出一套较优的目的蛋白的纯化流程。目前正采用座滴式气相扩散法对获取的蛋白样品进行晶体生长条件的筛选，以期获得高分辨率的蛋白晶体，进而解析其三维结构，为后期相关研究提供结构学参考和基础。

关键词：稻瘟菌；MoOsm1；蛋白纯化；晶体生长

* 第一作者：张圆圆，硕士研究生，植物保护专业：E-mail：zhangyuan1511@126.com

** 通信作者：张国珍，教授；E-mail：zhanggzh@cau.edu.cn；
刘俊峰，教授；E-mail：jliu@cau.edu.cn

甘青小麦条锈病菌对 UV-B 敏感性的测定*

赵雅琼[1**]，李婷婷[1]，马金星[1]，金社林[2]，姚　强[3]，马占鸿[1]，王海光[1***]

（1. 中国农业大学植物病理学系，北京 100193；2. 甘肃省农业科学院植物保护研究所，兰州 730070；3. 青海省农林科学院，农业部西宁作物有害生物科学观测实验站，青海省农业有害生物综合治理重点实验室，西宁　810016）

摘　要：小麦条锈病由条形柄锈菌小麦专化型（*Puccinia striiformis* f. sp. *tritici*）引起，是影响我国小麦安全生产的重要病害之一。小麦条锈病菌毒性发生变异是引起小麦抗条锈病抗性丧失的重要原因，是造成我国小麦品种大面积多次更换的重要因素，在我国小麦条锈病的发生流行中发挥着重要作用。紫外线对小麦条锈病菌的影响可能是条锈病菌自然发生变异的一个原因，尤其是在条锈病菌高空传播过程中或在紫外线辐射强度大的地区，紫外线对条锈病菌毒性变异的影响不容忽视。本研究对 2016 年采自甘肃省和青海省自然发病的小麦条锈病样品在人工气候室内进行了病原单孢子堆分离，共分离获得 25 个菌株，结合在甘肃和青海实际测定的 UV-B 辐射强度，选定 200μw/cm^{2}作为对分离菌株 UV-B 敏感性测定的照射强度。在此 UV-B 照射强度下，随机选择 3 个分离菌株，分别照射 30min、60min、90min、120min、150min 和 180min 后测定夏孢子相对萌发率，并进行方差分析，结果表明：在 UV-B 照射 120 min 情况下，3 个菌株间差异显著（$P<0.05$）。本研究中选定 120min 为照射时间对其他 22 个分离菌株进行 UV-B 敏感性测定。经 UV-B 照射处理后，在分离获得的 25 个菌株中，菌株夏孢子相对萌发率最低为 19.7%，最高为 59.3%，而且经 UV-B 照射后的夏孢子萌发存在一定的延迟。结果表明不同菌株之间对 UV-B 照射的敏感性存在较大差异，值得进一步研究分析潜在原因。本研究初步测定了甘肃、青海小麦条锈病菌分离菌株的 UV-B 敏感性，为进一步探究条锈病菌毒性变异机制和开展病害宏观防控奠定了一定基础。

关键词：小麦条锈病菌；夏孢子；UV-B 照射；敏感性；相对孢子萌发率

* 基金项目：国家重点基础研究发展计划（973）项目（2013CB127700）；国家自然科学基金项目（31471726）

** 第一作者：赵雅琼，女，山西阳泉人，博士研究生，主要从事植物病害流行学和宏观植物病理学研究；E-mail：18700944409@163.com

*** 通信作者：王海光，副教授，主要从事植物病害流行学和宏观植物病理学研究；E-mail：wanghaiguang@cau.edu.cn

胶孢炭疽菌转录因子 *CgbHLH6* 基因的分离与功能分析*

康浩**，苏初连，杨石有，刘晓妹，蒲金基***，张贺***
（中国热带农业科学院环境与植物保护研究所，海口 571101）

摘 要：胶孢炭疽菌是一类地理分布广泛、寄主范围广泛的病原真菌。胶孢炭疽菌在作物成熟前和成熟后都能侵染植株和果实，特别对成熟的热带作物为害极大，严重影响产品的出口质量。bHLH 转录因子（Basic helix-loop-helixtranscription factor）是真核生物中高度保守的一类转录因子，广泛存在动物、植物和真菌中，对生物的生长发育过程起着极其重要的调控作用。生物信息学分析表明，bHLH 转录因子具有两个 α 螺旋和一个短环。为了明确 bHLH 转录因子在胶孢炭疽菌中的作用，本实验以胶孢炭疽菌基因组 DNA 为模板，利用基因同源克隆技术扩增获得 *CgbHLH6*基因全长序列，该基因 DNA 和 cDNA 全长分别为 1 093 bp 和 999 bp，编码区含有一个内含子（94 bp），编码 333 个氨基酸，其分子量约为 46.02 ku，等电点为 6.42。分析其序列特征，预测其蛋白质功能域及结构域。通过潮霉素 B 抗性基因（*hygB*）插入突变和 PEG 介导的原生质体转化技术，获得 *CgbHLH6* 基因的敲除突变体并测定了其表型，为进一步分析 bHLH 转录因子家族在胶孢炭疽菌生长发育和致病过程的功能奠定基础。

关键词：胶孢炭疽菌；bHLH 转录因子；基因敲除；功能分析

---

* 基金项目：国家自然科学基金（31460455）；海南省自然科学基金项目（20163063）；海南省重大科技计划；中央级公益性科研院所基本科研业务费专项-中国热带农业科学院院级创新团队项目（17CXTD-09）

** 第一作者：康浩，男，在读硕士研究生，主要从事热带果树病理学研究；E-mail：kanghao0908@163.com

*** 通信作者：蒲金基，博士，研究员，主要从事热带果树病理学研究

张贺，硕士，助理研究员，主要从事热带果树病理学研究；E-mail：atzzhef@163.com

一种重楼新病害的病原菌鉴定*

钟　珊[1**]，丁万隆[2]，祁鹤兴[1]，张国珍[1***]

（1. 中国农业大学植物保护学院植物病理学系，北京　100193；
2. 中国医学科学院药用植物研究所，北京　100193）

摘　要：重楼为百合科植物，以根茎入药。近两年，甘肃省天水和云南省腾冲的重楼上发生了一种新病害。在3月春季重楼出苗期严重发生，死苗率最高可达50%。发病初期植株茎部出现一些小白点，1~2d后开始变大，病斑纵向扩展呈长梭形，中央凹陷并有开裂，多个病斑可汇合成片，致使茎部完全腐烂，倒伏，地上部枯死。也可从近土表处的茎基部开始发病向上蔓延扩展。或从叶柄基部开始发病，沿小叶柄向叶部和茎部扩展，叶片呈水渍状腐烂。湿度大时在病部可出现灰白色至灰黑色霉层。采用常规组织分离法分离病原菌，在PDA平板培养基上，20℃条件下培养。同时将消毒过的病组织进行保湿培养，1d后其表面覆盖大量灰色霉层。在PDA培养基上，初期菌落白色、平展，后期呈浅褐色，气生菌丝少，不产孢，可产较多的黑色菌核。病组织上的分生孢子梗直立，有分枝。分生孢子长卵圆形，（24.06~16.67）μm（21.15μm）×（13.38~9.69）μm（11.62μm）。从病斑上挑取单个分生孢子进行菌株纯化，从中选取3个代表菌株进行分子生物学鉴定，分别扩增了甘油醛-3-磷酸脱氢酶基因（*G3PDH*）、热激蛋白60基因（*HSP60*）和RNA聚合酶Ⅱ的第2亚基基因（*RPB2*）并测序。经BLAST比对分析，3个菌株均为葡萄孢属（*Botrytis* sp.）真菌。用PDA菌饼接种健康重楼的叶片，5d后叶片开始发病，症状与自然发病相同。结果表明：近两年在重楼出苗期发生的新病害是由葡萄孢属真菌引起的。鉴定结果为当地重楼新病害的防治提供了依据。

由于重楼分离菌株在培养性状、孢子形态以及所测3个基因的序列与灰葡萄孢（*B. cinerea*）均有明显差异，种的鉴定正在进行之中。目前，由葡萄孢属真菌引起的重楼灰霉病在国内外尚未见报道。

关键词：重楼；灰霉病；病原菌鉴定

---

* 基金项目：公益性行业（农业）科研专项（201303025）

** 第一作者：钟珊，在读硕士研究生，植物病理学专业；E-mail：zhongshan@cau.edu.com

*** 通信作者：张国珍，教授，主要从事植物病原真菌学的研究；E-mail：zhanggzh@cau.edu.cn

不同生育期禾谷丝核菌在小麦体内的分布*

金京京[1**]，李海燕[1]，马璐璐[1]，贾薇[1]，齐永志[1]，甄文超[2***]
(1. 河北农业大学植物保护学院，保定 071001；2. 河北农业大学农学院，保定 071001)

摘 要：为了明确禾谷丝核菌在小麦体内不同时期的分布情况，本试验用 QPCR 技术分析了从保定和辛集采集的小麦不同组织中禾谷丝核菌的生物量。结果表明：保定地区和辛集地区该菌的在小麦组织中的分布规律类似。从小麦的整个生育期来看，该菌的生物量在分蘖期最低，然后呈明显上升趋势；返青期与越冬前相比增加了 202.7%；该菌生物量在拔节期达到 3 909ng/g，比分蘖期提高了 13 倍。抽穗期植物组织老化程度加速严重抑制了禾谷丝核菌的扩展，导致该菌下降了 3%。另外，从每个时期不同组织中该菌的生物量的分布情况来看，分蘖期到越冬前，该菌主要分布在主茎第一叶叶鞘，占总植株菌量的 33%。返青期，由于缺少地上组织，病原菌主要侵染地中茎部分。起身期，病原菌主要侵染主茎和分蘖的第二片叶鞘。在拔节期，尽管有新生叶和节间生出，该菌还是主要分布在冬前叶中。抽穗期，有 21% 的菌转移到主茎的春生第三叶叶鞘和分蘖的春生第二叶叶鞘以及主茎和分蘖的第一节间上。

关键词：禾谷丝核菌的生物量；QPCR 技术；小麦组织；生育期

---

* 基金项目："十二五" 国家科技支撑计划粮食丰产科技工程课题（2011BAD16B08，2012BAD04B06，2013BAD07B05）；河北省现代农业产业技术体系小麦创新团队建设项目（HBCT2016010208）；河北省自然科学基金（C2016204211）；河北省高等学校科学技术研究重点项目（ZD2016162）

** 第一作者：金京京，在读博士生，主要从事植物生态病理学研究；E-mail：jingjing1809@ yeah. net

*** 通信作者：甄文超，博士，教授，博士生导师，主要从事农业生态学与植物生态病理学研究；E-mail：wenchao@ hebau. edu. cn

稻瘟菌效应子 AvrPib 的表达纯化，晶体生长和结构分析

张鑫，程希兰，杨俊，彭友良，刘俊峰
（中国农业大学植物保护学院植物病理系，北京 100193）

摘　要：由稻瘟病菌（*Magnaporthe oryzae*）引起的稻瘟病是为害水稻的主要病害。稻瘟病菌在侵染过程中分泌的无毒效应蛋白会被水稻中的抗病蛋白 R 蛋白直接或间接识别，并激发下游的抗病反应，抑制病原菌的进一步侵染。由于无毒效应蛋白在植物与病原物互作中的关键作用，稻瘟病菌侵染水稻过程中分泌的无毒效应子研究已成为研究的热点之一。例如 *AvrPib* 是最近通过图位克隆被克隆得到稻瘟病菌的无毒基因，与抗病基因 *Pib* 符合基因对基因假说，但无毒蛋白与抗病蛋白 Pib 互作模式尚不明确。

本研究通过解析 AvrPib 结构分析其功能和避免被寄主受体的识别机制，利用原核表达系统，将 *AvrPib* 基因克隆到原核表达载体中，在大肠杆菌中进行重组表达，通过亲和层析，离子交换层析和分子排阻层析等方法获得高纯度的目的蛋白，利用气相扩散法筛选不同晶体生长条件，获得目的蛋白晶体，优化后得到 1.8Å 衍射数据，利用 SAD（Single wavelength Anomalous Dispersion）求解相位，最终解析其晶体结构；与已报道的效应蛋白结构进行比对，发现具有保守的“β-sandwich”结构，同时也发现不同于其他效应子稳定二级结构的方式，解释田间分离菌株中 AvrPib 因关键氨基酸突变破坏自身结构稳定性而导致无毒功能丧失的结构机制，体外设计相关突变体进行验证，其他功能验证还在进行中。本研究为深入理解水稻与稻瘟病菌互分子机理提供新的线索和为培育高抗水稻品种提供重要理论基础。

关键词：稻瘟病菌；AvrPib；原核表达；晶体结构

马铃薯黑痣病生防芽孢杆菌筛选、防效及作用机理初探*

朱明明**，张岱，朱杰华，赵冬梅，潘阳，杨志辉***
（河北农业大学植物保护学院，保定 071001）

摘 要：马铃薯黑痣病是生产上一种重要的土传真菌病害，当前主要依靠沟施嘧菌酯等化学药剂进行防控。而化学药剂在杀死病原菌同时也会抑制或杀死土壤中的有益微生物，从而影响马铃薯的可持续发展。当前芽孢杆菌是生防菌中重要类群，具有广阔的发展和应用前景。本研究通过对马铃薯黑痣病生防芽孢杆菌的筛选、发酵条件优化、盆栽防效、促生作用和作用机理的初步研究，为马铃薯黑痣病优良芽孢杆菌生防菌剂的开发和应用奠定基础。

（1）筛选到对黑痣病菌抑制效果好的贝莱斯芽孢杆菌：通过土壤稀释法，从 44 份土壤样品中共分离到 142 株芽孢杆菌。经抑菌圈法和平板对峙法共得到 6 株对马铃薯黑痣病菌具有较强抑制效果的芽孢杆菌，其抑菌圈直径达 28.5mm 以上且抑菌带宽度大于 6.0mm。通过形态学观察及 gyrB 序列分析将其鉴定为贝莱斯芽孢杆菌（*Bacillus velezensis*）。

（2）优化出贝莱斯芽孢杆菌 HN-Q-8 的最佳培养条件：采用平板计数法比较各培养基的菌含量，获得了 HN-Q-8 的最佳培养基为 3.0%葡萄糖，0.5%酵母膏，0.3% NaCl，0.1% $MgSO_4$。采用比浊法比较各培养条件下的芽孢杆菌的数量，获得 HN-Q-8 的最佳培养条件为：温度 32~37℃，初始 pH 值 6.0，转速 200r/min，2%接种量，摇瓶最佳装液量为 40mL/250mL。

（3）测定了 HN-Q-8 对黑痣病的防效和促生作用：HN-Q-8 发酵液浸种和灌根处理均能够有效防治马铃薯黑痣病，防效分别为 52.72%和 61.22%，且该菌株能促进马铃薯植株的生长，尤其对马铃薯根部的促生作用明显，其增长率高达 77.45%。

（4）研究了 HN-Q-8 对植物抗逆防御酶活性的影响：将 HN-Q-8 发酵液施入马铃薯根际后显著提高了马铃薯叶片内 SOD、POD 和 CAT 活性，说明 HN-Q-8 能诱导马铃薯产生防卫反应。HN-Q-8 预处理后挑战接种病原菌明显提高了 POD 活性，说明 HN-Q-8 能在一定程度上诱导马铃薯产生抗病性。

（5）研究了 HN-Q-8 在马铃薯根际的定殖情况：采用绿色荧光蛋白（*GFP*）基因和氯霉素（*Cm*）抗性基因对 HN-Q-8 进行标记，将标记菌株的发酵液施入马铃薯根际土壤中，土壤中菌含量急剧增长，接种后 7d 达到最大值，之后迅速下降，28d 后趋于稳定，38d 检测到根际土壤中定殖量为 5.33×10^{3} CFU/g。此结果表明该菌株可稳定定殖于马铃薯根际土壤中。

关键词：立枯丝核菌；贝莱斯芽孢杆菌；生防机制；定殖

---

* 基金项目：现代农业产业技术体系建设专项资金资助（CARS-10-P12）；作物土传病害颉颃菌株的筛选及生防菌剂的研发（16S02）

** 第一作者：朱明明，硕士生，主要从事马铃薯黑痣病生物防治研究

*** 通信作者：杨志辉，教授，主要从事马铃薯病害和分子植物病理学研究；E-mail：bdyangzhihui@163.com

稻瘟菌海藻-6-磷酸合成酶 Tps1 及与相关配体复合物晶体生长条件的筛选

王珊珊*，赵彦翔，徐敏，彭友良，刘俊峰**
（中国农业大学植物保护学院植物病理学系，北京 100193）

摘　要：海藻糖是生物体内重要的还原型二糖，在生物体抗逆中具有重要的作用，由于海藻糖-6-磷酸合成酶（TPS）是真菌体内海藻糖合成的关键酶之一，在稻瘟菌中 MoTps1 是其致病所必需的，而且人和哺乳动物体内不存在 TPS 的同源蛋白，因而是理想的药物靶标。MoTps1 不仅是海藻糖合成的关键酶，而且可以感知 G6P 并与 NADPH 结合调控碳氮代谢以及下游致病基因的表达等，但发挥作用的结构基础不明确。本研究拟通过结构生物学手段获取 MoTps1 的 TPS 结构域自身及其与底物、NADPH 等配体的三维结构，分析它们在配体结合上的差异，确定发挥作用的关键位点，并分析其与已报道的 TPS 蛋白及复合物结构的差异，为新的环境友好型高效杀菌剂的开发提供结构基础。

首先，作者通过将 MoTps1 进行截短，在保证结构域完整的情况下对表达菌株及载体组合进行尝试，通过试亲和筛选到了较好的蛋白纯化条件，通过亲和层析，离子交换层析，凝胶过滤层析等方法对目标蛋白进行进一步的纯化，并最终获得了大量的纯度高且均一性好的蛋白，可用于后期的晶体生长。其次，作者通过利用气相扩散法对纯化好的目标蛋白开展了大量的晶体生长条件的筛选及优化，得到 MoTps1 截断体较好的晶体生长条件，获得母体晶体的 X-ray 衍射分辨率达到 2.7 Å。同时，作者继续对 MoTps1 截短体及其相关配体的晶体生长条件进行了筛选，最终得到其与 UDP 和 G6P 复合物晶体结构，X-ray 衍射分辨率达到 2.4 Å。数据利用同源模型进行结构解析，这些结果为进一步解释 MoTps1 的功能提供结构基础和为基于 MoTps1 结构的杀菌剂设计提供线索。

关键词：稻瘟菌；MoTps1；截断体；蛋白纯化；晶体生长

* 第一作者：王珊珊，硕士研究生，植物病理学专业；E-mail：15600912737@163.com
** 通信作者：刘俊峰，教授；E-mail：jliu@cau.edu.cn

芝麻球黑孢叶枯病田间流行动态的初步分析*

赵辉**，刘红彦***，刘新涛，倪云霞，文艺，刘玉霞，王飞，高素霞
（河南省农业科学院植物保护研究所，农业部华北南部作物有害生物综合治理重点实验室，
河南省农作物病虫害防治重点实验室，郑州 450002）

摘　要：芝麻球黑孢叶枯病是我国安徽、湖北和河南等芝麻主产区 2010 年以来出现的一种引起芝麻叶片枯萎的新病害，每年各产区约有 30%左右的地块发病，为害较重。为确定该病害的发生与时间、最高气温、有效积温和累积降雨量间的关系，明确其田间发生的最适发病条件，2016 年调查了河南省驻马店地区芝麻球黑孢叶枯病的自然发病情况，并通过 LINEAR（线性）、LOGLINEAR（对数线性模型）、INVERSE（倒数）、QUADRATIC（二次回归）、CUBIC（三次曲线模型）、COMPOUND（复合）、WEIGHTED EUCLIDEAN MODEL（加权回归）、S、GROWTH（增长）、EXPONENT（指数）和 LOGISTIC 等不同模型对田间流行动态进行了比对分析。初步明确了 2016 年平舆地区芝麻球黑孢叶枯病的发生与主要影响因素时间、最高气温、有效积温和累积降雨量之间关系可以用 QUADRATIC、CUBIC 和 S 模型模拟；通过模型模拟，确定在田间该病害发病初期为出苗后 45~52d（7 月下旬），最佳的发病温度为 31.67~33.33℃，最佳发病有效积温大于 1053.29~1252.67℃，发病的最佳累计降雨量大于 91.53~137.03mm。同时，这些时期也是防治球黑孢叶枯病的最佳时期。

关键词：芝麻球黑孢叶枯病；流行动态；预测模型

---

* 基金项目：农业部现代农业产业技术体系专项（CARS－15－1－05）；河南省农业科学院科研发展专项资金项目（YCY20167806）

** 第一作者：赵辉，男，副研究员，博士，主要从事植物病理学和植物病害生物防治研究；E-mail：zhaohui_ 0078@126. com

*** 通信作者：刘红彦，研究员，主要从事植物病理学和植物病害生物防治研究；E-mail：liuhy1219@163. com

水稻穗颈瘟发生的影响因素初探*

郭芳芳[1**]，汪锐辉[2]，王宁[1]，王世维[1]，杨仕新[1]，吴波明[1***]
（1. 中国农业大学植物保护学院，北京　100193；2. 江西省婺源县植保站，婺源　334000）

摘　要：为分析稻瘟病中各主要影响因子包括品种抗性、生育期以及环境条件的作用，在江西婺源县稻瘟病常发的两块水稻田进行田间试验，分析品种（蒙古稻、Y 两优 2 号及盐两优 2208 三个抗瘟性不同的水稻品种）、播期（自 5 月 10 日起每 5d 播种一批，共 5 批）和套袋时间（抽穗始期起，每 3d 一批，用硫酸纸袋保护稻穗，3d 后移去纸袋）对稻瘟病的影响。田间试验采用三因素裂区设计，主因素为品种，次要因素包括播期和套袋时间，每个小区面积为 1. 5m×5m。每个处理套 10 袋，每袋 5 穗，对照为不套袋。方差分析表明，穗颈瘟的发生受到品种、播期及套袋时间的显著影响。多重比较显示不同时间套袋保护下病情均显著低于对照不套袋处理下的病情。这一结果可能预示从抽穗期到蜡熟期，水稻穗部都可受到稻瘟病菌侵染。未来需要在不同地区进行更多的田间试验，以积累数据进一步分析天气条件的影响。

关键词：稻瘟病；影响因素；侵染时期

* 基金项目：国家重点研发计划“粮食丰产增效科技创新”专项
** 第一作者：郭芳芳，研究生，植物病害流行学；E-mail：guofangfang@ cau. edu. cn
*** 通信作者：吴波明，教授；E-mail：bmwu@ cau. edu. cn

中国香蕉枯萎病菌营养亲和群研究初报*

王维[1]，吴礼和[1]，付岗[1**]，杜婵娟[1]，杨迪，潘连富[1]，张晋[1]，叶云峰[2]

（1. 广西农业科学院微生物研究所，南宁　530007；

2. 广西农业科学院园艺研究所，南宁　530007）

摘　要：香蕉枯萎病是世界性分布的一种毁灭性土传病害，由尖孢镰刀菌古巴专化型（*Fusarium oxysporum* f. sp. *cubense*，Foc）侵染引起，病菌主要从香蕉根部侵入，蕉树一旦染病，则导致整株枯萎，直至病死，目前尚无有效治疗方法。选育抗病品种是最有应用前景的防治策略。然而，经过一个多世纪的变异和进化，Foc 现已分化出 4 个生理小种 24 个营养亲和群（Vegetative Compatibility Group，VCG）。明确香蕉枯萎病菌的群体多样性及其进化规律是进行有针对性的分子抗病育种的重要前提。

本研究从全国各产蕉省区共采集分离获得香蕉枯萎病菌 403 株，经致病性测定确定为病原菌。借助国际通用标准菌株对香蕉枯萎病菌进行了营养亲和群测试，结果表明：我国香蕉枯萎病菌的遗传多样性相当丰富，目前已鉴定出 7 个营养亲和群，分别为 VCG0123、VCG0124、VCG0125、VCG01213、VCG0120/15、VCG01213/16 和 VCG01218。在所有类群中，以 VCG01213/16 数量最多，占 41.7%；该群属于热带 4 号生理小种。香蕉枯萎病菌的群体多态性与地域分布密切相关，各省之间，不同 VCG 类群所占的比例存在很大差异。

关键词：香蕉枯萎病；营养亲和群；遗传多样性

---

* 基金项目：国家自然科学基金（31560006）；广西农业科学院科技发展基金（桂农科 2016JZ15，桂农科 2017JM33，2015YT79）

** 通信作者：付岗，博士，副研究员，主要从事植物病害及其生物防治研究；E-mail：fug110@gxaas.net

重庆道地药材黄连白粉病的发生情况及发病规律调查*

张永至[1**]，聂广楼[2]，孙现超[1***]
（1. 西南大学植物保护学院，重庆　400715；
2. 石柱农业特色产业发展中心，重庆　409100）

摘　要：黄连（学名：*Coptis chinensis* Franch.），别名：味连、川连、鸡爪连，属毛茛科、黄连属多年生草本植物。有清热燥湿，泻火解毒之功效。石柱县具有悠久的黄连种植生产历史，1989 年在山东泰安召开的全国道地药材学术研讨会上，石柱黄连被确认为道地黄连。黄连根茎生长 4~5 年采挖，较为缓慢，异地引种栽培困难。由于长期利用，野生黄连极为稀少，已被列为国家三级珍稀濒危植物。随着黄连需求量的不断增加和野生资源日益枯竭，2002 年石柱黄连产区被列为重庆市道地优势中药材 GAP 种植技术示范基地。2004 年石柱黄连获得国家地理标志产品保护，并通过中药材 GAP 认证。重庆石柱县作为黄连道地药材老产区，随着栽培面积的增加，生产中病害逐年加重，已经成为黄连规范化生产的重点和难点。其中黄连白粉病是黄连上常发病害之一。该病害在黄连生长过程中发生，主要为害叶片，可蔓延至叶柄和茎。发病初期叶背出现褪绿的黄褐色小斑点，病斑连片汇集后形成不规则褐色圆形病斑，而后霉斑逐渐增多连成一片，平铺在被害部位，发病盛期病斑上可见许多黑色小点，为颗粒状子囊壳，叶表多于叶背。发病由老叶渐向新叶蔓延，白粉也逐渐布满全株叶片，致使叶片焦枯死亡。重庆雨季的潮湿天气，白粉病发病严重，病株地上部整体呈水渍状暗绿色，叶片凋萎。轻者次年可生新叶，重者死亡缺株，造成减产减收。发病规律调查显示白粉病在 4 月底左右最初出现白粉病症状，然后发病率以较低的增长速度扩展，直到 6 月底。进入 7 月以后，各个田块发病率增长比较迅速，8 月下旬发病率增速达到高峰。

关键词：药材；黄连白粉病；发病调查

---

* 基金项目：重庆市社会事业与民生保障创新专项（cstc2016shmszx0368）；西南大学石柱科技创新基金项目：石柱黄连病害的无公害防

** 第一作者：张永至，硕士研究生，从事植物病理学研究

*** 通信作者：孙现超，博士，研究员，博士生导师，主要从事植物病毒学及植物病害生物防治研究；E-mail：sunxianchao@163. com

重庆地区苍术黑斑病发生情况及病原菌分离与鉴定*

张永至[1**]，曹浩[1]，胡晔[2]，孙现超[1***]

（1. 西南大学植物保护学院，重庆 400715

2. 重庆太极中药材种植开发有限公司，重庆 408000）

摘要：近年来苍术黑斑病严重发生，导致了苍术的产量和质量下降严重，并造成巨大经济损失。2016 年 6 月 21 日，课题组与重庆太极中药材种植开发有限公司合作，在城口和巫溪的太极中药材种植基地进行苍术黑斑病调查。该病在城口和巫溪苍术基地均是 2 年生苍术发病严重，同样地块，第一年苍术几乎没有任何病斑出现。两地发病初期均是 5 月连阴雨天气开始，随着雨水增多急剧加重。几天之内可沿茎基部蔓延至顶稍，严重时整株干枯死亡。为确定引起城口县苍术黑斑病的病原菌种类，为该病的病原菌寻找有效的防治措施打基础。从重庆市城口和巫溪苍术种植基地采集苍术黑斑病标样，而后采取常规组织分离法分离得到菌株，并依据柯赫氏法则对分离自发病叶片的病原菌菌株进行致病性测定，然后描述和测量了病原菌在 PDA 培养基上产生的分生孢子、分生孢子梗及喙的形态和大小，此外还对苍术黑斑病的生物学特性进行研究。最终，经过形态学鉴定、致病性测定及病原菌的 rDNA-ITS 序列分析研究，鉴定引起苍术黑斑病的病原为茄链格孢菌（*Alternaria solani*）。研究结果对城口县苍术黑斑病病害的防治和进一步研究具有重要的指导意义。

关键词：苍术；黑斑病；病菌分子鉴定；茄链格孢菌

---

* 基金项目：国家自然科学基金（31670148）；重庆市社会事业与民生保障创新专项（cstc2016shmszx0368）；中央高校基本科研业务费专项资金资助项目（XDJK2016A009，XDJK2017C015）

** 第一作者：张永至，男，硕士研究生，从事植物病理学研究

*** 通信作者：孙现超，博士，研究员，博士生导师，主要从事植物病毒学及生物防治研究；E-mail：sunxianchao@163.com

重庆石柱黄连根腐病根系真菌微生物的生物群落特性*

张永至[1**]，聂广楼[2]，孙现超[1***]
（1. 西南大学植物保护学院，重庆　400715
2. 石柱农业特色产业发展中心，重庆　409100）

摘　要：黄连是毛茛科的草本植物，其根茎可作为中药药材。据《中国药典》记载，常见的黄连药材有“味连、雅连、云连”，是我国常见中药材。中药药剂 3.2 万多方剂中，其中有 1 760种方剂中是包含有黄连的，占整个药剂的 5%。目前，由于栽培技术的成熟及产量高，适应性广，市场上主要的黄连药材为味连。味连主要的种植地为重庆、湖北、四川、陕西等省市。而重庆石柱则是味连的主要生产地，产量占全国的 60%，随着我国对中药材生产力度的投入，石柱地区黄连种植的面积将不断扩大，土地的重复耕作容易导致黄连病害的日益发生，其中黄连根腐病的发生尤为严重，对黄连的生产也是一个严重的危害，可以造成黄连的绝产。

黄连的根腐病主要为害黄连的根和根茎，可由多种细菌、真菌和病毒引起，病原通过植株根部的创口入侵引发病害发生，对于幼苗的侵染尤为严重。黄连喜阴，而根腐病在多雨、光照不足、湿度和气温较较高的条件下发病率较高，所以在黄连的种植过程中对根腐病的十分重要。通过对根系真菌的物种丰度的研究，了解根系真菌物种之间的差异，再结合未感病样品的土壤、感病样品的土壤当中的真菌成分，比较处理之间，对田间根系真菌种类影响黄连根腐病的发生做一个初步预测。

本次实验通过田间采取发病黄连及健康黄连，以及根际感病样品的土壤和健康样品的土壤，经过 Illumina Hiseq 2500 高通量测序平台测序检测，结果显示，感病植株中真菌的 OTU 丰度值均大于健康植株中的丰度，说明某些真菌的积累是黄连根腐病发生的一个重要原因；感病黄连的土壤中的 OTU 丰度也均大于黄连健康植株的土壤真菌 OTU 丰度。根据分类水平 PCA 图，得出健康根系和感病根系之间真菌种群存在着显著性的差异，其中说服性可以达到 33.3%；健康样品的土壤和感病样品的土壤之间真菌种群存在着显著性的差异，其中说服性可以达到 23.4%。进一步对感病黄连与健康黄连之间，健康黄连根际土壤与感病黄连根际土壤的物种对比分析发现，感病样品的土壤及感病黄连中 *Paraboeremia*、*Tetracladium*、柄孢壳属（*Podospora*）的含量均高于健康样品的土壤及感病黄连的含量；而在感病黄连根内与健康黄连根内的真菌种群对比中，健康植株中 *Llyonectriactri*、*Tetracladium* 的含量高于感病植株。这种真菌微生物的含量的差异是否是造成黄连根腐病发生的重要因素有待进一步的调查研究，且高通量测序技术对于研究植物内生真菌群落组成及多样性具有显著的优势。

关键词：黄连；根腐病；真菌；生物群落

---

* 基金项目：重庆市社会事业与民生保障创新专项（cstc2016shmszx0368）；西南大学石柱科技创新基金项目：石柱黄连病害的无公害防

** 第一作者：张永至，硕士研究生，从事植物病理学研究

*** 通信作者：孙现超，博士，研究员，博士生导师，主要从事植物病毒学及植物病害生物防治研究；E-mail：sunxianchao@163.com

稻瘟菌 MoHYR1 蛋白的原核表达与纯化*

钱恒伟[1**]，迟梦宇[1]，赵颖[1]，黄金光[1,2***]

（1. 青岛农业大学农学与植物保护学院分子植物病理与蛋白质科学研究中心，青岛 266109；
2. 山东省植物病虫害综合防控重点实验室，青岛 266109）

摘　要：稻瘟菌（*Magnaporthe oryzae*）是高等丝状子囊真菌，由其侵染水稻引发的稻瘟病是水稻第一大病害，生产上造成了巨大的经济损失。在植物与病原菌的互作过程中，植物会产生防御应答反应，其中包括活性氧自由基（reactive oxygen species，ROS）。ROS 在植物抵御病原菌的侵入以及病原菌在识别寄主后的过敏反应中发挥重要作用。稻瘟菌在侵染水稻早期能够通过 *MoHYR*1（MGG_ 07460）基因感知和调控 ROS 反应，确保病原菌成功侵染。*MoHYR*1 基因编码一种谷胱甘肽过氧化物酶（Glutathione peroxidase，GSH-Px），GSH-Px 能催化还原型谷胱甘肽（GSH）变为氧化型谷胱甘肽（oxidized glutathione），促进 H_2O_2的分解，使有毒的过氧化物还原成无毒的羟基化合物，从而保护细胞膜的结构和功能不受损害，该酶是机体内广泛存在的过氧化物分解酶。目前，在丝状真菌中该蛋白的三维结构未见报道，为进一步探究该蛋白的生化功能及蛋白质结构，作者开展了 *MoHYR*1 表达载体构建及重组蛋白 *MoHYR*1 表达纯化工作。以稻瘟菌 cDNA 为模板，通过 PCR 扩增 *MoHYR*1 基因片段，通过 PCR 扩增出 *MoHYR*1 基因片段并克隆到原核表达载体 pHAT2，利用 BL21（DE3）宿主菌在 IPTG 的诱导作用下表达。由于融合蛋白的 N 端带有 6×His 标签，故首先采用金属离子（Ni^{2+}）亲和层析纯化目的蛋白，结果表明：蛋白主要以可溶的形式存在，且产量高达 40mg/L；为了获得高纯度且均一性较高的蛋白，对该蛋白进一步进行凝胶层析，结果表明蛋白均一性较高，主要以单体的形式存在，且蛋白在纯化的过程中较为稳定，后期的稳定性试验也进一步验证了这一结果。该蛋白的成功表达与纯化为 MoHYR1 蛋白晶体的生长提供了条件，同时也为利用结构生物学的方法解析 MoHYR1 三维结构及分析其生化功能奠定基础。

关键词：*Magnaporthe oryzae*；*MoHYR*1；原核表达；亲和层析；凝胶过滤

---

* 项目基金：国家自然基金（31471735）

** 第一作者：钱恒伟，硕士研究生，研究方向为分子植物病理学；E-mail：qianhengweiqau@ 126. com

*** 通信作者：黄金光，博士，教授，研究方向蛋白结构生物学、分子植物病理学研究；E-mail：jghuang@ qau. edu. cn

马铃薯早疫病菌对苯醚甲环唑的抗性进化机制

李冬亮，周　倩，陈凤平，詹家绥
（福建农林大学植物病毒研究所，福州　350012）

摘　要：由链格孢属引起的马铃薯早疫病是马铃薯上为害严重的病害之一。本研究选用目前农业上广泛使用的 DMI 类药剂—苯醚甲环唑作为实验药剂，从全国 8 个省各挑选 20 株，共 160 株早疫病菌作为研究对象，研究其产生抗性的机制及进化机制，有望为生产中对马铃薯早疫病的防控及用药提供理论指导。前期研究发现，160 株早疫病菌对苯醚甲环唑的平均 EC_{50} 为 2. 157μg/mL，数值主要集中在 0. 25~2. 5μg/mL，频率为 0. 66。个体菌株间呈现明显的敏感性差异，EC_{50}最小值为 0. 043μg/mL，最大值为 65. 431μg/mL。省份间 EC_{50}平均值差异不大，最小值为河南省的 0. 980μg/mL，最大值为河北省的 2. 141μg/mL。进一步对苯醚甲环唑的靶标位点 CYP51 基因进行扩增测序，将 160 条 CYP51 基因序列进行比对发现，共有 47 个核苷酸变异位点，分为 23 个核苷酸单倍型。翻译成氨基酸后比对发现，第 48 位密码子由天冬酰胺替换为赖氨酸（N48K），第 361 位密码子由谷氨酰胺替换为组氨酸（Q361H）。查阅相关文献，并未报道相同的点突变，且发生突变的菌株并没有表现出对苯醚甲环唑有明显的敏感或抗性。采用实验室提供的 8 对引物，对 160 株早疫病菌进行 SSR 检测。共检测到 50 个等位基因，其中第二对引物 S2 包含 16 个等位基因，而最后一对引物 S8 只包含 2 个等位基因。将等位基因替换为数字代码后，共整理出 110 个 SSR 基因型，其中有 83 个基因型只出现一次。表明所测菌株的基因型多样性较丰富。目前正在进一步研究早疫病菌对苯醚甲环唑的其余的分子抗性机制，并结合基因型多样性、敏感性数据等做下一步研究。

关键词：链格孢属；苯醚甲环唑；EC_{50}；CYP51；抗性机制；SSR；相关性分析

Development of Simple Sequence Repeat Markers Based on Whole Genome Sequencing to Reveal the Genetic Diversity of *Glomerella cingulate* in China*

Liu Zhaotao[1], Lian Sen[1], Li Baohua[1**], Dong Xiangli[1],
Wang Caixian[1], Cho Wonkyong[2], Liang Wenxing[1]
(1. *College of Crop Protection, Qingdao Agricultural University, and Key Lab of Integrated Crop Pests Management of Shandong Province, Qingdao, Shandong* 266109, *People Republic of China*;
2. *Department of Agricultural Biotechnology, Research Institute of Agriculture and Life Sciences, and Plant Genomics and Breeding Institute, College of Agriculture and Life Sciences, Seoul National University, Seoul*, 08826, *Republic of Korea*)

Abstract: Glomerella Leaf Spot (GLS) has been in an outbreak since 2011 and severely affects apple production (*Malus domestica* Borkh.) in China. The pathogen that causes GLS is *Glomerella cingulata* (Stoneman) Spauld. & H. Schrenk, which is epidemic in regions with humid climates from May to October each year. GLS spreads more than 200 km per year, and most apple-producing areas have been invaded in the five years since 2011. To investigate the origin, evolution and diversity of *G. cingulata* for better GLSmanagement, we performed whole genome sequencing of *G. cingulata* isolate 030206 and developed 25 simple sequence repeat marker (SSR) markers according to the SSR motifs identified in the genome sequence. Using the 25SSRmarkers, we analyzed thegenetic diversity and population genetic structure of 96 *G. cingulata* isolates collected from eight prevalent GLS areas: Chiping, Feicheng, Haiyang, Muping, Qixia, Yishui, Yiyuan, and Zhanhua in Shandong province, China. The expected heterozygosities (H_E) for the populations were 0.351, 0.240, 0.471, 0.559, 0.556, 0.077, 0.452, and 0.036, respectively, with a mean of 0.343, and the Shannon indices (I) were 0.560, 0.385, 0.811, 1.100, 1.032, 0.140, 0.662, and 0.051, respectively, with a mean of 0.593. The gene flow index (Nm) was 2.937. The 96 *G. cingulata* isolates were divided into three subpopulations, denoted GC1, GC2 and GC3. GC1 contained 32 isolates, most of which were from the western areas of Shandong province, whereas GC2 contained 18 isolates and GC3 contained 46 isolates, most of which were from the eastern areas of Shandong province. In addition, the western genetic population, GC1, contained a few isolates from eastern collection sites, such as Haiyang, Muping and Qixia, whereas few isolates contained in GC2 and GC3 were from western areas. These results indirectly show that the GLS pathogens in western areas were originally from eastern areas of Shandong province.

Key words: *Glomerella cingulata*; Glomerella Leaf Spot; SSR marker; Population genetic

---

* Funding: This research was supported in part by the National Key Research and Development Program of China (2016YFD0201100); the Talents of High Level Scientific Research Foundation of Qingdao Agricultural University (6631116026); the China Agricultural Research System (CARS-28); and the Taishan Scholar Construction Project of Shandong Province

** Corresponding Author: Dr. Baohua Li; E-mail: baohuali@ qau. edu. cn

First Report of *Penicillium polonicum* Causing Blue Mold on Stored *Polygonatum cyrtonema* in China*

Liu Y. J. , Chen L. S. , Xu S. W. , Zhou H. B

(*Anhui Academy of Science and Technology*, *Hefei* 230031, *China*; *Anhui Academy of Applied Technology*, *Hefei* 230088, *China*)

Sealwort (*Polygonatum cyrtonema*) is a perennial herb belonging to the family lily, and its rhizome used for treatment of diabetes and asthma. In October 2015, decayedsealwort rhizomes with blue mold symptoms were found causing significant economic losses (at 20 to 30% incidence) at a storage facility in Hefei City, Anhui Province, China. The decayed area of the rhizome was circular, light brown, and-tissue was soft and watery. The lesions expanded rapidly and blue mold symptomswere apparent at room temperature. To isolate the causal organism, fragments (2mm×2mm) were excised from symptomatic rhizomes, surface-sterilized in 1. 5% sodiumhypochlorite for 1 min, rinsed three times with sterilized water, plated on potatodextrose agar (PDA) medium (Liu *et al.*, 2016), and incubated at 25℃ for 6 days. 20morphologically similar fungal isolates were recovered. Colony color was blue greenand reverse color was cream to yellow brown. Colony size was 36 mm in averagediameter at 7 days, plane, surface structure finely granular, marginal hyphae wascream and entirely subsurface. Conidiophores were terverticillate, stipes were septateand 167 to 300μm×3. 5 to 4μm in length with smooth to finely roughened walls, metulain verticils of 2 to 4 measuring 11 to 20μm×2. 8 to 3. 0μm, phialides were slender, ampulliform, and 8 to 10μm×2. 0 to 2. 5μm. Conidia were smooth walled, globose tosubglobose, and borne in long columns. Conidial measured 3. 3 to 4. 5 (3. 8) μm×3. 1 to 4. 0 (3. 5) μm ($n=50$). According to these cultural characteristics, the fungal isolateswere identified as *Penicillium polonicum* (Frisvad and Samson 2004). To validate thespecies identification, isolate HF-121 was selected for molecular identification byamplifying and sequencing the ribosomal DNA internal transcribed spacer (rDNA-ITS) region using the universal primer pair ITS1 and ITS4. 32 Sequencing of the PCR product revealed 99% similarity with ITS rDNA sequences of *P. polonicum* in GenBank (KY643770. 1, KX958077. 1, and KU847876. 1). The ITS sequence of the isolate was deposited in GenBank (MF135229). Pathogenicity was tested by woundinoculation with a sterilized needle onto disinfected rhizomes with a 20μL of conidialsuspension (5×10^5conidia/mL) from isolates grown on PDA. Ten asymptomaticrhizomes were inoculated with isolate HF-121. An equal number of control rhizomewas wounded and treated similarly with distilled water. The procedure was repeated three times, and rhizomes were incubated in plastic containers, under high humidity at 25℃ for ten days, typical symptoms of blue mold were developed on all inoculated rhizomes, while no symptoms were observed on control rhizomes. The fungus was recovered

* Funding: The research was supported by the science and technology key projects of Anhui Province, China (Grant No. 13010402103) and Anhui Province Postdoctoral foundation (2016B094)

from the rhizome symptoms and identified as *P. polonicum* byorphological characteristics, confirming Koch's postulates. This fungus has been previously reported as a causal agent of onion rot in Serbia (Duduk *et al.*, 2014), and is a pathogen on Garlic (*Allium sativum*) in Pakistan (Bashir *et al.*, 2017). To ourknowledge, this is the first report of *P. polonicum* causing blue mold decay on stored *P. cyrtonema* in worldwide. According to the literature, this fungus is an air borne fungus with a wide host range, so attention should be paid to the numerous other potentially susceptible host plants in the area where the disease was found.

***References*:**

Bashir, U., *et al.* 2017. Plant Dis. 101: 1037. 10. 1094/PDIS-07-16-1069-PDN.

Duduk, N., *et al.* 2014. Plant Dis. 98: 1440. 10. 1094/PDIS-05-14-0550-PDN.

Frisvad, J. C. and Samson, R. A. 2004. Stud. Mycol. 49: 1.

Liu, Y. J., *et al.* 2016. Plant Dis. 100: 1780. 10. 1094/PDIS-12-15-1433-PDN.

First Report of *Colletotrichum aenigma* Causing Anthracnose Fruit Rot on *Trichosanthes kirilowii* in China

Zhao Wei, Wang Tao, Chen Qingqing, Chi Yuankai, Qi Rrede

(*Institute of Plant Protection and Agro-products Safety, Anhui Academy of Agricultural Sciences, Hefei,* 230031 *China*)

Trichosanthes kirilowii Maxim. belonging to the genus Trichosanthes, family Cucurbitaceae, is an important medicinal plants, which roots, fruits and seeds can be used as Chinese herbalism. Furthermore, the seeds contain rich nutrient for edible. Since the potential big economic benefits, there are a lot of palaces to start growing *T. kirilowii* throughout China. However, heavy losses due to fruit rot by different disease have been damaging this industry. Over the last decade, water-soaked, dark brown-to-black, sunken lesions on the leaves and fruits of *T. kirilowii*, the typical anthracnose symptomshave been appearing, especiallyon the pre-mature fruits under high relative humidity, in Anhui Province, China. In 2015 and 2016, Small tissue pieces from the edges of disease lesions on fruits were disinfected in 70% (*v/v*) ethanol for 30s and in 2% (*v/v*) hypochlorous acid solution for 2 min, rinsed twice in sterilized water, air-dried, then incubated on potato dextrose agar (PDA) at 25℃. Six isolates of the pathogen were obtained from different diseased plants. The average diam of colonies grown from single conidia on Difco PDA were 28mm after 7 days. Aerial mycelium was sparse, fluffy, colourless edge to pale orange towards centre on Spezieller Nährstoffarmer Agar (SNA). The conidia were hyaline, aseptate, smooth, straight, cylindric with broadlyrounded ends, 13 to 16.5μm long and 5 to 6.5μm wide, L/W ratio≈2.4. These morphological characters are consistent with the description of *Colletotrichumaenigma* (Weir *et al.*, 2012). To confirm this identification, the internal transcribed spacer (ITS) rDNA regions, glyceraldehyde - 3-phosphate dehydrogenase (GADPH) gene, calmodulin (CAL) and β-Tubulin-2 (TUB2) gene were amplified with the primers ITS1/ITS4, GDF/GDR, CL1C/CL2Cand TUBT1/TUBT2 (Weir *et al.*, 2012), respectively. The consensus sequences of the ITS, GADPH, CAL and TUB2 were deposited in NCBI GenBank under the accession numbers KY130427, KX988006, KY197716 and KY197715, respectively. Compared with the sequence of strain ICMP 18608 of *C. aenigma* in GenBank (Weir *et al.*, 2012), the amplification products showed 100% homology with the ITS sequence (GenBank Accession JX010244), the GADPH sequence (JX010044), the CAL sequence (JX009683) and theTUB2sequence (JX010389), respectively. A pathogenicity test was performed by depositing 10μL droplets of a suspension (10^5 conidia/mL) on the surfaces of three artificially wounded fruits (prick a small hole on *T. kirilowii*fruit using needle). Two wounded fruits were inoculated with sterilized, distilled water as the control. After 7 days in a moist chamber at 25℃, symptoms similar to those observed on the original *T. kirilowii* plants had developed on the five woundedfruits that had been inoculated, and *C. aenigma*was re-isolated from the lesions. The control fruits remained healthy. *T. kirilowii* anthracnose disease caused by *C. gloeosporioides* was previously reported. Recently, *C. gloeosporioides* has been considered a species complex, and *C. aenigma*asone of the species within *C. gloeosporioides* species complex. To our knowledge, this is the first report of occurrence of anthracnose fruit rot caused by C. *aenigma* on *T. kirilowii* in China. This disease causes severe damage in the field and economic losses and there is need to develop effective control strategies in the *T. kirilowii* production in these regions.

A Class-Ⅱ Myosin Is Required for Growth, Conidiation, Cell Wall Integrity and Pathogenicity of *Magnaporthe oryzae* *

Guo Min[1], Tan Leyong[1], Nie Xiang[1**], Zhang Zhengguang[2]

(1. *Department of Plant Pathology, College of Plant Protection, Anhui Agricultural University, Hefei* 230036, *China*; 2. *Department of Plant Pathology, College of Plant Protection, Nanjing Agricultural University, Nanjing* 210095, *China*)

In eukaryotic organisms, myosin proteins are the major ring components that are involved in cytokinesis. To date, little is known about the biological functions of myosin proteins in *Magnaporthe oryzae*. In this study, insertional mutagenesis conducted in *M. oryzae* led to identification of *Momyo*2, a pathogenicity gene predicted to encode a class-II myosin protein homologous to *Saccharomyces cerevisiae-Myo*1. According to qRT-PCR, *Momyo*2 is highly expressed during early infectious stage. When this gene was disrupted, the resultant mutant isolates were attenuated in virulence on rice and barley. These were likely caused by defective mycelial growth and frequent emergence of branch hyphae and septum. The *Momyo*2 mutants were also defective in conidial and appressorial development, characterized by abnormal conidia and appressoria. These consequently resulted in plant tissue penetration defects that the wild type strain lacked, and mutants being less pathogenic. Cytorrhysis assay, CFW staining of appressorium and monitoring of protoplast release suggested that appressorial wall was altered, presumably affecting the level of turgor pressure within appressorium. Furthermore, impairments in conidial germination, glycogenmetabolites, tolerance to exogenous stressesand scavenging of host-derived reactive oxygen species were associated with defects on appressorium mediated penetration, and therefore attenuated the virulence of *Momyo*2 mutants. Taken together, these results suggestthat Momyo2 plays pleiotropic roles in fungal development, andis required for the full pathogenicity of *M. oryzae*.

---

* These authors have contributed equally to this work

** E-mail: Nie Xiang kandylemon@ 163. com; Zhang Zhengguang zhgzhang@ njau. edu. cn

不同寄主来源梨孢菌的系统发育及其致病性分析*

祁鹤兴**，赵　健，钟　珊，杨　俊，彭友良，张国珍***

（中国农业大学植物保护学院植物病理学系，北京　100193）

摘　要：稻瘟病的暴发对世界粮食作物的产量造成了持续的为害。引起水稻以及多种禾本科植物瘟病的病原菌是稻梨孢（*Pyricularia oryzae*，有性型为 *Magnaporthe oryzae*）。截至 2009 年，就已经发现了有 137 种禾本科植物能被梨孢菌侵染，包括小麦、大麦、谷子、燕麦等栽培作物，牛筋草、狗尾草、稗草、画眉草等杂草。2002 年 Couch 和 Kohn 将侵染马唐的梨孢菌的有性型定为 *M. grisea*，其他植物来源的梨孢菌的有性型定为 *M. oryzae*。*M. oryzae* 和 *M. grisea* 是近年来巨座壳属（*Magnaporthe*）中被研究最多的 2 个种，因为它们的易变性以及频繁的共进化特征使得它们能够侵染新的寄主植物。

本研究基于 *actin*、*β-tubulin* 和 *calmodulin* 基因，对分离自狗尾草、牛筋草、马唐、谷子和莎草的 48 株梨孢菌进行系统发育分析，发现 15 株马唐分离菌株中有 13 株为 *M. grisea*，而其余 2 株马唐分离菌株和其他寄主来源的菌株与水稻分离菌株同聚在 *M. oryzae* 这一分支，表明马唐分离菌株包括 2 个种，即 *M. oryzae* 和 *M. grisea*，此结果与 2002 年 Couch 和 Kohn 对 *M. grisea* 这个种的定义不完全一致。用孢子悬浮液划伤接种水稻丽江新团黑谷叶片，发现有 3 株狗尾草分离菌株能够引起发病，而分离自杂草上的其他 45 株菌株不能引起发病。推测稻田及周边常见杂草狗尾草有可能在稻瘟病的病害循环中发挥一定作用。

关键词：梨孢菌；系统发育；致病性

* 基金项目：粮食丰产工程项目（2016YFD0300700）；公益性行业（农业）科研专项（201203014）

** 第一作者：祁鹤兴，博士研究生，植物病理学专业；E-mail：hexqi@ cau. edu. cn

*** 通信作者：张国珍，教授；E-mail：zhanggzh@ cau. edu. cn

Integrated Transcriptome and Metabolome Analysis of *Oryza sative* L. Upon Infection with *Rhizoctonia solani*, the Causal Agent of Rice Sheath Blight*

Cao Wenlei[1], Xu Bin [1], Yuan Limin [2],
Chen Xijun [1], Pan Xuebiao [1], Zuo Shimin[1]
(1. *Jiangsu Key Laboratory of Crop Genetics and Physiology/Co-Innovation Center for Modern Production Technology of Grain Crops*, *Yangzhou University*, *Yangzhou* 225009, *China*; 2. *Testing Center of Yangzhou University*)

Abstract: Sheath blight (SB), caused by necrotrophic fungus *Rhizoctonia solani* Kühn (*R. solani*), gives rise to severely grain losses in rice. In this study, an integrated approach of transcriptomics and metabolomics was employed to try to uncover potential mechanism and to mine the candidate genes/pathways regulating rice SB resistance. YSBR1 is a novel germplasm with stable and high resistance to SB, which has been identified by using greenhouse and field trials. Histological analysis by SEM showed that the growth of *R. solani* mycelium was apparently suppressed in YSBR1 but not in Lemont (susceptible) after 60 HAI. *R. solani* infection was able to trigger substantialmetabolic reprogramming in rice, particularly in primary metabolismand energy metabolism. Phenylalanine and tyrosine, which have been confirmed to involve in plant-pathogen interaction via the biosynthesis of secondary metabolites, were specially increased in YSBR1 infected. Total of 332/62 (Lemont/YSBR1) and 2606/748 (Lemont/YSBR1) DEGsat 10 and 20 HAI were identified, respectively. A total of 247 unique DEGs were identified specifically in YSBR1, which were annotated to involve in several pathways, like amino acid metabolism, secondary metabolic, energy metabolism, photosynthesis, and sugar metabolism pathways. Through combining genetic analysis and omics evidence, 6 candidate genes localized within the known SBR QTL region (Lemont/YSBR1), encoding cellulose synthase-like family A, transcription factors TFⅡS, elongin A, EF hand family protein, calmodulin-related calcium sensor protein, ERF transcription factor, and thaumatin, have been identified.

Key words: *Rhizoctonia solani*; Rice; Metabolome; transcrptome; QTLs; Candidate genes/pathways

* Corresponding author: Zuo Shimin; E-mail: smzuo@ yzu. edu. cn

低温、冰冻、失水与苹果枝条对腐烂病菌敏感性*

林　晓**，李宝笃，王彩霞，练　森，李保华***
（青岛农业大学农学与植物保护学院/山东省植物病虫害综合防控重点实验室，青岛　266109）

摘　要：苹果树腐烂病（*Valsa mali*）是苹果树上的重要病害，我国各大苹果产区都有发生。苹果树腐烂病在春季发病严重，春季形成的病斑数占全年总病斑数的70%~80%。然而，对腐烂病春季高发的原因和机制还不明确，为了研究腐烂病春季高发的原因，于3—4月剪取苹果越冬后的枝条，截成30~40cm的茎段，经低温、冰冻等处理后，用菌饼接种枝条的芽眼处，每个枝段接种5个点，在25℃下保湿培养5d后，检查发病的芽眼数，依此推断低温、冰冻等因子对苹果枝条感病性的影响，结果如下。

在6个低温（-25℃、-18℃、-10℃、-7℃、0℃和5℃）处理中，-18℃处理的枝条发病点数最多，每个枝段平均1.7芽眼发病，其次为-25℃，每个枝段1.5个芽眼发病，0℃发病最轻，每枝段只有0.2个芽眼发病。将枝条在低温下处理不同时间后接种，发现处理96h枝条与处理10min枝条的芽眼发病率没有显著差异，表明低温处理时间对枝条的敏感性没有显著影响。将苹果枝条浸水1h，低温处理48h，取出后立即接种，浸水处理枝条与未浸水处理枝条芽眼发病率分别为1.2和0.7，浸水处理枝条芽眼的发病率显著高于未浸水处理枝条，表明结冰对枝条的敏感性有显著的影响。将苹果枝条浸水1h，低温处理48h，取出后在室温下让枝条失水不同时间后，再接种，失水处理72h的枝条，芽眼的发病率显著高于其他处理，每个枝段芽眼发病数高达3.1个，其次是48h和36h，每个枝条有1.6个和1.7个芽眼发病，失水处理2h的枝条发病最轻，每个枝段只有0.3个芽眼发病，表明低温后枝条失水对枝条的敏感性有显著的影响。在所有接种枝条中，1年生枝条和2年生枝条芽眼的发病率分别为1.3和0.7，1年生枝条芽眼的发病率显著高于2年生枝条；一年生枝条的梢部、中部和基部芽眼的发病率分别为1.2个、0.6个和0.4个病斑，三者差异显著，表明枝条的龄期对枝条的感病性有显著的影响。

关键词：苹果腐烂病；枝条敏感性；低温；失水

---

* 基金项目：国家苹果产业技术体系（CARS-28）；国家自然科学基金

** 第一作者：林晓，硕士研究生，研究方向为植物病害流行学；E-mail：1377189100@qq.com
*** 通信作者：李保华，教授，主要从事植物病害流行和果树病害研究；E-mail：baohuali@qau.edu.cn

核盘菌蛋白激发子 *Ss-Sm1* 基因克隆和功能分析*

魏君君，姚传春，任恒雪，高智谋，潘月敏**
（安徽农业大学植物保护学院，合肥　230036）

摘　要：核盘菌（*Sclerotinia sclerotiorum*）是在世界范围内都有分布的死体营养型病菌，其寄主广泛，可以侵染包括多种重要作物在内的 75 个科 400 多种植物。核盘菌侵染油菜和大豆产生的菌核病为害极大，在生产上造成了巨大的损失。激发子是一类能够引起寄主植物产生免疫反应的化合物，可以诱导植物产生乙烯、植保素、水杨酸、茉莉酸、病程相关蛋白等物质，提高寄主的抗病能力。本研究在核盘菌基因组中发现一个与木霉菌激发子蛋白 Sm1（small protein 1）具有较高同源性的蛋白，将该蛋白命名为Ss-Sm1。

为探索该蛋白在核盘菌生长发育过程中的功能，本实验构建了核盘菌蛋白激发子同源物 Ss-Sm1 的原核表达载体，将载体导入大肠杆菌 BL21，在 20℃条件下进行诱导表达，能够使目的蛋白在上清中大量表达。用 Ni 柱对目的蛋白进行纯化后，成功获得了纯化的可溶性目的蛋白。进一步用蛋白溶液注射浸润本氏烟叶片，发现该蛋白能够在本氏烟叶片上引起过敏性坏死反应（HR），因此确定了 Ss-Sm1 为核盘菌的蛋白激发子。通过使用 RNAi 技术和 PEG 介导的原生质体转化方法，获得了蛋白激发子 Ss-Sm1 编码基因的沉默转化子菌株。通过比较沉默转化子和野生型核盘菌菌株的生长发育及致病情况，发现沉默转化子的生长速度和致病力明显落后于野生型核盘菌。并且，沉默转化子与野生型菌株相比产生的侵染垫结构也大量减少，在 PDA 固体培养基上生长 3 周后，野生型核盘菌能够产生大量、形态正常的鼠粪状菌核，而沉默转化子只产生少量、畸形的菌核。对菌株进行的化学因子胁迫实验发现，转化子对 Sorbitol 和 SDS 的耐受力显著下降，沉默转化子在含有 1.2M Sorbitol 的 PDA 培养基上无法发育出成熟的菌核，而在含有 0.02% SDS 的 PDA 培养基上转化子菌丝不能长满整个平板。根据以上实验结果，作者推测该基因可能与核盘菌菌丝和菌核发育有关，但其具体的分子机制尚需进一步研究。

关键词：核盘菌；激发子；原核表达；RNAi

* 第一作者：魏君君，在读硕士研究生，植物病理学专业；E-mail：1903661675@ qq. com
** 通信作者：潘月敏，副教授；E-mail：panyuemin2008@ 163. com

苹果树腐烂病菌 β-葡萄糖苷酶基因的克隆及表达分析*

李　婷**，祝　山，徐静华，李保华，王彩霞***
（青岛农业大学植物保护学院，山东省植物病虫害综合防控重点实验室，青岛　266109）

摘　要：苹果树腐烂病（Apple valsa canker）是苹果生产上的三大重要病害之一，由黑腐皮壳菌 *Valsa mali* var. *mali* 引起，该病在我国各苹果产区发生普遍，可造成果树主干及整树死亡，甚至毁园。前期研究表明，腐烂病菌侵染致病过程中可产生的一系列的细胞壁降解酶：木聚糖酶、纤维素酶、β-葡萄糖苷酶、多聚半乳糖醛酸酶和果胶甲基半乳糖醛酸酶，是腐烂病菌重要致病因子之一。β-葡萄糖苷酶能够水解结合于末端非还原性 β-D-葡萄糖苷键，在协助纤维素酶降解纤维素的过程中起关键作用。

本研究通过 RT-PCR 与 TA 克隆技术获得了 β-glucosidase 基因的全长 cDNA 序列（GenBank 登录号 KY646110），命名为 *β-glucosidase* II，利用实时荧光定量 PCR 技术分析了 *β-glucosidase* II 在腐烂病菌侵染致病过程中的表达模式。序列分析结果显示：该基因包含 1 689bp 的开放阅读框，编码 562 个氨基酸，蛋白相对分子质量 62.7 ku。*β-glucosidase* II 与梨树腐烂病菌（*Valsa mali* var. *pyri*）*β-glucosidase*（KUI56905.1）氨基酸序列相似性最高为 88%，与向日葵茎溃疡病菌（*Diaporthe helianthi*）*β-glucosidase*（OCW40953.1）氨基酸序列相似性为 71%。qPCR 结果表明：腐烂病菌接种苹果枝条后，*β-glucosidase* II 表达量先下降后升高，接种后 48h，*β-glucosidase* II 的表达量比 12h 降低了 3 倍，接种后 120h 基因表达量为 48h 的 10 倍。苹果树腐烂病菌 *β-glucosidase* 基因的成功克隆和表达分析，为从分子水平上探究 β-葡萄糖苷酶在腐烂病菌致病过程中的作用奠定了基础。

关键词：苹果树腐烂病菌；β-glucosidase；基因克隆；表达分析

---

* 基金项目：国家自然科学基金（31272001 和 31371883）；现代农业产业技术体系建设专项资金（CARS-28）；大学生创新训练项目

** 第一作者：李婷，女，硕士研究生，研究方向果树病理学；E-mail：m18960878069@163.com

*** 通信作者：王彩霞，女，教授，主要从事果树病害研究；E-mail：cxwang@qau.edu.cn

云南省烟草靶斑病菌菌丝融合群及 ITS 序列分析*

侯慧慧**，赵秀香，吴元华***

（沈阳农业大学植物保护学院，沈阳　110866）

摘　要：2015—2016 年辽宁省、云南省、黑龙江省烟区均发生烟草靶斑病，其中我国云南省临沧市和普洱市烟草种植区为首次大面积发生，暴发面积高达 6 万亩，该病害主要为害叶片形成病斑，病斑坏死部分易碎形成穿孔，病斑易连片，造成经济损失严重。鉴此，本文针对云南烟草靶斑病菌（*Rhizoctonia solani* Kühn）的菌丝融合群和 ITS 序列进行了系统鉴定，研究结果如下：

1. 采集云南省的烟草靶斑病病叶标本 30 余份，采用常规组织分离法获得 36 个立枯丝核菌 *Rhizoctonia solani* 菌株。将所获菌株分别与立枯丝核菌标准融合群菌株 AG-1-IA、AG-2-1、AG-3、AG-4-HGI、AG-4-HGⅡ、AG-5、AG-6-HGI、AG-6GV、AG-8、AG-9 进行载玻片对峙培养，并进行菌丝融合观察，结果表明：36 个菌株均属于立枯丝核菌 AG-3 标准融合群，不与其他标准融合群发生融合反应，这表明我国云南省引起烟草靶斑病的病原菌 *Rhizoctonia solani* 均是 AG-3 融合群，是国内首次在云南省的罹病烟草植株上分离得到。

2. 随机选取云南省 9 个地区各 1 个代表性菌株，提取菌丝基因组 DNA。利用 5. 8S rDNA-ITS 区序列分析方法，采用 MEGA6. 0 软件对该地区烟草靶斑病菌不同地区代表菌株的系统演化关系进行了研究。结果表明：*Rhizoctonia solani* 菌株的 ITS 序列可明显的分成 3 个分支，同一菌株的不同 ITS 序列可分别存在于不同的分枝中，但均隶属于 AG-3 融合群，且隶属相同融合群的不同菌株之间其序列的一致性可高达 97%～99%。

关键词：烟草靶斑病；云南；立枯丝核菌；融合群；5. 8S rDNA-ITS 区序列分析

---

* 基金项目：中国烟草总公司科技重大专项项目（中烟办【2016】259 号）

** 第一作者：侯慧慧，女，硕士研究生，从事植物真菌病害研究；E-mail：1049346253@ qq. com

*** 通信作者：吴元华，博士，教授，主要从事植物病毒学和生物农药研究；E-mail：wuyh7799@ 163. com

苹果树腐烂病菌纤维素酶基因的克隆与原核表达*

王晓焕**，王彩霞，练 森，李保华***
（青岛农业大学农学与植物保护学院/山东省植物病虫害综合防控重点实验室，青岛 266109）

摘 要：苹果腐烂病（*Valsa mali*）是我国苹果上一种重要的枝干病害，主要造成死枝、死树，甚至毁园。腐烂病菌主要从剪锯口和各种伤口侵染，且能在木质部内生长扩展。腐烂病菌在木质部内生长扩展是导致腐烂病连年复发和剪锯口发病的主要原因。木质部的主要成分是木质素和纤维素，且腐烂病菌在对寄主植物侵染致病过程中可产生一系列的细胞壁降解酶：木聚糖酶、纤维素酶、β-葡萄糖苷酶、多聚半乳糖醛酸酶、果胶甲基半乳糖醛酸酶。

为了解析苹果树腐烂病菌纤维素酶基因的分子特征和表达，明确病原菌在木质部中生长扩展过程中的作用，通过 RT-PCR 技术从苹果腐烂病菌中克隆了 2 个纤维素酶基因，命名为 Cellulase Ⅰ、CellulaseⅡ，构建重组表达载体 pET32a-CellulaseⅠ，pET32a-CellulaseⅡ，将重组载体转入大肠杆菌 Rosetta 菌株，用 IPTG 诱导表达，并成功克隆。结果表明：CellulaseⅠ和 CellulaseⅡ的开放阅读框的长度分别为 1 008bp（编码 336 个氨基酸）、1 446bp（编码 482 个氨基酸），分子量分别为 34. 25ku、50. 66ku。融合蛋白的可溶性检测表明，CellulaseⅠ低温诱导下的蛋白主要以可溶性蛋白的形式存在，Western-blot 印迹证明 Rosetta 表达了一个分子量约为 66. 33ku 的蛋白质，与预测分子量大小不太一致。CellulaseⅡ低温诱导下的蛋白主要以包涵体蛋白的形式存在。本研究成功从苹果腐烂病菌中克隆到 2 个纤维素酶基因，并成功实现了其编码蛋白的原核表达，为后期探索腐烂病菌中的纤维素酶基因在木质部中生长扩展过程中的作用提供技术和方法。

关键词：苹果腐烂病菌；纤维素酶基因；基因克隆；原核表达

---

* 基金项目：国家苹果产业技术体系（CARS-28）；国家自然科学基金（6622313005）

** 第一作者：王晓焕，硕士研究生，研究方向为植物病害流行学；E-mail：15865521984@ 163. com

*** 通信作者：李保华，教授，主要从事植物病害流行和果树病害研究；E-mail：baohuali@ qau. edu. cn

我国梨树轮纹病和干腐病病原菌种内遗传多样性研究*

肖　峰，王国平**，洪　霓
（华中农业大学植物科学技术学院，农业微生物学国家重点实验室，武汉　430070）

摘　要：梨树轮纹病和干腐病病原菌是我国梨树枝干上常年发生的严重病害，造成树势减弱，严重威胁我国梨产区的经济发展。在我国已有的研究报道中，引起梨树轮纹病和干腐病的病原菌为葡萄座腔菌属（*Botryosphaeria*），目前发现的病原菌种类有 4 种，其中 *Botryosphaeria dothidea* 既能引起梨树枝干轮纹病斑又能引起梨树枝干干腐病斑；在分子鉴定方面，苹果轮纹病病原菌 *B. dothidea* 被鉴定出 4 种基因型，其中引起苹果轮纹病病原菌 *B. dothidea* 又被分成了两个种：*B. kuwatsukai* 和 *B. dothidea*。为明确我国梨树轮纹病和干腐病病原菌 *B. dothidea* 的种内遗传多样性，本研究以实验室保存的 146 株菌株为材料，开展了相关形态学观察和致病力分析。研究发现，在我国引起梨树轮纹病和干腐病病原菌 *B. dothidea* 存在两种基因型，Genotype Ⅰ 菌株有 24 株，Genotype Ⅱ 菌株有 122 株，在我国 15 个省份均有分布，其中 Genotype Ⅱ 菌株为优势菌株类群。在菌落形态上两种基因型菌株存在明显差异，Genotype Ⅰ 菌株在 PDA 培养基上菌丝生长旺盛，成灰绿色，生长速率较快，气生菌丝稀疏、竖起，易产生分生孢子器且产孢量大；Genotype Ⅱ 菌株在 PDA 培养基上菌丝生长密集、低隆，多为白色或灰黑色，生长速率次于 Genotype Ⅰ 菌株，不易产生分生孢子器且产孢量少。两种基因型菌株在 28℃ 培养环境中生长最快，37℃ 培养环境中生长受到抑制。在室外活体接种实验中发现，Genotype Ⅰ 菌株基本不能引起梨树枝干致病，而 Genotype Ⅱ 菌株在接种 25d 后在接种部位就能看到黑褐色凹陷病斑，接种 75d 后病斑开始明显扩展，并出现红褐色或黑褐色凹陷病斑，有的病斑四周出现开裂。本研究结果初步表明，引起我国梨树轮纹病和干腐病病原菌 *B. dothidea* 存在一定的分子分化，且在致病特性上表现一定的差异。

关键词：梨树轮纹病；干腐病；遗传多样性

---

* 基金项目：国家农业部公益性行业计划（201203076-03）；梨现代农业技术产业体系（nycytx-29-08）

** 通信作者：王国平；E-mail：gpwang@ mail. hzau. edu. cn

湿度对苹果轮纹病定殖与侵染的影响*

薛德胜**，李保华***
（青岛农业大学农学与植物保护学院/山东省植物病虫害综合防控重点实验室，青岛 266109）

摘　要：苹果轮纹病（*Botyosphaeria dothidea*）是苹果上的重要病害之一，主要为害枝干和果实，造成枝枯、死树、烂果，严重威胁苹果生产。轮纹病菌的孢子在果实和枝条表面萌发后，腐生于寄主表面，并在寄主表面生长发育。当腐生菌落扩大后，且条件适宜时，再陆续侵入寄主的活体组织，诱发病害。湿度对轮纹病菌孢子萌发，及在寄主表面定殖有重要影响，进而影响病菌的侵染发病率。为了解湿度对苹果轮纹病孢子萌发、定殖，及其对发病率的影响，本试验测试了不同保湿时间对轮纹病菌萌发、定殖和侵染的影响，结果如下。

用轮纹病菌的分生孢子以株为单位接种富士苹果果实，并将苹果树转移到雾室，模拟自然降雨过程，对接种树体喷淋 1h、2h、3h、4h、5h、6h、7h、8h、9h、10h、12h 和 24h 后，移到室外，使露珠自然晾干。待果实开始发病后每周定期检查病斑数量，共检查 4 周。结果表明：喷淋 1h 和 3h 的接种果实，未见发病病斑，喷淋 2h 的 36 个果实中，仅出现 1 个病斑。自 4h 后，随喷淋时间延长，接种果实上的病斑数逐渐增加，喷淋 12h 后，平均每个接种果实上的病斑数为 0.84 个。初步推断，在苹果生长季节，轮纹病菌孢子萌发并在果面定殖所需要的最短露时为 4h。果面结露时间越长，萌发和定殖孢子数量越多。

用分生孢子活体接种富士苹果当年生枝条，并用棉球包裹保湿，分别处理 1d、7d、14d、28d，待枝条有病瘤出现后开始调查发病情况，检查接种部位皮孔发病率。结果表明：枝条保湿 28d 后产生大量病瘤，接种皮孔发病率为 34.5%；保湿 7d 和 14d 枝条上皮孔的发病率显著低于保湿 28d 枝条上皮孔的发病率，接种皮孔发病率分别为 24.7%和 21.0%；保湿 1d 枝条上接种皮孔的发病率为 0.2%，显著低于其他保湿处理的皮孔发病率。结果表明：腐生于枝干表面的轮纹病菌在枝条湿润时才能生长扩展，才能有机会侵入寄主组织。枝条润湿时间越长，病菌的生长量越大，侵入寄主组织的数量越多。

关键词：轮纹病；湿度；果实；枝条

---

* 基金项目：国家苹果产业技术体系（CARS-28）；山东省农业科学院农业科技创新工程（CXGC2016B11）-主要农作物病虫害绿色防控关键技术

** 第一作者：薛德胜，硕士研究生，研究方向为植物病害流行学；E-mail：xuedesheng2008@126.com

*** 通信作者：李保华，教授，主要从事植物病害流行和果树病害研究；E-mail：baohuali@qau.edu.cn

核盘菌 *SsNR* 基因调节其生长发育及致病过程的功能研究*

姚传春，魏君君，任恒雪，高智谋，潘月敏**
（安徽农业大学植物保护学院，合肥 230036）

摘 要：核盘菌［*Sclerotinia sclerotiorum*（Lib.）de Bary］是一种为害作物和蔬菜的世界性的重要的植物病原菌，是典型的死体营养型同宗配合的子囊真菌。宿主范围十分广泛，能够在油菜、大豆、向日葵等多种油料经济作物上引起植物的菌核病，造成巨大的农业经济损失。随着核盘菌全基因组测序完成及 DNA 测序、RNA 干扰（RNA interference，RNAi）、高通量蛋白质组学和代谢组学等技术的快速发展，为全球植物病理学家开展核盘菌功能基因的研究提供了便利。

硝酸盐是高等植物、藻类、真菌、酵母和细菌等生物氮摄入的主要营养物质。硝酸还原酶（nitrate reductase，NR）是氮素同化过程中关键的酶，其在细菌、真菌和高等植物的多种器官和组织中均有分布，该酶的活力影响了硝酸盐同化成有机化合物的快慢，且与生物吸收利用氮素有密不可分的关系。

为探索该酶在核盘菌生长发育及致病过程中的功能，本研究克隆出核盘菌硝酸还原酶编码基因序列 SsNR，采用基因沉默构建了 *SsNR*-pSilent-1 表达载体，利用 PEG 介导的原生质体转化技术获得 *NR* 基因沉默突变体菌株。通过 HygB 抗性、表型生长和 RT-PCR 筛选出突变菌株 NR12 和 NR66 两株沉默抑制率较好的菌株。对 SsNR 在核盘菌生长、发育及致病过程中的作用进行了研究，主要结果如下：

（1）与野生型相比，SsNR 突变体菌落生长速度变慢，菌丝末端分支短而多，菌核发育形成量减少，侵染垫畸形，产草酸能力减弱。

（2）在 Congo Red、SDS、H_2O_2、Sorbitol 和 NaCl 逆境因子作用下，与野生菌株相比，突变体对 Congo Red、SDS、H_2O_2和 NaCl 高度敏感，菌落直径抑制明显，而仅对 Sorbitol 逆境胁迫不敏感。

（3）油菜与大豆离体叶片接种试验显示野生菌株病斑面积占整个叶片面积分别为 12.43%~13.66%和 55.72%~57.21%，而两个沉默突变体菌株病斑率分别为 3.28%~4.73%和14.39%~15.72%，突变体致病力明显下降。

（4）致病性相关基因的表达量分析表明沉默突变体 *Ggt*1、*Sac*1 和 *Smk*3 基因表达下调，而 *Ubq* 和 *Cyp* 基因表达上调。上述结果表明 SsNR 在核盘菌的营养生长、菌核发育形成、侵染垫的形成及致病过程中具有重要的作用，为了解核盘菌的致病因子提供信息和理论依据。

关键词：核盘菌；硝酸还原酶；基因沉默；生长发育；致病性

* 第一作者：姚传春，硕士研究生，植物病理学专业；E-mail：1536864236@ qq. com
** 通信作者：潘月敏，副教授；E-mail：panyuemin2008@ 163. com

Functional characterization of a wheat lipid transfer protein TaLTP as potential target of rust effector PNPi*

Bi Weishuai[1], Gao Jing[1], Dubcovsky Jorge[2,4], Yu Xiumei[3],
Yang Wenxiang[1], Liu Daqun[1*] and Wang Xiaodong[1*]
(1. *College of Plant Protection, Biological Control Center for Plant Diseases and Plant Pests of Hebei, Agriculture University of Hebei, Baoding, Hebei* 071000, *P. R. China*; 2. *Department of Plant Science, University of California, Davis, CA* 95616, *USA.*; 3. *College of Life Science, Agriculture University of Hebei, Baoding, Hebei* 071000, *P. R. China.* 4. *Howard Hughes Medical Institute (HHMI), Chevy Chase, MD* 20815, *USA.*)

Abstract: In Arabidopsis, NPR1 is a key transcriptional co-regulator of systemic acquired resistance (SAR). In barley plants, NPR1 homolog showed a similar function in regulating the expression of pathogenesis-related (*PR*) genes during acquired resistance (AR) triggered by *Pseudomonas syringae* pv. *tomato* DC3000. In our previous research, we have found a virulent effector from *Puccinia striiformis* f. sp. *tritici* directly targeting wheat TaNPR1 protein, designated as *Puccinia* NPR1 interactor (PNPi). NPR1-mediated *PR* inductions weresuppressed by PNPi possibly via its competitive effect on the protein interaction between NPR1 and TGA2. Here in this study, ascreen of a yeast two-hybrid (Y2H) library from wheat leaves infectedwith *Puccinia striiformis* f. sp. *tritici* was carried outusing PNPi as bait. Among all the potential protein interactors, a wheat lipid transfer protein TaLTP showed a strong interaction with PNPi in our subsequent Y2H validation. Expression profile of *TaLTP*measured by qRT-PCR indicates a possible involvement of this gene in wheat resistance to rust infection. Moreover, a significant up-regulation of its barley homolog gene *HvLTP* during AR triggered by DC3000 in a *NPR*1-independent manner was observed. Further functional characterization of TaLTP as potential target of rust effector PNPi during both wheat resistance to rust and AR will be carried out.

* Corresponding authors: Liu Daqun; E-mail: liudaqun@ caas. cn
Wang Xiaodong; E-mail: zhbwxd@ hebau. edu. cn

麦根腐平脐蠕孢分泌蛋白基因 *CsSP*3 的功能研究*

张一凡，王利民，丁胜利**，李洪连**
（河南农业大学植物保护学院，郑州　450002）

摘　要：麦根腐平脐蠕孢（*Bipolaris sorokiniana*），有性态为禾旋孢腔菌（*Cochliobolus sativus*），是小麦上一种可以引起根腐、叶斑和黑胚等多种症状的重要病原真菌，广泛分布于世界各地，可为害多种禾谷类作物及杂草。麦根腐平脐蠕孢引起的小麦根腐病在英国、法国、德国、印度以及北美的加拿大、美国等地发生严重为害，造成小麦大量减产，也是我国小麦根腐病的主要病原之一。近年，小麦根腐病在我国主要小麦主产区发生呈现不断加重趋势，部分田块为害减产达 10%～20%，严重地块达 30%以上，对小麦生产造成了一定危害。

本课题组在麦根腐平脐蠕孢侵染小麦过程的转录组数据中，筛选了 4 个表达模式相似的高表达基因 *CsSP*1-4，其编码的蛋白在 N 端都具有经典信号肽序列，初步推定为分泌蛋白。其中 *CsSP*3 基因含有 294 个核苷酸，编码的蛋白含有 18 个氨基酸的信号肽，8 个 Cys，1 个糖基化位点，2 个丝氨酸和 2 个赖氨酸磷酸化位点；软件分析表明，*CsSp*3 为胞外分泌；BLAST 分析在其他真菌中没有找到同源类似物。

采用 split-marker 基因敲除方法，成功获得 Δ*cssp*3 缺失突变体。Δ*cssp*3 突变体在 CM 平板培养基上生长速率较野生型菌株变慢，洋葱内表皮侵染实验显示突变体侵染菌丝发育延迟，对小麦的致病性降低。初步研究结果表明：*CsSP*3 参与麦根腐平脐蠕孢的营养生长和侵染菌丝发育过程。相关的功能互补和编码蛋白的亚细胞定位等实验正在进行中。

关键词：麦根腐平脐蠕孢；分泌蛋虫；基因

---

* 基金项目：国家公益性行业（农业）科研专项（201503112）；河南农业大学人才引进项目（30600861）和河南省科技计划项目（152300410073）

** 通信作者：丁胜利，男，校特聘教授；E-mail：shengliding@ henau. edu. cn
李洪连，男，教授，博导；E-mail：honglianli@ sina. com

假禾谷镰刀菌非核糖体多肽合成酶基因 *FpgNPS9* 的功能研究*

康瑞姣，席靖豪，杜振林，王利民，丁胜利**，李洪连
（河南农业大学植物保护学院，郑州 45002）

摘 要：镰刀菌茎基腐病（Fusarium Crown Rot，FCR）是小麦生产上的重要病害，目前已在世界上多个国家报道发生，给小麦生产造成了严重的经济损失。2012 年，李洪连等在我国首次报道假禾谷镰刀菌造成的小麦茎基腐病以来，已在河南、河北、山西、山东等省发现该病菌的为害，严重地块的发病率达 70%以上，该病菌不仅可以引起茎基腐病，还可以造成穗腐（赤霉病），对小麦生产威胁很大。关于该病菌致病的分子机理研究报道很少。过去研究发现，非核糖体（nonribosomal peptides，NRPs）可影响一些植物病原真菌的致病性，如玉米小斑病菌 NPS6 参与铁代谢，与其致病性和抗氧化压力（oxidative stress）有关，而且在水稻胡麻斑病菌、禾谷镰刀菌和甘蓝链格孢等子囊菌中功能非常保守。敲除禾谷镰刀菌的 NRPS7 和假禾谷镰刀菌的 NRP32 分别会影响环脂肪肽类 fusaristatin A 及 W493 A 和 B 的合成。环脂肽具有抗菌、抗肿瘤和抗炎等广泛生物学活性，该类抗生素已经应用于临床。假禾谷镰刀菌 PKS-NRPS 基因簇与禾谷镰刀菌有很大的不同，许多 *NRPS* 基因的功能有待研究开发。

本课题组经过转录组分析，在假禾谷镰刀菌侵染小麦 5d 和 15d 时非核糖体肽合成酶 *NPS* 基因显著表达，并且该基因保守性高，因此推测 *NPS* 可能影响该菌对小麦的侵染过程。作者通过 *Split-Marker* 基因敲除方法获得了 *FpgNPS9* 基因的缺失突变体，通过孢子悬浮液侵染小麦胚芽鞘和盆栽土壤接种侵染小麦试验测定了突变体和野生型菌株在致病性上的差异，发现突变体接种发病明显减轻；利用菌丝体在不同的温度和摇培时间获得粗过滤液处理发芽的小麦种子，突变体粗滤液对小麦胚根生长的抑制能力明显低于野生型菌株，具体代谢产物的差异需要进一步测定。

研究结果初步表明，非核糖体肽合成酶 *NPS* 基因对假禾谷镰刀菌的致病性有一定影响，并可以调控产生影响小麦根生长的代谢产物。其影响的分子机理还在进一步研究中。

关键词：假禾谷镰刀菌；合成酶基因；小麦

---

* 基金项目：国家公益性行业科研专项（201503112）；河南省科技计划项目（152300410073）和河南农业大学人才引进项目（3600861）

** 通信作者：丁胜利，教授，主要从事植物病原菌致病分子机理研究；E-mail：shengliding@ nenau. edu. cn

假禾谷镰刀菌促分裂原蛋白激酶 FpPMK1 及其下游转录因子 *FpFST*12 基因的功能研究*

王利民**，张一凡**，杜振林，张银山，丁胜利***，李洪连
（河南农业大学植物保护学院，郑州　450002）

摘　要： 小麦茎基腐病是假禾谷镰刀菌、禾谷镰刀菌和黄色镰刀菌等多种病原物引起的一种土传病害。国内外研究表明，假禾谷镰刀菌是小麦茎基腐病优势病原菌。目前对假禾谷镰刀菌致病的分子机理研究很少。过去的研究显示，MAPK 信号途径对多种病原真菌的菌丝生长、繁殖和发育过程具有关键的调控作用，这些信号途径中的关键基因在病原真菌的致病过程中起着必不可少的作用。尤其是 MAP 激酶及其调控的下游转录因子。因此，了解 MAPK 信号途径在假禾谷镰刀菌中的作用，能够更好地了解该菌的致病分子机理。

信息素是 MAPK 信号途径中重要的传导途径之一。作者根据假禾谷镰刀菌侵染小麦的转录组数据分析，发现了多个关于信息素信号途径（*Fus*3/*Kss*1 signaling pathway）的基因差异表达，但没有 *PMK*1 同源基因的转录数据，却含有其磷酸化底物 *FST*12 同源基因上调表达，反转录克隆了假禾谷镰刀菌 *PMK*1 同源基因的全长 cDNA，*FpPMK*1 含有 1 244个核苷酸，有 3 个内含子，编码 355 个氨基酸；*FpFST*12 含有 2 193个核苷酸，有 2 个内含子，编码 695 个氨基酸。所以选择了 MAPK 途径中的 *Fus*3 信息素信号途径作为研究的重点。

采用反向遗传学的研究思路，利用 split-marker 的方法来构建含有 HYG（潮霉素磷酸转移酶基因）的基因敲除盒，再通过 PEG 介导的原生质体转化和体内同源重组得到抗潮霉素的转化子。用四对引物对得到的转化子进行 PCR 检测初步筛选，然后再进 Southern 验证进一步确认转化子，从而得到真正 Δ*fppmk*1、Δ*fpfst*12 缺失突变体。*FpPMK*1 基因的缺失突变体菌落生长速度稍为减慢，气生菌丝减少，后期会出现自溶现象，分生孢子产量下降约 98%，对 SDS 敏感，但对 H_2O_2 不敏感。通过小麦胚芽鞘和盆栽接种致病性测定，发现该突变体致病力完全丧失。*FpFST*12 基因的缺失突变体菌落表型与野生型没有明显差别，产孢量下降 50%，对 H_2O_2 和 SDS 都不敏感，Δ*fpfst*12 致病性显著下降。与野生型假禾谷镰刀菌菌株相比，两个基因的缺失突变体均丧失对玻璃纸的穿透能力。综合上述结果，MAPK 信号途径在调控假禾谷镰刀菌侵染过程中发挥着重要作用，其中 MAP 激酶 *FpPMK*1 和其调控的下游转录因子 *FpFST*12 两个关键基因是致病力重要因子，在假禾谷镰刀菌侵染初期发挥决定性作用。

关键词： 假禾谷镰刀菌；转录因子；基因功能

---

* 基金项目：国家公益性行业（农业）科研专项（201503112）；河南农业大学人才引进项目（30600861）和河南省科技计划项目（152300410073）

** 第一作者：王利民，男；张一凡，女，在读硕士研究生

*** 通信作者：丁胜利，男，校特聘教授；E-mail：shengliding@ henau. edu. cn

油菜菌核病菌弱致病力菌株的生物学性状研究*

吴　亚，赵振宇，王　琴，王伟燕，张华建，潘月敏，高智谋**
（安徽农业大学植物保护学院，合肥　230036）

摘　要：核盘菌（*Sclerotinia sclerotiorum*）是一种破坏性和腐生性很强的土传性植物病原真菌，由该病菌侵染油菜引起的油菜菌核病在我国的各油菜种植区均能发生，尤其严重威胁长江流域及东南沿海地区的油菜产区，使油菜产量和品质严重下降。为了探讨油菜菌核病菌的致病力分化及弱致病力形成机制，作者对油菜菌核病菌弱致病力菌株的生物学性状进行了研究，主要结果如下。

选取安徽省不同地区来源的油菜菌核病菌株，利用其作为研究对象，采用离体叶片法，测定了不同菌株间的致病力。结果表明：同一菌株接种 4 个不同油菜品种产生的病斑大小不同；同一油菜品种被 25 个不同菌株接种产生的病斑大小也不同，进而说明油菜菌核病菌菌株之间存在明显致病力分化。根据测定结果，选取有代表性的 6 个油菜菌核病菌菌株（强致病力菌株 WJB-4、TC-7、SX-2，弱致病力菌株 LA3、WW2、SX-6-1）为供试菌株。

对弱致病力菌株 LA3、WW2、SX-6-1 进行了分子鉴定，ITS 序列分析结果得出，弱致病力菌株与已知核盘菌的序列的同源性均高达 100%。由此可见，弱致病力菌株为核盘菌。进一步研究了 6 个供试油菜菌核病菌菌株（强致病力菌株 WJB-4、TC-7、SX-2，弱致病力菌株 LA3、WW2、SX-6-1）的生物学性状。结果表明：WJB-4 为快生长类型，生长速度为 4. 20cm/d，菌落扩展均匀且生物量较高；而弱致病力菌株中 SX-6-1 为慢生长类型，生长速度仅为 2. 43cm/d，菌落扩展不均匀且生物量较低。

为了分析弱致病力菌株的生化机制，比较其在致病因子方面是否存在明显差异，分别测定了供试菌株的果胶酶和纤维素酶活性、草酸产生量，并检测了侵染垫形成数量。结果表明：与强致病力菌株相比，弱致病力菌株草酸产生量较低，果胶酶活性较弱，侵染垫形成较少。

此外，从内生细菌鉴定和 dsRNA 病毒检测两方面初步探索弱致病力菌株形成的可能原因。结果表明：自弱致病力菌株体内未分离到内生细菌，说明内生细菌不是引起弱致病力菌株的形成原因。同时，在弱致病力菌株中没有提取到 dsRNA 病毒，进而说明 dsRNA 病毒也不是引起供试弱致病力菌株的形成原因。

关键词：油菜菌核病菌；致病力；生物学

* 基金项目：公益性行业（农业）科研专项（201103016）

** 通信作者：高智谋，教授；E-mail：gaozhimou@ 126. com

油菜菌核病菌（*Sclerotinia sclerotiorum*）营养生理研究*

张茜茹，王　姣，宋延香，潘月敏，张华建，高智谋**
（安徽农业大学植物保护学院，合肥　230036）

油菜菌核病是我国油菜上主要病害。为了明确其病原菌——核盘菌（*Sclerotinia sclerotiorum*）的营养生理，从而为核盘菌的相关研究及油菜菌核病的“营养疗法”提供试验依据，作者研究了碳氮源营养及 B、Zn、Se 等微量元素对油菜核盘菌生长速率及菌核产生量的影响，取得的主要实验结果如下。

1　不同培养基对油菜核盘菌生长速率及菌核产生量的影响

以采自安徽油菜上的 12 个核盘菌（*Sclerotinia sclerotiorum*）菌株为供试菌株，测定了 PDA、PSA、胡萝卜培养基、燕麦培养基、甘薯培养基 5 种培养基对核盘菌生长速率及菌核产生量的影响，供试菌株在 5 种培养基上培养 2d 后测量菌丝直径，7d、14d、30d 记录菌核数目。实验结果表明：大多数油菜核盘菌菌株的菌丝生长以在 PSA 培养基上最好，菌核形成数量也以在 PSA 培养基上最多；也即 PSA 培养基对于大多数核盘菌菌株的菌丝生长和菌核形成是最佳的。

2　不同碳氮源营养对核盘菌生长速率及菌核产生量的影响

以查式培养基为基础，分别以供试碳源、氮源代替培养基中的蔗糖、硝酸钠配制成不同碳氮源培养基，将供试的 12 个核盘菌菌株接种到上述不同碳氮源培养基上，置 25℃黑暗培养。2d 后测量菌落直径，第 7d、14d、30d 时记录菌核数目，测定分析不同碳氮源营养对核盘菌生长速率及菌核产生量的影响。实验结果表明：对于核盘菌菌丝生长和菌核形成而言，最佳碳源为蔗糖，最佳氮源是蛋白胨。

3　硼元素对核盘菌菌丝生长速率及孢子萌发的影响

在熔化的 PDA 培养基中加入硼酸钾（$K_2B_4O7.5H_2O$），配制成硼酸钾含量分别为 0.01mg/L，0.05mg/L，0.10mg/L，0.5mg/L，1mg/L，2mg/L，4mg/L，8mg/L，16mg/L 的 PDA 培养基，将供试的 12 个核盘菌菌株接种到上述不同含硼培养基上，置 25℃黑暗培养，以不加硼酸钾的 PDA 培养基为对照。接种 2d 后测量菌落直径，第 7d、14d、30d 时记录菌核数目。测定结果表明：硼在低浓度（0.01mg/L、0.05mg/L）时，对菌丝生长有促进作用；而在较高浓度（0.5~16mg/L）时，对菌丝生长有抑制作用，且随着硼元素浓度增高，对菌丝的抑制作用就越明显。硼在一定浓度下对油菜核盘菌孢子萌发有抑制作用，在硼离子浓度为 2~16mg/L 时，随着浓度的增加，孢子萌发抑制率增加。

4　锌元素对核盘菌菌丝生长速率及孢子萌发的影响

测定方法同上，只是用硫酸锌（$ZnSO_4$）配制成不同浓度的含锌培养基。实验结果表明：当

* 基金项目：公益性行业（农业）科研专项（201103016）

** 通信作者：高智谋，教授；E-mail：gaozhimou@126.com

锌离子的浓度为 0.01~4.00mg/L 时，对核盘菌菌丝生长总体有明显促进作用。当锌离子的浓度为 0.01~4.00mg/L 时，各浓度处理对核盘菌孢子萌发率均低于空白对照，说明锌对核盘菌孢子萌发有明显抑制作用。

5 硒元素对核盘菌生长速率及菌核产生量的影响

在熔化的 PDA 培养基中加入亚硒酸钠（Na_2SeO_3）配制成不同含硒培养基，将供试的 12 个核盘菌菌株接种到上述不同含硒培养基上，置 25℃黑暗培养，以不加亚硒酸钠的 PDA 培养基为对照。实验结果表明：加入了低浓度的亚硒酸钠（<0.01mg/L）的培养基，其核盘菌的生长与对照组进行对比，可显著看到菌丝生长浓密，有一定的促进核盘菌生长的作用，菌核生长速率较快，7d 即可形成。而在培养基含硒浓度较高（10~100mg/L）时，硒对核盘菌的菌丝生长有明显的抑制作用，菌丝生长稀疏，且随着浓度的升高，菌丝生长抑制率增加。此外，菌丝形态观察发现，在硒浓度>10mg/L 时，核盘菌菌丝形状变化明显，与空白对照（CK）相比，菌丝较短且较粗，菌丝扭曲，呈现畸形。

此外，测定结果还表明，在试验设置的培养基含硒浓度（0.01~100mg/L）范围内，硒元素对菌核形成有明显的抑制作用，且当硒元素浓度达到>80mg/L 时，菌核便不再形成。

水稻纹枯病菌细胞壁降解酶与致病力的相关性研究*

张　优**，王海宁，莫礼宁，杨晓贺，罗文芳，李思博，张照茹，魏松红***
（沈阳农业大学植物保护学院，沈阳　110866）

摘　要：水稻纹枯病在我国南北方水稻种植区普遍发生，近年来在东北地区发病趋势严重，已成为限制水稻高产和稳产的主要因素之一。本试验 2015 年从辽宁省、吉林省和黑龙江省多个水稻种植区采集病样，分离纯化出 132 株水稻纹枯病菌，经鉴定主要为茄丝核菌（*Rhizoctonia solani*）和稻丝核菌（*Rhizoctonia oryzae*），对不同地理来源的水稻纹枯病菌的致病力差异进行了研究，发现两种病原菌的致病性差异较大，茄丝核菌菌株均表现出较强致病性，稻丝核菌菌株均表现出弱致病性。

细胞壁降解酶是水稻纹枯病菌重要的致病因子，水稻纹枯病菌在改良的 Marcus 培养液中能产生多聚半乳糖醛酸酶（PG）、果胶甲基半乳糖醛酸酶（PMG）、纤维素酶（Cx）、多聚半乳糖醛酸反式消除酶（PGTE）、果胶甲基反式消除酶（PMTE）5 种细胞壁降解酶。本试验研究了东北三省不同来源的水稻纹枯病菌对水稻的致病力与各种细胞壁降解酶活性之间的关系。结果表明：不同碳源培养液对细胞壁降解酶影响活性存在差异，在果胶培养液中 PG 活性最高，在纤维培养液中 Cx 活性最高而 PGTE 和 PMTE 活性均较低。供试菌株 PG、PMG 和 Cx 活性与致病力强弱之间具有较强的相关性，其中 PG 与致病力的相关性最强。强致病性菌株的 PG 酶活性较高，弱致病性菌株的酶活性较低；茄丝核菌菌株的细胞壁降解酶活性均明显高于稻丝核菌菌株。

关键词：水稻纹枯病菌；细胞壁降解酶；致病力

* 基金项目：国家水稻产业技术体系项目（CARS-01）；辽宁水稻产业技术体系项目（辽农科<2013>271 号）

** 第一作者：张优，硕士研究生，研究方向为植物病理学；E-mail：2385073088@qq.com

*** 通信作者：魏松红，博士，教授，主要从事植物病原真菌学及水稻病害研究；E-mail：songhongw125@163.com

稻瘟病菌氧胁迫应答机制的转录组分析*

魏松红**，刘　伟，王海宁，李　帅，孔令春，许　月，张照茹
（沈阳农业大学植物保护学院，沈阳　110866）

摘　要：植物病原真菌的氧化应激反应由活性氧（Reactive oxygen species，ROS）超负荷引起，是细胞对逆境条件防御和适应性反应中最重要的一种。植物病原真菌在萌发、侵染、温度胁迫、农药胁迫及营养匮乏时均会受到活性氧的胁迫。对氧胁迫防御能力的提高会导致植物病原真菌致病力上升、危害地区扩大、越冬存活率提高及对杀菌剂敏感性下降等。所以研究植物病原真菌对氧胁迫的抗逆机制对于植物病害的综合防治具有重要的意义。

本研究采用 RNA-seq 技术，将稻瘟病菌在不同氧胁迫条件下（PDA 培养基、CM 培养基、H_2O_2 0. 5mmol/L、紫外光照 25 min、pH 值=6）进行培养，提取各处理的总 RNA，委托北京百迈客生物科技有限公司进行 cDNA 文库的构建和 I11umina 测序，对测序得到的原始 reads 进行数据评估，然后组装得到稻瘟病菌的 Unigene Library，再基于 Unigene 库进行基因结构注释、基因表达分析、基因功能注释等，筛选出稻瘟病菌识别和防御氧胁迫的相关基因。结果表明：

（1）差异基因的表达量、数量和类群存在异同。作为低氧胁迫的 MoPDA 处理组与其他 4 个高氧胁迫的处理组的差异表达基因的总数为 5 154个，其中上调表达基因数 3 541个，下调基因表达数为 1 613个；在 4 个高氧胁迫处理组中，共有相同表达基因 3 314个，其中上调基因表达数 2 086个，不同基因表达数 1 228个。其中参与氨基酸的转运和代谢、碳水化合物的转运和代谢等的基因最多，其次为参与抗逆胁迫响应、基础代谢、细胞功能、蛋白修饰、信号转导、离子运输、氧化还原等生物代谢进程。

（2）GO 富集的基因种类和功能存在差异。将不同氧胁迫处理组中，上述备选基因分别富集在 36 条 GO-item 中，包括生物进程、分子功能和细胞成分 3 个方面。根据本研究目的对活性氧的形成和清除、生物大分子的损伤修复、代谢途径的适应性调整等方面进行专项整理，主要包括活性氧解毒与氧化还原平衡、碳水化合物的转运和代谢、氨基酸和蛋白质的合成、核苷酸的转运与代谢、信号转导等方面。

（3）KEGG Pathway 的数量和种类存在差异。从 5 个不同氧胁迫处理组的转录组数据中富集得到 35 个代谢通路，其中下列 5 个代谢通路占据的数目最多，分别为代谢途径、次生代谢产物的生物合成、氨基酸的生物合成、信号转导和糖代谢。该研究可为阐明稻瘟病菌抗氧胁迫的调控机制提供一定理论基础。

关键词：稻瘟病菌；氧胁迫；转录组

---

* 基金项目：国家水稻产业技术体系项目（CARS-01）；辽宁水稻产业技术体系项目（辽农科<2013>271 号）

** 第一作者：魏松红，教授，研究方向为植物病原真菌学及水稻病害研究；E-mail：songhongw125@ 163. com

稻瘟菌氧固醇结合蛋白激发子功能的初步研究*

陈萌萌**，房雅丽，范 军***
（中国农业大学植物病理学系，北京 100193）

摘 要：病原菌中能够诱导植物免疫反应的激发子的分离、鉴定及功能分析对于研究植物先天免疫反应机理具有重要意义。本文利用经改造的激活标记双元载体构建的稻瘟菌基因组文库，通过农杆菌在烟草上瞬时表达稻瘟菌基因，筛选出了稻瘟菌的氧固醇结合蛋白，它能够引起烟草叶片出现类似过敏性反应的细胞死亡。利用5-氨基水杨酸作为反应底物，笔者检测到原核表达纯化的稻瘟菌氧固醇结合蛋白能够触发烟草悬浮细胞活性氧累积；此外，接种实验结果表明：稻瘟菌氧固醇结合蛋白处理拟南芥叶片可导致植株对丁香假单胞菌番茄致病变种的抗性增强，表现为病原细菌的生长量明显减少。这些结果暗示稻瘟菌氧固醇结合蛋白可能是一种激发子，能够触发植物的先天免疫反应。这对进一步解析该蛋白诱导植物先天免疫反应的分子机制奠定了基础。

关键词：氧固醇结合蛋白；稻瘟菌；植物先天免疫；激发子

---

* 基金项目：教育部博士点基金（20130008110005）

** 第一作者：陈萌萌，女，博士研究生，主要研究方向植物病理学

*** 通信作者：范军，男，教授，主要研究方向植物病理学；E-mail：jfan@ cau. edu. cn

Isolation and Identification of *Alternaria* Speices on Compositae Plants in China*

Luo Huan**, Jia Guogeng, Liu Haifeng, Pei Dongfang, Deng Jianxin***

(*College of Agriculture*, *Yangtze University*, *Jingzhou*, *Hubei* 434025, *China*)

Abstract: *Alternaria* species are widely spreaded all over the world with a variety of ecological niches. In order to know the diversity of *Altenaria*species from Compositae plants in China, specimens were collected from 15 geogrophy areas in 7 provinces. 160 isolates were obtained bysingle spore isolation. Their pathogenicity on host was tested on detached leaves accdoring to the Koch's rule. Based on the morphological characteristics and phylogenetic analysis of a combined gene sequence dataset (*rDNA-ITS*, *ATPse* and *GPDH*). 43 strains were selected and determined in the study. Four new recorded species (*A. argyranthemi*、*A. cinerariae*、*A. calendulae*and *A. helianthinficiens*) and one new species, were reported and identified. Meanwhile, the diversity and phylogenetic relationship of *Altenaria* species on Compositae plants in China were determined.

Key words: *Alternaria*; Compositae; Morphology; Phylogenetic

* 基金项目：国家自然科学基金项目（31400014）；长江大学青年人才基金项目（2015cqr17；2016cqr08）

** 第一作者：罗欢，硕士，主要从事植物病理研究；E-mail：luohuan_ 0813@ 163. com

*** 通信作者：邓建新，男，副教授，主要从事植物真菌病害研究；E-mail：djxin555@ hotmail. com

广西防城港西番莲茎基腐病发生为害调查与病原鉴定

陈　星[1*]，高淑梅[2]，李迎宾[1]，张治萍[1]，蒋　娜[1]，罗来鑫[1]　李健强[1**]
（1. 中国农业大学植物病理系/种子病害检验与防控北京市重点实验室，北京　100193；
2. 北京社会管理职业学院，河北燕郊　101601）

摘　要：西番莲（*Passiflora edulia*），因其果汁散发菠萝、香蕉、石榴、杧果等多种水果香气又称为百香果，具有很高的食用价值与药用价值，在我国广东、广西、福建、台湾、云南等地均有种植。2016 年广西防城港西番莲种植产区大面积发生茎基腐病害，造成严重经济损失。为掌握该地区西番莲病害的发生为害及探究病原种类，对当地发病较重的那屋背村等西番莲种植基地进行了病害调查、病样采集和病原菌分离鉴定。结果表明：该地区西番莲种植基地茎基腐病发生较普遍，发生区域和严重程度不一，生产区病害发生率为 20%左右，重病区发生率达 60%～93%，甚至造成毁园。对采集的发病植株根、茎、叶、果实和根围土壤进行病原菌分离，共获得 43 个真菌分离物，综合形态学和 ITS 序列鉴定结果，确定分离物中包括毛色二孢属（*Lasiodiplodia*）、镰刀菌属（*Fusarium*）、黑孢属（*Nigrospora*）、刺盘孢属（*Colletotrichum*）、新拟盘多毛孢属（*Neopestalotiopsis*）、假拟盘多毛孢属（*Pseudopestalotiopsis*）、拟茎点霉属（*Phomopsis*）、漆斑菌属（*Myrothecium*）和木霉属（*Trichoderma*）共 9 个属的真菌。综合文献报道，对分离物镰刀菌 F-1 进行进一步的分子鉴定，镰刀菌特异性引物 EF-1/EF-2 扩增测序结果表明 F-1 菌株为尖孢镰刀菌（*Fusarium oxysporum*）。致病性试验显示，采用 5mL 浓度为 10^6 个/mL 的 F-1 菌株孢子悬浮液接种西番莲幼苗（90d 苗龄），60d 后接种处理组植株叶片出现褪绿、萎蔫和茎基部变褐、脱水症状，后整株枯死，表明菌株 F-1 对西番莲幼苗具有致病性。关于该病害的防控药剂筛选和田间防病试验正在进行中。

关键词：西番莲茎基腐病；病原鉴定；ITS；尖孢镰刀菌；致病性

* 第一作者：陈星，硕士研究生，主要从事种传病害研究；E-mail：chenxing2028@163.com
** 通信作者：李健强，教授，主要从事种子病理学和杀菌剂药理学研究；E-mail：lijq231@cau.edu.cn

魔芋块根腐烂病原真菌的分离及其致病性研究

李迎宾[1*]，暴晓凯[1]，万 琪[1]，徐文果[2]，高 凡[3]，杨明文[4]，谢世清[5]，李健强[1**]
（1. 中国农业大学植物保护学院/种子病害检验与防控北京市重点实验室，北京 100193；
2. 云南省德宏州农科所，潞西 678400；3. 云南省西双版纳州农科所，景洪 666100；
4. 云南省临沧市农技推广站，临沧 677000；5. 云南农业
大学农学与生物技术学院，昆明 650201）

摘　要：魔芋属于天南星科魔芋属多年生草本植物，具有重要的经济价值和营养价值，在中国西南地区栽培面积较大。魔芋软腐病是魔芋生产中的重要病害，以云南昭通为例，平均发病率在 15%~30%，严重时可达 80%。引起魔芋软腐病的病原菌已有报道主要为胡萝卜软腐果胶杆菌胡萝卜亚种（*Pectobacterium carotovora* subsp. *carotovora*）和菊果胶杆菌（*P. chrysanthemi*）。目前，鲜有关于病原真菌引起魔芋块根腐烂的报道。

（1）病原菌分离、纯化与鉴定：基于传统分离培养的方法，从采自云南的 2 个种共计 20 个具有软腐症状的魔芋块根样品的病健交界处进行病原真菌的分离；挑取单孢进行纯化，共获得 2 株纯培养物，编号分别为 MY-1 和 MY-3；形态学鉴定结果显示，在 PDA 培养基上培养 3d 后，分离物 MY-1 菌落呈白色绒毛状，边缘整齐，以大型分生孢子为主，孢子呈镰刀形，无色，具 1~3 个分隔；分离物 MY-3 菌落呈淡紫色，边缘不整齐，以小型分生孢子为主，孢子呈卵圆形。ITS 和 tef 两对引物分别进行 PCR 扩增测序结果显示，MY-1 与 *Fusarium solani* 相似度高达 100%，MY-3 与 *F. oxysporum* 相似度高达 100%。

（2）致病性测定：选择“珠芽黄魔芋”（*Amorphophallus muelleri* Bulbifer）和“花魔芋”（*Amorphophallus konjac* K. Koch）两个种的块根，分别接种 MY-1 和 MY-3。结果表明：接种“花魔芋”6d 后，*F. solani* 与 *F. oxysporum* 均可引起致病；而接种“珠芽黄魔芋”后，仅观察到 *F. solani* 可引起致病。从接种后的两种魔芋上，又可以再次分离得到与接种体同样的真菌。

何雯等 2016 年在魔芋根域分离获得了 *F. solani* 与 *F. oxysporum*，通过离体接种毒素粗提液，证明其可引起魔芋块根腐烂。本研究获得的 2 株镰刀菌与其报道的菌落形态存在差异，同时采用离体菌丝块接种方法，发现 *F. oxysporum* 对不同种魔芋块根的致病力存在一定差异。

关键词：魔芋；腐烂；镰刀菌；致病性测定

* 第一作者：李迎宾，博士研究生，主要从事种子病理学和病害生物防治研究；E-mail：lyb_ cau_ 2014@ sina. com
** 通信作者：李健强，教授，博士生导师，主要从事种子病理及杀菌剂药理学研究；E-mail：lijq231@ cau. edu. cn

High throughput sequencing reveals endophytic fungal communityin field-grown soybean of Huang-huai-hai region of China

Yang Hongjun, Ye Wenwu, Ma Jiaxin,
Zeng Dandan, Wang Yuanchao, Zheng Xiaobo
(*Plant Pathology Department*, *Nanjing Agricultural University*, *Nanjing*, *China*, 210095)

Abstract: Plants depend upon beneficial interactions between roots and endophytic fungal for growth promotion, disease suppression, and stress tolerance. In this study, we analyzed the endophytic fungal community harbored in the root of soybean plants grown in field in Huang-huai-hai region, one of the main producing areas of soybeans in China. Based on ITS amplicon sequencing for soybean samples collected from Shandong, Jiangsu, and Anhui province, we found the majority of endophyte taxa were Ascomycota (87%), followed by Basidiomycota (10%) and Zygomycota (1%). The species composition and abundanceof endophytic fungi were significantly dependent on the geographic locality and the plant developmental stage. The deep sequencing results were confirmed by traditional methods for isolation of fungi from corresponding tissues. The most frequently isolatesin genus were *Haematonectria* (60%), *Phoma* (10%), *Fusarium* (8%), *Rhizoctonia* (6), *Clonostachys* (6), *Macrophomina* (3), and *Ceratobasidium* (2%). Many phytopathogenic species associated with soybean disease were found; however, it remains to be determined whether these endophytes are latent pathogens or non-pathogenic forms that benefit the plant. The results have revealed the endophytic fungal community in field-grown soybean of Huang-huai-hai region, providing insight for further study of plant-endophyte interaction.

Key words: Endophytic fungi; Soybean; ITS; Plant pathogens; Microbiome

The ArfGAP protein MoGlo3 regulates the development and pathogenicity of *Magnaporthe oryzae**

Zhang Shengpei[1], Liu Xiu[1], Li Lianwei[1], Yu Rui[1], He Jialiang[1], Zhang Haifeng[1], Zheng Xiaobo[1], Wang Ping[2], and Zhang Zhengguang[1]

(1. *Department of Plant Pathology, College of Plant Protection, Nanjing Agricultural University, and Key Laboratory of Integrated Management of Crop Diseases and Pests, Ministry of Education, Nanjing* 210095, *China*; 2. *Departments of Pediatrics, and Microbiology, Immunology, and Parasitology, Louisiana State University Health Sciences Center, New Orleans, Louisiana* 70112, *USA*)

Abstract: The ADP ribosylation factor (Arf) and the coat protein complex I (COPI) are both involved in vesicle transport. Together with GTPase-activating proteins (ArfGAPs) and guanine exchange factors (ArfGEFs) that regulate the activity of Arf, they govern vesicle formation, COPI trafficking, and the maintenance of the Golgi complex. In an ongoing effort to study the role of membrane trafficking in pathogenesis of the rice blast fungus *Magnaporthe oryzae*, we identified MoGlo3 as an ArfGAP proteinthat is homologous to Glo3p of the budding yeast *Saccharomyces cerevisiae*. As suspected, MoGlo3 partially complements the function of yeast Glo3p. Consistent with findings in *S. cerevisiae*, MoGlo3 is localized to the Golgi and that the localization is dependent on the conserved BoCCS domain. We found that MoGlo3 is highly expressed during conidiation and early infection stages, and is required for vegetative growth, conidial production, and sexual development. We further found that the Δ*Moglo3* mutantis defective in endocytosis, scavengingof the reactive oxygen species (ROS), and in the response to endoplasmic reticulum (ER) stress. The combined effects result in failed appressorium function and decreased pathogenicity. Moreover, we provided evidence showing that the domains including the GAP, BoCCS and GRM are all important for normal MoGlo3 functions. Our studies further illustrate the importance of normal membrane trafficking in the physiology and pathogenicity of the rice blast fungus.

Key words: *Magnaporthe oryzae*; Endocytosis; Development; Pathogenicity

* Corresponding author: Zhang Zheng-guang; E-mail: zhgzhang@ njau. edu. cn

四川盆地小麦条锈病菌小种鉴定及其寄生适合度测定*

王树和**，初炳瑶，马占鸿***
（中国农业大学植物病理系，农业部植物病理学重点开放实验室，北京 100193）

摘 要：为了解四川盆地春季菌源区小麦条锈病菌的毒性结构及优势小种寄生适合度，于2014年和2015年在四川盆地绵阳市涪城区和盐亭县连续两年种植不同抗性小麦品种，分离小麦条锈病菌，应用中国鉴别寄主进行生理小种鉴定，并对优势小种采用综合病情指数法测定其相对寄生适合度。2014年和2015年的鉴定结果均显示贵农22类群（G22）处于第一位，出现频率分别为38.3%和38.2%；其次是水源11类群和杂46类群。对优势小种G22-83、CYR32和CYR33采用综合病情指数法在田间分小种接种四川省109个小麦品种（系）测定其相对寄生适合度。在群体水平，致病类型G22-83、CYR32和CYR33对四川省109个小麦品种（系）平均相对寄生适合度值分别为0.1930、0.0560和0.0379，致病类型G22的平均寄生适合度显著高于CYR32和CYR33（$P<0.01$）。新致病类型G22在四川省近年小麦生产品种和后备品种上的寄生适合度较高，在今后一段时间内可能在该地区流行为害。

关键词：小麦条锈病；寄生适合度；四川省

* 基金项目：国家重点基础研究发展计划（2013CB127700）；国家重点研发计划项目（2016YFD0300702、2016YFD0201302）；国家自然科学基金（31371881）

** 第一作者：王树和，男，甘肃人，博士研究生，主要从事植物病害流行学研究；E-mail：wang_shuhe@163.com
*** 通信作者：马占鸿，教授，主要从事植物病害流行和宏观植物病理学研究；E-mail：mazh@cau.edu.cn

黄淮部分地区玉米穗粒腐病致病镰孢菌种类研究*

席靖豪，林焕洁，赵清爽，袁虹霞，李洪连**
（河南农业大学植物保护学院，郑州 450002）

摘 要：穗粒腐病在世界各玉米种植区广泛发生，田间损失一般为 5%～10%。穗粒腐病不仅造成玉米产量损失，也会因致病菌在玉米籽粒中产生多种毒素，作为食物或饲料被食用后对人畜健康会产生严重影响。黄淮夏玉米种植区是我国玉米主产区之一，近年来穗粒腐病发生逐年加重，给玉米安全生产造成较大威胁。为了明确该地区玉米穗粒腐病主要病原菌的种类及优势种，为该病害的防治工作提供理论依据，本课题组于 2016 年 9 月中下旬在河南、山东、安徽、江苏 4 个省的 25 个区县采集玉米穗腐样品，经组织分离和单孢纯化，共获得 282 个病原分离物，根据形态学初步鉴定，其中 214 个分离物属于镰孢菌，占 75.89%，说明镰孢菌为黄淮地区玉米穗粒腐病的主要病原菌。

进一步采用分子生物学方法（rDNA-ITS 和 Ef-1α 基因序列分析）对不同分离物进行种类鉴定。鉴定结果表明：本地区玉米穗粒腐病的病原镰孢菌种类比较复杂，包括拟轮枝镰孢（*Fusarium verticillioide*）、禾谷镰孢（*F. graminearum*）、层出镰孢（*F. proliferatum*）、尖镰孢（*F. oxysporum*）、木贼镰孢（*F. equiseti*）和新知镰孢（*F. andiyazi*）等。其中拟轮枝镰孢（*Fusarium verticillioide*）的平均分离频率最高，达 38.2%，层出镰孢（*F. proliferatum*）次之，平均分离频率为 31.2%；禾谷镰孢平均分离频率较低，只有 6.38%。同时研究发现，不同采样点病原镰孢菌种类差别较大。初步研究结果发现，拟轮枝镰孢（*F. verticillioide*）和层出镰孢（*F. proliferatum*）为该地区玉米穗粒腐病的优势致病菌，部分地区禾谷镰孢（*F. graminearum*）占有较高比例。但由于气候、品种和地域差异，不同年份和地区玉米穗粒腐病病原菌种类变化较大，应当扩大采样范围，持续进行该地区玉米穗粒腐病病原菌的种类监测。

关键词：玉米穗粒腐病；病原菌；监测

* 基金项目：河南省玉米产业技术体系植保岗位科学家科研专项（S2015-02-G05）
** 通信作者：李洪连，教授，主要从事植物病害监测和防控技术研究；E-mail：honglianli@ sina. com

α-1，3-甘露糖转移酶基因 *FpALG3* 在假禾谷镰刀菌中的功能分析*

杜振林，王利民，张一凡，丁胜利**，李洪连
（河南农业大学植物保护学院，郑州 450002）

摘　要：小麦茎基腐病是一种由多种病原引起的世界性土传病害，近年在我国主要小麦产区黄淮麦区的发生普遍，为害逐年加重，已经成为小麦生产上的严重问题。经本课题组系统研究发现，假禾谷镰刀菌（*Fusarium pseudograminearum*）分离频率高，致病力强，是黄淮麦区小麦茎基腐病的优势病原菌。在侵染过程中，假禾谷镰刀菌孢子发芽形成芽管，顶端膨大形成类似稻瘟菌的足型附着孢和细侵染钉，侵染菌丝在寄主细胞内扩展定殖，同时在小麦茎秆中产生高浓度的类似于禾谷镰孢菌的 DON 毒素，目前，国内外对该病原菌的致病分子机理研究较少。

甘露糖转移酶是一类参与蛋白质 N-糖基化途径的关键酶，一些研究发现 N-糖蛋白不仅影响多肽的机构、生物活性、限定糖基结合物与其他蛋白、大型分子复合物以及细胞等物质的黏膜性状，还参与细胞识别和分子识别。在稻瘟菌中，敲除 α-1，3-甘露糖转移酶严重抑制次生侵染菌丝的发育，突变体在侵染寄主时引起寄主强烈的活性氧生成，扩展定殖受抑制，致病性显著降低，由其介导的效应子糖基化作用是突破寄主防御反应的关键。其在尖孢镰刀菌甜瓜专化型中的同源基因因 T-DNA 插入突变体则完全丧失对寄主的致病性（未发表）。但甘露糖转移酶在假禾谷镰刀菌中生理功能及在致病过程中的作用尚不清楚。

作者基于假禾谷镰孢菌侵染小麦的转录组数据，分析了编码 α-1，3-甘露糖转移酶的同源基因 *FpALG*3 侵染诱导上调表达，推测其可能在侵染过程中发挥相似的功能。采用 Split-Marker 基因敲除方法，获得了假禾谷镰孢菌 Δ*fpalg*3 缺失突变体。与野生型 WZ-8A 和异位插入突变体菌株比较，突变体 Δ*fpalg*3 在 PDA 平板上菌落形态不规则，色素沉淀呈黄色。不仅菌丝的扩展状态不同，而且生长速率上存在极显著差异。进一步采用菌饼直接接种大麦叶片、分生孢子悬浮液定量接种小麦胚芽鞘和小米培养物土壤接种方法测定了该突变体的致病性，发现突变体 Δ*fpalg*3 的致病性显著下降。关于突变体 Δ*fpalg*3 的互补、*FpALG*3-GFP 融合蛋白的定位和 FpAlg3 糖基化靶标的筛选和鉴定等实验正在进行当中。

关键词：假禾谷镰刀菌；功能分析；小麦茎基腐病

---

* 项目基金：国家公益性行业科研专项（201503112）；河南省科技计划项目（152300410073）和河南农业大学人才引进项目（3600861）

** 通信作者：丁胜利，教授，主要从事植物病原菌致病分子机理研究；E-mail：shengliding@ henau. edu. cn

玉米叶片中多堆柄锈菌潜伏侵染的 real-time PCR 检测和应用[*]

张克瑜[**]，马占鸿[***]
（中国农业大学植物病理系，农业部植物病理学重点开放实验室，北京 100193）

摘 要：玉米南方锈病是由多堆柄锈菌（*Puccinia polysora* Underw.）引起的一种世界范围内毁灭性的气传病害。快速及时诊断与定量检测潜伏状态下的玉米多堆柄锈菌，对玉米南方锈病流行侵染的预测预报，以及制定合理的防治措施均具有重要意义。本研究利用 real-time PCR 技术建立了一种简单易行且灵敏准确的玉米多堆柄锈菌潜育检测体系，并应用此方法对人工接种和田间的潜育期玉米南方锈病进行检测，建立病情预测方程。

根据多堆柄锈菌的 ITS 序列和玉米 *Actin* 2 基因序列，分别设计了特异性引物和 TaqMan 探针：PpoF/PpoR 和 PpoP（多堆柄锈菌）、ZmF/ZmR 和 ZmP（玉米）。经 real-time PCR 扩增对 2 对引物的特异性和灵敏性进行了测定，结果表明：2 对引物和探针对各自靶标片段均具有良好的特异性，且最低可检测出的多堆柄锈菌 DNA 浓度和玉米 DNA 浓度分别为 10^{-3} ng/μL 和 10^{-2} ng/μL。通过梯度稀释模板 DNA 建立了多堆柄锈菌和玉米各自的标准曲线，能够准确定量检测潜伏状态下的多堆柄锈菌以及玉米叶片的 DNA 含量。应用 2 对特异性引物和探针建立了玉米南方锈病潜育期的 real-time PCR 检测体系。对人工接种后不同潜育期天数的玉米叶片中病原菌的 DNA 量和叶片的 DNA 量进行 real-time PCR 检测，并在发病后不同天数调查病情，计算分子病情指数（MDI）和病情指数（DI），其中分子病情指数定义为叶片中病原菌的 DNA 量（pg）与叶片的总 DNA 量（ng）的比值，将 MDI-AUDPC 与 DI-AUDPC 进行线性回归分析，建立了病情预测方程：$y=0.7472x+212.4$（$R^2=0.8175$，$P=0.0351$）。采用同样的方法，本研究还通过对田间玉米南方锈病潜育期的分子病情检测与发病后病情的实际调查，建立了田间病情预测方程：$y=31.797x+22.501$（$R^2=0.7256$，$P=0.0017$）。为玉米南方锈病的早期检测和预测预报提供了科学合理的技术手段，并能够为该病害的早期防治提供参考依据。

关键词：玉米南方锈病；潜伏侵染；Real-time PCR

---

* 基金项目：国家重点研发计划项目（2016YFD0300702、2016YFD0201302）

** 第一作者：张克瑜，女，硕士研究生，研究方向为植物病害流行学；E-mail：zhangkeyu1994@126.com

*** 通信作者：马占鸿，男，教授，研究方向为植物病害流行学与宏观植物病理学；E-mail：mazh@cau.edu.cn

香蕉枯萎病菌分泌蛋白质的差异表达分析*

李云锋[1]，周　淦[1]，周玲菀[1]，王振中[1]，聂燕芳[2**]
（1. 华南农业大学农学院；2. 华南农业大学材料与能源学院，广州　510642）

摘　要：香蕉枯萎病是香蕉生产上最重要的病害之一，其病原菌为尖孢镰刀菌古巴专化型（*Fusarium oxysporum* f. sp. *cubense*，Foc）。Foc 有 4 个生理小种，其中，1 号小种（Foc1）和 4 号小种（Foc4）在我国分布最为广泛。分泌蛋白作为一类重要的致病因子，在 Foc 侵染香蕉过程中起着重要作用。

以 Foc1 和 Foc4 为研究对象，通过巴西蕉组织提取物的诱导，采用非标记蛋白定量技术（Label-free）分析了 2 个小种的菌丝分泌蛋白质表达变化，并结合生物信息学等方法对差异表达的分泌蛋白进行了功能预测分析。结果表明：采用 Label-free 的蛋白质组学技术共鉴定了 2 573个 Foc 分泌蛋白质。与对照相比，诱导条件下的 Foc1 有 42 个分泌蛋白质差异表达，其中经典分泌蛋白有 30 个，非经典分泌蛋白有 12 个。与对照相比，诱导条件下的 Foc4 有 57 个分泌蛋白质差异表达，其中经典分泌蛋白有 43 个，非经典分泌蛋白有 14 个。

对诱导条件下 Foc1 和 Foc4 间差异表达的分泌蛋白质进行了比较分析，结果表明在 92 个差异表达的分泌蛋白中，其功能主要涉及到细胞壁降解、蛋白修饰、氧化胁迫反应过程、氧化还原过程和能量代谢等过程。其中，谷氨酰胺合成酶、丝氨酸-苏氨酸蛋白磷酸酶等仅在 Foc1 中差异表达；磷酸化激酶、角质酶、1，6-β-葡糖苷酶、1，3-β 葡糖苷酶、几丁质脱乙酰酶、羧肽酶、天冬氨酸蛋白酶等仅在 Foc4 中差异表达。根据生物学功能预测分析，推测这些分泌蛋白的表达差异可能与 2 个小种的致病力差异相关。

关键词：香蕉枯萎病菌；蛋白质；差异表达

* 基金项目：国家自然科学基金（31600663）和广东省科技计划项目（2016A020210098）

** 通信作者：聂燕芳；E-mail：yanfangnie@ scau. edu. cn

稻曲病菌中稻绿核菌素合成基因簇的初步研究*

李月娇，王 明，方安菲，张 楠，郑馨航，邱姗姗，孙文献
（中国农业大学植物保护学院，北京 100193）

摘 要：稻曲病已成为我国水稻生产上的三大病害之一。该病的症状是在稻穗上形成体积数倍于稻粒的稻曲球，造成严重的减产。此外，稻曲球中含有大量对人畜有害的毒素，其中包括一类有色的吡喃酮的衍生物——稻绿核菌素。目前，在稻绿核菌素的分离、鉴定以及生物活性的测定方面取得了重要进展。但是，对稻曲病菌中稻绿核菌素的生物合成途径及其调控机理尚不明确。本研究对通过生物信息学预测的稻绿核菌素生物合成基因簇进行了初步功能鉴定。首先，基于同源重组的正负向筛选敲除体系以及农杆菌介导转化法获得了稻曲病菌 P1 菌株的 *UV_* 2080、*UV_* 2083、*UV_* 2086、*UV_* 2090 和 *UV_* 2091 等 5 个基因的敲除突变体；其次，对这些基因敲除体进行生长表型和毒素合成能力的测定，发现 Δ*UV_* 2086、Δ*UV_* 2090 和 Δ*UV_* 2091 丧失了稻绿核菌素的合成能力，但在 Δ*UV_* 2080 和 Δ*UV_* 2083 中仍能检测到稻绿核菌素。Δ*UV_* 2083 较野生型菌株生长速率显著增加；Δ*UV_* 2086 相较于野生型菌株在固体培养基上颜色明显变浅，而 Δ*UV_* 2080、Δ*UV_* 2083、Δ*UV_* 2090 和 Δ*UV_* 2091 菌株颜色相对于野生型菌株有不同程度的变化，推测可能是该毒素生物合成过程中积累了有色的中间产物。后续将对稻绿核菌素的生物合成路径以及合成基因的表达调控进行深入研究。

关键词：稻绿核菌素；基因敲除；毒素检测

---

* 基金项目：国家自然科学基金项目（No. 31471728）

Title: Population Genetic Study on *Puccinia striiformis* f. sp. *tritici* in Ethiopia and Kenya

Jiang Shuchang, Zhan Gangming, Meng Yan, Huang Lili, Kang Zhensheng

(*State Key Laboratory of Crop Stress Biology for Arid Areas*, *College of Plant Protection*, *Northwest A&F University*, *Yangling* 712100, *Shaanxi*, *China*)

Abstract: Stripe rust or yellow rust of wheat, caused by *Puccinia striiformis* f. sp. *tritici* (*Pst*), is an important disease in Ethiopia and Kenya. The high frequency of wheat stripe rust epidemics caused severe loss in yield and quality under favorable environmental conditions. To investigate the virulent component and population structure of the pathogen, 96 single-pustule isolates collected from Ethiopia and Kenya were tested on 24 Near Isogenic Lines (NILs) differentials and 5 Chinese supplementary differentials and genotyped at 12 simple-sequence repeat loci. Virulence complexity of the isolates ranged from 5 to 22 with a mean of 15. 3. In total, virulence frequency was higher than 80% to *Yr*2 (100%), *Yr*6 (100%), *Yr*7 (100%), *YrA*, *Yr*9, *Yr*17, *Yr*25, *Yr*27 and *YrExp*2. Virulence frequency was lower than 40% to *Yr*5, *Yr*15, *YrSp*, *YrTye*, *Yr*3, *Yr*4, *Yr*10, *Yr*24, *Yr*32, and *YrSu*. Virulence and SSR data had no correlation. Both virulence and molecular data revealed that the genetic diversity gradually decreased from north to south in East Africa. In other words, genetic diversity of *Pst* in Ethiopia was higher than Kenya. According to the Wei' s θ, wheat stripe rust populations in four regions (Gonder, Shoa, Bale and Nairobi) were different significantly. AMOVA analysis revealed that the majority (99. 0%; $P=0.012$) of the molecular variation was within the geographic populations. The asexual reproductive mode was indicated by the excess of observed heterozygosity.

Key words: Wheat stripe rust; Genetic diversity; AMOVA; Asexual reproduction

Basidiomycete-specific *PsCaMKL*1 Encoding a CaMK-like Protein Kinase is Required for Full Virulence of *Puccinia striiformis* f. sp. *tritici*

Jiao Min, Kang Zhensheng, Guo Jun

(*State Key Laboratory of Crop Stress Biology for Arid Areas*, *College of Plant Protection*, *Northwest A&F University*, *Yangling* 712100, *Shaanxi*, *China*)

Abstract: Calcium/calmodulin-dependent kinases (CaMKs) are Ser/Thr protein kinases (PKs) that respond to changes in cytosolic free Ca^{2+} and play diverse roles in eukaryotes. In fungi, CAMKs are generally classified into four families CAMK1, CAMKL, RAD53, and CAMK-Unique. Among these, CAMKL constitutes the largest family. In some fungal plant pathogens members of the CaMKL family have been shown to be responsible for pathogenesis. However, little is known about their role (s) in rust fungi. In this study, we functionally characterized a novel PK gene, *PsCaMKL*1, from *Puccinia striiformis* f. sp. *tritici* (*Pst*). *Ps*CaMKL1 belongs to a group of PKs that is evolutionarily specific to basidiomycete-ous fungi. *Ps*CaMKL1 shows little intra-species polymorphism and cytoplasmic localization in wheat protoplasts. *PsCaMKL*1 transcripts are highly elevated at early infection stages, whereas gene expression is down-regulated in barely germinated urediospores under KN93 treatment. Overexpression of *PsCaMKL*1 in fission yeast increased resistance to environmental stresses. Knock down of *PsCaMKL*1 using host-induced gene silencing (HIGS) reduced the virulence of *Pst* accompanied by reactive oxygen species (ROS) accumulation and a hypersensitive response. These results suggest that *Ps*CaMKL1 is a novel pathogenicity factor that exerts it virulence function by regulating ROS production in wheat.

Key words: CaMKs; *Pst*; *PsCaMKL*1; HIGS; ROS

中国小麦条锈菌 CYR32 和 CYR33 的毒性和遗传多样性分析

王洁荣，詹刚明，卢 霞，杨彩柏，黄丽丽，康振生
（旱区逆境生物学国家重点实验室，西北农林科技大学植物保护学院，杨凌 712100）

摘 要：由担子菌条形柄锈菌（*Puccinia striiformis* Westend f. sp. *tritici*）引发的小麦条锈病是小麦种植区的毁灭性病害。根据在中国鉴别寄主不同小麦品种上的致病特征，将小麦条锈菌命名为致病类型或生理小种。至今，中国依次命名了 34 个生理小种，即条中 1 号–条中 34 号（CYR1-CYR34）。自 2000 年以来，CYR32 和 CYR33 一直是我国出现频率最高的条锈菌生理小种，然而，多种证据表明中国小麦条锈菌存在有性过程，生理小种毒性存在频繁变异，进而说明这套中国鉴别寄主对小种的区分力度不足，未能及时反映条锈菌新的毒性变异，不能及时为抗病育种提供有效信息。当前，国际小麦条锈菌毒性鉴定中，一套以 Avocet 为背景的近等基因系鉴别寄主除了不能区分一些不侵染感病对照 Avocet S 的菌系外，经证明可用来有效区分病原菌小种。目前国际上倾向于用这套有区分力度的近等基因系鉴别寄主来鉴定条锈菌的毒性。为了明确当前主要流行生理小种 CYR32 和 CYR33 的毒性和遗传分化情况，从中国 11 个省采取的小麦条锈菌标样在中国鉴别寄主上进行毒性鉴定后，每个省分别随机选取生理小种为 CYR32 和 CYR33 的菌株，共挑选 29 个 CYR32 和 39 个 CYR33 个菌株。用近等基因系对其进行毒性分析，发现供试的 CYR32 和 CYR33 菌系分别有 17 种和 18 种毒性表现型，并且都分别在含单个抗病基因 *Yr*2、*Yr*17、*Yr*27、*Yr*32、*Yr*43、*YrSp*、*YrExp*2、*Yr*28、*YrV*23 小麦材料上发生了分化，毒性多样性指数（*K*）和 Nei´s 多样性指数（*H*）分别是 0. 089、0. 075 和 0. 101、0. 081。当相似性系数为 0. 93 时，CYR32 被聚为 5 个毒性类群，CYR33 被聚为 8 个毒性类群。再用 SSR 分子标记对其遗传结构进行分析，发现 CYR32 和 CYR33 菌系的 Shannon 信息指数（*I*）分别是 0. 44 和 0. 45，当相似性系数为 0. 81 时，CYR32 被聚为 9 个基因型类群，CYR33 被聚为 12 个基因型类群。上述结果表明这两个优势生理小种 CYR32 和 CYR33 内部存在高度分化，因此在条锈菌菌系毒性鉴定中有必要用近等基因系鉴别寄主进行进一步区分，以便在系统了解小麦条锈菌毒性结构基础上，采取针对性的抗性基因合理布局，以延长抗病品种的使用年限，在全面“双减”的基础上，有效防控小麦条锈病，达到稳产、增产的目的。

关键词：小麦锈病；CYR32；CYR33；毒性鉴定；多样性分析

Function of the Ss-SFH1 Transcription Factor in *Sclerotinia sclerotiorum*[*]

Liu Ling, Wang Qiaochu, Pan Hongyu[**]

(*College of Plant Sciences*, *Jilin University*, *Changchun*, 130062, *China*)

Abstract: *Sclerotinia sclerotiorum* (Lib.) de Bary is a necrotrophic plant pathogen with a worldwide distribution. Here, a gene named *Ss-Sfh*1 was cloned in *S. sclerotiorum*, which encodes a snf5-box-containing protein. The predicted Ss-SFH1 protein also has a GATA Zn-finger domain in a C-terminal region. To investigate the role of Ss-SFH1 in *S. sclerotiorum*, the partial sequence of Ss-SFH1 was cloned and RNA interference (RNAi) -based gene silencing method was employed to alter the expression of Ss-SFH1. RNA silenced mutants Ss-SFH1 RNA levels reduced exhibited slow hyphal growth. Appressoria and oxalic acid accumulation were reduced in *Ss-Sfh*1 RNAi mutants. Disease assays demonstrated that pathogenicity in RNAi-silenced strains was significantly compromised with the development of a smaller infection lesion on soybean leaves. In addition, *Ss-Sfh*1-silenced transformants were more tolerant to oxidative stress compared with the wild-type strain. Furthermore, the expression levels of putative probably involved in ROS production-related genes were significantly different in silenced strains. All the results suggest that Ss-SFH1 is involved in hyphal growth, virulence and the tolerance to oxidative stress in *S. sclerotiorum*.

* Funding: This work was financially supported by the National Natural Science Foundation of China (31271991, 31471730)

** Corresponding author: Pan Hongyu, E-mail: panhongyu@ jlu. edu. cn

Gene Differential Expression Analysis of the Interaction between *Sclerotinia sclerotiorum* and *Glycine max*[*]

Wang Qiaochu, Cheng Hailong, Liu Ling, Liu Jinliang,
Zhang Yanhua, Zhang Xianghui, Pan Hongyu[**]
(*College of Plant Sciences, Jilin University, Changchun, China*)

Abstract: *Sclerotinia sclerotiorum* (Lib.) de Bary is a filamentous pathogenic fungus. It has broad host range and causes substantial losses of important crops. In this study, a transcriptional factor Ss-Nsd1 gene from *S. sclerotiorum* was cloned, then *Ss-Nsd*1 gene knock-out mutant was obtained. Infection cushion was not found in *Ss-Nsd*1 knock-out mutant, meanwhile the mutant lost the ability to invade the host plant. However the secretion of toxins such as oxalic acid was not affected.

Toillustrate the forming pathway of infection cushion and find functional genes in primary infection, RNA-seq technique was used to search differentially expressed genes. Fourteen soybean varieties were screened, and susceptible variety Soybean No. 1 and the resistant variety Jennon 28 were selected as experimental hosts for this study. In initial formation period the infection cushion was found in 24h after the infection. The strains of wild type UF−70 and the mutant ΔSs-Nsd1 infected two soybean varieties in 24h and 48h respectively, the RNA-sequencing was proceeded.

The results of transcriptional analysis showed that there were 516 differentially expressed genes in the initial infection stage of *S. sclerotiorum*, in which 293 genes related to molecular function. Among 293 genes, there are 23 transmembrane transporters, 28 zinc ion bindings, 26 dehydrogenases, 3 ligases, 29 transferases, 2 peroxidase, 14 transcription factors and so on; Among 516 differentially expressed genes, there are 278 biological process related genes, 245 metabolic related genes, 106 cellular components related genes including 42 membrane composition related genes. On the basis of these results, virulent genes of *S. sclerotiorum* were deeply excavated from the hydrolytic enzymes, detoxification, secondary metabolites biosynthesis, oxalic acid production, and generation of reactive oxygen species.

* Funding: This work was financially supported by the National Natural Science Foundation of China (31471730)

** Corresponding author: Hongyu Pan, E-mail: panhongyu@ jlu. edu. cn

Screening and Identification Proteins that Interact with Transcription Factor SsMCM1 in *Sclerotinia sclerotiorum*[*]

Xu Tingtao, Liu Xiaoli, Liu Ling, Wang Qiaochu,
Liu Jinliang, Zhang Xianghui[**], Pan Hongyu, Zhang Yanhua
(*College of Plant Sciences*, *Jilin University*, *Changchun* 130062, *China*)

Abstract: The *Sclerotinia sclerotiorum* (Lib.) de Bary is one of agriculture's most devastating necrotrophic fungal plant pathogens. It infects 408 different plant species from 278 genera encompassing 75 plant families and causes unrestricted lesion development and tissue maceration of its hosts.

One family of transcription factors that are well conserved in eukaryotic organisms is MADS-box proteins. Its founding members are MCM1 (yeast), Agamous (plant), Deficiens (Drosophila) and serum response factor (SRF, human). The putative *SsMCM*1 gene was cloned in our lab, and it was highly similar to the orthologues *S. cerevisiae* Mcm1, including a conserved DNA-binding domain. SsMCM1 function was investigated using RNA interference. Our results suggest that the MADS-box transcription factor SsMCM1 is involved growth and virulence in *S. sclerotiorum*.

Considering its importance, SsMCM1 may also interact and form heterodimers with other protein in *S. sclerotiorum* to regulate growth and virulence. Yeast two-hybrid and Bimolecular Fluorescent Complementary have been applied to identify proteins that physically associate with protein SsMCM1 in *S. sclerotiorum*. 12 putative SsMCM1 interacting proteins were identified. SsIP37 contains the Pmp3 family domain; SsIP43 contains the PGM3 family domain; SsIP44 contains the Mpv17-PMP22 domain; SsIP86 contains the Catalase domain; SsIP98 contains the Cyclin domain; SsIP106 contains the RGL11 domain. The other interacting proteins function are unknown. SsCdc28 (SSIG_ 02296) was cloned which highly homologous to yeast Cdc28. Yeast two-hybrid assays indicating that Cdc28 physically interacts with SsIP86. This study clarified the interaction between SsMCM1-SsIP86-SsCdc28. It will be important to reveal the regulatory network and regulation mechanism of SsMCM1 transcription factor in *S. sclerotiorum*.

* Funding: This work was supported in part by the National Natural Science Foundation of China (No. 31101394, 31471730)

** Corresponding author: ZHANG Yanhua, E-mail: yh_ zhang@ jlu. edu. cn

西藏的木霉多样性初探

王洪凯

（浙江大学生物技术研究所，杭州　310058）

摘　要：笔者从西藏采集到土样200份，地点分布在西藏的7个市（州）。从土样中分离到木霉168株，通过形态学和分子生物学相结合的方法进行鉴定，发现有18个种，分别是 *Trichoderma afroharzianum*, *T. alni*, *T. asperellum*, *T. citrinoviride*, *T. erinaceus*, *T. hamatum*, *T. harzianum*, *T. koningii*, *T. koningiopsis*, *T. longibrachiatum*, *T. paraviridescens*, *T. polysporum*, *T. saturnisporum*, *T. stramineum*, *T. tomentosum*, *T. velutinum*, *T. viride*, *T. viridescens*。其中 *T. harzianum* 和 *T. koningii* 是优势种。此结果为进一步研究开发这些木霉菌打下了基础。

关键词：西藏；木霉；多样性

第二部分　卵　菌

2016年福建福清致病疫霉群体遗传多样性SSR分析

谢家慧*，沈林林，姜艳坤，王　甜，詹家绥**
（福建农林大学植物病毒研究所，福州　350012）

摘　要：致病疫霉（*Phytophthora infestans*）引起的晚疫病是马铃薯的毁灭性灾害，明确致病疫霉的遗传多样性能为马铃薯晚疫病的防治提供理论基础。本试验采用8对SSR引物对2016年4月初采自福清4个田块的182个致病疫霉样本进行遗传多样性分析，实验结果检测出共有25个等位基因和42个基因型。平均期望杂合度为0.414 1，平均香农指数为0.652 9，总体遗传多样性水平中等。其中，C田块的平均期望杂合度最高，为0.464 3，说明遗传多样性也较高于其他田块。

关键词：致病疫霉；SSR；遗传多样性

---

* 第一作者：谢家慧，女，硕士研究生，主要从事分子植物病理学研究；E-mail：bbxjh1994@163.com
** 通信作者：詹家绥，教授，主要从事群体遗传学；E-mail：Jiasui.zhan@fafu.edu.cn

A *Phytophthora capsici* virulent RXLR effector targets plant PP2a isoforms that confer Phytophthora blight resistance

Chen Xiaoren[1,2,3*], Li Yanpeng[1], Xing Yuping[1], Zhai Yi[2,3], Ma Wenbo[2,3]

(1. *Department of Plant Protection*, *Yangzhou University*, *Yangzhou*, *China* 225009; 2. *Department of Plant Pathology and Microbiology*, *University of California*, *Riverside*, *CA* 92521; 3. *Center for Plant Cell Biology*, *University of California*, *Riverside*, *CA* 92521)

Abstract: Plant pathogens deliver an array of effectors to alter host physiology and defense responses. To understand the molecular basis underlying these, it is important to identify and characterize the target proteins of pathogen effectors in plants. Here we show that the RXLR effector PcAvh1 of *Phytophthora capsici* is highly conserved across *Phytophthora* genus. The encoding gene *PcAvh*1 is upregulated during the plant infection stages versus barely expressed during the developmental stages. The effector is important for the pathogen virulence on pepper and *Nicotiana benthamiana* via activity in the host nucleus and cytoplasm and of RNA silencing suppression. PcAvh1 interacts with two host protein phosphatase 2a (PP2a) structural isoforms. Silencing of the *PP2a* isoforms facilitates pathogen infection. Furthermore, silencing of the isoforms together severely compromises the growth of plants. These results demonstrate that host PP2a activity is required for the plant resistance against the pathogen. Taken together, we conclude that PcAvh1 is an important virulence determinant that targets host PP2a isoforms to attenuate plant resistance.

* 通信作者：陈孝仁，副教授，主要从事植物病原真菌学研究；E-mail：xrchen@ yzu. edu. cn

Pectin Acetylesterase is Required for the Virulence of Oomycete Plantpathogen *Peronophythora litchii*

Kong Guanghui, Wan Lang, Yang Wensheng, Li Wen, Jiang Liqun, Situ Junjian, Li Minhui, Xi Pinggen, Jiang Zide

(*Department of Plant Pathology/Guangdong Province Key Laboratory of Microbial Signals and Disease Control*, *South China Agricultural University*, *Guangzhou* 510642, *China*)

Abstract: Pectin is a major component of the primary cell wall of higher plants. Some galacturonyl residues in the backbone of pectinaceous polysaccharides are of ten *O*-acetylated, and the resulting acetylesters change dynamically. This processes involve both enzymatic acetylation and deacetylation. Our study found pectin acetylesterases (PAEs), which can remove of acetyl-substituents, are widespread in the plant pathogenic oomycete and showed sequence and transcriptional polymorphism. We also characterized the functions of *PlPAE*4 and *PlPAE*5 from *P. lithii* and found that *PlPAE*5 knockout mutantsinoculated into lychee were less invasive than the wild-type *P. litchii* strain SHS3, demonstrating PAE5 functions in the *P. litchii* infection. Ectopic expression of PAE5 on *Nicotiana benthamiana* could promote the infection of *Phytophthora capsici*. This study firstly reported that PAE protein is involved in the infection of oomycete plant pathogen.

马铃薯晚疫病 LAMP 检测方法研究[*]

林桐司骐[**]，苏　瑞，谢欣娱，叶芯妤，王伟伟[***]，唐　唯[***]
（云南师范大学马铃薯科学研究院，昆明　650500）

摘　要：马铃薯晚疫病（Potato late blight）是一种为害马铃薯生产的毁灭性病害，病原为 *Phytophthora infestans*，属异宗配合卵菌。此病害造成的经济损失巨大，因此，对该病原的前期检测非常重要。本研究基于 LAMP（Loop-mediated isothermal amplification）环介导等温扩增技术，用马铃薯晚疫病菌 510bp ITS（Internal Transcribed Spacer）区域（KU992300. 1）片段，设计出 1 对 LAMP 引物，包括一对外引物（F3 和 B3）和一对内引物（FIP 和 BIP）。引物经特异性检测后，对 LAMP 反应条件进行了优化，最终确定为：25μL LAMP 反应体系在 62℃ 条件下反应 45min。产物直接加入稀释 100 倍的 SYBR Green I 荧光染料，可立即在自然光或紫外光下对 LAMP 反应产物进行可视化检测。若荧光染料呈现绿色，则为阳性，若荧光染料呈现橙黄色，则为阴性。本研究所用 LAMP 技术与普通 PCR 扩增检测方法相比，灵敏度较高，DNA 的检测下限为 8×10^{-5}ng。在 2016 年 7—9 月，对云南省丽江市太安地区随机调查晚疫病发病情况。在田间未见晚疫病症状时，采集叶片进行 LAMP 检测，实验结果表明：LAMP 检测率和采集后第 7d、第 15d 的实际发病率与检测率的相关性（R^2）分别为 0. 670 6和 0. 783 2。因此该方法可用于快速预测田间马铃薯晚疫病发病趋势，也有助于检测种薯储藏期晚疫病害的预期发生情况。因 LAMP 仅需恒温条件，其相关适用于田间操作的简易 DNA 提取方法，以及移动式 LAMP 反应平台的搭建正在进一步完善中。

关键词：马铃薯；ITS；晚疫病；LAMP

---

* 基金项目：云南省大学生科技创新项目（田间快速检测马铃薯晚疫病菌的新方法）

** 第一作者：林桐司骐，生物科学专业三年级本科生；E-mail：even. lintong@ foxmail. com

*** 通信作者：王伟伟，在读硕士研究生，主要从事马铃薯晚疫病研究；E-mail：1174043860@ qq. com
唐唯，博士，讲师，主要从事马铃薯晚疫病防治研究；E-mail：4497049@ qq. com

Enhancement of Tomato Resistance to *Phytophthora infestans* Via Silencing of Tomato MiR482b With Short Tandem Target Mimic[*]

Jiang Ning[1**], Meng Jun[2***], Cui Jun[1], Yang Guanglei[1], Xiao Yu[1], Han Lu[1], Hou Xinxin[1], Luan Yushi[1***]

(1. *School of Life Science and Biotechnology, Dalian University of Technology, Dalian*, 116024, *China*; 2. *School of Computer Science and Technology, Dalian University of Technology, Dalian* 116024, *China*)

Abstract: Small Tandem Target Mimic (STTM) developed in our lab is a highly effective technology that can target and degrade/block miRNAs, and concomitantly up-regulate all target genes by expression of a simple small artificial RNA structure. This approach is especially useful in revealing the overall functions of specific miRNAs that have multiple target genes. In our previous work, using miRNA and degradome datasets, the tomato miR482b targeting mRNAs of *NBS-LRR* disease-resistance genes were validated. Overexpression tomato miR482b and NBS-LRR gene enhanced the tomato susceptibility and resistance to *Phytophthora infestans* infection, respectively. In this study, the STTM482b was used to investigate the role of miR482b in tomato−*P. infestans* interaction. STTM482b was transformed into tomato and three transgenic tomato lines were selected by kanamycin and validated by PCR. Compared to the wild-type tomato plant, the miR482b was significantly in less abundant in STTM482b transgenic tomato plant, while the expression of its target gene *NBS-LRR* was significantly increased. In STTM482b transgenic tomato plant, the genes encoding proteins related to plant-pathogen interaction and biosynthesis of archetypal defense hormones were in increased expression, suggesting that silencing of miR482b might enhance the resistance of tomato to *P. infestans*. Transgenic tomato plants that overexpressed the STTM482b displayed less serious disease symptoms than wild-type tomato plants after infection with *P. infestans*, as shown by fewer number of necrotic cells, shorter lesion sizes and less disease index. Our results further provide insight into miR482b involved in the response of tomato to *P. infestans* infection, demonstrate that silencing miR482b by STTM increases the expression of genes associated with defense response and defense-related hormones biosynthesis to enhance the resistance of tomato to *P. infestans*, and provide candidates for breeding to enhance biotic stress-resistance in tomato.

---

* Funding: National Natural Science Foundation of China (Nos. 31471880 and 61472061)

** First author: Jiang Ning, Ph. D candidate; E-mail: jiangning@mail.dlut.edu.cn

*** Corresponding authors: Luan Yushi; E-mail: luanyush@dlut.edu.cn; Meng Jun; E-mail: mengjun@dlut.edu.cn

大豆疫霉对噻唑菌胺抗性风险评估*

彭 钦**，方 媛，薛昭霖，刘西莉***
（中国农业大学植物病理学系，北京 100193）

摘 要：噻唑菌胺（ethaboxam）是韩国 LG 生命科学公司（原 LG 化学有限公司）开发并于 1998 年首先在韩国获准登记的噻唑酰胺类杀菌剂，作用机制尚不明确。其防治对象主要是致病疫霉（*Phytophthora infestans*）、葡萄霜霉（*Plasmopara viticola*）、辣椒疫霉（*Phytophthora capsici*）和古巴假霜霉（*Pseudoperonospora cubensis*）等植物病原卵菌。本研究建立了大豆疫霉对噻唑菌胺的敏感基线，评估了大豆疫霉对噻唑菌胺的抗性风险，为生产中进行抗性监测和药剂合理使用提供了参考。

研究中采用菌丝生长速率法，测定了采自全国不同省区的 112 株大豆疫霉对噻唑菌胺的敏感性。结果表明：供试大豆疫霉对噻唑菌胺的 EC_{50} 分布于 0.018～0.049μg/mL，平均 EC_{50} 值为（0.033±0.007）μg/mL。研究显示，试验获得的大豆疫霉对噻唑菌胺的敏感性频率分布呈单峰曲线，未发现敏感性下降的亚群体。因此，可将其作为大豆疫霉对噻唑菌胺的敏感性基线，该结果为田间抗药性监测提供了参考和依据。

采用药剂驯化的方法筛选获得了 6 株大豆疫霉对噻唑菌胺的抗性突变体。突变体抗性倍数介于 6.67～20.62 倍，经 10 代继代培养后，突变体抗性倍数稳定。生物学性状研究表明，突变体的菌丝生长速率、产孢能力和致病力等显著低于亲本，表明突变体的生存适合度较低。同时，交互抗药性研究结果表明：噻唑菌胺与苯酰菌胺具有正交互抗药性，与氟吗啉、甲霜灵、霜脲氰和氟噻唑吡乙酮没有交互抗药性。结合大豆疫霉的传播方式，抗性菌株的突变频率以及生存适合度，综合评估大豆疫霉对噻唑菌胺具有低等抗性风险。

---

* 基金项目：公益性行业（农业）科研专项（编号：201303023）

** 第一作者：彭钦，在读博士研究生；E-mail：pengqin1991@126.com

*** 通信作者：刘西莉，女，教授，主要从事植物病原菌与杀菌剂互作研究；E-mail：seedling@cau.edu.cn

Transcription Profiling and Identification of Infection-Related Genes in *Phytophthora cactorum*

Chen Xiaoren *, Huang Shenxin, Zhang Ye, Sheng GuiLin,
Zhang Boyue, Li Qiyuan, Zhu Feng, Tong Yunhui, Xu Jingyou
(*College of Horticulture and Plant Protection*, *Yangzhou University*, *Yangzhou* 225009, *China*)

Abstract: *Phytophthora cactorum* is an economically important oomycete that infects an extremely wide range of hosts within several plant families. To gain insight into the repertoire of the infection-related genes used by *P. cactorum*, we performed massively parallel sequencing of cDNA derived from three important life cycles of *P. cactorum*, mycelia (MY), zoospores (ZO) and germinating cysts with germ tubes (GC). From over 9.8 million Illumina reads for each cDNA library, 18 402, 18 569 and 19 443 distinct genes were identified for MY, ZO and GC libraries respectively. Comparative analysis between two samples showed major differences between the expressed gene content of MY, ZO and GC stages. A large number of genes associated with specific stages and pathogenicity were identified, including 166 predicted effector genes. Of them, 14genes encoding small cysteine-rich (SCR) secretory proteins showed differential expression during *P. cactorum* developmental stages and *in planta*. Ectopic expression in the Solanaceae indicated that SCR113 and one elicitin PcINF1 can trigger plant cell death on *Nicotiana benthamiana*, *N. tabacum* and *Solanum lycopersicum*. No conserved domain or homologues of SCR113 in other organism can be identified. Our data provide the most comprehensive sequence resource available for *P. cactorum* study and will help define the pathogenicity mechanism of this broad-host-range pathogen.

* 通信作者：陈孝仁，男，副教授，主要从事植物真菌病害研究；E-mail：xrchen@ yzu. edu. cn

外源甾醇对辣椒疫霉生长发育的作用*

王为镇**，李腾蛟，刘西莉***
（中国农业大学植物保护学院植物病理系，北京　100193）

摘　要：辣椒疫霉（*Phytophthora capsici*）是一种重要的植物病原卵菌，其寄主范围广泛，可对农业生产造成严重的经济损失。已有研究表明，疫霉属卵菌因其基因组缺乏甾醇合成必要的基因，自身不能合成甾醇。也有研究表明疫霉可以利用外源的甾醇类物质。本课题组利用不含有任何甾醇的疫霉基础培养基进行辣椒疫霉的培养，并在收集菌丝后进行甾醇类物质的提取，通过 GC-MS 进行检测，也未能检测到任何甾醇类物质。

基于上述研究背景，本研究通过外源添加的方法研究了 4 种外源甾醇对辣椒疫霉生长发育的作用。研究结果表明：辣椒疫霉在不添加甾醇的基础培养基上依然可以进行营养生长，表明辣椒疫霉的营养生长不依赖甾醇类物质。但胆固醇、β-谷甾醇、豆甾醇在供试浓度范围内对辣椒疫霉的菌丝生长具有明显的促进作用，在一定范围内该促进作用随外源甾醇浓度的提高而增强；麦角甾醇对辣椒疫霉菌丝生长的作用表现为低浓度（<1μg/mL）促进，高浓度（>10μg/mL）抑制。另外，辣椒疫霉在不添加外源甾醇的基础培养基上，不能产生游动孢子，而 4 种外源甾醇均可以显著促进游动孢子的产生，其中以胆固醇促进作用最弱，而 β-谷甾醇、豆甾醇和麦角甾醇对辣椒疫霉游动孢子产生的促进作用相当，均明显优于胆固醇。

辣椒疫霉的寄主体内含有甾醇类物质，推测辣椒疫霉在侵染寄主时可利用寄主体内的植物甾醇，从而调控其自身的生长发育。

关键词：辣椒疫霉；外源甾醇；生长

---

* 基金项目：公益性农业行业科研专项（201303023）

** 第一作者：王为镇，博士研究生；E-mail：wzwangyx163@163.com

*** 通信作者：刘西莉，教授，主要从事植物病原菌与杀菌剂互作研究；E-mail：seedling@cau.edu.cn

Network and role Analysis of Autophagy in *Phytophthora sojae**

Chen Linlin[1], Zhang Xiong[2], Wang Wen[2], Dou Daolong[2], Li Honglian[1]

(1. *Department of Plant Pathology*, *Henan Agricultural University*, *Zhengzhou* 450002, *China*;
2. *Department of Plant Pathology*, *Nanjing Agricultural University*, *Nanjing* 210095, *China*)

Abstract: Autophagy is an evolutionarily conserved mechanism in eukaryotes with roles in development and the virulence of plant fungal pathogens. However, few reports on autophagy in oomycete species have been published. Here, we identified 26 autophagy-related genes (*ATGs*) belonging to 20 different groups in *Phytophthora sojae* using a genome-wide survey, and core ATGs in oomycetes were used to construct a preliminary autophagy pathway model. Expression profile analysis revealed that these ATGs are broadly expressed and that the majority of them significantly increase during infection stages, suggesting a central role for autophagy in virulence. Autophagy in *P. sojae* was detected using a GFP-PsAtg8 fusion protein and the fluorescent dye MDC during rapamycin and starvation treatment. In addition, autophagy was significantly induced during sporangium formation and cyst germination. Silencing *PsAtg6a* in *P. sojae* significantly reduced sporulation and pathogenicity. Furthermore, a *PsAtg6a*-silenced strain showed haustorial formation defects. These results suggested that autophagy might play essential roles in both the development and infection mechanism of *P. sojae*.

* 基金项目：国家公益性行业科研专项（201503112）

大豆疫霉菌效应分子 Avh5 的植物靶标筛选

靳雨婷，刘美彤，王群青
（山东农业大学植物保护学院，泰安 271018）

摘 要：大豆疫霉菌（*Phytophthora sojae*）引起的大豆根茎腐病是大豆生产上的毁灭性病害，每年在全球范围内造成的经济损失高达数十亿美元。大豆疫霉是兼性寄生卵菌，在侵染前期营寄生生活，亲和互作条件下侵入寄主约 24h 之后转入腐生生活。在寄生阶段，大豆疫霉通过分泌效应分子蛋白干扰植物免疫反应，抑制植物的程序性细胞死亡（Programmed Cell Death，PCD），同时依靠吸器从寄主活体细胞中获取营养和水分，建立寄生关系。因此，抑制植物的 PCD 反应是大豆疫霉菌在侵染前期的主要策略。研究发现以 Avh5 为代表的大豆疫霉菌 RXLR 效应分子家族普遍具有一个 W-domain，与抑制植物的 PCD 有关。生物信息学分析发现 W-domain 的蛋白结构由两个 α 螺旋组成，通过实验证明位于第二个 α 螺旋上的 4 个疏水性氨基酸对于 Avh5 抑制植物 PCD 的功能起关键作用。三维蛋白结构分析显示 Avh5 蛋白 C 端 W-domain 及周围的氨基酸共同形成了一段钳形的内嵌结构，这种结构很有可能与其靶标直接互作有关。因此本研究通过酵母双杂交筛选 Avh5 的植物靶标，共转化 Avh5 和烟草 cDNA 文库到酵母细胞，在高严谨性筛选平板上共获得了 6 个阳性克隆，其中在两个克隆中鉴定出了一个 bHLH（basichelix-loop-helixprotein）转录因子。bHLH 是真核生物中存在最广泛的一大类转录因子，其通过特定的氨基酸残基与靶基因相互作用，进而调节相关基因的表达，对于真核生物的正常生长及发育是必不可缺的。已经有报道在绒毡层中特异性表达的 bHLH 转录因子 EAT1 参与调节植物发育性的程序性细胞死亡的新机制，因此效应分子的 W-domain 极有可能通过干扰 bHLH 转录因子的功能达到抑制植物 PCD 的目的。本研究将进一步解析疫霉菌 RXLR 效应分子 W-domain 参与抑制植物 PCD 的机制，为制定疫霉菌防治策略和针对性的开发农药靶标提供理论基础。

关键词：大豆疫霉菌；植物靶标；筛选

Transcriptome Signatures in Tomato Leaf Allows Identification of Transcription Factor *SpWRKY3* Conferring Resistance to *Phytophthora infestans* in Tomato*

Cui Jun[1**], Meng Jun[2], Li Jingbin[1], Jiang Ning[1], Luan Yushi[1***]

(1. *School of Life Science and Biotechnology*, *Dalian University of Technology*, *Dalian*, 116024, *China*; 2. *School of Computer Science and Technology*, *Dalian University of Technology*, *Dalian* 116024, *China*)

Abstract: Transcription factors (TF) play crucial roles in the plant response to various pathogens. In this present study, we used comparative transcriptome analysis of tomato inoculated with and without *Phytophthora infestans* to identify 1103 differentially expressed genes. After GO enrichment analysis, 7 GO terms (level 4) associated with plant resistance to pathogen were identified. After expression pattern analysis, it was found that 34 selected transcription factor (TF) genes from GO enriched term, sequence-specific DNA binding transcription factor activity were induced by *P. infestans*. Of these TFs, a homologous gene of *WRKY* named *SpWRKY3* was isolated form *P. infestans*-resistant tomato, *Solanum pimpinellifolium* L3708. *SpWRKY3*expression was significantly induced by treatment with salt, drought, salicylic acid, methyl jasmonate and abscisic acid. Overexpression of *SpWRKY3* in tomato positively modulated *P. infestans* defense response as shown by decreased number of necrotic cells, lesion sizes and disease index, while the resistance was impaired after *SpWRKY3* silencing. After *P. infestans* inoculation, the expression levels of *PR* genesin transgenic tomato plants that overexpressed *SpWRKY3*were significantly higher than those in WT, while number of necrotic cells and ROS accumulation were fewer and lower. This result suggests that SpWRKY3 induces *PR* gene expression and reduces ROS accumulation to protect against cell membrane injury, leading to enhanced resistance to *P. infestans*. Our results provide insight into *SpWRKY3* as a positive regulatorinvolved in the response of tomato to *P. infestans* infection.

---

* Funding: National Natural Science Foundation of China (Nos. 31471880 and 61472061)

** First author: Cui Jun, Ph. D candidate; E-mail: cuijun@ dlut. edu. cn

*** Corresponding authors: Luan Yushi; E-mail: luanyush@ dlut. edu. cn; Meng Jun; E-mail: mengjun@ dlut. edu. cn

安徽省大豆疫霉对氟吗啉的抗药性监测及遗传研究*

刘腾飞，丁　旭，刘　枭，赵振宇，陈方新，潘月敏，高智谋**
（安徽农业大学植物保护学院，合肥　230036）

摘　要：大豆疫霉（*Phytophthora sojae*）是引致大豆疫病的重要卵菌。氟吗啉是我国自主研发的用于防治卵菌病害的有效杀菌剂。迄今关于大豆疫霉对氟吗啉的抗药性监测和遗传研究较少。为了阐明大豆疫霉对氟吗啉的抑制机理及遗传规律，为生产上合理使用氟吗啉防治大豆疫病及对氟吗啉的抗药性治理提供试验依据，作者对安徽省大豆疫霉菌株对氟吗啉的抗药性进行了监测，并进一步研究了大豆疫霉菌株对氟吗啉的抗药性遗传及其稳定性，主要结果如下。

通过对安徽省 43 个大豆疫霉菌株进行了敏感性检测，结果表明：43 个大豆疫霉菌株的 EC_{50} 值的范围在 0. 075 8～0. 183 9μg/mL，平均值为 0. 1213μg/mL，均小于 1μg/mL。可以看出，安徽省的 43 个大豆疫霉菌对氟吗啉是敏感的，没有抗药性的菌株。抗性监测结果显示，安徽省的 42 株大豆疫霉菌株，根据采集地的不同，对氟吗啉的敏感性程度也是有一定差异。来自太和地区的 TH 系列菌株对氟吗啉最敏感，其 EC_{50} 值为 0. 0809μg/mL；而来自安徽泗县 SX 系列菌株及萧县 XX 系列菌株对氟吗啉产生了耐药性其 EC_{50} 值分别为 0. 1590μg/mL 和 0. 1478μg/mL。

通过对安徽省的 43 个大豆疫霉菌株进行高浓度药剂诱变，发现有部分菌株会出现角变区。将角变区在 20μg/mL 的含药培养基上继续培养，只有 1 个角变区能够快速生长，最终得到 1 个抗药性突变菌株 Fr。通过研究突变菌株 Fr 的抗药性水平，结果发现，突变菌株 Fr 的抗药水平比亲本敏感性的菌株高了 1 023倍。同时试验结果显示，突变菌株 Fr 较亲本敏感性菌株在以下生物学性状方面没有显著差异，主要体现为：①突变菌株 Fr 的生长速率和亲本无显著差异；②Fr 菌株的游动孢子囊的产量及大小与亲本敏感性菌株无显著差异；③突变菌株 Fr 在排孢孔的大小及卵孢子大小与亲本敏感性菌株无显著差异，但卵孢子产量较亲本低。试验还发现，突变菌株 Fr 与亲本敏感性菌株有着相似的致病力，都能快速侵染大豆植株。

通过建立抗药性突变菌株 Fr 的单游动孢子系 D1 代及 D2 群体及单卵孢系的 G1 代及 G2 代群体，研究该抗性突变菌株的遗传特性。结果表明：大豆疫霉对氟吗啉抗性遗传至少存在 2 种类型：①抗药性突变菌株 Fr 在继代培养时，即丧失抗性；该性状可能由细胞质中不稳定遗传的线粒体基因控制。②抗性突变菌株 Fr，其抗药性在无性繁殖及有性繁殖后代均是稳定的，但来自同一突变菌株的后代类型的不同单孢个体间对氟吗啉的抗性程度有所差异。

通过对该突变菌株 Fr 后代的生长速率及致病力的研究发现，突变菌株 Fr 的无性及有性繁殖后代，在生长速率及致病性方面性状不能够稳定遗传。

---

* 基金项目：公益性行业（农业）科研专项（201303018）

** 通信作者：高智谋，教授；E-mail：gaozhimou@ 126. com

大豆疫霉生物学性状在单游动孢子后代的遗传研究*

郑　婷，王艺烨，王伟燕，胡九龙，潘月敏，高智谋**
（安徽农业大学植物保护学院，合肥　230036）

摘　要：大豆疫霉（*Phytophthora sojae*）对大豆的生产为害极大，目前仍被列为我国的进境植物检疫对象。随着全球化发展的进程，各国均对外开放，进出口产品也越来越多，大豆进口数量每年都在增长。这也说明国外大豆疫霉传入我国可能性越来越高。目前，安徽省沿淮和淮北地区的大豆疫病的发生比较普遍。除了用农药治理，最好的便是种植抗病品种，利用植株抗病性控制大豆疫病，然而随着大豆疫霉新的毒力类型增长越来越快，这在一定程度上使得抗病品种极易丧失。为了探讨大豆疫霉毒性遗传规律及新的毒力类型产生机制，进而为大豆抗病育种和抗病品种的合理布局提供必要依据，作者对安徽省部分地区的大豆疫霉菌株的毒性及菌落形态、生长速率等其他生物学特性在其无性后代的遗传进行了研究，主要结果如下。

（1）以菌株 HY12 和 MC48 为亲本分别建立其单游动孢子第一代和第二代菌株群体，第一代共分离 63 个单孢菌株，第二代共 20 个单孢株，分别接种到 14 个含有单抗基因的鉴别寄主上。结果表明：大豆疫霉的毒性在单游动孢子后代的遗传不稳定，其致病型的遗传也是如此。

（2）从亲本分离出来的单游动孢子后代，通过测量它们的菌丝生长速率和观察菌落形态。结果发现，无论是第一代还是第二代，它们的菌落形态基本相同，都呈现近似圆形，菌落白色，气生菌丝呈现棉絮状，比较多，菌落边缘光滑，很整齐，质地比较光滑，菌丝生长速率和亲本无显著性差异。这表明它们之间基本没有差异，可稳定遗传。

（3）分别用菌株 HY12 和 MC48 的单游动孢子第一代和第二代菌株，分别接种合丰 35 与合丰 47 两个大豆品种叶片，结果表明：大豆疫霉对大豆的致病力在无性后代遗传不稳定，与亲本相比较，后代菌株的致病力有的变强，有的变弱。

关键词：大豆疫霉；生物学；遗传

---

* 基金项目：公益性行业（农业）科研专项（201303018）
** 通信作者：高智谋，教授；E-mail：gaozhimou@ 126. com

大豆疫霉与大豆叶片互作代谢组学研究*

刘　冬，卓　新，左荣华，胡九龙，潘月敏，高智谋**
（安徽农业大学植物保护学院，合肥　230036）

摘　要：近年来，病原菌和植物互作发展成为当今植物病理学研究热点之一。代谢组学是研究互作的有效技术手段，气质联用是代谢组学分析中较为普遍的一种方法。本文利用气质联用的代谢物的分析方法研究大豆疫霉（*Phytophthora sojae*）与大豆叶片互作代谢组学，利用气质联用分析大豆疫霉与大豆叶片互作 0~48h 产生的挥发性代谢物。结果表明：大豆叶片中都含有醛类挥发性代谢物质乙醛，且乙醛的含量随互作时间减少；互作中挥发性代谢物的种类增多与对照中代谢物均有差异。大豆叶片与大豆疫霉互作 6h 后主要挥发性代谢物有 7 类，增加了烷烃类和环烷烃两大类代谢物。大豆叶片与大豆疫霉互作 24h 后同种挥发性代谢物乙醛和 1-辛烯-3 醇的含量减少，6h 和 24h 对比大豆叶片同种挥发性代谢物环柠檬醛的含量减少。48h 出现了代谢物含量较少，峰面积值不显示，同对照相比没有酚类代谢物，增加了酮类和环烷烃类化合物，与 24h 对比醛类挥发性代谢物十三醛含量减少。以上结果说明大豆疫霉与大豆叶片互作过程中大豆叶片的代谢物发生了变化，大豆叶片受大豆疫霉影响代谢物种类增多，同时相同代谢物的含量减少。

关键词：大豆叶片；大豆疫霉；互作；代谢组学；气质联用

---

* 基金项目：公益性行业（农业）科研专项（201303018）
** 通信作者：高智谋，教授；E-mail：gaozhimou@ 126. com

辣椒疫霉菌蛋白分泌途径相关基因 *PcLHS*1 和 *PcSec*6 的功能分析[*]

陈国梁[**]，刘裴清，王荣波，李本金，翁启勇，陈庆河[***]
（福建省作物有害生物监测与治理重点实验室，福建省农业科学院植物保护研究所，福州　350003）

摘　要：本研究通过分子生物学、分子遗传学、生物化学以及植物病理学等研究方法研究辣椒疫霉菌 *PcLHS*1 在致病过程及其生长发育中的作用。研究结果表明：*PcLHS*1 的沉默致使辣椒疫霉菌生长速率降低，菌丝分叉明显增多且形态畸形，孢子囊产量显著减少；同时对过氧化氢等细胞壁敏感抑制剂的敏感性增强，进一步研究发现 *PcLHS*1 沉默突变体对内质网 ER 抑制剂更加敏感；在进行胞外酶活性实验中，*PcLHS*1 的沉默影响某些胞外酶（漆酶）的分泌；在致病性测定中发现，沉默突变体对感病辣椒品种的致病性减弱，不能激发本氏烟产生 HR 反应，ROS 明显减少。游动孢子侵染洋葱表皮实验中，沉默突变体侵染结构的产生时间要明显长于野生型。

此外，笔者采用同样的研究方法完成了对蛋白分泌相关的同源蛋白 Sec6 的部分研究工作，研究发现 *PcSec*6 虽然对辣椒疫霉的生长速率不产生影响，但是 *PcSec*6 沉默突变体的菌丝产生畸形且分叉增多；与野生型相比致病性也有明显的减弱。

综上所述，辣椒疫霉菌 *PcLHS*1 与蛋白分泌以及生长发育有关，参与了效应蛋白和胞外酶的分泌过程，影响辣椒疫霉的致病性。本研究为进一步研究辣椒疫霉菌效应蛋白的分泌及其致病机制提供一定的理论依据。

关键词：辣椒疫霉；蛋白分泌；*PcLHS*1；*PcSec*6

---

* 基金项目：公益性行业（农业）科研专项（201303018）；福建省属公益类科研专项（2016R1023-1）

** 第一作者：陈国梁，福建三明人，研究实习员

*** 通信作者：陈庆河，研究员；E-mail：chenqh@ faas. cn

辣椒疫霉中 3 个 RxLR 效应子的功能研究*

蒋　玥**，陈国梁，王荣波，刘裴清，李本金，翁启勇，陈庆河***
（福建省作物有害生物监测与治理重点实验室，福建省农业
科学院植物保护研究所，福州　350003）

摘　要：辣椒疫霉菌（*Phytophthora capsici*）是农业生产上极为重要的植物病原菌之一，其寄主范围广，除了侵染辣椒外，还可以侵染茄科、葫芦科和豆类等多种植物，引起的作物疫病是作物生产上的毁灭性病害，造成巨大的经济损失，同时还能侵染模式植物拟南芥和本氏烟，是卵菌研究的重要模式材料。疫霉菌分泌大量的效应子，其中 RxLR 效应子是卵菌中一类胞内效应子，其能干扰寄主植物细胞的正常生理代谢和功能。

通过生物信息学分析，在辣椒疫霉菌 LT1534 基因组中鉴定了 371 个候选的 RxLR 效应子。为了明确这些候选效应子在抑制植物防卫反应中的作用及 RxLR 效应子的功能，本研究首先通过农杆菌介导的瞬时表达方法对 67 个 RxLR 效应子进行筛选，筛选出 3 个效应子 Avh108、Avh110 和 Avh121 能够引起烟草叶片的 HR 反应同时还能够抑制 BAX 引起的 PCD 反应。进一步通过 PEG 介导的原生质体转化的方法进行获得了 Avh121 的沉默突变体。通过对比沉默突变体和野生型菌株的菌丝形态发现，突变体的菌丝出现畸形；通过洋葱表皮接种实验检测菌丝和游动孢子的侵染情况，结果发现突变体菌丝的侵染能力明显低于野生型，并且突变体游动孢子的生活力也明显低于野生型。

关键词：辣椒疫霉；RxLR 效应子；筛选；基因沉默

* 基金项目：公益性行业（农业）科研专项（201303018）；福建省属公益类科研专项（2016R1023-1）

** 第一作者：蒋玥，河南新乡人，研究生

*** 通信作者：陈庆河，研究员；E-mail：chenqh@ faas. cn

Transcriptional Programming of *Phytophthora sojae* for Organ-Specific Infection

Ye Wenwu, Wang Yang, Lin Long, Wu Jiawei,
Chen Han, Lin Yachun, Wang Yuanchao
(*Department of Plant Pathology*, *Nanjing Agricultural University*, *Nanjing*, *China* 210095)

Abstract: The different organs of a plant are significantly different in structure, metabolism, and defense response, however, the mechanisms on how plant pathogens regulate for organ-specific infection were largely unknown. Effector proteins are secreted by many plant pathogens topromote infection, e. g. , the genome of the soybean root rot pathogen *Phytophthora sojae* contains hundreds of effector genes. We have found that different effector genes could be transcriptionally programmed following infection, and the mechanism facilitated the effectors target different functional branches of the plant defense response. To learn our next hypothesis that *P. sojae* transcriptionally reprogram effector genes to facilitate its organ-specific infection in soybean, we recently compared the transcriptomes of *P. sojae* during the early stage of its infection in soybean roots and leaves. We identified the differentially expressed genes and found many enriched pathogenicity-related gene families were likely associated with the response of pathogen against different plant cell structure components and defense stresses. Functions of several candidate genes were further studied based on CRISPR/CAS9-mediaed gene knock out. A bZIP transcription factor of *P. sojae* was revealed to play a specific role in the infection of soybean roots. The results revealed that transcriptional programming are important for successful infection and host environment adaptability of *Phytophthora* pathogens.

Key words: *Phytophthora*; Organ-specific infection; Transcription; Effectors; CRISPR

A bHLH Transcription Factor, Associated with PsHint1, May be Required for the Chemotaxis and Pathogenicity of *Phytophthora sojae*

Qiu Min, Zhang Baiyu, Li Yaling, Ye Wenwu, Wang Yuanchao
(*Plant Pathology Department*, *Nanjing Agricultural University*, *Nanjing*, *China*, 210095)

Abstract: Zoospore chemotaxis and early infectionare key stages in the interaction between *Phytophthora sojae* and its host soybean. Transcriptional regulation is required for precise gene expression during a series of biological processes including these stages. Our previous studies show that a G-protein α subunitencoded by PsGPA1 regulates the chemotaxis and pathogenicity of *P. sojae*. In addition, PsHint1, a histidine triad (HIT) domain-containing protein orthologous to human HIT nucleotide-binding protein 1 (HINT1), interacts with PsGPA1 and exhibited a PsGPA1-like biological function in *P. sojae*. Recently, we noticed that Hint1 could usually inhibit bHLH family transcription factors (TFs) in mammals. We found there were 6 bHLH TFs in *P. sojae*, and PsbHLH1showed higher transcription levels during stages of zoospores, cysts, and cystgermination. Interestingly, PsbHLH1 interacts with PsHint1 in the GST-Pull down assay. On the basis of CRISPR/Cas9-mediaed gene knock-out, weare focusing on phenotypic analysis for the obtained mutants. Preliminary results showed that PsbHLH1may also be required for the chemotaxis and pathogenicity of *P. sojae*, playing as a downstream transcriptional regulator in the PsGPA1-associated signaling pathway.

Key words: *Phytophthora sojae*; bHLH transcription factor; Chemotaxis; Pathogenicity

河南省烟草疫霉菌交配型与抗药性的初步研究*

高鹏飞，胡艳红，崔林开**
（河南科技大学林学院，洛阳 471023）

摘 要：从河南省宜阳、洛宁、渑池、卢氏、邓州、郏县、舞阳、襄城、鹿邑、确山、睢阳、济源、登封和嵩县等地烟田分离到30株烟草疫霉菌，采用直接配对法测定了这些菌株的交配型，并进一步采用菌丝生长速率法测定了这些菌株对甲霜灵和烯酰吗啉的敏感性。结果表明：30个菌株均为A2交配型，未发现其他类型的交配型；甲霜灵对烟草疫霉菌的EC_{50}值在0.08～2.82μg/mL，平均EC_{50}值为1.12μg/mL；烯酰吗啉对烟草疫霉菌的EC_{50}值在0.07～0.59μg/mL，平均EC_{50}值为0.36μg/mL。这些结果说明河南省的烟草疫霉菌交配型较为单一，以无性繁殖为主；河南省的烟草疫霉菌对甲霜灵已经产生了一定的抗性，而对烯酰吗啉还较为敏感。这些结果为河南省烟草黑胫病的防治提供了一定的理论依据。

关键词：烟草疫霉；交配型；敏感性；甲霜灵；烯酰吗啉

---

* 基金项目：国家自然科学基金项目（U1404318）；河南科技大学青年基金项目（2012QN001）；河南科技大学培育基金项目（2013ZCX013）

** 通信作者：崔林开，副教授，主要从事植物疫霉病害研究；E-mail：cuilk@haust.edu.cn

第三部分　病　毒

The Movement Protein of *Barley Yellow Dwarf Virus* Targets AtVOZ1 and AtVOZ2 to Delay Flowering in Arabidopsis

Huang Caiping, Tao Ye, Chen Yujia, Wu Yunfeng

(*State Key Laboratory of Crop Stress Biology for Arid Areas and Key Laboratory of Crop Pest Integrated Pest Management on the Loess Plateau of Ministry of Agriculture*, *College of Plant Protection*, *Northwest A&F University*, *Yangling*712100, *Shaanxi*, *China*)

Abstract: The movement protein (MP) of barley yellow dwarf virus-GAV (BYDV-GAV) was previously considered to bind the viral RNA and transport viral genome, however, a new function of MP of BYDV-GAV on plant physiology was found in this study. Transgenic *Arabidopsis thaliana* plants that constitutively expressed MP were produced and exhibited a late-flowering phenotype. Two key proteins involved in a signal transduction pathway about flowering in Arabidopsis, vascular plant one-zinc protein 1 (AtVOZ1) and vascular plant one-zinc protein 2 (AtVOZ2), were found to interact with MP by yeast two-hybrid assay and bimolecular fluorescence complementation experiment. Additionally, N-terminal of the MP was confirmed to be essential for the interaction. The result of quantitative real-time PCR (qRT-PCR) revealed that MP up-regulated the expression of*AtVOZ*1 and *AtVOZ*2 and also affected other genes in phyB signal transduction pathway. Overall, our work indicated that MP of BYDV-GAV can delay flowering in Arabidopsis through interacting with AtVOZ1 and AtVOZ2. These findings provided new insights into the molecular mechanisms by which MP influence plant growth and development.

Key words: Barley yellow dwarf virus; Movement protein; AtVOZ1; AtVOZ2; Late-flowering

黄瓜花叶病毒致病性研究*

邱艳红[1**]，雷　荣[1]，王超楠[1,2]，朱水芳[***]
（1. 中国检验检验科学研究院植物检疫研究所，北京　100176；
2. 中国农业大学植物保护学院，北京　100193）

摘　要：黄瓜花叶病毒（*Cucumber mosaic virus*，CMV）寄主广泛，为害严重。研究发现病毒编码蛋白与核基因编码的叶绿体蛋白存在互作，并影响叶绿体的发育，引起类囊体基粒片层大幅减少，导致光系统Ⅱ（PSⅡ）中多个蛋白含量降低。此外，转录组数据表明病毒侵染可影响烟草体内多个代谢通路，尤其是叶绿体相关代谢途径。病毒小干扰 RNA（small interfering RNA，siRNA）研究发现 CMV 的 siRNA 可靶标于烟草的多个基因。研究将进一步探索 CMV—寄主因子互作的具体机制，以及病毒 siRNA—寄主靶基因—症状三者之间的相互关系，深入阐述病毒致病相关机制。

关键词：黄瓜花叶病毒；致病研究；叶绿体；病毒小干扰 RNA

---

* 基金项目：主要入侵生物的生物学特性研究（2016YFC1200600）

** 第一作者：邱艳红，助理专家，主要从事植物病毒研究；E-mail：qiuyh@ caiq. gov. cn

*** 通信作者：朱水芳，研究员，主要从事植物病毒研究；E-mail：zhusf@ caiq. gov. cn

Complete Genome Sequence of Two *Strawberry Vein Banding Virus* Isolates From China*

Li Shuai, Jiang Xizi**, Zuo Dengpan,
Zhang Xiangxiang, Hu Yahui, Jiang Tong
(*School of Plant Protection, Anhui Agricultural University, Changjiang west Road No.* 130, *Hefei* 230036, *China*)

Abstract: It was rarely reported about *strawberry vein banding virus* (SVBV) genome sequence in China and most countries worldwide. In the present study, we have determined two complete nucleotide sequences of SVBV China isolates, designated SVBV-AH (accession number: KX787430) and SVBV-BJ (accession number: KR080547), which were obtained in naturally infected strawberry samples from Anhui province and Beijing city of China, respectively. The complete genomes of SVBV-AH and SVBV-BJ were 7 862 nucleotides (nts) and 7 863 nts long, respectively, and both component with seven genes typical of the genus. Alignment of complete nucleotide sequences showed that SVBV-AH and SVBV-BJ shared the significant higher nucleotide sequence similarity (97.7%) with each other, and had 85.7% and 86.0% sequence similarity related to SVBV from the United States (SVBV-US), respectively. Phylogenetic analysis of the two Chinese isolates of SVBV aligned with other caulimoviruses indicated that two SVBV China isolates formed a separate branch and most closely connected with SVBV-US, and have relatively distant genetic relationships with other caulimoviruses.

Key words: Strawberry vein banding virus; Sequence analysis; Genetic characterization

* Funding: National Natural Science Foundation of China (31671999, 31371915)

** Corresponding author: Jiang Tong; E-mail: jiangtong4650@ sina. com

应用自然诱发观察萍乡市主栽水稻品种对南方水稻黑条矮缩病发生状况

龚航莲[1*]，龚朝辉[2**]，敖新萍[3]

（1. 江西省萍乡市植保站，萍乡　337000；2. 江西省萍乡市农技站，萍乡　337000；
3. 江西省萍乡市芦溪县银河镇国家农业科技示范园，337251）

摘　要：南方水稻黑条矮缩病（Southern rice black-stveaked dwarf，SRBSDV）属呼肠孤病毒科（Reociridae）斐济病毒属（Fijirus）。2001 年在广东阳西首次发现，目前该病毒的发生地涉及到国内的海南、两广、湖南、江西、江浙以及国外的越南等。据萍乡市植保植检站统计：萍乡市 2009 年水稻只零星发病，发病面积 670hm^2，2010 年该病在萍乡市大暴发，发病面积达 3 万 hm^2。2011—2015 年每年属偏轻发生，发病面积保持在 210～350hm^2，2016 年略有回升，发病面积 510hm^2。

2010—2016 年坚持应用水稻秧苗期测定入迁白背飞虱带病率预测南方水稻黑条矮缩病发生趋势。其方法是：温室培育水稻无菌秧苗单株接入白背飞虱带毒率。结果：2010 年通过 5 月 6 日、5 月 18 日、5 月 25 日、6 月 15 日、6 月 23 日 5 次入迁萍乡市的白背飞虱带毒率分别为 0%、2. 17%、6. 52%、39. 96%、30. 43%。2011 年通过 5 月 7 日、5 月 18 日、6 月 10 日、6 月 17 日、6 月 21 日 5 次入迁白背飞虱带毒率分别为 0%、0%、0%、0、2. 2%。2012 年通过 5 月 5 日、5 月 8 日、5 月 31 日、6 月 9 日、6 月 13 日 5 次入迁白背飞虱带毒率分别为 0%、0%、4. 4%、2. 1%、6. 7%。2013 年通过 5 月 9 日、5 月 18 日、5 月 31 日、6 月 9 日、6 月 13 日 5 次入迁白背飞虱带毒率分别是 0%、0%、4. 34%、2. 17%、6. 25%。2014 年通过 5 月 6 日、5 月 10 日、5 月 12 日、5 月 20 日、5 月 28 日、6 月 2 日 5 次入迁白背飞虱带毒率分别是 0%、4. 44%、0%、2. 22%、6. 67%。2015 年通过 5 月 7 日、5 月 10 日、5 月 20 日、5 月 29 日、6 月 9 日 5 次入迁白背飞虱带毒率分别是 0%、4. 54%、2%、2. 27%、2. 27%。2016 年通过 5 月 4 日、5 月 13 日、5 月 30 日、6 月 5 日、6 月 14 日 5 次入迁白背飞虱带毒率分别是 0%、0%、2. 17%、4. 34%、8. 69%。6 年的苗期测定其结论是入迁白背飞虱 SRSDV 带毒率达到 4%～8%，当年南方水稻黑条矮缩病为轻发生，而其带毒率达到 30%以上，就有可能大暴发的趋势。只有 2010 年带毒率达到 30%以上而造成萍乡市水稻南方水稻黑条矮缩病的大暴发。

2010—2016 年应用自然诱发观察萍乡市主栽水稻品种对南方水稻黑条矮缩病发生状况，2010 年在萍乡市连陂病虫观察区自然诱发鉴定圃中，供试主栽水稻品种 25 个，病株率在 0. 83%～2. 15%有 13 个品种，占 52%；病株率在 3. 2%～5. 6% 7 个品种占 28%，病株率在 11. 37%～21% 3 个品种占 12%，病株率 100%（绝收）1 个品种占 4%，只有 Y 两优 1 号一个品种未发病占 4%。2011 年同样的试验方法，只有培优 1993 及宜香 725 两个品种病株率 4%，其他 23 个品种均未发病。2012—2015 年 4 年中主栽的水稻品种病株率均控制在 0. 38%～2. 3%，即所有的主栽水稻均发病。2016 年主栽的水稻品种有 4 个品种病株率上升 0. 12%～0. 93%，病株率达到 0. 5%～3. 16%。略有回升，无绝收品种及不发病品种。

---

* 通信作者，龚航莲，高级农艺师，研究方向：中长期病虫测报；E-mail：ghl1942916@ sina. com

** 第一作者：龚朝辉，男，农艺师，研究方向：病虫综合治理；E-mail：26537383@ qq. com

2014—2016 年将江西东乡野生稻 10 个植株，海南崖城野生稻 10 个植株，萍乡本地高粱 10 个植株，华农 18 玉米 10 个植株，李氏禾 10 个植株，杂草稻 10 个植株，李氏禾、杂草稻在试验圃繁育的苗期植株移栽在萍乡市连陂病虫观察区自然诱发鉴定圃中，60d 后（即 7 月下旬末）观察鉴定结果：2014 年只有杂草稻有 4 个植株出现典型南方水稻黑条矮缩病症状。其他处理未出现症状。2015—2016 年，江西东乡野生稻有 2 个植株，海南崖城野生稻有一个植株。萍乡本地高粱有 3 个植株，华农 18 玉米有 4 个植株、李氏禾有 2 个植株、杂草稻有 6 个植株出现典型病状。

应用水稻秧苗期及自然诱发鉴定圃能有效地监测南方水稻黑条矮缩病在萍乡市水稻上为害程度，发生态势，流行趋势及禾本科植物感病程度，为大面积防治提供依据。

关键词：自然诱发；南方水稻黑条矮缩病；水稻；发生状况

小麦黄花叶病毒 3′末端 poly（A）长度变化对体外翻译与复制的影响*

耿国伟**，于成明，原雪峰***

（山东农业大学植物保护学院植物病理系，山东省农业微生物重点实验室，泰安 271018）

摘 要：小麦黄花叶病毒（*Wheat yellow mosaic virus*，WYMV）是小麦土传花叶病的主要病原之一，对我国多个地区冬小麦的生长、发育构成严重危害。WYMV 粒子呈弯曲线状，长度分为 300nm 和 600nm 两种。WYMV 为二分体基因组，由两条线性的单链正义 RNA 组成。RNA1 全长约 7. 6kb，RNA2 全长约 3. 6kb，分别编码一个 270ku 和 100ku 的多聚蛋白。WYMVRNA 的 5′端有共价结合的 VPg，3′端有一个 poly（A）尾。

通过对山东省泰安大汶口地区感病冬小麦 WYMV 的 3′RACE 发现，WYMV RNA1 在体内存在 3′末端没有 poly（A）的现象，随机选取 100 个克隆统计发现该 WYMV 分离物 RNA1 3′末端的 poly（A）的长度存在两极分化现象。在体外翻译系统中，利用含有 WYMV RNA 5′UTR 与 3′UTR 的 Fluc 载体，分析有无 poly（A）对翻译的调控。实验表明，同时含有 5′UTR 和 3′UTR 的 Fluc 载体相对于空载体，翻译效率提高 20 倍；加入 poly（A）后则增加到空载体的 40 倍，说明 WYMV RNA 3′末端的 poly（A）在体外翻译中起正调控作用。另外，利用 WYMV NIb 介导的体外复制体系分析了有无 poly（A）对体外复制的效应。发现加入 poly（A）的 WYMV RNA 的体外复制效率仅为不加 poly（A）的 WYMV RNA 的 30% 左右。说明 WYMV RNA 3′末端的 poly（A）在体外复制中起到抑制作用。以上实验结果侧面说明了 WYMV RNA 3′末端在体内存在 poly（A）长度变化的生物学意义，不同的生物学进程（如翻译、复制）可能利用不同形式的 WYMV RNA 模板［有无 poly（A）］。

关键词：小麦黄花叶病毒；poly（A）；翻译；复制

---

* 基金项目：国家自然科学基金资助项目（31670147；31370179）；山东省自然科学基金资助项目（ZR2013CM015）

** 第一作者：耿国伟，山东菏泽人，硕博连读研究生，主要从事分子植物病毒学研究；E-mail：guowgeng@ 163. com

*** 通信作者：原雪峰，教授，博士生导师，主要从事分子植物病毒学研究；E-mail：snowpeak77@ 163. com

一个新的烟草丛顶病毒（TBTV）不依赖帽子翻译元件-TSS*

王德亚**，于成明，刘珊珊，窦宝存，原雪峰***
（山东农业大学植物保护学院植物病理系，山东省农业微生物重点实验室，泰安 271018）

摘　要：烟草丛顶病毒（*Tobacco bushy top virus*，TBTV）属于番茄丛矮病毒科幽影病毒属。TBTV 基因组由一条正义单链 RNA（+ssRNA）组成，编码 4 个 ORF，5′端缺乏帽子结构（m7GPPPN），3′末端也没有 poly（A）尾。p35（ORF1）蛋白是通过不依赖帽子翻译机制而表达，已报道的不依赖帽子翻译元件为类 BTE 元件。通过萤火虫荧光素酶（Fluc）载体和 TBTV 全长 RNA 的体外翻译分析，表明类 BTE 元件的核心组件为可能结合 eIF4E 的 SL-I 以及与 5′UTR 形成远距离 RNA-RNA 互作的 SL-III B；并且其结构稳定性对于 p35 的不依赖帽子翻译也至关重要。

TBTV 3′UTR 长 645nt，类 BTE 元件位于 3 724~3 838位，长约 115nt，在如此冗长的 3′UTR 中是否存在除类 BTE 元件之外的翻译调控元件？通过对 TBTV3′UTR 区域的系列缺失实验，发现 TB-D9（4 040~4 099）和 TB-D10（4 093~4 152）的缺失会明显降低 p35 蛋白的表达，通过 Mfold 预测发现 4 003~4 152形成独立结构域，并且利用 SHAPE 技术分析发现 D9 和 D10 的突变并没有影响类 BTE 元件的活性结构。反式竞争实验也证实了 4 003~4 152区对于 p35 不依赖帽子翻译的调控作用。因此，在 TBTV 的 3′UTR 还存在除类 BTE 元件之外的不依赖帽子翻译调控元件。利用 In line-probing 技术体外分析 4003－4152 的 RNA 结构，表明其包含 5 个发夹结构和 3 个假结点。此结构与 TCV、CMoV 以及 PEMV 的 3′UTR 区的 TSS 结构类似。因此，以上结果表明：TBTV 的 3′UTR 存在类 BTE 元件之外的新的不依赖帽子翻译元件——TSS。

关键词：烟草丛顶病毒；不依赖帽子翻译；类 BTE 元件；TSS 元件

---

* 基金项目：国家自然科学基金资助项目（31370179）；山东省自然科学基金资助项目（ZR2013CM015）
** 第一作者：王德亚，在读博士生，主要从事分子植物病毒学研究；E-mail：wangdeyasdny@163.com
*** 通信作者：原雪峰，教授，博士生导师，主要从事分子植物病毒学研究；E-mail：snowpeak77@163.com

我国部分烟区烟草病毒病的病原分析*

刘珊珊**，原雪峰***

（山东农业大学植物保护学院植物病理系，山东省农业微生物重点实验室，泰安 271018）

摘 要：烟草病毒病是目前烟草生产上分布最广、发生最为普遍的一类病害，对我国部分烟区造成严重损失。世界各烟区为害烟草的病毒病已报道有 20 多种，我国已发现的烟草病毒病有 17 种。为给烟草病毒病防治提供参考，对来自山东省、云南省、贵州省 3 个省份 6 个地区的 64 份烟草样品进行了 20 种烟草候选病毒的检测并对发病严重的烟草病毒痁进行研究分析。64 份烟草样品中共检测出 7 种烟草病毒，其中黄瓜花叶病毒（*Cucumber mosaic virus*，CMV）、烟草普通花叶病毒（*Tobacco mosaic virus*，TMV）、马铃薯 Y 病毒（*Potato virus Y*，PVY）检出率较高，分别为 34.38%、15.62%和 9.38%；两种及两种以上的病毒复合侵染检出率达 18.75%。另外，不同烟区的病毒种类组成有差异。通过对三省份不同地区烟草病毒病检测，基本掌握了现阶段我国常见及多发性的烟草病毒病种类，为烟草病毒病的防治提供了数据支持。

关键词：烟草病毒病；黄瓜花叶病毒；烟草普通花叶病毒；马铃薯 Y 病毒

---

* 基金项目：国家自然科学基金资助项目（31670147；31370179）

** 第一作者：刘珊珊，硕士研究生，主要从事分子植物病毒学研究；E-mail：shansd1218@126.com

*** 通信作者：原雪峰，教授，博士生导师，主要从事分子植物病毒学研究；E-mail：snowpeak77@163.com

烟草丛顶病毒 RdRp 蛋白-1 位移码的调控机制*

于成明**，王德亚，王国鲁，逯晓明，原雪峰***
（山东农业大学植物保护学院植物病理系，山东省农业微生物重点实验室，泰安　271018）

摘　要：烟草丛顶病毒（*Tobacco bushy top virus*，TBTV）属于番茄丛矮病毒科幽影病毒属。TBTV 基因组是一条（+）ssRNA，由 4152 个核苷酸组成，有 4 个开放阅读框（ORF），编码 4 个蛋白，5′末端缺乏帽子结构，3′末端也不带 poly（A）尾。RNA 依赖的 RNA 聚合酶（RdRp）是 ORF1 编码的 p35 蛋白通过-1 位移码机制表达，分子量等于 ORF1 和 ORF2 的分子量之和。

本研究通过体外翻译、RNA 结构体外分析等技术研究 TBTV RdRp 的-1 位移码机制，定位影响移码的关键性元件并解析其发挥调控作用的分子基础。首先，通过序列保守型分析和突变分析，定位了移码必需七核苷酸滑动序列，位于 946~952 位的 GGAUUUU；根据 RNA 结构预测发现，七核苷酸滑动序列存在一个大的茎环结构，突变分析表明此结构距离七核苷酸滑动序列的距离以 6~9nt 为宜，否则影响移码效率；进一步的 RNA 结构体外分析和突变分析显示，七核苷酸滑动序列下游调控移码的茎环结构可能有 2~3 个。通过 EMSA 分析了基因组范围内的 RNA-RNA 远距离互作，其中 3′末端的 Pr 环碱基与七核苷酸滑动序列下游的茎环结构区存在 RNA 互作；然后分别通过 In-line probing 定位了 Pr 环中的互作碱基；通过 SHAPE 技术分析七核苷酸滑动序列下游的茎环结构区的互作碱基，位于 1 023~1 025位的 AGU。本研究确定了 TBTV RdRp-1 位移码机制的 3 个层次调控元件：七核苷酸滑动序列、下游茎环结构和远距离 RNA-RNA 互作；进一步的调控细节研究进行中。

关键词：烟草丛顶病毒；移码；调控元件；RNA 结构

---

* 基金项目：国家自然科学基金资助项目（31670147；31370179）；山东省自然科学基金资助项目（ZR2013CM015）

** 第一作者：于成明，山东平阴人，在读博士生，主要从事分子植物病毒学研究；E-mail：ycm2006. apple@ 163. com
*** 通信作者：原雪峰，教授，博士生导师，主要从事分子植物病毒学研究；E-mail：snowpeak77@ 163. com

马铃薯 Y 病毒多基因系统发育分析及其株系快速鉴定方法的建立

邹文超[1]，沈林林[1]，沈建国[2]，蔡　伟[3]，詹家绥[1]，高芳銮[1]
（1. 福建农林大学植物病毒研究所/福建省植物病毒学重点实验室，福州　350002；
2. 福建出入境检验检疫局检验检疫技术中心/福建省检验检疫技术研究重点实验室，福州　350001；3. 福清出入境检验检疫局，福州　350300）

摘　要：马铃薯 Y 病毒（*Potato virus Y*，PVY）是世界范围内为害烟草与马铃薯的重要植物病毒之一。本文旨在建立快速准确的多基因联合体系，实现 PVY 常见株系的快速鉴定。以 PVY 的 P1、HC-Pro、VPg 和 CP 4 个基因为研究对象，根据基因的不同组合建立 5 个不同数据集，分别进行系统发育分析，并通过贝叶斯标签关联显著性（Bayesian tip-association significance，BaTS）分析各数据集中代表分离物与株系的关联性，以确定实现 PVY 快速鉴定的最佳组合。不同数据集的系统发育及 BaTS 分析结果显示，除了联合 P1、VPg 和 CP 3 个基因数据集外，其他 4 个数据集均无法实现 PVY 常见株系的准确鉴定。采用不同建树方法对联合 P1、VPg 和 CP 3 个基因数据集比较分析显示，基于 ML 法和 NJ 法的系统发育树在拓扑结构上基于一致，均优于基于贝叶斯算法的最大分支置信（Maximum clade credibility，MCC）树；同时以 HLJ26 分离物为研究对象，对建立的多基因联合体系进行实际应用，结果显示该分离物与 PVY^{NTN-NW} 株系的 3 个分离物 SYR-Ⅱ-2-8、SYR-Ⅱ-Be1 和 SYR-Ⅱ-DrH 以高置信值聚为一亚簇，表明该分离物可能属于 PVY^{NTN-NW} 株系（SYR-Ⅱ型）。重组分析显示，HLJ26 基因组存在 4 个潜在的重组信号，分别位于 P1、HC-Pro/P3、VPg 和 CP 的 5′-末端，与 PVY^{NTN-NW} 株系（SYR-Ⅱ型）的重组位点相一致，表明其属于 PVY^{NTN-NW} 株系（SYR-Ⅱ型）。同时，应用多重 RT-PCR 成功扩增出约为 1 000bp 和 400bp 的 2 个特异性片段，与 PVY^{NTN-NW} 株系（SYR-Ⅱ型）的特异条带大小相一致。这些结果进一步支持了多基因联系的鉴定结果。联合 P1、VPg 和 CP 3 个基因数据集系统发育分析，可以实现 PVY 常见株系的准确鉴定。HLJ26 分离物属于 PVY^{NTN-NW} 株系（SYR-Ⅱ型），这是 PVY 马铃薯上该株系 SYR-Ⅱ型全因组序列的首次报道。

关键词：马铃薯 Y 病毒；多基因系统发育分析；系统发育与性状关联分析；SYR-Ⅱ型

Sequence analyses of an RNA virus from the oomycete *Phytophthora infestans* in China*

Zhan Fangfang[1**], Wang Tian[1], Zhan Jiasui[1,2***]

(1. *Fujian Key Laboratory of Plant Virology*, *Institute of Plant Virology*, *Fujian Agriculture and Forestry University*, *Fuzhou*, *Fujian* 350002, *China*; 2. *Key Lab of Biopesticide and Chemical Biology*, *Ministry of Education*, *Fujian Agriculture and Forestry University*, *Fuzhou*, *Fujian* 350002, *China*)

Abstract: Eleven RNA virus 4 (PiRV-4) were detected from the 42 *Phytophthora infestans* isolates sampled from Shanxi Province, China and sequenced. All of the eleven viral genomes consist of 2 904 nucleotides, encoding an RdRp protein of 962 amino acids (aa). The genomes share 98%~99% nucleotide and aa sequence identities with the three reference PiRV-4 sequences (FLa2005, US970001 and MX980317) downloaded from GenBank. Phylogenetic analysis shows that the 11 sequences from the current study clustered them together with the references in the *Narnavirus* clade, suggesting that PiRV-4 is likely a species in the genus *Narnavirus*.

Key words: *Phytophthora infestans*; RNA virus; Sequence analyses

* Funding: The project was supported by the Modern Agricultural Industry and Technology System (No. CARS-10)

** First author: Zhan Fangfang, research assistant, major in plant pathology; E-mail: zhanfangfang@fafu.edu.cn

*** Corresponding author: Zhan Jiasui, mainly engaged in population genetics; E-mail: Jiasui.zhan@fafu.edu.cn

Strain Composition of *Potato virus Y* in Fujian Province Detected with the Concatenated Sequence Approach*

Shen Linlin, Zou Wenchao, Xie Jiahui, Wang Tian,
Chen Meiling, Zhan Jiasui**

(*Institute of Plant Virology, Fujian Agriculture and Forestry University/Key Laboratory of Plant Virology of Fujian Province, Fuzhou* 350002)

Abstract: *Potato virus Y* (PVY) is one of the most destructive pathogens constraining sustainable development of potato industry. The objective of this study is to develop a fast, easy-to use and accurate approach to timely detect PVY strains and apply the approach to investigate the occurrence, distribution and composition of PVY strains in Fujian Province. Three pairs of degenerate primers designed from the conserved regions of P1, VPg and CP genes in the referencePVY sequences downloaded from GenBank were used to amplify the positive samples by ELISA test. Individual gene analyses showed that P1, VPg and CP sequences inFujian isolates shared 72%~99%, 85%~99% and 88%~99% nucleotide identity with the reference strains, respectively. Concatenated sequence analysis showed that FQ01 and FQ08 isolates shared the highest sequence identity with $PVY^{N\text{-}Wi}$ and PVY^{E}, respectively. Recombination signals were identifiedin P1 and VPg genes but not in CP of PVY. Phylogenetic tree revealed that isolates CL01, CL02, CL05 and FQ13 were grouped with $PVY^{NTN\text{-}NW}$ (SYR-I) and isolates CL03, CL04, FQ02, FQ06, FQ09, FQ11 and FQ12 were clustered with the $PVY^{NTN\text{-}NW}$ (SYR-II), whereas isolate FQ01 was grouped with $PVY^{N\text{-}Wi}$ and isolate FQ08 was clustered with thePVY^{E}, respectively. PVY occurs frequently in Changle and Fuqing cities in Fujian Province andrecombinant strains, particularly $PVY^{NTN\text{-}NW}$ are dominant in the PVY isolates from the province.

Key words: *Potato virus Y* (PVY); ELISA; Concatenated sequence; Phylogeny; Recombinant strain

---

* First author: Shen Linlin, Master student, research direction for potato late blight; E-mail: 18649708815@163.com

** Corresponding author: Zhan Jiasui, Professor, research interests for population genetics; E-mail: Jiasui.zhan@fafu.edu.cn

甘蔗新品种花叶病病原检测及其系统进化分析*

王晓燕**，李文凤，张荣跃，单红丽，尹　炯，罗志明，黄应昆***
（云南省农业科学院甘蔗研究所/云南省甘蔗遗传改良重点实验室，开远　661699）

摘　要：甘蔗花叶病是一类广泛发生在世界各大甘蔗产区的重要病毒病害，常造成巨大的经济损失。近年来，在甘蔗生产快速发展中，甘蔗品种改良更新加快，种植制度多样化，而甘蔗花叶病的分布、发生为害和各甘蔗新品种的感病程度等情况均不清，有效防控甘蔗花叶病缺乏可靠依据。为此，笔者对云南开远、弥勒和元江蔗区繁殖示范甘蔗新品种的花叶病发生情况进行了调查和田间采样，采用 RT-PCR 技术对其病原进行检测，并测定各花叶病病原的 *CP* 基因序列，系统进化分析其遗传多样性、类群构成以及地理相关性等，以期为针对性地选育和利用抗病品种，在甘蔗生产上进行品种合理布局，有效防控甘蔗花叶病提供科学依据。甘蔗花叶病病原检测结果表明：77 份甘蔗新品种中共检出 SCSMV、SrMV 和 SCMV 3 种病毒，其中 SCSMV 的检出率最高为 100%，SrMV 次之为 27.3%，SCMV 最低为 1.3%，首次检测发现 SCSMV、SrMV 和 SCMV 3 种病毒复合侵染甘蔗新品种黔糖 5 号。对 SCSMV 和 SrMV 代表性样品 *CP* 基因序列进行克隆测序，并与已公布的相应 *CP* 基因序列构建 ML 树进行系统进化分析。结果表明：SCSMV 系统进化树有明显地理差异性，所有中国 SCSMV 分离物都位于 China 组；不同甘蔗新品种 SCSMV 分离物分布在 China 组各亚组间或形成独立分支。所有 SrMV 分离物 *CP* 基因序列可分为两个组，组内可分为不同的亚组；不同甘蔗新品种 SrMV 分离物在不同组和亚组间交叉分布，无明显地理差异性。

关键词：甘蔗新品种；甘蔗花叶病；病原检测；系统进化

* 基金项目：国家现代农业产业技术体系建设专项资金资助（CARS-20-2-2）；云南省现代农业产业技术体系建设专项资金资助

** 第一作者：王晓燕，女，副研究员，主要从事甘蔗病害研究；E-mail：xiaoyanwang402@ sina. com
*** 通信作者：黄应昆，研究员，从事甘蔗病害防控研究；E-mail：huangyk64@ 163. com

一种同步检测 3 种菠萝病毒的多重实时荧光定量 RT-PCR 方法*

罗志文[1**]，胡加谊[1,2]，范鸿雁[1]，李向宏[1]，陈业光[1]，余乃通[3]，
郭利军[1]，胡福初[1]，何 凡[1]，刘志昕[3]，张治礼[1***]

（1. 海南省农业科学院热带果树研究所/农业部海口热带果树科学观测实验站/海南省热带果树生物学重点实验室/海南省热带果树育种工程技术研究中心，海口 571100；
2. 黄埔出入境检验检疫局，广州 510730；3. 中国热带农业科学院热带生物技术研究所/农业部热带作物生物学与遗传资源利用重点实验室，海口 571101）

摘 要：菠萝凋萎病是菠萝生产中的一种重要病害，目前尚难防控。为实现菠萝植株病毒快速定量检测和实时监测，笔者根据 NCBI 报道的 3 种菠萝凋萎相关病毒（PMWaV－1、PMWaV-2、PMWaV-3，PMWaV-1、-2、-3）的外壳蛋白基因（Coat Protein gene，CP gene）保守序列，分别设计合成 3 种病毒的特异性引物和 TaqMan 探针，通过多重 RT-qPCR 反应体系优化，进而以 3 种病毒 *CP* 基因目的片段的重组质粒为标准品绘制标准曲线，初步建立了针对 3 种目标病毒的三重 RT-qPCR 检测体系。试验结果表明：该体系对 PMWaV-1、-2、-3 的扩增效率分别为 101.7%、97.1%和 106.7%，线性相关系数 R^2 为 0.996~0.999，可信度高；对PMWaV-1、-2、-3 的最低检测线分别为 99 拷贝、98 拷贝和 868 拷贝；重复性实验结果显示，该体系下 PMWaV-1、-2、-3 三重 RT-qPCR 扩增曲线的标准差均不超过 0.26，变异系数均不超过 1.44%。这表明本研究针对 PMWaV-1、-2、-3 初步建立多重 RT-qPCR 检测体系是一种能够快速、灵敏、稳定地同步定量检测目标病毒的方法，这为菠萝凋萎相关病毒（Pineapple Mealybug Wilt-associated Viruses，PMWaVs）的定性定量研究和未来开展复合侵染机制研究提供了技术支撑。

关键词：菠萝凋萎相关病毒（PMWaV）；菠萝病毒病；实时荧光定量 RT-PCR；多重检测体系；同步检测

---

* 基金项目：海南省重点研发计划项目（ZDYF2016035）；公益性行业（农业）科研专项经费项目（201203021）；海南省农业科学院农业科技创新专项经费项目（CXZX201410）

** 第一作者：罗志文，男，硕士，助理研究员，研究方向：果树植物病理学；E-mail：zhiwenluo@163.com

*** 通信作者：张治礼；E-mail：zzl_haas@163.com

Selection, evaluation and validation of reference genes for expression analysis of miRNAs in cucumber under virus stress*

Liang Chaoqiong[2,3,4**], Luo Laixin[1,2], Barbara Baker[3,4], Li Jianqiang[1,2***]

(1. *Department of Plant Pathology/Key laboratory of Plant Pathology, China Agricultural University, Yuanmingyuan West Road No2., Beijing, China*; 2. *Beijing Engineering Research Center of Seed and Plant Health (BERC-SPH) /Beijing Key Laboratory of Seed Disease Testing and Control (BKL-SDTC), Yuanmingyuan West Road No2., Beijing, China*; 3. *Plant Gene Expression Center, United States Department of Agriculture, Agricultural Research Service*, 800 *Buchanan St. Albany, CA, USA*; 4. *Department of Plant and Microbial Biology, University of California, Berkeley, Berkeley, CA, USA*)

Abstract: Real-time quantitative PCR (RT-qPCR) is a feasible tool for determining gene expression profiles, but the accuracy and reliability of the results depends on the stable expression of selected reference genes in different samples. Recent studies in various plant species have established microRNAs play a crucial role in the regulation of key biological processes such as development, signal transduction, abiotic and biotic stresses, including viral stress. Studies on expression characteristics of plant microRNAsare an important aspect of elucidating functions and regulatory mechanisms and require particularly careful selection of suitable reference genes which are not affected by the plant growth and development nor bythe environmental factor. Therefore, the objective of this study was to select the most reliable reference genes for expression analysis of microRNAsin leaf, stem and root tissues of CGMMV (*Cucumber green mottle mosaic virus*) -infected cucumberby RT-qPCR. The seven commonly used internal controls (*actin*, *tubulin*, *EF−1a*, *18S rRNA*, *ubiquitin*, *GAPDH* and *cyclophilin*) for gene expression studies of cucumber were used in thisstudy. The results showed that stability of reference gene expression significantly varied depending on cucumber growth stages and tissue types. *EF−1a*, *cyclophilin* and *ubiquitin* are the most appropriate reference genes for normalization of expression of microRNAsin leaf, stem and root samples using the comparative delta-Ct, geNorm, Norm Finder, Best Keeper and Ref Finder methods. The combination of two reference genes was needed to obtain highly reliable results of microRNAs expression. *Ubiquitin* and *EF−1a* were found to be the most suitable for normalization of expression of microRNAsin all tested cucumber samples. The conclusions provide the appropriate reference genes for characterizing cucumber microRNAs responding to CGMMV infection and facilitate expression and functional studies of cucumber microRNAs.

Key words: Reference genes; Cucumber; MicroRNAs; Expression normalization; Real-time quantitative polymerase chain reaction; Virus stress

---

* Funding: National Natural Science Foundation of China (NSFC) (Project No. 31371910) and Chinese Scholarship Council

** First author: Liang Chaoqiong, PhD student, mainly focused on the study of plant virus and host interactions; E-mail: lcq19880305@126.com

*** Corresponding author: Li Jianqiang, research field focused on Seed Pathology and Fungicide Pharmacology; E-mail: lijq231@cau.edu.cn

Molecular detection of *Potato virus X* (PVX), *Potato virus Y* (*PVY*), *Potato leaf roll virus* (PLRV) and *Potato virus S* (PVS) from Bangladesh*

M. Rashid**, Zhang Xiaoyang, Wang Ying, Han Chenggui***

(*Department of Plant Pathology and State Key Laboratory for Agro-Biotechnology, China Agricultural University*, *Beijing* 100193, *China*)

Abstract: Potato (*Solanum tuberosum*) is considered as the fourth most important staple food source after rice (*Oryza sativa*), maize (*Zea mays*) and wheat (*Triticum aestivum*) which is the most popular and important vegetable in Bangladesh. But it is frequently affected by different viruses and cause a huge yield loses. Viruses known to infect potato in Bangladesh include Potato virus X (PVX), Potato virus Y (PVY), Potato leaf roll virus (PLRV) and Potato virus S (PVS). Though the increasing incidence of PVX, PVY and PLRV in main potato growing areas of Bangladesh is getting an alarming position andcaused significant yield losses, no molecular study had done before this work. In this research, molecular detection of potato viruses was done from the potato tuber collected from Munshiganj and Jessore districts of Bangladesh. To detect the plant viruses, total RNAs were extracted from the virus affected potato tubers and complementary cDNAs were synthesized. Reverse transcriptase polymerase chain reaction (RT-PCR) based detection conditions were optimized by using coat protein (CP) gene specific primers. After RT-PCR we found the amplified fragments of expected length for PVX, PVY, PLRV and PVS that were confirmed by subsequent sequencing. This is the first report for PVS in Bangladesh. Our results will be helpful for further molecular study on PVX, PVY, PLRV and PVS and their management.

Key words: Molecular detection; Potato virus X (PVX); Potato virus Y (PVY); Potato leaf roll virus (PLRV); Potato virus S (PVS); Bangladesh

---

* Funding: China Scholarship Council, Ministry of Education, China

** First author: M. Rashid, PhD student, Major in Plant Pathology, China Agricultural University; E-mail: mamun_ 1961@ yahoo. com

*** Corresponding author: Han Chenggui; E-mail: hanchenggui@ cau. edu. cn

香橼中柑橘裂皮类病毒子代种群的分子变异*

王亚飞**，周常勇***，曹孟籍***
（西南大学柑桔研究所，重庆　400712）

摘　要：柑橘裂皮类病毒（*Citrus exocortis viroid*，CEVd）属于马铃薯纺锤块茎类病毒科（Pospiviroidae）马铃薯纺锤块茎类病毒属（*Pospiviroid*），是大小约为370nt，不编码任何多肽的闭合环状的单链小分子RNA。其基因组包含5个功能区，分别为左手末端区，致病区，中央区，可变区以及右手末端区。CEVd为害以枳和枳橙为砧木的柑橘植株，能引起砧木明显的裂皮症状，造成严重为害。'Arizona 861-S1'香橼常用来作为CEVd的指示鉴定植物，本研究借助从柑橘品种'眉山9号'中获取的CEVd主序列通过侵染性克隆构建和体外转录成功侵染香橼植株，待香橼植株出现明显症状后对CEVd子代种群分子变异情况进行分析，旨在了解CEVd在香橼中的子代种群信息，为防治柑橘裂皮病提供科学依据。

笔者利用CEVd的特异性引物从接种香橼中扩增到371bp的特异条带，然后通过分子克隆及序列测定建立CEVd子代种群，进而对其遗传多样性和变异水平进行分析。结果表明：子代种群突变克隆百分比为41.6%，所得77个克隆中有32个克隆发生突变，种群碱基突变频率为1.79×10^{-3}。子代种群共检测到33个碱基替代突变，15个碱基缺失突变和3个碱基插入突变。碱基突变以碱基替代突变为主，其中有8个C→U突变，8个U→C突变，7个A→G突变，4个G→C突变，3个G→A突变，2个G→U突变，1个C→A突变，碱基转换与碱基颠换比值为26∶7，碱基替代突变中碱基转换占比达78.9%。3个碱基插入突变分别为1个碱基C插入，1个碱基G插入，1个碱基A插入，没有发现碱基U的插入。15个碱基缺失突变中有11个碱基G缺失，3个碱基A缺失，1个碱基C缺失，没有发现碱基U的缺失。对突变区域进一步分析发现，突变主要集中在左手末端区，右手末端区以及可变区，而在致病区及中央区发现的碱基突变较少。

关键词：柑橘裂皮类病毒（*Citrus exocortis viroid*，CEVd）；香橼；种群；分子变异

---

* 基金项目：国家自然科学基金（31501611）

** 第一作者：王亚飞，博士研究生，分子植物病理学方向；E-mail：776545177@qq.com

*** 通信作者：周常勇，博导，研究员，主要从事分子植物病毒学研究；E-mail：zhoucy@cric.cn
曹孟籍，副研究员，主要从事植物病理学研究；E-mail：caomengji@cric.cn

江苏省牡丹病毒病原分子检测与鉴定

贺 振*，陈春峰，陈孝仁
（扬州大学园艺与植物保护学院，扬州 225009）

摘 要：牡丹（*Paeonia suffruticosa*）隶属于毛茛科（Ranunculaceae）芍药属（*Paeonia*），落叶小灌木。牡丹雍容华贵，国色天香，在中国有悠久的栽培历史和巨大的栽培规模，被推崇为“花王”。病毒病是牡丹上的一种重要病害，典型症状包括叶片上产生大小不一的环斑、轮斑或者不规则病斑，部分叶片具有黄化、花叶和斑驳等症状。目前，已报道的病原物主要有三种，包括烟草脆裂病毒（*Tobacco rattle virus*，TRV）、苜蓿花叶病毒（*Alfalfa mosaic virus*，AMV）和番茄斑萎病毒（*Tomato spotted wilt virus*，TSWV）。国际上，1976 年，TRV 首次在日本牡丹上发现；2001 年，在意大利帕尔玛大学植物园发现 AMV 可以侵染日本牡丹；2009 年，首次在美国阿拉斯加州牡丹上发现了 TSWV。在国内，牡丹病毒病于 1990 年被首次报道，经过生物学鉴定为烟草脆裂病毒的牡丹分离物（TRV-Pa），但直到目前仍缺少分子检测鉴定方面的确切报道。

本研究在扬州地区采集到 14 个具有典型花叶、斑驳症状的牡丹样品，经 RT-PCR 和 ELISA 法检测发现多数样品 TRV 呈阳性，检出率为 85.7%，未发现 AMV 和 TSWV 存在。在 TRV 阳性样品中，任意选取一个样品，对其 ORF1 部分保守区序列克隆测序，经 BLAST 比对发现，该分离物与已报道 TRV-ppk20 具有 99%的一致率。通过本研究，笔者在分子角度上明确扬州地区牡丹病毒病病原为 TRV，为牡丹病毒病害的有效防控提供了理论依据。

关键词：牡丹；病毒；分子检测

* 通信作者：贺振，讲师，主要从事分子植物病毒学研究；E-mail：hezhen@ yzu. edu. cn

CGMMV侵染后的西瓜果实转录组初步分析

李晓冬*，夏子豪，安梦楠，吴元华**
（沈阳农业大学植物保护学院，沈阳　110866）

摘　要：黄瓜绿斑驳花叶病毒（*Cucumber green mottle mosaic virus*，CGMMV）是烟草花叶病毒属成员之一，在世界范围内严重为害葫芦科作物生产。CGMMV侵染会导致西瓜（*Citrullus lanatus*）植株叶片产生典型的花叶症状，果实往往会发生腐烂、酸化等果质败坏现象。目前已有一些研究对葫芦科植物寄主响应CGMMV侵染的miRNA及vsiRNA展开了探讨，而关于CGMMV侵染对西瓜果实的寄主转录组表达的影响鲜见报道。因此，本文利用高通量测序技术对接种及未接种CGMMV的西瓜果实的转录谱进行了RNA-seq测序及差异分析。测序结果发现共有1 621个基因（DEGs）在两组果实样品的转录谱中显著差异表达，其中1 052个DEGs上调表达，569个DEGs下调表达；仅在接种CGMMV的样品中表达的DEGs有45个，且均为上调。整体上西瓜对CGMMV的侵染表现为积极响应。利用GO、COG、KEGG等数据库对DEGs的生物学功能进行注释。GO分析表明，共有683个DEGs参与分子功能（MF），1 562个参与生物过程（BP），1 187个参与细胞组分（CC）。GO富集结果表明：DEGs主要富集于结合（binding）和催化活性（catalytic activity）等MF功能，代谢过程（metabolic process）和细胞过程（cellular process）等BP功能，及细胞（cell）和细胞部分（cell part）等CC功能。COG分析表明，共有828个DEGs被归至23个分类中，主要涉及信号转导机制（signal transduction mechanisms）、转录后修饰、蛋白翻转及伴随蛋白（posttranslational modification，protein turnover，chaperones）、翻译、核糖体结构及生物起源（translation，ribosomal structure and biogenesis）、碳水化合物运输及代谢（carbohydrate transport and metabolism）、氨基酸运输及代谢（amino acid transport and metabolism）、转录（transcription）等。KEGG分析表明，DEGs共涉及296条代谢通路，主要参与代谢途径（metabolic processes）、次级代谢产物生物合成（biosynthesis of secondary metabolites）、核糖体（ribosome）、碳循环（carbon metabolism）、淀粉和蔗糖代谢（starch and sucrose metabolism）等。本研究对西瓜果实响应CGMMV侵染后的转录谱进行测定与分析，能够为CGMMV的致病机理及其与寄主互作等研究提供有力参考。

关键词：黄瓜绿斑驳花叶病毒（CGMMV）；西瓜（*Citrullus lanatus*）；转录组测序（RNA-seq）；生物信息学分析

* 第一作者：李晓冬，女，在读博士，研究方向：植物病毒学；E-mail：lxdong1221@126.com
** 通信作者：吴元华，男，教授，博士生导师，研究方向：植物病毒学和生物农药；E-mail：wuyh09@vip.sina.com

Study on Temperature Sensitivity of *Chinese Wheat Mosaic Furovirus* (CWMV) by Its Infectious Clones*

Yang Jian, Zhang Fen, Li Jing, Chen Jianping**, Zhang Hengmu**

(*State Key Laboratory Breeding Base for Zhejiang Sustainable Pest and Disease Control*; *Key Laboratory of Plant Protection and Biotechnology*, *Ministry of Agriculture*, *China*; *Zhejiang Provincial Key Laboratory of Plant Virology*, *Institute of Virology and Biotechnology*, *Zhejiang Academy of Agricultural Sciences*, *Hangzhou* 310021, *China*)

Abstract: *Chinese wheat mosaic virus* (CWMV) is one of recognised members within the genus *Furovirus* (family *Virgaviridae*) which naturally infects cereal plants and transmitted by *Polymyxa graminis*. Infected plants display typical symptoms, including light chlorotic streaking on the young leaves and bright yellow chlorotic streaking or even purple chlorotic stripes on old leaves. The virus has rigid rod-shaped particles with predominant lengths of 140 ~ 160nm and 260 ~ 300nm and contains a bipartite single-strand positive RNA (ssRNA) genome, which encodes three proteins, required for viral replication and movement, on RNA1 and four proteins, a major coat protein (CP), two minor CP-related proteins, and a RNA silencing suppressor, on RNA2. In the past years, we firstly produced the infectious full-length cDNA clones of CWMV and initiated its reverse genetics in wheat and *N. benthamiana* plants. Previous studies have shown that CWMV infects wheat and *N. benthamiana* at 16℃ but not at 24℃ and that the optimal temperature for replication of *Soil-borne wheat mosaic virus* (SBWMV, another member of genus *Furovirus*) was 17℃ inbarley protoplasts. However, the mechanism of temperature dependence remains largely unknown. To characterizethe biological properties of CWMV, here its infectious full-length cDNA clones were applied to analyze the viral temperature sensitivity. The infectious constructs were agro-infiltrated onto *N. benthamiana* plants grown at a range of temperatures (including 12℃, 15℃, 17℃, 20℃, and 25℃). Two weeks after infiltration, the inoculated and upper leaves were tested by northern blotting assays. Interestingly, in the inoculated leaves the accumulation of viral RNAs was greatest at 12℃, next at 15℃ and much less at 25℃. However, in the upper leaves, the accumulation of viral RNAs was greatest at 17℃, next at 15℃ and there were almost no signals of viral RNA accumulation at lower (12℃) or higher temperatures (20℃ and 25℃). These results indicated that CWMV replicates best at low temperatures but that the optimal temperature for systemic infection was 17℃, putatively reflecting the adaption of the virus to plants that grow in a cool climate.

* Funding: China Agriculture Research System (CARS-3-1) from the Ministry of AgricultureandProject of New Varieties of Genetically Modified Wheat of China (2016ZX08002001)

** Corresponding authors: Zhang Hengmu; E-mail: zhhengmu@ tsinghua. org. cn

Chen Jianping; E-mail: jpchen2001@ 126. com

Interaction between a Furoviral Replicase and Host HSP70 Promotes the Furoviral RNA Replication*

Yang Jian, Zhang Fen, Cai Nianjun, Wu Ne, Chen Xuan, Li Jing,
Chen Jianping**, Zhang Hengmu**

(*State Key Laboratory Breeding Base for ZhejiangSustainable Pest and Disease Control*; *Key Laboratory of Plant Protection and Biotechnology*, *Ministry of Agriculture*, *China*; *Zhejiang Provincial Key Laboratory of Plant Virology*, *Institute of Virology and Biotechnology*, *Zhejiang Academy of Agricultural Sciences*, *Hangzhou* 310021, *China*)

Abstract: RNA viruses are ubiquitous biotic stresses on both unicellular and multicellular organisms and include many important pathogens of humans, domestic and farm animals and of crops. Their successful infection and symptom development depends on complex molecular interactions between viral and host factors. In recent years, many host factors have been identified that interact with viral genes and are involved in viral infection cycles and symptom expression. Among these, heat shock proteins (HSPs), highly conserved and ubiquitous cytoprotective proteins, often play crucial roles in important processes, including replication, virion assembly, and intracellular movement. However, although furoviruses cause important diseases of cereals worldwide, no host factors have yet been identified that interact with furoviral genes or participate in the viral infection cycle. *Chinese wheat mosaic virus* (CWMV) is a member of the genus *Furovirus*, family *Virgaviridae*, and causes a damaging disease of cereal plants in China. Recently, its infective full-length cDNA clones have been developed. In this study, both TaHSP70 and NbHSP70 were up-regulated in plantsagro-infiltrated with its infectious cDNA clones. Their overexpression and inhibition were correlated with the accumulation of viral genomic RNAs, suggesting that the HSP70 genes could be necessary for CWMV infection. The subcellular distributions of TaHSP70 and NbHSP70 were significantly affected by CWMV infection or by infiltration of RNA1 alone. Further assays showed that the viral replicase encoded by CWMV RNA1 interacts with both TaHSP70 and NbHSP70 in *vivo* and *vitro* and that its region aa167-333 was responsible for the interaction. Subcellular assays showed that the viral replicase could recruit both TaHSP70 and NbHSP70 from the cytoplasm or nucleus to the granular aggregations or inclusion-like structures on the intracellular membrane system, suggesting that both HSP70s may be recruited into the viral replication complex (VRC) to promote furoviral replication. This is the first host factor identified to be involved in furoviral infection, which extends the list and functional scope of HSP70 chaperones.

* Funding: China Agriculture Research System (CARS-3-1) from the Ministry of AgricultureandProject of New Varieties of Genetically Modified Wheat of China (2016ZX08002001)

** Corresponding authors: Zhang Hengmu; E-mail: zhhengmu@ tsinghua. org. cn
Chen Jianping; E-mail: jpchen2001@ 126. com

Characterization of interaction between *southern rice black-streaked dwarf virus* (SRBSDV) minor core protein P8 and a rice zinc finger transcription factor*

Li Jing[1], Xue Jin[1,2], Zhang Hengmu[1**], Yang Jian[1],
Li Xie[1], Chen Jianping[1**]
(1. *State Key Laboratory Breeding Base for Zhejiang Sustainable Pest and Disease Control*; *Key Laboratory of Plant Protection and Biotechnology*, *Ministry of Agriculture*, *China*; *Zhejiang Provincial Key Laboratory of Plant Virology*, *Institute of Virology and Biotechnology*, *Zhejiang Academy of Agricultural Sciences*, *Hangzhou* 310021, *China*; 2. *College of Bio-Safety Science and Technology*, *Hunan Agricultural University*, *Changsha* 410128, *China*)

Southern rice black-streaked dwarf virus (SRBSDV) (or Rice black-streaked dwarf virus 2, RBSDV-2) is a recently-described member of the genus *Fijivirus* within the family *Reoviridae* and causes one of the most serious viral diseases of rice in China and Vietnam. Its virion consists of multiple genomic dsRNA segments and structural proteins, making it an excellent model for complex macromolecular interaction. The completion of SRBSDV genomic sequences and construction of its full-length cDNA libraries provides a practical base for further investigation of molecular interactions. Previous studies showed that the complex homologous or heterologous interactions between viroplasm proteins P5-1, P6 and P9-1, were necessary for viroplasm-like structure (VLS) formation. To better understand the molecular basis of SRBSDV infection, a yeast two-hybrid screen of a rice cDNA library was carried out using P8, a minor core protein of SRBSDV, as the bait. A rice Cys2His2-type zinc finger protein (OsZFP) was found to interact with SRBSDV P8. A strong interaction between SRBSDV P8 and OsZFP was then confirmed by pull-down assays, and bimolecular fluorescence complementation assays then showed that the in vivo interaction was specifically localized in the nucleus of plant cells. Using a series of deletion mutants, it was shown that both the NTP-binding region of P8 and the first two zinc fingers of OsZFP were crucial for their interaction in plant cells. The localization in the nucleus and activation of transcription in yeast supports the notion that OsZFP is a transcription factor. SRBSDV P8 may play an important role in fijiviral infection and symptom development by interfering with the transcription activity of OsZFP in host plants.

---

* Funding: National Science and Technology Support Program (2012BAD19B03) and the State Basic Research Program of China (2014CB160309)

** Corresponding authors: Zhang Hengmu; E-mail: zhhengmu@ tsinghua. org. cn
Chen Jianping; E-mail: jpchen2001@ 126. com

茉莉 C 病毒外壳蛋白的原核表达和特异性抗血清的制备*

陈梓茵，江朝杨，陆承聪，韩艳红**
（福建农林大学，海峡联合研究院，福州 350002）

摘 要：茉莉 C 病毒（*Jasmine virus C*，JaVC）是侵染茉莉（*Jasminum sambac*）的一种病毒，该病毒属于乙型线性病毒科（Betaflexiviridae），香石竹潜隐病毒属（*Carlavirus*），首先发现于台湾地区。现有研究表明，JaVC 可能是茉莉黄化嵌纹病的病原之一；本实验室尚未发现 JaVC 单独侵染的茉莉植株。为了便于进一步的研究，根据茉莉 C 病毒台湾分离物（JaVC-TW）（GenBank No. NC_ 030926. 1）和本实验室测定的序列，采用引物 carCPF：5′-AAGAAGGAGATATACCATGAGTTCCACTTCAGACGACAATTC-3′（下划线为载体序列）和 carCPR：5′-TGGTGGTGGTGGTGGTGCTCTCTTTTCCAAATCCCCCGTTC-3′（下划线为载体序列）扩增获得 JaVC 外壳蛋白（CP）基因的全长片段。利用 Gibson 组装（NEB # E5510S）的方法将该片段克隆到 pET-28a（+）载体上，获得重组质粒 pET-28aCP。SDS-PAGE 分析显示，pET-28aCP 在大肠杆菌 Rosetta 中，经 1mmol/L IPTG 诱导后，可特异性的表达分子量大小约为 33ku 的融合蛋白，但是该蛋白只存在于包涵体中。将包涵体蛋白溶于高浓度尿素溶液中，经 Ni-NTA Agarose 亲和层析柱纯化后，获得高纯度的目的融合蛋白。Western blot 检测纯化后的蛋白，结果表明该融合蛋白可被 His 抗体识别，证实了目的蛋白成功表达并且融合表达了 His 标签。用纯化后的 CP 融合蛋白作为抗原免疫健康白兔，制备了特异性的抗血清。

关键词：茉莉；C 病毒；抗血清

---

* 基金项目：本研究得到福建省茶产业重大农技推广服务试点建设项目（KNJ-151053）资助

** 通信作者：韩艳红；E-mail：yan-hong@ fafu. edu. cn

一种侵染茉莉的番茄丛矮病毒科病毒的发现和分子鉴定*

朱丽娟，陆承聪，江朝杨，陈梓茵，韩艳红**
（福建农林大学，海峡联合研究院，福州　350002）

摘　要：茉莉（*Jasminum sambac*）又名茉莉花，属于木樨科（Oleaceae）素馨属（*Jasminum* Linn.），原产于印度、巴基斯坦，我国早已引种，且具有重大的经济和文化价值。之前茉莉病毒病的研究，因为电镜观察等的局限性多为具有线性病毒粒子的病毒。已报道的茉莉病毒有马铃薯 Y 病毒属的茉莉 T 病毒（*Jasmine virus* T）和香石竹潜隐病毒属（*Carlavirus*）的茉莉 C 病毒（*Jasmine virus* C）。本实验室采集疑似病毒病茉莉样品，利用小 RNA 二代测序和拼接发现并克隆（GenBank No. KX897157）了一种新的侵染茉莉的番茄丛矮病毒科（Tombusviridae）病毒的全序列，并暂命名为茉莉 H 病毒（Jasmine virus H，JaVH）。该病毒的基因组全长序列为 3 867nt，软件预测含有 5 个开放阅读框。该病毒的基因组序列同番茄丛矮病毒科其他属病毒的相似性在 41.48%~58.43%，与天竺葵环斑病毒属（*Pelarspovirus*）的相似度最高，因此认为该病毒为天竺葵黄斑病毒属成员。分析比较复制酶和外壳蛋白序列的相似性，得出相同的结论。复制酶和外壳蛋白与该属成员玫瑰叶畸形病毒（*Rosa rugose leaf distortion virus*，RrLDV）的相似度最高，分别为 58.36% 和 52.80%，远低于成立该属新成员的标准，因此 JaVH 为天竺葵黄斑病毒属的一个新成员。

此外，分析茉莉样品小 RNA 序列，结果表明来源于 JaVH 的小 RNA 主要大小为 20nt、21nt 和 22nt，Northern 杂交也验证了这一现象。不同于其他多数正单链 RNA 病毒，病毒小 RNA 主要来源于病毒的正链，JaVH 病毒小 RNA 来源于正链和负链的比例接近，且偏向性很小。

关键词：茉莉；番茄丛矮病毒；分子鉴定

---

* 基金项目：本研究得到福建省茶产业重大农技推广服务试点建设项目（KNJ-151053）资助

** 通信作者：韩艳红；E-mail：yan-hong@ fafu. edu. cn

ToMV 侵染可引起 IP-L 表达量上调*

彭浩然**，蒲运丹，薛　杨，叶思涵，青　玲，孙现超***
（西南大学植物保护学院，重庆　400715）

摘　要：前期研究显示 ToMV 外壳蛋白（coat protein，CP）可以与烟草蛋白 L（CP-interacting protein-L，IP-L）相互作用，本研究旨在明确 ToMV CP 与 IP-L 共定位情况和在 ToMV 侵染条件下烟草中 IP-L 及其编码蛋白的表达变化情况。为进一步明确 IP-L 的功能提供依据。通过双酶切法从本实验室构建保存的 pGBKT7-CP 和 pGADT7-IP-L 重组载体上切下 ToMV CP 和 IP-L 目的片段，构建融合蛋白植物表达载体 pPZP-IP-L-N-EGFP 和 pPZP-CP-N-DsRed 及原核表达载体 pEGX-IP-L。通过热激法将融合表达的 EGFP 和 DsRed 植物表达载体转化至农杆菌 EHA105，在烟草表皮细胞瞬时表达 IP-L-N-EGFP 和 CP-N-DsRed，共聚焦荧光显微镜下观察 IP-L-N-EGFP 和 CP-N-DsRed 共定位情况。用实时荧光定量 RT-PCR 分析 IP-L 在本氏烟各组织的表达量。原核表达载体 pEGX-IP-L，转化原核表达菌 BL21，优化 GST-IP-L 可溶性表达条件后大量表达蛋白，用 GST 亲和层析法纯化获得可溶性 GST-IP-L 蛋白，免疫家兔制备多克隆抗体。ELISA 测定抗体效价，Western 印迹法明确抗体的特异性后，利用该抗体分析 ToMV 侵染条件下番茄 IP-L 蛋白表达情况，用实时荧光定量 RT-PCR 检测 IP-L 表达变化与蛋白水平是否一致。结果显示：ToMV CP 在本氏烟叶片表皮细胞的细胞质和细胞质膜以及叶绿体均有分布，IP-L 只在本氏烟叶片表皮细胞质膜表达，二者共定位在细胞质膜。IP-L 在本氏烟叶片中特异高表达，显著高于茎、根和花中的相对表达量。在温度为 30℃，IPTG 诱导浓度为 0.3mmol/L 条件下表达出大小为 42.8ku 的可溶性 GST-IP-L 融合蛋白。纯化获得约 4.2mg 可溶性蛋白，免疫家兔制备了效价为 1/6 400的多克隆抗体。Western 印迹结果表明：该抗体可以与 IP-L 的原核产物特异性结合。Western 印迹分析表明 IP-L 在 ToMV 侵染本氏烟第 7d 后再在叶片内表达量明显上调，实时荧光定量 RT-PCR 检测结果显示 IP-L 在 ToMV 侵染本氏烟第 7d 后的叶片内表达量是健康对照的 3 倍多，差异达到显著水平。

---

* 基金项目：国家自然科学基金（31670148）；重庆市社会事业与民生保障创新专项（cstc2015shms-ztzx80011，cstc2015shms-ztzx80012）；中央高校基本科研业务费专项资金资助项目（XDJK2016A009，XDJK2017C015）

** 第一作者：彭浩然，男，硕士研究生，从事植物病理学研究

*** 通信作者：孙现超，博士，研究员，博士生导师，主要从事植物病毒学及植物病害生物防治研究；E-mail：sunxianchao@163.com

本氏烟中瞬时表达番茄 SYTA 可以促进 TMV 侵染移动*

潘　琪**，彭浩然，薛　杨，吴根土，青　玲，孙现超***
（西南大学植物保护学院，重庆　400715）

摘　要：Synaptotagmin 最早是在大型致密核心囊泡（large dense core vesicle，LDCV）和突触囊泡的膜表面上发现的抗原，1990 年被命名为 Synaptotagmin。Synaptotagmin 作为植物中第一个被发现的参与细胞膜修复的蛋白质，能够在植物细胞中广泛表达，并且定位于细胞质膜上，为了克隆获得番茄 SYTA（*S. lycopersicum* STYA，S. l SYTA），分析基因序列生物信息学特征和预测蛋白的结构特征，明确 S. l SYTA 亚细胞定位和组织表达，并分析其在绿色荧光蛋白（Green Fluorescence Protein，GFP）标记的烟草花叶病毒（*Tobacco mosaic virus*，TMV）侵染下的表达变化及其对 TMV 移动的影响。笔者根据番茄基因组含有的 SYTA 同源基因序列，利用 Primer Premier 5. 0 软件设计克隆引物，采用 RT-PCR 技术克隆 *S. l SYTA* 全长序列；应用生物信息学方法分析该基因的序列特征；使用 MEGA7. 0 对 S. l SYTA 蛋白序列及其同源序列进行多序列比对，并构建同源物种间系统进化树；通过与 GFP 蛋白融合进行亚细胞定位；利用实时荧光定量 PCR（qRT-PCR）检测番茄各个部位 *S. l SYTA* 的表达量以及在 TMV 胁迫的番茄中 *S. l SYTA* 的表达变化；构建植物瞬时表达载体 pCV-SYTA-mGFP，通过农杆菌介导在本氏烟草中瞬时表达，TMV-GFP 攻毒，ELISA 检测在本氏烟中瞬时表达 S. l SYTA 时 TMV-GFP 的积累和移动情况。结果显示克隆得到 1 620bp 的 *S. l SYTA* 基因开放阅读框全长，序列比对及生物信息学分析表明，其编码的氨基酸序列具有 SYT 家族的典型特征，含有 N 端的跨膜区、胞间连接区和 C 端的两个 C2 结构域；多序列对比及系统进化树分析发现，与茄科林生烟草、绒毛状烟草等植物亲缘关系较近，与黄瓜较远；亚细胞定位显示 S. l SYTA 定位在细胞质膜。在番茄的根、茎和叶中 *S. l SYTA* 的表达量从高到低依次为根>叶>茎；TMV-GFP 侵染番茄导致其 *S. l SYTA* 表达量在接种后第 1d 显著上调，在第 3d 下降至正常水平。在 S. l SYTA 瞬时表达的本氏烟叶片部位接种 TMV-GFP，TMV-GFP 在接种第 5d 时已经到达新叶，而接种部位仅表达空载体对照的本氏烟新叶中未观察到 TMV-GFP，且接种第 5d 时 TMV-GFP 在接种叶和新叶中的积累量均明显高于其在空载体对照处理的叶片。该结果表明在本氏烟中瞬时表达 S. l SYTA 有利于 TMV-GFP 侵染初期的积累和移动。

---

* 基金项目：国家自然科学基金（31670148）；重庆市社会事业与民生保障创新专项（cstc2015shms-ztzx80011，cstc2015shms-ztzx80012）；中央高校基本科研业务费专项资金资助项目（XDJK2016A009，XDJK2017C015）

** 第一作者：潘琪，女，硕士研究生，从事植物病理学研究

*** 通信作者：孙现超，博士，研究员，博士生导师，主要从事植物病毒学及植物病害生物防治研究；E-mail：sunxianchao@163. com

芸薹黄化病毒含有 GFP 标记基因侵染性 cDNA 克隆的构建[*]

陈清华[**]，赵添羽，张晓燕，王　颖，张宗英，李大伟，于嘉林，韩成贵[***]
（中国农业大学作物有害生物监测与绿色防控农业部重点实验室，北京　100193）

摘　要：芸薹黄化病毒（Brassica yellows virus，BrYV）是本实验室发现的一种新的马铃薯卷叶属病毒，与芜菁黄化病毒的亲缘关系较近。BrYV 是一种正单链 RNA 病毒，在我国分布广泛，主要侵染十字花科植物，能够引起植株叶片黄化和卷曲的症状。BrYV 基因组 RNA 含有 7 个 ORFs，从 5′端通过基因组翻译的 ORFs 依次产生 P0、P1 和 P2，通过亚基因组策略翻译的 ORFs 依次产生 P3、P3a、P4 和 P3～P5，根据 BrYV 基因组 5′端的序列特点可分为 A、B、C 三种基因型。本研究在已经构建 BrYV-A 型侵染性 cDNA 克隆的基础上，通过利用同源重组的方法进一步进行带有标记基因的侵染性 cDNA 克隆的改造修饰。

本研究成功构建了 3 种含有 GFP 标记基因的 BrYV 侵染性 cDNA 克隆。（1）在 BrYV 基因组的 ORF2 后插入 2A-sGFP，该侵染性 cDNA 克隆简称为 P2-2A-sGFP。（2）在 BrYV 基因组的 ORF5 后插入 2A-sGFP，该侵染性 cDNA 克隆简称为 P5-2A-sGFP。（3）将 BrYV 基因组的 ORF5 缺失，并在 ORF3 后插入 2A-sGFP，该侵染性 cDNA 克隆简称为 ΔP5-2A-sGFP。通过农杆菌浸润注射的方法将这 3 种载体浸润注射到本生烟叶片上，注射浓度 $OD_{600}=0.5$，注射 3 d 后，运用 Western blot 方法分别检测病毒的外壳蛋白 CP 和 GFP 的表达量。结果表明：P2-2A-sGFP 的 CP 表达量最少，GFP 的表达量最多；P5-2A-sGFP 与 ΔP5-2A-sGFP 的 CP 表达量正常，但前者 GFP 的表达量要多于后者。对于上述 3 种侵染性 cDNA 克隆是否能够系统侵染本生烟并表达 GFP，尚需进一步对系统叶进行检测研究。本研究为深入研究 BrYV 的基因功能提供了有效 GFP 标记的病毒载体工具。

关键词：芸薹黄化病毒；cDNA 克隆；GFP 标记基因

致谢：感谢王献兵教授和张永亮副教授对本研究工作的建议。

* 基金项目：国家自然科学资金面上项目（批准号 31371909 和 31671995）

** 第一作者：陈清华，硕士生，主要开展植物病毒与寄主互作研究；E-mail：chenqinghua@cau.edu.cn
*** 通信作者：韩成贵，教授，主要从事植物病毒学与抗病毒基因工程；E-mail：hanchenggui@cau.edu.cn

玉簪提取物生物碱抗烟草花叶病毒（TMV）转录组分析*

陈雅寒[1**]，谢咸升[2]，翟颖妍[1]，成巨龙[3]，张 强[3]，安德荣[1***]

（1. 西北农林科技大学植物保护学院，旱区作物逆境生物学国家重点实验室，杨凌 712100；
2. 山西省农业科学院小麦研究所，临汾 041000；3. 陕西省烟草公司烟草所，西安 7100068）

摘 要：玉簪中富含生物碱类化学成分，抗病毒效果显著。本试验于本氏烟 5~6 叶期设喷清水对照（T01）、喷生物碱（T02）、接 TMV 前 24h 喷生物碱（T03）、接种 TMV（T04）和接 TMV 后 24h 喷生物碱（T05）5 个处理，3 次重复，处理 72h 后采摘各处理顶部第 2 片烟叶经液氮速冻后干冰送样，混样提取总 RNA，构建 RNA 文库，经 HiSeq 2500 测序，读长为双末端 PE125。原始 reads 经过滤及去除过长、过短序列获得 clean reads 数据，与本氏烟参考基因组对比获得 Mapped Data，比较分析了基因表达差异。试验结果表明：样品 T05、T03、T04 基因表达水平比清水对照 T01 增加；接 TMV 前 24h 喷生物碱（T02）与接种 TMV（T04）处理之间存在大量显著性差异的基因表达；喷生物碱后 24h 接 TMV（T03）的差异基因数比接 TMV 后 24h 喷生物碱（T05）的少。因此，在病毒胁迫时喷施生物碱能起到调控基因表达的作用，免疫调节作用明显，而且预防作用大于治疗作用。

关键词：烟草花叶病毒；玉簪；生物碱；转录组分析

---

* 基金项目：国家“863”计划（2012AA101504）；国家公益性行业（农业）科研专项（201203021）；国家自然科学基金（31471816）

** 第一作者：陈雅寒，女，在读博士，主要从事植物保护研究；E-mail：562991357@qq.com

*** 通信作者：安德荣，男，教授，博士生导师，主要从事微生物资源利用及植物病毒学研究；E-mail：anderong323@163.com

北京地区不同草莓品种5种主要病毒的检测*

褚明昕[1**]，魏　然[1]，席　昕[2]，卢　蝶[1]，邢冬梅[3]，尚巧霞[1***]

（1. 农业应用新技术北京市重点实验室，植物科学技术学院，北京农学院，北京　102206；
2. 北京市植物保护站，北京100029；3. 北京市昌平区植保植检站，北京　102200）

摘　要：草莓（*Fragaria ananassa* Duch.）是蔷薇科草莓属多年生草本植物，其产量在世界小浆果中居首位。无性繁殖的草莓植株一旦受到病毒侵染，病毒会随着无性繁殖材料传播扩散。经过几年的带毒栽培，引进品种的品质变劣、退化现象普遍发生，造成草莓严重减产，1年内可达30%~80%，且危害逐年递增。目前北京地区草莓生产上造成严重损失的主要有5种病毒，包括草莓轻型黄边病毒（*Strawberry mild yellow edge virus*，SMYEV）、草莓斑驳病毒（*Strawberry mottle virus*，SMoV）、草莓镶脉病毒（*Strawberry vein banding virus*，SVBV）和草莓皱缩病毒（*Strawberry crinkle virus*，SCV）和黄瓜花叶病毒（*Cucumber mosaic virus*，CMV）。CMV是于2014年首次报道的可侵染草莓植株的病毒种类。为了明确不同草莓品种病毒病的发生情况，本文对北京地区栽培的21个草莓品种进行了主要病毒病发生情况调查和病毒检测。

从北京地区各草莓种植区采集有典型病毒症状或表现正常的草莓叶片样品进行以上5种病毒的RT-PCR检测，主要调查的品种有红颜、章姬、圣诞红、甜查理等21种草莓品种。SMYEV检测引物为YT_1（CCGCTGCAGTTGTAGGGTA）和Y_2（CATGGCACTCATTGGAGCTGGG），退火温度60℃，目的片段为861bp；SMoV检测引物为6126F1（GGTTTGAAGGAATAGGGTTGTTGG）和6732R1（CAGGTTACTCTAGTACGTCACCAC），退火温度58℃，目的片段为606bp；SVBV检测引物为SV5508F（TCGGGAAYTTGCAGGWAAAACATAG）和6606R（TACTCGTGATTCTCAGG-TAGATTGG），退火温度56℃，目的片段为1 098 bp；SCV检测引物为deta（CATTGGTG-GCAGACCCATCA）和detb（TTCAGGACCTATTTGATGACA），退火温度58℃，目的片段为345bp；CMV检测引物1F（GSAGGTGGTTAACGGTCTTT）和1R（TGCATCGTCTTTTGAATACACGA），目的片段大小为910bp，退火温度为56℃。

检测结果显示，北京地区草莓品种以红颜为主，其病毒侵染率为56.89%；从病毒侵染种类来看，SVBV能够侵染15种草莓品种，SMoV侵染5种草莓品种，SMYEV侵染4种草莓品种，SCV仅侵染红颜，CMV侵染红颜和章姬2种品种；5种被测草莓品种存在复合侵染情况。草莓受到病毒侵染后，从无明显症状到矮化、丛簇、花叶、畸形、斑驳、皱缩等症状表现多样，与感染的病毒种类间的关系并不明确。

关键词：草莓；病毒；检测

---

* 基金项目：北京市农业科技项目（20160127）

** 第一作者：褚明昕，硕士研究生，植物保护；E-mail：1184207948@qq.com

*** 通信作者：尚巧霞，副教授，研究方向：果蔬植物病毒学与果蔬病害防治研究；E-mail：shangqiaoxia@bua.edu.cn

CYVCV-TGB 基因的生物信息学分析及亚细胞定位研究[*]

崔甜甜[1**]，王艳娇[1]，李中安[1]，周常勇[1]，宋　震[1***]
（西南大学柑桔研究所，中国农业科学院柑桔研究所，国家柑桔工程技术研究中心，重庆　400712）

摘　要：柑橘黄化脉明病毒（*Citrus yellow vein clearing virus*，CYVCV）是印度柑橘病毒属（*Mandarivirus*）的一种正义单链 RNA 病毒，其基因组含有 6 个开放阅读框（ORF）。其中，ORF2、ORF3 和 ORF4 组成了三基因连锁结构（triple gene block，TGB）。已有研究表明 CYVCV 的 TGB 参与病毒在胞间的运动，但关于柑橘黄化脉明病毒 TGB 的基本性质、结构和具体作用机制尚不清晰。为进一步解析 TGB 相关信息，本研究以感染 CYVCV 的尤力克柠檬［*Citrus limon*（L.）Burm. f.］植株的总 RNA 为模板，扩增和克隆了 CYVCV 的 *TGB* 基因，并开展了其编码蛋白的理化性质和分子特性等生物信息学分析预测；进而通过 In-Fusion ©技术构建了 TGB 各基因的亚细胞定位载体，经农杆菌介导转化洋葱表皮细胞并在荧光显微镜下进行亚细胞定位观察。生物信息学分析结果表明：CYVCV-TGB 的理化性质和分子特性与 *Potexvirus* 属病毒相似，即 CYVCV-TGB 的运动需要外壳蛋白的参与。亚细胞定位结果显示：TGBp1 定位于细胞壁（膜）上，TGBp2 和 TGBp3 则少量呈星点状位于胞内，大部分位于细胞壁（膜）上。研究结果为揭示 CYVCV 在寄主中的运动机理奠定了基础。

关键词：柑橘黄化脉明病毒；生物信息学分析；三基因连锁结构；亚细胞定位

---

* 基金项目：国家公益性行业（农业）科研专项（201203076）；重庆市两江学者计划项目；重庆市基础及前沿研究项目（CSTC2014jcyjA80033）；中央高校基本科研业务专项（XDJK2017B024）

** 第一作者：崔甜甜，女，在读硕士研究生；E-mail：15736032158@163.com

*** 通信作者：宋震；E-mail：songzhen@cric.cn

柑橘黄化脉明病毒胶体金免疫层析检测试纸条的研制*

宾　羽**，宋　震，李中安，周常勇***
（中国农业科学院/西南大学柑桔研究所，国家柑桔工程技术研究中心，重庆　400712）

摘　要：柑橘黄脉病是由柑橘黄化脉明病毒（*Citrus yellow vein clearing virus*，CYVCV）引起的一种新发柑橘病毒病，随着该病的发生范围不断扩大，给柠檬产业带来极大损失。为更快速、简便地检测 CYVCV，对其进行有效的监测防治，本试验对 CYVCV 胶体金免疫层析检测试纸条进行研制和评估。选择 30nm 胶体金标记 CYVCV 单克隆抗体（1E1），将羊抗鼠 IgG 和 1E1 包被在硝酸纤维素膜上，分别作为质控线（C）和检测线（T），试纸条组装后对其进行评估。结果显示，该试纸条与柑橘衰退病病毒（*Citrus tristeza virus*，CTV）等 8 种柑橘常见病原物无交叉反应，特异性强；样品检测的最低稀释限度为 1：320［W/V，叶片组织重（g）/缓冲液（mL）］，灵敏度与 dot-ELISA 相当；检测结果于加样后 5～10min 即可获得。本实验研制的 CYVCV 胶体金免疫层析检测试纸条具有特异性强、快速、简便、低成本等优点，可推广用于田间柑橘黄脉病早期快速检测，进而指导病害防治。

关键词：胶体金免疫层析试纸条；柑橘黄化脉明病毒；快速检测

* 基金项目：农业部公益性行业（农业）科研专项（201203076）；两江学者计划；中央高校基本科研专项（XDJK2017B024）

** 第一作者：宾羽，女，博士研究生

*** 通信作者：周常勇，男，研究员，主要从事柑橘病毒类病害研究和无病毒繁育体系建设；E-mail：zhoucy@swu.edu.cn

广西主栽水稻品种抗水稻齿叶矮缩病鉴定*

谢慧婷**，崔丽贤，李战彪，秦碧霞**，陈锦清，蔡健和***

（广西壮族自治区农业科学院植物保护研究所/广西作物病虫害生物学重点实验室，南宁 530007）

摘　要：水稻齿叶矮缩病是东南亚、东亚和南亚一些国家的重要水稻病害，其病原水稻齿叶矮缩病毒（*Rice ragged stunt virus*，RRSV）属于呼肠孤病毒科（Reoviridae）稻病毒属（*Oryzavirus*），主要由褐飞虱（*Nilaparvata lugens*）以持久增殖型方式传播，不经卵传。发病田块一般减产10%~20%，严重的减产80%~100%，是影响水稻生产的主要病毒病害之一。近年来，水稻齿叶矮缩病在水稻种植区的发病率呈上升趋势，2006 年该病害在越南大面积暴发，2007 年以来，在我国福建、云南、海南和广西等地相继造成严重危害，水稻齿叶矮缩病的发生流行对水稻生产和粮食安全构成潜在威胁，开展水稻品种抗水稻齿叶矮缩病鉴定，对挖掘抗病品种材料、开展抗病性遗传分析和抗病基因定位研究及病害综合防控有重要意义。本研究采用苗期集团接种法对广西主栽水稻品种进行抗水稻齿叶矮缩病人工接种鉴定，根据发病症状并结合 RT-PCR 检测结果，调查接种品种和感病对照的发病率，明确了 45 份广西主栽水稻品种对水稻齿叶矮缩病的抗性。抗性鉴定结果表明：测试的 45 个主栽水稻品种人工接种后发病率 42.31%~96.55%，根据水稻病毒病抗性分级标准进行抗病性评价，表现高感的品种 41 个，占测试品种的 91.11%，表现中感的品种 4 个，占测试品种的 8.89%，在广西主栽水稻品种中未发现抗水稻齿叶矮缩病的品种。在后续研究中仍需加大对水稻品种和育种材料等的抗病鉴定筛选力度，以期为水稻齿叶矮缩病综合防控品种布局和抗病育种等研究储备抗性材料资源。

关键词：水稻品种；水稻齿叶矮缩病；抗性鉴定；人工接种

* 基金项目：广西科学研究与技术开发计划项目（桂科重 14121001-1-6）；广西农业科学院基本科研业务专项项目（桂农科 2014YZ26、桂农科 2016YM48）；广西作物病虫害生物学重点实验室基金项目（14-045-50-ST-09）；公益性行业（农业）科研专项经费项目（项目编号：201403075）

** 第一作者：谢慧婷，助理研究员，主要从事植物病理学研究；E-mail：huitingx@163.com

*** 通信作者：秦碧霞，副研究员，主要从事植物病毒学研究；E-mail：qbx33@163.com

蔡健和，研究员，主要从事植物病毒学研究；E-mail：caijianhe@gxaas.net

海南胡椒花叶病发生及为害现状*

车海彦**，刘培培，曹学仁，罗大全***
（中国热带农业科学院环境与植物保护研究所/
农业部热带作物有害生物综合治理重点实验室/
海南省热带农业有害生物监测与控制重点实验，海口 571101）

摘 要：胡椒（*Piper nigrum* L.）是一种重要的热带香辛作物，素有“香料之王”的美誉，经济寿命长达20~30年。目前我国胡椒收获面积和产量分别居世界第6位和第5位，海南胡椒生产占全国90%以上。胡椒花叶病是生产上的重要病害之一，在海南危害日趋严重，然而对胡椒花叶病在海南的发生分布等还缺乏系统研究。本研究通过对海南琼海、文昌、海口、万宁、定安等胡椒主栽区田间初步调查发现，胡椒花叶病在以上市县均有发生，主要症状为花叶、斑驳，叶片变小、变狭、卷曲皱缩，节间变短，生长缓慢等。调查的大部分胡椒园发病率在10%~20%，发病严重的胡椒园可高达87.5%。各市县中琼海的发病率最高达36%，发病最轻的为万宁7.5%，整个海南胡椒主栽区的发病率为14.4%。另外，调查中发现与槟榔、荔枝、柠檬等套种的胡椒园发病率为9.9%，而单一模式的胡椒园发病率为16.8%。通过PCR法检测发现，海南胡椒花叶病样品中存在黄瓜花叶病毒（Cucumber mosaic virus，CMV）和胡椒黄斑驳病毒（Piper yellow mottle virus，PYMoV）两种病原。

关键词：海南；胡椒花叶病；为害情况

---

* 基金项目：国家公益性行业（农业）科研专项（201303028）
** 作者简介：车海彦，副研究员，从事热带作物病毒与植原体病害研究；E-mail：chehaiyan2012@126.com
*** 通信作者：罗大全，研究员；E-mail：luodaquan@163.com

核盘菌低毒相关内源病毒 SsEV2 的分子特性研究*

成淑芬[1,2**]，李　波[1,2]，谢甲涛[1,2]，陈　桃[1]，程家森[1]，付艳苹[1]，姜道宏[1,2***]
（1. 湖北省作物病害监测和安全控制重点实验室（华中农业大学），武汉　430070；
2. 农业微生物学国家重点实验室（华中农业大学），武汉　430070）

摘　要：由核盘菌（*Sclerotinia sclerotiorum*）引起的作物菌核病是一种重要真菌病害，严重影响油菜、大豆等作物的产量和品质。病原真菌低毒相关病毒防治真菌病害是一种潜在的防治措施。前期研究发现核盘菌中蕴含丰富的低毒力相关病毒资源。核盘菌弱毒菌株 SX276 中含有 9 种不同的病毒，其中内源病毒 Sclerotinia sclerotiorumendornavirus 2（SsEV2）的 cDNA 全长 12661bp，推测仅含有一个 ORF，编码一个推定的多聚蛋白，该多聚蛋白具有 5 个保守结构域，分别为甲基转移酶（MTR），DExD RNA 解旋酶，病毒解旋酶（HEL），S7 核心保守结构域和复制酶（RdRp），此外，病毒 SsEV2 编码蛋白含有一个保守的半胱氨酸富集区 CRR（cysteine-rich region）。基于病毒保守的 RdRp，MTR 和 HEL 序列进行系统发育进化分析，发现虽然 SsEV2 与侵染丝状真菌的内源病毒 SsEV1、GaBRV-XL1 和 GaBRV-XL2 亲缘关系较近，但在系统发育进化树上独属一支，且与报道的 SsEV1 等病毒编码蛋白相似性低于 26%，因此，推测 SsEV2 是内源病毒科（Endornaviridae）科 *Endornavirus* 属一种新的病毒。通过原生质体再生技术，获得仅含有 SsEV2 的再生菌株，且该再生菌株呈现弱毒特性；通过病毒水平传播等研究证实了 SsEV2 可引致核盘菌生长速度减慢，菌落形态呈现扇形，丧失对寄主的致病力。因此，SsEV2 与核盘菌致病力衰退相关，是一种新的潜在的生物防治资源。

关键词：核盘菌；病毒；分子特性

* 资助基金：油菜现代产业技术体系岗位科学家科研专项资金（CARS-13）

** 第一作者：成淑芬，女，硕士研究生，主要从事植物病害生物防治（真菌病毒）相关研究；E-mail：chengshufen2012@163. com

*** 通信作者：姜道宏，教授，主要从事分子植物病理学及生物防治的相关研究；E-mail：daohongjiang@ mail. hzau. edu. cn

甜菜坏死黄脉病毒 P14 蛋白的原核表达纯化*

侯春香**，祖力亚夏尔·玉山诺，姜 宁，张宗英，韩成贵，王 颖***
（中国农业大学作物有害生物监测与绿色防控农业部重点实验室，北京 100193）

摘 要：由甜菜坏死黄脉病毒（*Beet necrotic yellow vein virus*，BNYVV）引起的甜菜丛根病（Rhizomania）是世界主要甜菜产区普遍发生的病毒病害，由根部专性寄生菌甜菜多粘菌（*Polymyxa betae*）传播。BNYVV 是甜菜坏死黄脉病毒属（*Benyvirus*）的代表种，含有 4～5 个 RNA 组分。

P14 蛋白是 RNA2 编码的最后一个蛋白，由亚基因组 c 翻译表达。该蛋白富含半胱氨酸，其“锌指”结构可以在体外与锌离子结合，属于富半胱氨酸结构。由 BNYVV 编码的 P14 蛋白目前已被证明是一种基因沉默抑制子，为了给研究该蛋白的生物学功能提供实验材料，将 BNYVV 的 P14 蛋白克隆到 pGEX-KG 表达载体上，通过转化大肠杆菌 BL21 菌株，结果显示，在 18℃ 下，终浓度 0.2mmol/L IPTG 诱导 14h 后，p14 融合蛋白可在上清中大量表达，进一步通过 GST 亲和层析柱纯化得到目的蛋白。纯化的 BNYVV P14 可用于多克隆抗体的制备，为甜菜坏死黄脉病毒的检测及该蛋白抑制基因沉默的特点和机制研究打下了材料基础。

关键词：甜菜坏死黄脉病毒；P14；原核表达；抗血清制备

致谢：感谢于嘉林、李大伟、王献兵教授和张永亮副教授对本研究的建议。

---

* 基金项目：国家自然科学基金项目（31401708）和高校基本业务费（2016QC014）资助

** 第一作者：侯春香，硕士生，主要从事甜菜病毒的研究；E-mail：houchunxiang@cau.edu.cn

*** 通信作者：王颖，副教授，主要从事植物病毒学研究；E-mail：yingwang@cau.edu.cn

葫芦科种子携带黄瓜绿斑驳花叶病毒检测*

秦碧霞**，崔丽贤，谢慧婷，李战彪，邓铁军，蔡健和***

（广西壮族自治区农业科学院植物保护研究所/广西作物病虫害生物学重点实验室，南宁 530007）

摘　要：黄瓜绿斑驳花叶病毒（*Cucumber green mottle mosaic virus*，CGMMV）是严重威胁葫芦科作物生产的重要种传病毒，在英国、希腊、罗马尼亚、巴西、日本、伊朗、印度、韩国等及中国均有分布，是我国农业植物检疫性有害生物和进境检疫性有害生物。近几年该病毒先后在辽宁、北京、甘肃、四川、云南、广东、湖北、浙江、山东、广西等省（自治区）部分地区田间造成危害，严重威胁葫芦科作物尤其是西瓜嫁接苗的安全生产。据报道，CGMMV 葫芦种传率 1%~5%，西瓜种传率 0%~3%，且极易通过汁液摩擦、农事操作和嫁接等方式传播。由于西瓜产区普遍采取嫁接方式集中培育西瓜苗，种子带毒是嫁接苗育苗环节扩散传播的最初毒源，砧木或接穗种子携带病毒，极易通过“摘除砧木生长点→插竹签→削接穗→拔竹签→插接穗”等嫁接操作使该病害在苗床扩散传播并蔓延到田间，加强葫芦科种子携带病毒检测，使用无毒健康种子和种苗是防治该病害的重要措施。

2013—2016 年笔者利用 DAS-ELISA、RT-PCR、IC-RT-PCR 等检测方法，对广西各地植保植检部门、在广西销售种子的各种苗公司、育苗户等送检的葫芦、西瓜、南瓜、甜瓜等葫芦科种子进行 CGMMV 检测，了解市场售卖种子的带毒情况，根据检测结果指导育苗户购买和使用不携带病毒的种子，为西瓜嫁接苗生产保驾护航。送检的葫芦种包括韩砧一号、丰砧一号、丰乐五号、安生、金砧一号、强力一号、强力二号、亲抗水瓜、本地葫芦及来自山东、甘肃、安徽等地的散装葫芦种，西瓜种包括正大小麒麟、农优新一号、珠农大美人、珠农黑美人、甜王三号、甜王五号、大美龙、嘉丽、野生西瓜种等，南瓜种包括雪藤木二号、壮士、金威龙、雪砧金刚 F1、东顺砧木、壮根等，甜瓜种有甜蜜蜜、美浓、蜜玉、白玉等，黄瓜种是太湖溪 1 号。4 年合计检测葫芦种子 1 860份，626 份检出 CGMMV，送检种子带毒率 33.66%，各年份送检葫芦种带毒率分别为 36.64%、40.30%、27.80%、36.70%，检出 CGMMV 的葫芦种涉及散装和部分包装的种子。4 年共检测西瓜种子 267 份、南瓜种子 181 份、甜瓜种子 10 份、黄瓜种子 16 份，均未检出 CGMMV。检测结果表明葫芦砧木种子携带病毒是目前西瓜嫁接苗生产出现 CGMMV 暴发的主要原因，根据笔者几年的跟踪调查，凡是按规定进行种子送检并采用健康种子生产的西瓜嫁接苗基本解决了黄瓜绿斑驳花叶病毒病的问题，因此加强对育种、制种基地的检疫监管，加大对有关技术人员和西瓜育苗户进行黄瓜绿斑驳花叶病毒病防控技术培训力度，广泛宣传，提高认识，做好种子检测、消毒等处理措施，就能从源头上切断病毒的传播，促进我国葫芦科作物产业的可持续健康发展。

关键词：黄瓜绿斑驳花叶病毒；葫芦科种子；带毒率；检测

---

* 基金项目：广西特色水果创新团队建设专项（nycytxgxcxtd-04-19-2）；河南省果树瓜类生物学重点实验室项目(HNS-201508-10)；广西作物病虫害生物学重点实验室基金项目（13-051-47-ST-4）；广西农业科学院基本科研业务专项（2015YT42）

** 第一作者：秦碧霞，副研究员，主要从事植物病毒学研究；E-mail：qbx33@163.com

*** 通信作者：蔡健和，研究员，主要从事植物病毒学研究；E-mail：caijianhe@gxaas.net

番茄褪绿病毒的 RT-PCR 检测及 SYBR Green I 实时荧光定量方法的建立与应用*

孙晓辉**，高利利，王少立，王雪雨，杨园园，竺晓平***
（山东农业大学植物保护学院；山东省蔬菜病虫生物学重点实验室，泰安　271018）

摘　要：为了建立检测番茄中番茄褪绿病毒（*Tomato chlorosis virus*，ToCV）的实时荧光定量方法。根据 ToCV 热激蛋白 70（HSP70）的基因序列，设计了 ToCV 实时荧光定量 RT-PCR 的特异性引物，获得了 190bp 大小的片段，并将小片段克隆到 pMD18-T 载体上，鉴定结果显示，PCR 扩增获得片段与 ToCV HSP70 基因部分片段同源性高，表明设计的引物特异性好且该片段已经成功连接至载体，获得了重组质粒 ToCV-1。通过对退火温度和引物浓度的优化，最终确定退火温度为 63℃，引物浓度为 0.2μmol/L。选择 5 个梯度稀释的 ToCV-1 进行 SYBR Green I 实时荧光定量 PCR 反应，获得的标准曲线线性关系良好，相应的回归方程为 Ct=-3.227Log（copy number）+49.508。熔解曲线为特异性的单峰，且没有明显的非特异性扩增或引物二聚体的峰出现，证明该方法特异性良好。通过与普通 PCR 的灵敏度比较发现建立的荧光定量 PCR 比普通 PCR 灵敏 100 倍。同时与瓜类褪绿黄化病毒（*Cucurbit chlorotic yellows virus*，CCYV）、黄瓜花叶病毒（*Cucumber mosaic virus*，CMV）、番茄花叶病毒（*Tomato masaic virus*，ToMV）和番茄斑萎病毒（*Tomato spotted wilt virus*，TSWV）均无交叉反应；组间和组内变异系数均小于 5%，表明该方法具有良好的重复性和稳定性。针对不同感病植株不同部位病毒相对含量比较，茎的韧皮部中含量最高，老叶次之，然后是根部，而上部新叶病毒量最低；不同品种番茄供试样品中粉果番茄品种抗病性比红果番茄品种抗病性差；采用粉虱接种法对健康植株接种番茄褪绿病毒，在第四周病毒含量大幅度增加。研究建立的基于 SYBR Green I 实时荧光定量 PCR 技术的快速检测方法速度快、特异性强、灵敏度高、重复性好。

关键词：番茄褪绿病毒；SYBR Green I 实时荧光定量 PCR；病毒检测

* 基金项目：国家公益性（农业）行业科研专项资助项目（201303028）；山东省科技发展计划资助项目（2014GNC111008）

** 第一作者：孙晓辉，男，在读硕士研究生，主要从事植物病理学研究；E-mail：ddxiaohuifly@163.com
*** 通信作者：竺晓平，教授；E-mail：zhuxp@sdau.edu.cn

Complete Genome Sequences of Two Divergent Isolates of *Citrus Tatter Leaf Virus* Infecting Citrus in China

Li Ping*, Li Min, Wang Jun, Tian Xin, Li Zhongan**, Cao Mengji**

(*National Citrus Engineering Research Center*, *Citrus Research Institute*, *Southwest University*, *Chongqing*, *China*, 400712)

Abstract: Two novel isolates of *Citrus tatter leaf virus* (CTLV) was identified from citrus in China. The complete sequences of isolate CTLV-HJY1 and CTLV-HJY2 both were 6 496 nucleotides, excluding the poly (A) tail, encode two open reading frames (ORFs) and the genomic structure of the two isolates were similar to previously reported isolates of CTLV. The complete nucleotide sequences of CTLV–HJY1 and CTLV-HJY2 shared 81%~86% and 82%~98% identities with other available *Apple stem grooving virus* (ASGV) or CTLV isolates in GeneBank, respectively. Coat protein (CP) gene is the most conserved gene of ASGV. The CP gene of CTLV-HJY1 and CTLV-HJY2 had the highest identity with ASGV-HPKu (LT160740) isolate (94.3%) and with CTLV–Kumquat (AY646511) isolate (99.0%), respectively. A phylogenetic tree based on the complete genome sequence of available *Capilloviruses* showed that the two isolates of CTLV were grouped in two separate clusters. Recombination analysis indicated CTLV-HJY1 was a potential major parent of the recombinant ASGV-FKSS2 (LC143387).

Key words: *Citrus tatter leaf virus*; *Apple stem grooving virus*; Genome sequence; *Capillovirus*

* First author: Li Ping; E-mail: 183757352202@163.com

** Corresponding authors: Li Zhongan; E-mail: zhongan@cric.cn

Cao Mengji; E-mail: caomengji@cric.cn

马铃薯 Y 病毒分离物基因的全序列比较分析

白艳菊[1*]，孙旭红[1]，高艳玲[2]，韩树鑫[3]，张　威[2]，范国权[2]，
申　宇[2]，邱彩玲[2]，张　抒[2]，毛艳芝[1]，尚　慧[3]
（1. 黑龙江省农业科学院克山分院，齐齐哈尔　4233246；2. 黑龙江省农业科学院植物脱毒苗木研究所，哈尔滨　150086；3. 东北农业大学，哈尔滨　150030）

摘　要：针对 PVY 病毒血清学为 O 的 7 个样品进行全序列分析，经与 Genebank 序列比对，SC-1-1-2，HLJ-9-4，HLJ-6-1 与 HN2（GQ200836）的序列相似度达到 99%，而与其他株系分离物的相似度均小于 99%。HLJ-C-429，HLJ-BDH-2 与 Mb112（AY745491）和 SGS-AG（JQ924288）的相似度均达到了 98%。INM-W-369-12 与 SYR-II-2-8（AB461451）的序列相似度也达到了 99%。最后，HLJ-30-2 的序列则与 Mb112 的相似度最高，达到 99%。将 7 个全序列与 Genebank 中 39 个来自世界各国的 PVY 全序列进行比较，通过系统进化树分析，41 个 PVY 不同分离物共形成 15 个簇，HLJ-6-1 与 HLJ-9-4 分离物在系统发育关系上与 GQ200836 最近。同时，INM-W-369-12 与 SC-1-1-2 分离物以置信值（54%）与 INM-W-369-12、SC-1-1-2 及 GQ200836 聚一个大的簇，说明 INM-W-369-12 与 SC-1-1-2 分离物在系统发育上，与 HLJ-6-1 和 HLJ-9-4 分离物有一定的距离。另外，在 N-Wi 及 N：O 两个大簇中，HLJ-C-429 及 HLJ-BDH-2 两个分离物与其他的 N-Wi 分离物以高置信值（89%）聚为一个大簇，说明它们与 N-Wi 株系在系统发育分析的角度上看是最为接近的。而 HLJ-30-2 以 100%的高置信值与 N：O 株系的聚为一大簇，则说明其与 N-Wi 的关系最为密切。将测试材料接种到烟草和马铃薯敏感品种克新 13 上，症状表现有明显差异，其中 HLJ-9-4 和 HLJ-6-1 症状相同，HLJ-C-429 和 HLJ-BDH-2症状相同，SC-1-1-2 和 INM-W-369-12 症状相同，HLJ-30-2 的症状与其他的分离物均不相同。INM-W-369-12 及 SC-1-1-2 所造成的危害最为严重，接种后 25d 内出现叶脉坏死，主茎条斑坏死，新叶花叶等症状，并最终造成植株死亡。而 HLJ-C-429 和 HLJ-BDH-2 则造成的危害最轻，仅仅能产生花叶症状。剩余 3 种分离物的症状则介于以上 4 个分离物之间，其中 HLJ-30-2 造成的症状要比另外两个分离物严重。本研究所筛选材料的遗传背景和生物学信息为 PVY 株系深入研究提供了可参考的理论依据。

关键词：马铃薯 Y 病毒；分离物；基因

* 通信作者：白艳菊，研究员；E-mail：yanjubai@163.com

Complete Genome Sequence of a Divergent Strain of *Lettuce Chlorosis Virus* from Periwinkle in China

Tian Xin, Shen Pan, Zhang Song, Zhou Changyong and Cao Mengji
(*National Citrus Engineering Research Center*, *Citrus Research Institute*, *Southwest University*, *Chongqing*, *China*, 400712)

Abstract: A novel strain of *Lettuce chlorosis virus* (LCV) was identified from periwinkle (PW) with foliar interveinal chlorosis and plant dwarf in China. Complete nucleotide (nt) sequences of genomic RNA1 and RNA2 of the virus are 8 602nt and 8 456nt, respectively. The genomic organization of LCV-PW resembles that of a California isolate (LCV-CA), but its RNA2 only encodes 8 open reading frames. The LCV-PWRNA2 is shorter than that of LCV-CA, mainly due to two large deletions. Phylogenetic analyses were performed using the amino acid sequences of the RdRp (ORF1b), HSP70h, coat protein (CP) and minor coat protein (CPm) of the two LCV isolates and representative members of the genus *Crinivirus*. Neighbor-joining trees show that the two LCV isolates are grouped together in a cluster containing *Cucurbit chlorotic yellows virus* (CCYV), *Bean yellow disorder virus* (BnYDV), *Cucurbit yellow stunting disorder virus* (CYSDV) and *Tetterwort vein chlorosis virus* (TVCV). LCV-PWis different from LCV-CA by 84% (RNA1) and 89% (RNA2) at the genomic sequences and the bean isolate by 84. 0% at the 421-nt HSP70h sequence (KC602375, RNA2), respectively. More recently, a tomato isolate of LCV (KX158863) was identified from Shandong province. Comparison of a 446-nt partial 1a polyprotein sequence of the two Chinese isolates showed that they were 99. 8% identical withone another, suggesting that they might bethe same variant.

Key words: Lettuce chlorosis virus; Genome sequence; *Crinivirus*; Periwinkle

SRBSDV 编码的非结构蛋白 P7-1 与介体白背飞虱的互作研究*

王　甄[1,2**]，贾东升[2]，韩　玉[2]，毛倩卓[2]，陈红燕[2]，魏太云[2***]
（1. 恩施州农业科学院，恩施　445000；2. 福建农林大学植物病毒研究所，福州　350002）

摘　要： 近几年由南方水稻黑条矮缩病毒（*Southern rice black-streaked dwarf virus*，SRBSDV）引起的南方水稻黑条矮缩病在我国南方稻区流行，对水稻生产造成了严重的损失。SRBSDV 由白背飞虱（*Sogatella furcifera*，Horváth）以持久增殖型方式传播。已知 SRBSDV 编码的非结构蛋白 P7-1 形成包裹病毒粒体的管状结构，是病毒在介体白背飞虱体内突破各种屏障并快速扩散的重要通道，而病毒的扩散必然需要介体因子的参与才能完成，因此筛选与 P7-1 互作的介体蛋白将有助于研究 SRBSDV 与白背飞虱高度亲和的分子机制。

本研究首先采用酵母双杂交系统，以 SRBSDV 非结构蛋白 P7-1 为诱饵，初步从带毒白背飞虱 cDNA 文库中筛选出与 P7-1 互作的介体因子 63 个。经过序列分析及功能预测，挑选出 6 个可能与 P7-1 互作的候选蛋白进一步验证。通过双分子荧光互补实验在共聚焦显微镜下观察到B-cell receptor-associated protein 31-like（BAP31）、Ly-6/neurotoxin superfamily member 1（Neurotoxin）、DNA-J/hsp40 protein（DNA J）、ubiquitin-protein E3 ligase（E3）和 myosin RLC2（Myosin）这 5 个候选蛋白与 P7-1 在本氏烟中互作产生黄色荧光，skeletal muscle growth protein 5（Skeletal）与 P7-1 不产生荧光，两者不发生互作。利用昆虫杆状病毒表达系统在 Sf9 昆虫细胞中分别单独表达和共表达 P7-1 与 BAP31、Neurotoxin、DNA J、E3 和 Myosin 5 个候选蛋白，观察到蛋白单独表达与共表达时的形态和位置均有变化，说明在 Sf9 细胞中 P7-1 与 5 个候选蛋白也可以发生互作。本研究通过酵母双杂交筛选与 SRBSDV 编码的 P7-1 管状结构互作的候选蛋白，并通过 BiFC、杆状病毒表达系统证明了 5 个候选蛋白（BAP31、Neurotoxin、DNA J、E3 和 Myosin）与 P7-1 存在互作关系，为解析 SRBSDV 与介体昆虫间的互作机制奠定了基础。

关键词： 南方水稻黑条矮缩病毒；白背飞虱；P7-1 管状蛋白；蛋白互作

---

* 基金项目：国家自然科学基金项目（31400136）；福建省自然科学基金面上项目（2017J06011）

** 第一作者：王甄，女，硕士，农艺师，主要从事水稻病毒和马铃薯病害防治与遗传育种研究

*** 通信作者：魏太云，研究员，主要从事水稻病毒与介体昆虫的互作关系；E-mail：weitaiyun@ fafu. edu. cn；weitaiyun@ 163. com

感染病毒条件下砂梨 *Pp*miR397a 的鉴定及表达分析*

杨岳昆[1,2]，王国平[1,2]，洪　霓[1,2]，王利平[1,2**]

（1. 华中农业大学植物科学技术学院，湖北省作物病害监测与安全控制重点实验室；
2. 华中农业大学，农业微生物学国家重点实验室，武汉　430070）

摘　要：苹果茎沟病毒（*Apple stem grooving virus*，ASGV）是我国栽培梨上普遍携带的一种病毒。miRNA 是一类非编码 21-24nt 小 RNA，在转录水平或翻译水平上通过调控靶基因进而对植物生长发育、信号转导、抗病抗逆中起重要调控作用。前期研究经小 RNA 高通量测序发现并获得了 ASGV 感染条件下砂梨（*Pyrus pyrifolia*）茎尖部位表达的一些保守的 miRNAs 以及梨特有的新的 miRNAs。已有文献报道 miR397a 通过靶标漆酶基因在维管束细胞壁形成，木质素合成中发挥功能。为探究病毒感染条件下 *Pp*miR397a 在梨中的功能，本研究选取了砂梨“金水二号”小 RNA 文库中生物信息学分析存在的 miR397a（命名为 *Pp*miR397a），采用茎环引物反转录法（Stem-loop RT-PCR）对 *Pp*miR397a 成熟体进行了克隆鉴定。预测 *Pp*miR397a 前体在梨基因组 DNA 上的位置，克隆获得相应的前体序列，对其进行克隆，构建了 *Pp*miR397a 超量表达载体，为探究 *Pp*miR397a 在梨中基因功能提供材料。进而采用 RLM-RACE 法验证了 *Pp*miR397a 与梨漆酶家族成员 *Lac*4 基因（命名为 Pp-*Lac*4）在 miRNA 结合处的第 10 和第 11 位碱基间存在靶标切割关系，克隆获得了 Pp-*Lac*4 基因的 CDS 序列，明确其在漆酶家族中的分类地位。结合有毒和无毒梨离体植株茎尖和基部的横截石蜡切片显微观察，结合 qRT-PCR 对病毒感染条件下 *Pp*miR397a 及其对应的靶基因 Pp-*Lac*4 在梨不同组织部位进行表达量联合分析，结果显示，无毒梨植株茎尖和茎基部部位木质化程度均高于感染 ASGV 病毒植株部位，且 Pp-*Lac*4 基因不同部位表达量与木质化程度正相关，*Pp*miR397a 与 *PpLac*4 存在负调控关系。本研究结果初步表明携带 ASGV 与无病毒梨植株的木质化程度差异与 *Pp*miR397a 对靶基因 Pp-*Lac*4 的调控存在一定的相关性，相关研究还在进一步试验中。

关键词：砂梨；miRNA；苹果茎沟病毒；茎环引物反转录法

---

* 基金项目：国家自然科学基金（31201488）和国家梨产业技术体系（CARS-29-10）

** 通信作者：王利平，副教授，研究方向为果树病理学；E-mail：wlp09@mail. hzau. edu. cn

一种侵染梨干腐病菌的单组分 dsRNA 病毒的研究*

杨萌萌，洪　霓，王国平**
（华中农业大学植物科学技术学院，农业微生物学国家重点实验室，武汉　430070）

摘　要：本实验室前期研究证实梨干腐病与梨轮纹病是由 *Botryosphaeria dothidea* 导致的同一病害的两类症状表现，本研究从来源于山东省烟台市表现干腐症状的 *B. dothidea* 菌株 MSD53 中分离出一种新的 dsRNA 病毒，暂命名为 Botryosphaeria dothidea victoria-like virus 1（BdVLV1）。

BdVLV1 病毒中包含一条线性 dsRNA，大小为 5 090bp，该基因组包含两个开放阅读框，ORF1 和 ORF2 可能分别编码病毒的 CP 和 RdRp。两个开放阅读框有 8 个碱基的重叠部分，ORF1 终止密码子上游第 3～5 个碱基为 ORF2 起始密码子位置，之间由两个腺嘌呤核苷酸（AA）隔开，这种结构与全病毒科（Totiviridae）的大多数病毒的移码翻译模式不同，是 BdVLV1 翻译的一个关键 Motif。从 BdVLV1/MSD53 中提取病毒，负染后透射电镜观察到直径约为 35 nm 的球形粒体。从病毒粒子中提取的 dsRNA，其带型和大小与从菌丝中提取的一致。SDS-PAGE 电泳分析显示，BdVLV1 病毒粒体中可能含有两种结构蛋白，大小分别为 80ku 和 90ku。根据 BdVLV1 所预测 RdRp 的氨基酸序列，采用临接法进行系统发育分析。结果表明：BdVLV1 与全病毒科（Totiviridae）维多利亚病毒属（*Victorivirus*）的病毒聚在同一分支中。

关键词：梨干腐病菌；dsRNA 病毒；单组分

---

* 基金项目：梨现代农业技术产业体系（CARS-29-10）；国家农业部公益性行业计划（201203034-02）
** 通信作者：王国平；E-mail：gpwang@ mail. hzau. edu. cn

Identification and Characterization of miRNAs and Their Targets in Maize in Response to *Sugarcane Mosaic Virus* Infection by Deep Sequencing*

Xia Zihao[1**], Zhao Zhenxing[2], Jiao Zhiyuan[2],
Zhou Tao[2], Wu Yuanhua[1], Fan Zaifeng[2***]
(1. *College of Plant Protection*, *Shenyang Agricultural University*, *Shenyang* 110866;
2. *Department of Plant Pathology*, *China Agricultural University*, *Beijing* 100193)

Abstract: MicroRNAs (miRNAs) are endogenous non-coding small RNAs that play essential regulatory roles in plant development and environmental stress responses. Maize (*Zea mays* L.) is a global economically important food and forage crop. To date, a number of conserved and non-conserved miRNAs of maize have been identified as being involved in plant development and stress responses. However, the miRNA-mediated gene regulatory networks responsive to virus infections in maize remain largely unknown. In this study, the profiles of small RNAs in buffer-and *Sugarcane mosaic virus* (SCMV) -inoculated maize plants at 8 days post-inoculation (dpi) were obtained by deep sequencing, respectively. A total of 154 known miRNAs (belonging to 28 miRNA families) and 354 novel miRNAs (of which 38 novel miRNAs were induced by SCMV infection) were profiled. Most of the miRNAs identified were differentially expressed and almost all miRNAs were up-regulated after SCMV infection. Northern blotting and quantitative real-time PCR were performed to determine the accumulation of several miRNAs and their targets in maize plants inoculated with either buffer or SCMV at 4, 8 and 12 dpi, respectively. These results provide new insights into the regulatory networks of miRNAs and their targets in maize plants responsive to SCMV infection.

Key words: *Sugarcane mosaic virus*; Maize; miRNAs; Targets; Deep sequencing

---

* 基金项目：农业生物技术国家重点实验室（2013SKLAB1-3）
** 第一作者：夏子豪，讲师，主要从事小 RNA 和植物病毒相关研究；E-mail：zihao8337@ 126. com
*** 通信作者：范在丰，教授，主要从事植物与病毒互作相关研究；E-mail：fanzf@ cau. edu. cn

柑橘黄化脉明病毒和衰退病毒的二重 RT-PCR 检测体系的建立与应用*

赵恒燕**，关桂静，王洪苏，周常勇，李中安，刘金香***
（西南大学柑桔研究所/中国农业科学院柑桔研究所，国家柑桔工程技术研究中心，重庆 400712）

摘　要：柑橘黄化脉明病毒（*Citrus yellow vein clearing virus*，CYVCV）和柑橘衰退病毒（*Citrus tristeza virus*，CTV）均由昆虫传播，并常在田间混合感染。本研究选用 CYVCV 和 CTV 两种病毒外壳蛋白保守序列及内参基因 Ubiquitin 的特异性引物，优化了影响二重 RT-PCR 反应的 Mg^{2+}浓度、dNTPs 浓度、引物浓度和退火温度，建立了针对 CYVCV 及 CTV 两种病毒的一步法二重 RT-PCR 检测体系。二重 RT-PCR 获得 CYVCV、CTV 及 *Ubiquitin* 的特异性片段大小分别为 614bp、373bp 及 194bp，经克隆测序和序列对比，结果显示它们与已报道的病毒序列具有较高的同源性。该体系最低能从 40ng/μL 总核酸中检测出 CYVCV，从 4ng/μL 总核酸中检测出 CTV，其灵敏度与单一 RT-PCR 检测灵敏度一致。利用该体系对 33 份田间样品进行检测，CYVCV 和 CTV 感染率分别为 54.5%和 66.7%，复合侵染率高达 36.4%。本研究建立的一步法二重 RT-PCR 技术体系可用于大量田间样品中 CYVCV 和 CTV 的快速同步检测。

关键词：柑橘黄化脉明病毒；柑橘衰退病毒；二重 RT-PCR；检测

* 基金项目：西南大学基本科研业务费（XDJK2014C131）；果树病毒病防控技术研究与示范（201203076）项目资助
** 第一作者：赵恒燕，女，在读硕士研究生；E-mail：zhy938756108@163.com
*** 通信作者：刘金香，女，副研究员，主要从事分子植物病毒学研究；E-mail：Ljinxiang@126.com

浙江柑橘产区柑橘黄脉病毒发生与分子特性研究

张艳慧*，刘莹洁，王　琴，李雪燕，周　彦**
（西南大学柑桔研究所，重庆　400712；重庆市农业信息中心，重庆　400015）

摘　要：柑橘黄脉病毒（*Citrus yellow vein clearing virus*，CYVCV）是一种柑橘新病毒，其在中国柑橘产区分布较为广泛，但尚未有其在浙江发生的报道。本研究运用 RT-PCR 技术对采自浙江 12 个柑橘产地的 181 份柑橘样品进行检测，首次从浙江的红美人和佛手上检测出 CYVCV，其检出率分别为 42.7%和 68.0%。选取 4 个 CYVCV 毒株与 11 个已知 CYVCV 毒株进行全序列分析的结果显示，CYVCV 序列保守性较高，所有 15 个 CYVCV 毒株的核苷酸和氨基酸序列相似性分别为 96.9%~99.8%和 97.9%~99.4%。虽然采自浙江的 4 个 CYVCV 毒株在进化树中聚成单独的簇，但 CYVCV 毒株间的亲缘关系可能不与其采样地存在相关性。

关键词：柑橘黄脉病毒；检测；序列分析；进化树

---

* 第一作者：张艳慧，女，在读硕士研究生；E-mail：zhangyanhui0323@163.com
** 通信作者：周彦，研究员；E-mail：zybook1@163.com

山东省辣椒主要病毒种类的分子检测与鉴定*

王少立[1**]，谭玮萍[1]，杨园园[1]，代惠洁[2]，孙晓辉[1]，乔　宁[1,2]，竺晓平[1***]
（1. 山东省蔬菜病虫生物学重点实验室，山东农业大学植物保护学院，
泰安　271018；2. 潍坊科技学院，寿光　262700）

摘　要：辣椒（*Capsicum annuum*）是茄科（Solanaceae）辣椒属（*Capsicum*）植物，是一种重要的蔬菜作物。随着辣椒种植规模的逐步扩大，辣椒病毒病也愈发严重，已成为危害我国辣椒生产中的重要病害。本研究对山东省 10 个市区的辣椒病毒病的发生情况进行了调查，通过分子生物学方法对引起辣椒病毒病的病原进行了初步的检测与鉴定。

本研究对采集山东省 10 个市区 253 份疑似感病的辣椒植株叶片，提取叶片总 RNA 和总 DNA，利用双生病毒的通用引物（PA/PB）、马铃薯卷叶病毒属通用引物（POL-F/POL-R）及已报道侵染辣椒的主要病毒的检测引物对样品进行 PCR、RT-PCR 分子检测与鉴定。结果表明：在山东地区采集的 253 份辣椒样品可被 15 种病毒侵染，病毒复合侵染率高达 89.92%，最多可同时感染 6 种病毒，辣椒轻斑驳病毒（*Pepper mild mottle virus*，PMMoV）、黄瓜花叶病毒（*Cucumber mosaic virus*，CMV）为主要病毒种类。PMMoV、CMV 总检出率均在 60%以上，分别为 61.66%、60.08%；烟草轻型绿花叶病毒（*Tobacco mild green mosaic virus*，TMGMV）、蚕豆萎蔫病毒 2 号（*Broad bean wilt virus* 2，BBWV-2）、甜菜西方黄化病毒（*Beet western yellows virus*，BWYV）、烟草花叶病毒（*Tobacco mosaic virus*，TMV）发生也比较普遍，检出率是 41.90%、34.78%、33.20%和 24.90%；辣椒潜隐病毒 2 号（*Pepper cryptic virus* 2，PCV-2）、番茄花叶病毒（*Tomato mosaic virus*，ToMV）、番茄黄化曲叶病毒（*Tomato yellow leaf curl virus*，TYLCV）、马铃薯 Y 病毒（*Potato virus Y*，PVY）、甜瓜蚜传黄化病毒（*Melon aphid-borne yellows virus*，MABYV）、辣椒脉黄化病毒（*Pepper vein yellows virus*，PeVYV）、辣椒潜隐病毒 1 号（*Pepper cryptic virus* 1，PCV-1）、辣椒脉斑驳病毒（*Chilli veinal mottle virus*，ChiVMV）、苜蓿花叶病毒（*Alfalfa mosaic virus*，AMV）侵染现象较少，检出率分别为 11.86%、9.88%、9.09%、6.72%，5.53%、3.56%、3.16%、0.79%和 0.40%。首次在中国辣椒上检测到 MABYV，在山东首次检测到了 PeVYV、BWYV、PCV-1、PCV-2。

本研究基本摸清了山东省主要辣椒产区病毒病发生情况以及病毒种类的主次关系，为山东省辣椒病毒病的防控提供了依据。

关键词：辣椒病毒病；辣椒轻斑驳病毒；甜菜西方黄化病毒；甜瓜蚜传黄化病毒；复合侵染

---

* 基金项目：公益性行业（农业）科研专项（201303028）；山东省科技发展计划项目（2014GNC111008）
** 第一作者：王少立，女，硕士研究生，植物病理学专业；E-mail：wangshaolisdau@163.com
*** 通信作者：竺晓平，教授；E-mail：zhuxp@sdau.edu.cn

番茄褪绿病毒自然侵染豇豆的首次报道*

王雪雨**，冯　佳，臧连毅，闫亚莉，杨园园，竺晓平***
（山东省蔬菜病虫生物学重点实验室，山东农业大学植物保护学院，泰安　271018）

摘　要：番茄褪绿病毒（*Tomato chlorosis virus*，ToCV），隶属于长线形病毒科（Closteroviridae）毛形病毒属（*Crinivirus*），是一种由粉虱传播的 RNA 病毒，可以侵染番茄、辣椒、茄子及南瓜等常见蔬菜，已在世界多地造成了严重危害。2014 年和 2016 年的夏天，在山东省聊城市番茄感染 ToCV 的塑料温室中，观察到与番茄间作的豇豆植株叶片表现出脉间褪绿黄化，叶脉深绿色，叶缘卷曲，并伴有大量烟粉虱。番茄植株的发病率约为 87%，豇豆植株的病害发病率约为 5%。为了鉴定病原体，通过 TRizol 试剂盒从 6 个有症状的豇豆叶片和 8 个番茄叶片中提取总 RNA。利用 ToCV*CP* 和 *HSP*70 基因的特异性引物分别进行 RT-PCR 扩增，获得了具有预期大小的扩增片段。通过 Q 型烟粉虱进行虫传实验，将无毒烟粉虱在微虫笼中取食毒源 48h，然后用接毒后的粉虱传毒到三叶期的豇豆幼苗 24h。5 周后，虫传的豇豆植株，在幼苗期后症状明显。利用 ToCV 外壳蛋白 *CP* 基因和热激蛋白 *HSP*70 基因的特异性引物分别进行 RT-PCR 扩增分析，经 NCBI BLAST 比对发现，聊城分离物的 *CP* 和 *HSP*70 基因与 GenBank 上传的 ToCV 典型分离物序列的同源性均达 98%以上。据报道，ToCV 可以感染豌豆（*Pisum sativum*）、菜豆（*Phaseolus vulgaris*）和蚕豆（*Vicia faba*）但没有关于 ToCV 感染豇豆植物的相关报道。据笔者所知，这是豇豆（*Vignaun guiculata*）作为 ToCV 的天然宿主的首次报道。这对 ToCV 寄主范围的界定及病害防治提供了理论依据。

关键词：番茄褪绿病毒；豇豆；粉虱传毒

* 基金项目：公益性行业（农业）科研专项（201303028）；山东省科技发展计划项目（2014GNC111008）
** 第一作者：王雪雨，硕士研究生，从事植物病理学研究；E-mail：wangxysdau@163.com
*** 通信作者：竺晓平，教授，从事植物病理学研究；E-mail：zhuxp@sdau.edu.cn

BNYVV p42 蛋白的原核表达及纯化*

姜　宁**，杨　芳，张宗英，韩成贵，王　颖***
（中国农业大学作物有害生物监测与绿色防控农业部重点实验室，北京　100193）

摘　要： 甜菜坏死黄脉病毒（*Beet necrotic yellow vein virus*，BNYVV）由甜菜多黏菌（*Polymyxa betae*）传播，能够引起甜菜丛根病（Rhizomania），对世界各地甜菜生产造成极大破坏。BNYVV 是一种多分体正单链 RNA 病毒，其基因组由 4~5 条 RNA 组成，根据核苷酸的长度依次命名为 RNA1-RNA5。RNA1 与 RNA2 编码“持家基因”，是 BNYVV 侵染寄主的必要组分，其中 RNA1 编码复制酶，与病毒基因组复制相关，RNA2 编码 6 个蛋白，依次是外壳蛋白 CP、CP 通读蛋白 p75、三联基因区蛋白（Triple gene block，TGB）以及基因沉默抑制子 p14。RNA3、RNA4、RNA5 编码一些小分子蛋白，与病毒的致病性及菌传效率等相关。

三联基因区由 p42、p13、p15 组成，这 3 个蛋白共同参与病毒的胞间运动，缺少这 3 个蛋白中任何一个，病毒均不能在昆诺藜上进行胞间运动。结构分析表明 p42 有至少两个功能域，分别是位于蛋白 N 端的核酸结合区以及位于蛋白 C 端的依赖 ATP/GTP 的解旋酶保守功能域，这两个功能域对 p42 的活性均有影响。亚细胞定位实验表明 p42 定位于胞间连丝和胼胝质。

为了更方便地检测 BNYVV 的积累量和进一步研究病毒的运动及 p42 的功能，笔者将 p42 基因构建到了原核表达载体 pHAT2 上，转化大肠杆菌表达菌株 BL21（DE3），经 0.2 mmol/L IPTG 诱导后进行检测，发现 6XHis-p42 融合蛋白大量存在于上清中，进一步通过镍柱亲和纯化融合蛋白并通过不同浓度的咪唑洗脱，得到纯化的目标蛋白。

关键词： 甜菜坏死黄脉病毒；p42；原核表达

致谢：感谢于嘉林、李大伟、王献兵教授和张永亮副教授对本研究的建议。

---

* 基金项目：国家自然科学基金项目（31401708）和高校基本业务费（2016QC014）资助

** 第一作者：姜宁，博士生，主要从事甜菜病毒的研究；E-mail：jiangning90y@cau.edu.cn

*** 通信作者：王颖，副教授，主要从事植物病毒学研究；E-mail：yingwang@cau.edu.cn

玉米黄花叶病毒河北分离物的初步鉴定*

李　畅**，孙　倩，李源源，王　颖，张宗英，李大伟，于嘉林，韩成贵***
（中国农业大学作物有害生物监测与绿色防控农业部重点实验室，北京　100193）

摘　要：马铃薯卷叶病毒属（*Polerovirus*）病毒在世界上分布广泛，可为害多种作物。玉米黄花叶病毒（Maize yellow mosaic virus，MaYMV）是 2016 年报道的 *Polerovirus* 新成员，其基因组 RNA 长 5 642nt，为正义单链 RNA，包含 7 个可读框：ORF0（44～838nt）、ORF1（174～2 042 nt）、ORF2（1 602～3 316nt）、ORF3（3 510～4 103nt）、ORF4（3 529～4 107nt）、ORF5（4 103～5 504nt）、ORF3a（3 392～3 526 nt，由 AUA 起始翻译），3 个非编码区：5′UTR、3′UTR、intergenic UTR（位于 ORF2 和 ORF3a 之间）。目前报道的云南和贵州分离物，基因组序列一致性达 98%。ORF0 所编码 P0 蛋白具有基因沉默抑制子活性，ORFs1-2 编码 RdRp 融合蛋白（1047 aa），ORF3 编码外壳蛋白 CP（197 aa），ORF4 编码运动蛋白 MP（192 aa），ORF3a 编码 P3a（44 aa），构建的全长 cDNA 克隆 Pcb301-MaYMV 农杆菌浸润接种本生烟（*Nicotiana benthamiana*）表明能够系统侵染。

本文对采自河北张家口市蔚县小五台山和廊坊市的玉米样品，利用 polerovirus 简并引物 PoconF（GAYTGYTCYGGTTTTGACTGG）/PocoCPR（CGTCTACCTATTTSGGRTTN）进行 RT-PCR 检测，结果均为阳性。扩增产物经过克隆、序列测定和 blastn 确定为 MaYMV。核苷酸序列比对分析发现与 MaYMV 一致性为 99%；推导的氨基酸序列比较表明，MaYMV 河北分离物与已报道的云南分离物同源性较高：CP 氨基酸同源性为 99%，MP 氨基酸同源性为 99%，P3a 氨基酸同源性为 100%，RdRp C 端（849～1 047aa）氨基酸同源性为 100%。检测结果为深入研究该病毒在我国的发生与分布提供了新的数据。

关键词：玉米；玉米黄花叶病毒；河北分离物；RT-PCR

致谢：感谢吴波明教授和蔡青年教授协助采样，感谢王献兵教授和张永亮副教授对本研究工作的建议。

---

* 基金项目：粮食作物重要病毒病防控技术研究与示范子课题（201303021）和粮食主产区主要病虫草害发生及其绿色防控关键技术子课题（2016YFD0300710-1）

** 第一作者：李畅，硕士生，主要开展植物病毒病害防控研究；E-mail：lich@ cau. edu. cn

*** 通信作者：韩成贵，教授，主要从事植物病毒学与抗病毒基因工程；E-mail：hanchenggui@ cau. edu. cn

重组酶聚合酶扩增技术（RPA）检测葡萄卷叶伴随病毒2号*

张永江[1**]，魏　霜[2]，黄　帅[2]，刘建华[3]，任　娇[2]，乾义柯[4]，林春贵[2]
（1. 中国检验检疫科学研究院，北京　100176；2. 广东出入境检验检疫局，广州　510632；3. 国家质检总局标准与技术法规研究中心，北京　100028；4. 伊犁出入境检验检疫局，伊宁　835221）

摘　要：利用重组酶聚合酶扩增技术检测葡萄卷叶伴随病毒2号（Grapevine leaf-roll-associated virus 2，GLRaV-2）。根据GLRaV-2的HSP70基因保守序列，设计RPA特异性引物，建立GLRaV-2的RPA检测方法；对供试样品GLRaV-2、GLRaV-3、GRSPaV、GFkV、GVA共5种病毒进行检测，以验证RPA方法的特异性；将GLRaV-2带毒株的总RNA制备的cDNA设计6个浓度梯度，利用RPA检测方法与普通PCR方法比较，以验证所建立的RPA方法的灵敏性。结果显示，建立的RPA检测方法能够从GLRaV-2带毒株中检测到约284 bp的特异性条带，仅需在37℃下恒温反应40min，不需要特殊的仪器设备；除GLRaV-2能检测到特异性条带以外，其他4种病毒GLRaV-3、GRSPaV、GFkV及GVA均未检测到条带，证明该方法特异性好；RPA和PCR两种方法检测GLRaV-2的灵敏度一致，均达到10^{-4}稀释度，说明所建立的RPA方法灵敏性好。综上所述，本研究建立的RPA方法检测GLRaV-2特异性强、灵敏度高，无须特殊的仪器设备，适合实验室快速检测。

关键词：葡萄卷叶伴随病毒2号；RPA；检测

Detection of Grapevine leafroll-associated virus 2 by Recombinase Polymerase Amplification

Zhang Yongjiang[1], Wei Shuang[2], Huang Shuai[2], Liu Jianhua[3], Ren Jiao[2], Qian Yike[3], Lin Chungui[2]
(1. *Chinese Academy of Inspection and Quarantine*, *Beijing* 100176, *China*; 2. *Guangdong Entry-Exit Inspection and Quarantine Bureau*, *Guangzhou* 510632, *China*; 3. *Standard and Technical Regulation Research Center of AQSIQ*, *Beijing* 100028; 4. *Yili Entry-Exit Inspection and Quarantine Bureau*, *Yining* 835221, *China*)

Abstract: Todevelop a recombinase polymerase amplification (RPA) assay for the detection of Grapevine leafroll-associated virus 2 (GLRaV-2), specific RPA primers were designed based on HSP

* 基金项目：国家重点研发计划《高频跨境细菌和病毒高精准检测技术研究》（2016YFF0203203）
** 第一作者兼通信作者：张永江，研究员，从事植物病毒检测研究；E-mail：zhangyjpvi@ yeah. net

70 (heat shock protein 70) gene of GLRaV-2 and a RPA assay was developed for the detection of GLRaV-2 after optimization reaction condition. Specificity of the RPA assay was tested by GLRaV-3, GRSPaV, GFkV and GVA. Six gradient cDNA concentrations, which from total RNA preparation of GLRaV-2 strains, were designed to verify the sensitivity between the established RPA and ordinary PCR method, The result showed that only a 284 bp fragment was detected about GLRaV-2 by RPA assay. The RPA reaction condition was just to keep the temperature at 37℃ about 40 min without using other special equipments. Besides other 4 virus samples, GLRaV-3, GRSPaV, GFkV and GVA were negative reaction by this assay, which can prove the method is good specificity. The sensitivity of RPA was 10^{-4} degrees cDNA dilution, which was consist with PCR. This RPA assay proved to be rapid, sensitive and specific for the detection of GLRaV-2, and is suitable for the rapid screening of GLRaV-2..

Key words: Grapevine leafroll-associated virus 2; RPA; Detection

葡萄卷叶伴随病毒（Grapevine leaf-rol-associated virus，GLRaV）是仅次于葡萄扇叶病毒的分布广泛、危害最为严重的葡萄病毒之一，在国内分布较为普遍、危害大，在北京、河北、天津、山东及宁夏地区均有发生。田间感病率较高，对葡萄产量和品质影响很大，且引起该病的病毒由多种病毒单独或复合侵染造成[1-3]。现已报道的有 11 种葡萄卷叶伴随病毒：分别为GLRaV-1~9，GLRaV-Dr，GLRaV-De[4-5]，而我国已鉴定的共有 6 种，即 GLRaV-1~5 和 7[6-7]。GLRaV 具有半潜隐性，导致葡萄植株生长衰弱，果粒大小不均，果穗着色不良，果实成熟期推迟，含糖量减少，产量下降，发病严重的葡萄园产量损失超过 45%，对葡萄的生长发育有显著影响。其中葡萄卷叶伴随病毒 2 号（GLRaV-2）最早由 Gugerli 等在 1984 年发现，GLRaV-2 为长线形病毒属（Closterovirus）成员，可随苗木运输传播，且发病率高、危害大[8]。因此，研究 GLRaV-2 的检测方法对该类病毒检验检疫和葡萄无毒苗木生产具有重要意义。

目前，指示植物法、酶联免疫吸附法（ELISA）[9]、反转录聚合酶链式反应（RT-PCR）[10]、免疫捕捉 RT-PCR[11]、实时荧光 RT-PCR[12]和探针杂交[13-14]以及 LAMP 方法[15]等技术已用于植物病毒的检测中。尤其是 PCR 技术由于检测灵敏度高，适用范围广，操作简便等原因应用尤为广泛，但常见的 PCR 检测技术检测程序复杂、需要精密的仪器、检测时间较长，不利于非实验室环境下现场检测以及基层实验室的推广应用。RPA（Recombinase polymerase amplifcation）是一项新型的等温扩增技术[16]，该技术扩增是在单链 DNA 结合蛋白（single-stranded DNA binding protein，SSB）的作用下，使模板 DNA 解链，并在 DNA 聚合酶的作用下，引物与模板正确配对形成复合体，然后 DNA 聚合酶延伸引物生成新的 DNA 互补链。该方法主要具备以下优点：①不需要高低温度循环实现核酸解链和退火，只需要 37℃下恒温反应即可；②只需要 1 对引物即可在 37℃恒温进行模板核酸的扩增；③反应时间短，仅需 40 min；④结果容易辨识，不像 LAMP 产物的弥散带，RPA 扩增产物根据引物设计位点具有特定大小的条带[17]，而且 RPA 产物可测序进一步确认。

目前国内利用 RPA 技术检测葡萄卷叶伴随病毒 2 号尚未见报道，本研究拟根据 GLRaV-2 的 HSP70 基因（Heat shock protein 70）保守序列，设计 RPA 检测引物，建立 GLRaV-2 的 RPA 检测方法，并验证其特异性和灵敏度，建立一种快速检测 GLRaV-2 的简便高效、特异性强、灵敏度高的适于现场检测的方法。

1 材料与方法

1.1 材料与试剂

葡萄卷叶伴随病毒 2 号（GLRaV-2）、葡萄卷叶伴随病毒 3 号（GLRaV-3）、沙地葡萄茎痘伴随病毒（Grapevine rupestris stem pitting associated virus，GRSPaV）、葡萄斑点病毒（Grapevine fleck virus，GFkV）及葡萄病毒 A（Grapevine virus A，GVA）的葡萄叶片采自新疆伊宁，同时设置阴性对照（DNA 模板为脱毒葡萄苗），样品经 PCR 检测和序列测定确认后-80℃冻干保存。植物总 RNA 提取试剂盒购自天根生物科技有限公司；反转录试剂盒 Prime Script RT-PCR Kit 为 TaKaRa 公司产品。

1.2 主要仪器与设备

PCR 扩增仪（TC3000G，英国 TECHNE 公司）、微量紫外分光光度计（ND-1000，Nano Drop 公司）、电泳仪（600SI，上海博彩生物科技有限公司）、凝胶成像系统（GBOX-F3，GENE 公司）。

1.3 方法

1.3.1 总 RNA 提取及 cDNA 的合成

参照试剂盒操作，提取感病葡萄叶片的总 RNA。并参照反转录试剂盒（Prime Script RT-PCR Kit，TaKaRa）的操作，将提取的 RNA 制备成 cDNA，-20℃保存备用。

1.3.2 引物设计

参考 RPA 引物设计的要求，根据 GLRaV-2 的 HSP70 基因的保守序列，设计检测葡萄卷叶伴随病毒 2 号的 RPA 检测引物，扩增条带 284 bp。上游引物 GLRaV-2-F：5′-CGGTCTCGT-CATAACCGACGCTTCTAGT-TTGAATG-3′；下游引物 GLRaV-2-R：5′-TGAATCTATTAAGAGGT-GCGCACCCTTCAAGCACT-3′。引物由上海生工生物工程有限公司合成。

1.3.3 RPA 检测

以 1.3.1 中制备的 cDNA 为模板，采用 GLRaV-2-F 和 GLRaV-2-R 引物进行 RPA 扩增，同时设置阴性对照。RPA 扩增体系的配制方法如下：向含有冻干酶粉的 0.2mL Twist Amp 反应管（Twist AmpBasic kits，Twist）中加入再水化缓冲液（Rehydration Buffer）29.5μL，去离子水 12.5μL，上、下游引物各 2μL（终浓度为 0.4μmol/L），模板 cDNA 1μL，最后再加入醋酸镁溶液 2.5μL（280mmol/L）。将 RPA 扩增体系充分混匀，置于 37℃的金属浴上反应 40min。RPA 反应结束后，向 RPA 扩增产物中加入 50μL 苯酚/氯仿溶液，充分混匀后 12 000r/min 离心 2min，取上清液 5μL 于 1.5%琼脂糖凝胶电泳，在凝胶成像系统上观察结果。

1.3.4 RPA 检测方法特异性评价

按照 1.3.3 的 RPA 检测反应体系，对供试样品 GLRaV-2、GLRaV-3、GRSPaV、GFkV 和 GVA 共 5 种病毒的 cDNA 进行检测，同时用脱毒葡萄苗的 cDNA 做阴性对照，对所建立的 RPA 检测方法特异性进行评价。

1.3.5 RPA 检测方法灵敏度评价

将 GLRaV-2 带毒株的总 RNA 制备的 cDNA 进行 10 倍梯度稀释，共 6 个稀释度，以梯度稀释的 cDNA 为模板，按照 1.3.3 所示的 RPA 检测反应体系进行 RPA 灵敏度实验，并参照已报道的 GLRaV-2 检测引物 V2dCPf2/V2CPr1 进行 PCR 灵敏度实验，引物序列 V2dCPf2：5′-ACGGT-GTGCTATAGTGCGTG-3′；V2CPr1-：5′-GCAGCTAAGTACGAATCTTC-3′，扩增产物长度为 515 bp，比较两种方法检测灵敏度。

2 结果

2.1 GLRaV-2 的 RPA 检测方法的建立

以含 GLRaV-2 的葡萄叶片的 cDNA 为模板，并以脱毒葡萄苗的 cDNA 为阴性对照，建立了 RPA 检测方法检测 GLRaV-2（图 1）。GLRaV-2 扩增条带与预期目的条带大小一致约 284 bp，阴性对照脱毒葡萄苗没有条带，建立的 RPA 检测体系能够准确检测到 GLRaV-2。

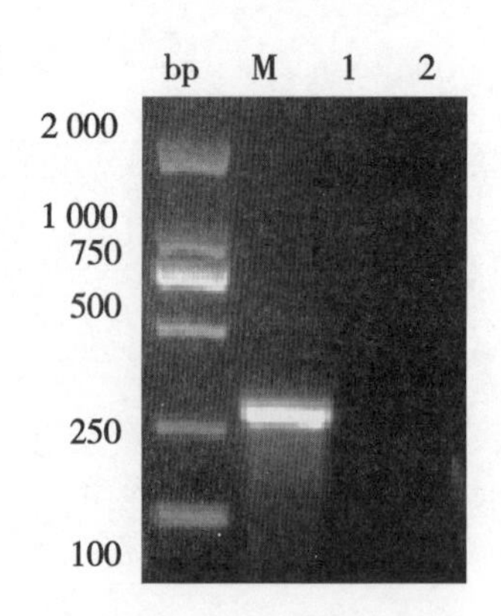

图 1　RPA 技术检测 GLRaV-2

M：Marker DL 2000；1：GLRaV-2；2：阴性对照

2.2　RPA 检测方法的特异性评价

GLRaV-2 的 RPA 特异性评价结果显示，GLRaV-2 样品能够检测到 284 bp 的特异性条带，其他 4 种病毒样品 GLRaV-3、GRSPaV、GFkV 及 GVA 均未检测到条带，证明该方法具有较强的特异性（图 2）。

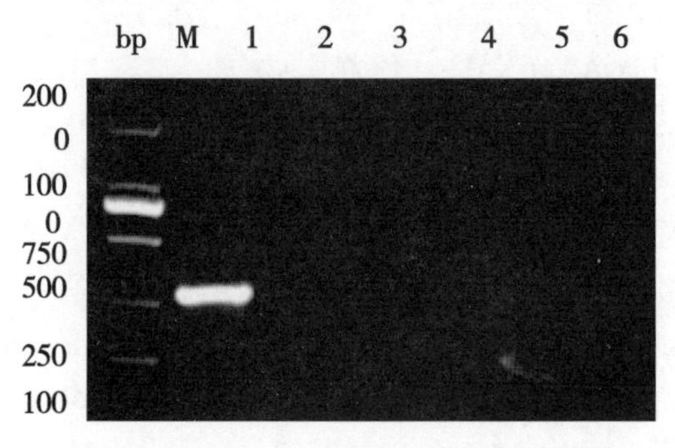

图 2　GLRaV-2 的 RPA 特异性检测结果

M：Marker DL 2000；1：GLRaV - 2；2：GLRaV - 3；3：GRSPaV；4：GFkV；5：GVA；6：阴性对照

2.3　RPA 检测方法与普通 PCR 方法的灵敏度比较

对比 GLRaV-2 的 RPA 检测方法与普通 PCR 方法灵敏度，检测结果显示（图 3、图 4），当普通 PCR 反应设置 30 个循环，稀释度为 10^{-4}时，仍可以检测到约 515bp 的目的条带；当 RPA 检测方法反应时间设置 40min，稀释度为 10^{-4}时，也可以检测约 284bp 的目的条带，当稀释度为 10^{-5}时；两种方法均未能扩增到目的条带，说明两种方法灵敏度相当。

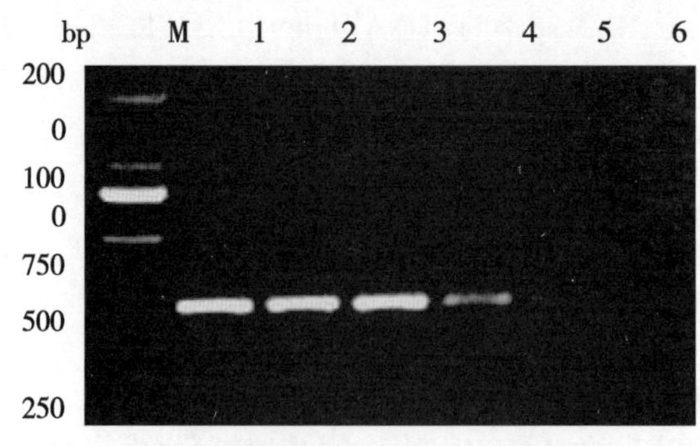

图 3　GLRaV-2 的 RPA 检测灵敏度

M：Marker DL 2000；1-6：cDNA 稀释度依次为 10^{0}、10^{-1}、10^{-2}、10^{-3}、10^{-4}、10^{-5}

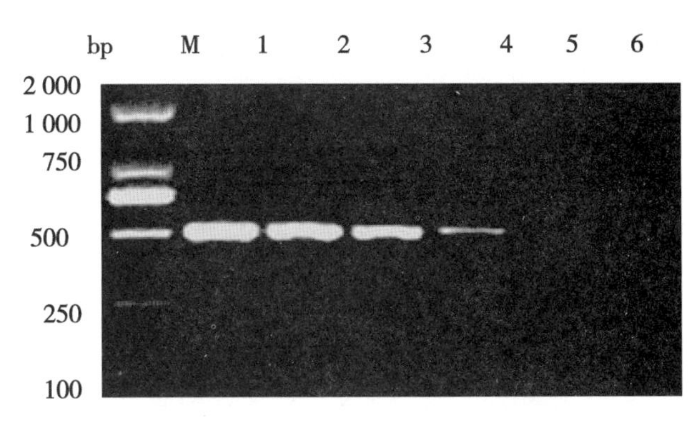

图4　GLRaV-2的PCR检测灵敏度

M：Marker DL 2000；1-6：cDNA 稀释度依次为 10^0、10^{-1}、10^{-2}、10^{-3}、10^{-4}、10^{-5}

3　讨论

RPA技术是近几年兴起的常温扩增技术，目前在医学病原菌及转基因的检测中应用比较广泛[18-22]，但该技术在植物病毒检测中应用较少。本研究建立了GLRaV-2的RPA检测方法，并对该方法的特异性进行验证，同时将该技术的灵敏度与普通PCR进行了对比，结果显示该方法特异性强，灵敏度与普通PCR相当，且反应快速、对仪器设备依赖性低，适合基层单位和现场检测使用，为葡萄病毒病害的田间诊断、预测预报和葡萄脱毒苗的生产等提供了更为简便高效的技术方法。

RPA技术最大的优点在于反应条件在37℃恒温反应，不需要PCR仪等热循环设备，并且这一温度接近人体温度，因此，有学者利用RPA的这一优点，建立了一种不利用任何仪器设备，仅仅利用人体体温即能完成DNA扩增的RPA检测方法[23]，这大大增强了该技术的适用范围。

灵敏度是一项检测方法的关键，本研究建立的RPA技术用于检测GLRaV-2方法，其灵敏度和普通PCR的结果相当，与已报道的LAMP检测技术灵敏度（质量分数0.1%~0.01%）也相当[24-25]。但是，RPA技术相比普通PCR，反应更为快速（仅需40min），而且无须PCR仪器；RPA技术相比LAMP，仅需1对引物即可完成扩增，而LAMP需要4对引物，引物设计难度加大，而且LAMP电泳结果为弥散带，无法进一步确认产物，而RPA根据引物设计位点有确定大小的条带，产物可测序进一步确认。这说明，RPA技术相比PCR和LAMP技术，在不降低灵敏度的情况下同时获得了许多优点，具有较高的实际应用价值。

当然RPA作为一种新兴起的检测技术，还有许多不足等待完善，首先是RPA体系对于蛋白活性要求比较高，必须保证体系中任意一种酶均有活性，否则将导致RPA体系停止工作；其次，RPA扩增体系中存在大量蛋白酶，使得RPA扩增产物必须除去蛋白后才能电泳或进行后续试验[17]；最后，RPA技术所需引物（30~35bp）较长，在短序列核酸检测上不适用。但RPA技术作为一种新的核酸扩增技术，具有特异性强、灵敏度高、操作快速便捷等优点，被誉为“可替代PCR”的技术，已逐渐应用于病原体检测及食品安全等众多领域，相信随着RPA技术反应体系和条件的不断优化，该技术的应用范围也将越来越广。

参考文献

［1］　Dovas C I, Katis N I. A spot multiplex nested RT-PCR for the simultaneous and generic detection of viruses involved in the aetiology of grapevine leafroll and rugose wood of grapevine [J]. Journal of Virological Methods, 2003, 109 (2): 217-226.

[2] 王升吉，尚佑芬，赵玖华，等．葡萄卷叶病毒主要检测技术比较 [J]．山东农业科学，2008，35 (3)：95-98.

[3] 侯义龙．葡萄卷叶病毒 RT-PCR 检测技术 [J]．植物保护学报，2003，30 (3)：305-308.

[4] Maliogka V I, Dovas C I, Katis N I. Evolutionary relationships of virus species belonging to a distinct lineage within the ampelovirus genus [J]. Virus Research, 2008, 135 (1): 125-135.

[5] Ghanem-Sabanadzovic N A, Sabanadzovic S, Uyemoto J K, *et al.* A putative new ampelovirus associated with grapevine leafroll disease [J]. Archives of Virology, 2010, 155 (11): 1871-1876.

[6] 裴光前，董雅凤，张尊平，等．4 种葡萄卷叶伴随病毒多重 RT-PCR 检测 [J]．植物病理学报，2010，40 (1)：21-26.

[7] 范旭东，董雅凤，张尊平，等．葡萄卷叶伴随病毒 4 和 5 简并引物和多重 PCR 检测 [J]．植物保护，2012，38 (6)：95-97.

[8] 刘广穗．葡萄卷叶伴随病毒-2CP 基因转化本氏烟的初步研究 [D]．武汉：华中农业大学，2008.

[9] 陈建军，刘崇怀，古勤生，等．DAS-ELISA、RT-PCR 和 IC-RT-PCR 检测葡萄卷叶病毒 3 的比较研究 [J]．果树学报，2003，20 (3)：173-177.

[10] Thompson J R, Fuchs M, Fischer K F, *et al.* Macroarray detection of grapevine leafroll-associated viruses [J]. Journal of Virological Methods, 2012, 183 (2): 161-169.

[11] Engel E, Girardi C P, Arredondo V, *et al.* Genome analysis and detection of a chilean isolate of grapevine leafroll associated virus-3 [J]. Virus Genes, 2008, 37 (1): 110-118.

[12] Osman F, Rowhani A. Application of a spotting sample preparation technique for the detection of pathogens in woody plants by RT-PCR and real-time PCR (TaqMan) [J]. Journal of Virological Methods, 2006, 133 (2): 130-136.

[13] Fatima O, Christian L, Deborah G, *et al.* Comparison of low-density arrays, RT-PCR and real-time TaqMan; RT-PCR in detection of grapevine viruses [J]. Journal of Virological Methods, 2008, 149 (2): 292-299.

[14] Saldarelli P, Minafra A, Martelli G P, *et al.* Detection of grapevine leafroll-associated closterovirus iii by molecular hybridization [J]. Plant Pathology, 1994, 43 (1): 91-96.

[15] 范旭东，董雅凤，张尊平，等．沙地葡萄茎痘相关病毒的 RT-LAMP 检测方法 [J]．植物病理学报，2013，43 (3)：286-293.

[16] Lutz S, Weber P, Focke M, *et al.* Microfluidic lab-on-a-foil for nucleic acid analysis based on isothermal recombinase polymerase amplification (RPA) lab chip [J]. Lab on A Chip, 2010, 10 (7): 887-893.

[17] 张娜，乾义柯，魏霜，等．基于重组酶聚合酶扩增技术（RPA）的葡萄卷叶伴随病毒 3 号检测方法 [J]．新疆农业科学，2016，53 (2)：302-308.

[18] Xu C, Li L, Jin W, *et al.* Recombinase polymerase amplification (RPA) of camv-35s promoter and nos terminator for rapid detection of genetically modified crops [J]. International Journal of Molecular Sciences, 2014, 15 (10): 18, 197-198, 205.

[19] 邓婷婷，黄文胜，程奇，等．重组酶聚合酶扩增技术检测转基因水稻中的 Cry1Ab/c 基因 [J]．中国食品学报，2015，15 (3)：187-193.

[20] Euler M, Wang Y, Nentwich O, *et al.* Recombinase polymerase amplification assay for rapid detection of rift valley fever virus [J]. Journal of Clinical Virology the Official Publication of the Pan American Society for Clinical Virology, 2012, 54 (4): 308-312.

[21] Milena E, Yongjie W, Peter O, *et al.* Recombinase polymerase amplification assay for rapid detection of francisella tularensis [J]. Journal of Clinical Microbiology, 2012, 50 (7): 2, 234-238

[22] Boyle D S, Lehman D A, Lorraine L, *et al.* Rapid detection of hiv-1 proviral dna for early infant diagnosis using recombinase polymerase amplification [J]. Mbio, 2013, 4 (2): 49-52.

[23] Zachary Austin C, Brittany R, Rebecca R K. Equipment-free incubation of recombinase polymerase amplification reactions using body heat [J]. PLoS ONE, 2014, 9 (11): 112, 146.

[24] Kiddle G, Hardinge P, Buttigieg N, *et al.* GMO detection using a bioluminescent real time reporter (BART) of loop mediated isothermal amplification (LAMP) suitable for field use [J]. BMC Biotechnology, 2012, 12 (1): 15.

[25] Rostamkhani N, Haghnazari A, Tohidfar M, *et al.* Rapid identification of transgenic cotton (gossypium hirsutum l.) plants by loop-mediated isothermal amplification [J]. Czech Journal of Genetics&Plant Breeding, 2011, 47 (4): 140-148.

Development the GFP Expression Vector of *Cucurbit chlorotic yellows virus* *

Yan Shi**, Sun Xinyan, Wei Ying, Han Xiaoyu,
Chen Linlin, Wang Zhenyue
(*College of Plant Protection*, *Henan Agricultural University*, *Zhengzhou* 450002, *China*)

Abstract: *Cucurbit chlorotic yellows virus* (CCYV), the recently found cucurbit-infecting crinivirus in the family *Closteroviridae* is among the largest single stranded positive-sense RNA viruses. The bipartite RNA genome is comprised of 8607-nucleotide [nt] (RNA1 and 8041-nt RNA2). Recent studies on CCYV were hampered by the lack of reverse genetic tools. Construction of full-length infectious cDNA clones will facilitate the investigation of viral determinants in virus replication and movement, as well as the interactions between viral proteins and host factors. Our previous study developed two sets of full-length CCYV cDNA clones under the control of the T7 RNA polymerase and 35S promoters. Here we constructed CCYV GFP expression vector using "add a gene" strategy according to the previous study on the closterovirus *Citrus tristeza virus* suggesting the "add a gene" strategy is better than ORF substitution and ORF fusion, and tested the GFP fluorescence on systemic leaves of cucumber plants. Twenty-five days after agroinfiltration of pCCYVGFP together with pCBCCYVRNA1 and pCBP1/HC-Pro in *Cucumis sativus*, GFP fluorescence was observed using epifluorescence microscope. It was worth noting that the GFP fluorescence was not only detectable in the leaf veins but also in the surrounding mesophyll tissues which is consistent with the previous study on the CCYV localization in leaf lamina using immunoblots. Comparison of the three different constructs, it was found that $pCCYVGFP_{CGC}$ inoculated cucumber leaves showed the cell spread at 25 dpi, while the other two constructs were mainly in isolated cells. Besides, the fluorescence cell number of $pCCYVGFP_{CGC}$ was significantly higher than $pCCYVGFP_{CGS}$.

Key words: CCYV; Infectious clone; GFP expression

---

* 基金项目：河南省科技攻关项目（162102110102）

** 通信作者：施艳，副教授，博士，主要从事园艺植物病毒学研究；E-mail：shiyan00925@126.com

柑橘黄化脉明病毒侵染对柠檬植株基因表达的影响*

牛炳棵[1,2]，洪　霓**，王国平[1,2]
（1. 华中农业大学农业微生物国家重点实验室；
2. 华中农业大学植物科技学院，武汉　430070）

摘　要：柑橘黄化脉明病毒（*Citrus yellow clearing vein virus*，CYVCV）是近些年在我国新发生的一种 RNA 病毒，在柠檬植株上可引起明显的黄斑、脉明和叶片畸形等症状，严重影响柠檬植株的生长。为明确该病毒侵染对寄主植物基因表达及相关的生理代谢影响，本研究以柠檬 1 年生实生苗为寄主植物，对嫁接接种 CYVCV 的柠檬植株叶片和模拟接种的柠檬植株叶片分别进行转录组测序，以甜橙基因组作为参考进行序列比对，分别获得 33 911 975和 24 792 524条读长的序列。与模拟接种的无毒柠檬样品相比，CYVCV 侵染的柠檬样品中共有 1 316个差异表达基因，上调和下调表达基因分别为 679 个和 637 个。对差异表达基因进行功能注释，注释到 GO 数据库的差异表达基因共 1 114个，注释到 KEGG 数据库的共 453 个。KEGG 通路分析结果表明：这些基因可注释到的通路有 5 类，其中代谢通路所占比例最大，为 377 个，其他通路分别有 12 个、11 个、39 个和 14 个。KEGG 富集分析显示，光合作用、叶绿素生物合成与代谢、碳代谢、光合作用中的碳固定等代谢途径受到显著影响，共有 68 个差异表达基因注释到光合作用、叶绿素生物合成与代谢、植物病原互作途径，推测这些代谢途径的基因表达变化可能与柠檬受 CYVCV 侵染后叶片表现的叶片褪绿及黄化等症状相关。选取涉及 4 个代谢通路的 11 个差异表达较明显的基因进行 qRT-PCR 验证，确认这些基因的表达变化与测序结果一致，但存在时间动态的变化。本研究为深入探讨 CYVCV 与柠檬植株的互作提供了重要参考信息。

关键词：柑橘黄化脉明病毒；柠檬；通路分析；转录测序

---

* 基金项目：国家自然科学基金资助项目（No. 30871684）

** 通信作者：洪霓，教授，研究方向为果树病毒；E-mail：whni@ mail. hzau. edu. cn

灰葡萄孢 HBstr-470 中真菌病毒的克隆及相关生物学分析[*]

贺国园，周梓良，张　静，杨　龙，李国庆，吴明德[**]
（华中农业大学植物科学技术学院，武汉　430070）

摘　要：灰葡萄孢菌株 HBstr-470 是从湖北省分离得到的一株弱致病力菌株，具有生长缓慢、不产菌核，而气生菌丝及分生孢子产量多等特性。前期研究表明，该菌株中至少含有 5 条 dsRNA 条带，进一步检测确定该菌株中含有 7 条 dsRNA 条带，大小分别约为 15. 0kb、10. 0kb、5. 0kb、4. 5kb、4. 0kb、3. 5kb、2. 0kb，通过序列克隆分别获得其各自全长 cDNA 序列。其中，15. 0-kbdsRNA 全长 cDNA 长度为 15 222bp（除 polyA 外），其编码的蛋白与 *Hypoviridae* 的病毒成员 SsHV2/SX247 的氨基酸同源性为 95. 88%，命名为 *Sclerotinia sclerotiorum hypovirus* 2/HBstr-470（SsHV2/HBstr-470）；10. 0-kbdsRNA 全长 cDNA 长度为 10 217bp（除 polyA 外），Blast 分析与 *Hypoviridae* 的病毒成员同源性较高，命名为 *Botrytis cinereahypovirus* 1/HBstr-470（BcHV1/HBstr-470）；5. 0-kbdsRNA 全长 cDNA 长度为 5 228bp，与 *Botryotinia fuckeliana totivirus* 1（BfTV1）的 RdRp 氨基酸比对同源性为 91. 25%，系统发育分析与 *Totiviridae* 中 *Victorivirus* 病毒属的成员聚为一支，命名为 *Botrytis cinereavictorivirus* 1/HBstr-470（BcVV1/HBstr470）；4. 5-kb、4. 0-kb 和 3. 5-kbdsRNA 全长分别为 4 996bp、4 263bp 和 3 931bp，其末端非编码区的同源性高，因而认为该 3 条 dsRNA 为同一病毒的三条片段，基于 RdRp 构建系统发育树，分析表明该病毒与 Amasya cherry disease-associated mycovirus 和其他几种病毒形成一大支，自展支持率为 100%，可能为一种新的真菌病毒，暂命名为 *Botrytis cinerea RNA virus* 2/HBstr-470（BcRV2/HBstr-470）；2. 0-kbdsRNA 其中含有两种片段，它们的大小分别为 1 909bp 和 1 883bp，分别编码同一病毒的 RdRp 和 CP。基于 RdRp 构建系统发育树分析表明，该病毒和 *Partitiviridae* 的病毒成员构成一大支，命名为 *Botrytis cinerea partitivirus* 2/HBstr-470（BcPV2/HBstr-470）。通过供体菌株 HBstr-470 分别与受体菌株 HeBtom-25、HNstr-2 及 B05. 10 进行水平传染，获得 SsHV2/HBstr-470 单独感染的衍生菌株 ST25-1（HeBtom-25 为亲本菌株），SsHV2/HBstr-470 和 BcHV1/HBstr-470 共侵染的衍生菌株 SS2-2（HNstr-2 为亲本菌株）及 BcPV2/HBstr-470 单独感染的衍生菌株 S0-36（B05. 10 为亲本菌株）。所有衍生菌株的分生孢子产量均较其各自的亲本菌株显著增加，无菌核形成。菌株 ST25-1 的菌丝生长速度和致病力显著降低，菌株 S0-36 的生长速度正常但致病力显著降低，菌株 SS2-2 的生长速度与致病力较亲本菌株相比无明显变化。此外，菌株 SS2-2 和 S0-36 的菌落产生大量气生菌丝。因此，灰葡萄孢菌株 HBstr-470 菌落的异常及致病力的衰退可能与真菌病毒的复合感染有关，SsHV2/HBstr-470、BcHV1/HBstr-470 和 BcPV2/HBstr-470 可能共同导致菌株 HBstr-470 的产孢量显著增多，SsHV2/HBstr-470 和 BcPV2/HBstr-470 可能造成了菌株致病力的衰退。然而，BcVV1/HBstr470 和 BcRV2/HBstr-470 对灰葡萄孢的影响还待进一步研究。

关键词：灰葡萄孢；真菌病毒；dsRNA

---

* 基金项目：行业专项“保护地果蔬灰霉病绿色防控技术研究与示范”（201303025）

** 通信作者：吴明德，博士；E-mail：mingde@ mail. hzau. edu. cn

柑橘病毒与类病毒病害及其脱毒技术研究进展*

李双花**，易　龙***，姚林建，夏宜林
（赣南师范大学国家脐橙工程技术研究中心/生命与环境科学学院，赣州　341000）

摘　要：柑橘病毒与类病毒病害严重威胁着柑橘产业的发展，生产中无病化苗木是保障柑橘产业健康发展的首要措施。本文围绕柑橘衰退病、碎叶病、裂皮病3种病害的病原性质及为害进行阐述，并重点分析了目前脱毒技术在柑橘苗木脱毒上的应用及发展。

关键词：柑橘；病毒与类病毒病；脱毒技术

柑橘是世界三大水果产品之一，是经济价值最高且极具发展力的果树之一。我国是主要的柑橘生产国，总产量仅次于巴西和美国排第3位。目前，柑橘栽培方式主要采用无性繁殖，在长期的生长过程中很容易感染病害，其中多数病害能随传播介质或苗木运输而进行传播且多数病害在柑橘树上前期很难察觉，极易暴发，从而导致重大的经济损失。因此，柑橘病害是威胁世界柑橘产业发展的重要因素之一，且目前尚无有效的防治药剂。据不完全统计柑橘病害有100余种，其中世界上曾经报道过的病毒与类病毒病害就有80多种，其中柑橘衰退病、裂皮病、碎叶病等对柑橘生产造成严重的影响[1-2]。因此对柑橘良种进行脱毒处理获得无病毒苗木有着非常重要的意义，本文以近年来对我国柑橘产业造成严重危害的病毒与类病毒病害的病原性质及病毒脱除技术进行综述。

1　我国柑橘主要病毒与类病害

柑橘衰退病（Citrus Tristeza Virus，CTV）病原是已知植物病毒基因组中最大的病毒，属于长线型病毒（Closterovirus）的一种正义单链RNA病毒。CTV存在复杂的株系分化现象，根据不同的CTV分离物在指示植物显示的不同症状，可以将CTV分为3个株系：衰退株系（decline inducing，DI）、茎陷点株系（stem pitting，SP）、苗黄株系（seeding yellow，SY）。其中茎陷点株系对我国柑橘产业危害最为严重，它不仅能侵染以枳壳为砧木的柑橘类果树，还导致植株矮化、果实小、树势弱等症状，且能通过蚜虫和嫁接快速传播。随着甜橙、柚类等感病品种的增加，柑橘衰退病的危害日益加剧[3]。

柑橘碎叶病（Citrus tatter leaf，CTLV）病原为发状病毒属（Capillovirus），与苹果茎沟病毒（Apple stem grooving virus，ASGV，苹果的一种潜隐性病毒）有很近的亲缘关系。柑橘碎叶病在诸多柑橘种植国家暴发过，我国和日本发生最为频繁，2000年湖南安化县唐溪园艺场栽种的3 780株早津温州蜜柑因感染柑橘碎叶病全部死掉[5]。柑橘碎叶病传播方式为嫁接和污染的工具。柑橘碎叶病主要为害的对象是以枳壳或枳壳杂交品种为砧木的柑橘类果树，其大体表现为：嫁接部位肿大、嫁接口易折断、树冠叶片黄化等症状[6]。

柑橘裂皮病（Citrus exocortis viroid，CEVD）病原是一种低分子量的闭合环状RNA病毒，属

* 基金项目：江西省高等学校科技落地项目（KJLD13079）；江西省重点研发项目（20161BBF60070）

** 第一作者：李双花，女，在读硕士研究生，主要从事柑橘病害防治工程研究；E-mail：676966032@qq.com
*** 通信作者：易龙，男，教授，博士，主要从事柑橘病害防控研究；E-mail：yilongswu@163.com

于马铃薯纺锤块茎类病毒属（*Pospiviroid*），主要通过嫁接传播。以枳壳为砧木的柑橘极易感染 CEVD，该病引起柑橘砧木部分外皮开裂、树体矮化；叶片小、新梢少等症状，从而导致植株生长障碍，严重时整个植株死亡[7]。20 世纪 60 年代，我国四川、广西、浙江等地相继报道 CEVD 的发生，之后又在湖南、福建等地暴发，成为影响我国柑橘产业发展的重要因素之一。

2 病毒与类病毒病害的脱除技术

柑橘病毒类病毒病害对柑橘产业造成的危害日益严重，因此越来越受广大病害研究者的重视。柑橘脱毒技术的研发，有效的缓解了柑橘病害对柑橘产业造成的影响。目前，主要的脱毒方法有：珠心培养法、热处理脱毒法、茎尖培养法以及茎尖嫁接等。但多数方法都存在着相应的缺陷，如珠心培养法的童期较长，在长时间的过程中容易产生变异，且一些种传的病毒不能够通过这种方法脱毒；热处理脱毒法受脱毒对象的限制，像裂皮病、碎叶病这样的病害不能够通过热处理脱除[8]；茎尖组织培养法缺点在于柑橘组培苗生根困难，脱毒苗在后期的生长过程中存在很多问题。

2.1 茎尖嫁接脱毒法

近年来，被誉为最有脱毒效果方法茎尖嫁接技术[9-12]；1972 年 Murashinge 等[13]首先提出茎尖嫁接技术，使茎尖分生组织嫁接到试管中经脱毒培养的黄化苗上，继续培养成一个完整的个体。1975 年 Navarro 等[14]加以实施和改进，利用甜橙、克里曼丁桔的茎尖组织进行了嫁接，获得了无病毒苗木。茎尖嫁接能解决茎尖培养困难、生根艰难和移栽成活率低，将实生砧木培养于培养基中，再从成年树上取茎尖作为接穗嫁接到试管中的幼砧木上，从而获得无病毒苗木。另外采用茎尖嫁接技术培育的苗木有童期缩短、无不良变异等优点。已有用此方法获得无病毒苗木的例子，我国相关报道有通过茎尖嫁接脱除柑橘黄龙病、柑橘裂皮病的成功事例[15-16]。

影响茎尖嫁接成活的因素有：砧木品种、接穗、茎尖大小、嫁接方法等。砧木与接穗的亲缘性是影响茎尖嫁接成活的重要因素，1976 年 Navarro 等[17]研究发现，嫁接使用特洛伊枳橙和柠檬作砧木的成活率相对较高，而用莱蒙和香橼成活率极低。1984 年 Edriss & Burger[28]将同一柑橘品种嫁接到不同的砧木上，结果表明不同砧木影响茎尖嫁接成活率。之后有 Fifaei 等[19]、Lahoty 等[20]、徐雪荣等[21]都进行了相关实验并与 Navarro 等[17]实验结果相近。

大量试验说明不同接穗品种使茎尖嫁接的成活率有明显差异。罗君琴等[22]用同一嫁接方式茎尖嫁接每 15 个茎尖嫁接一个品种，共 5 个品种（温州蜜柑、本地早蜜橘、秋辉、杂柑、黄皮）；其后培养 20d 观察结果：5 个品种茎尖嫁接存活率分别是 60%、46.7%、20%、26.7%、90%。李丽[23]、祁鹏志[24]都做了相关研究实验结果与罗君琴等[22]试验结果相近。

茎尖的大小和茎尖嫁接后是否成活、与脱除病毒程度都有着重要的关系。陈如珠等[25]试验用同一柑橘品种砧木（酸橙）嫁接不同大小的茎尖（1 个叶原基、2 个叶原基、3 个叶原基、4 个叶原基）；其成活率分别为：0%、12%~19.2%、76.5%、100%。Navarro 等[12,15]实验结果表明只有在 3 个叶原基（0.18mm）大小以下的茎尖才是无毒。因此茎尖大小与嫁接成活率呈正相关，与脱毒程度呈负相关。

嫁接方法多种多样，不一样的嫁接方法影响茎尖嫁接的成活率。Murashinge 等[13]、Rangan 等[26]运用的顶接法和倒 T 法都存在明显的不足：顶接法的接穗不易成活容易枯水干瘪；倒 T 法的接穗会因为生长旺盛的愈伤组织抑制。1989 年马凤桐和 Murashinge[27]创建了点接法：试验用无病枸橼的种子进行试管实生黄化苗的培育作为砧木备用，粗柠檬为接穗（无菌条件下窃取 0.1~0.2mm 茎尖），距离顶部横切面 1~2mm 处，轻切出一小切口后迅速将切好的茎尖放入切口处，继续在适宜的条件下培养。点接法茎尖嫁接成活率明显提高，达到 90%，成苗率达到 77%~85%。

2.2　茎尖嫁接与热处理组合脱毒方法

采集含有 HLB、CTV、CTLV 和 CEVd 病原的柑橘病苗进行茎尖嫁接，实验结果茎尖嫁接脱除 HLB 的脱除率为 100%，CTV 和 CEVd 的平均脱除率为：80.6%、54.1%。CTLV 脱除率为 0。单凭茎尖嫁接不能完全脱除柑橘病原，常与其他脱毒手段结合，从而提高脱毒率。热处理与茎尖嫁接的结合促使脱毒效率快速提升，宋瑞琳等[28]经过前期试验发现只用茎尖嫁接处理 CTLV 病苗，无脱毒效果。随后试验利用已感染 CTLV 的特洛亚枳橙（14 株）、鲁斯克枳橙（16 株）、芦柑（22 株）和本地早（30 株）4 种柑橘品种（每品种设置 3 组试验组）经 30℃、60℃、90℃预热处理，处理（昼处理：40℃，16h；夜处理：30℃，8h）时间在 30~90d。结果表明处理时间 ≥30d 的脱毒成效显著，达到 90d 的可完全脱毒。周常勇等[29]采集含 SDV 的 2 种柑橘品种（枳砧兴津温州蜜柑、枳砧宫本温州蜜柑）幼苗作为试验材料，经热处理：昼处理 40℃，夜处理 30℃，均为 12h，7~40d；嫁接进行的条件为相对湿度为 80%。茎尖嫁接了 72 株苗全部脱毒。

3　展望

随着我国柑橘产业迅速发展，病毒与类病毒病害危害也日益严重，对柑橘的品质与产量都造成了严重的影响。为克服或缓解柑橘病毒与类病毒病害带来的危害并促进柑橘产业的长足发展，柑橘无病毒苗木繁育体系是柑橘产业健康发展的重要保障，而母本树及苗木的无病化是关键技术，但柑橘的脱毒效果受柑橘品种、病害类型以及脱毒方法等多种因素的影响，单一的脱毒方法或多或少存在着一些缺点，不能令人满意，通过茎尖嫁接和茎尖嫁接结合热处理技术是目前柑橘脱毒技术中最有效的方法。

参考文献

[1]　李红叶，陈力耕，周雪平．柑橘病毒与类似病毒分子生物学和抗病毒基因工程研究进展［J］．果树科学，2000，17（2）：131-137.

[2]　周常勇，赵学源．柑桔良种无病毒苗木繁育体系建设［J］．广西园艺，2004，15（04）：11-17.

[3]　周彦．交叉保护防治柑橘衰退病研究进展［J］．园艺学报，2014，41（9）：1793-1801.

[4]　Zanutto C A，Corazza M J，de Carvalho Nunes W M，*et al.* Evaluation of the protective capacity of new mild Citrus tristeza virus（CTV）isolates selected for a preimmunization program［J］. Scientia Agricola，2013，70（2）：116-124.

[5]　刘干生，刘震，刘尚泉，等．柑桔碎叶病危害的调查［J］．中国南方果树，2003，32（2）：22.

[6]　Ito T，Ieki H，Ozaki K. Simultaneous detection of six citrus viroids and Apple stem grooving virus from citrus plants by multiplex reverse transcription polymerase chain reaction.［J］. Journal of Virological Methods，2002，106（2）：235-239.

[7]　Semancik J S，Szychowski J A，Rakowski A G，*et al.* A stable 463 nucleotide variant of citrus exocortis viroid produced by terminal repeats［J］. Journal of General Virology，1994，75（Pt 4）（4）：727.

[8]　SchwarzRE，and GC，Green. Heat requirement for symptom suppression and inactivation of the greening pathogen. In：Proe. 5th Conf. Intl. Organ. Citrus.（Eds.）：W. C. Price. University of Florida Press，Gainesville，1 972：44-51.

[9]　Navarro L. Application of shoot-tip grafting in vitro to woody species［J］. Acta Horticulturae，1998，227（227）：43-45.

[10]　贺和初．果树无病毒栽培［J］．经济林研究，1995，13（3）：45-46.

[11]　Chand L，Sharma S，Dalal R，*et al.* In vitro shoot tip grafting in citrus species-a review.［J］. AgriculturalReviews，2013，34（4）：279-287.

[12]　Juárez J，Aleza P，Navarro L，*et al.* Applications of citrus shoot-tip grafting in vitro［J］. Acta Horticulturae，2015（1065）：635-642.

[13] Murashinge T, *et al*. A technique of shoot apex grafting and its utilization towards recovering virus-free citrus clones [J]. Hortiscience, 1972, 7: 118-119.

[14] Navarro L, *et al*. Improvement of shoot-tip grafting in vitro for virus-free citrus [J]. Journal of American Society for Horticultural Science, 1975, 100 (5): 471-479.

[15] 蒋元晖，等．通过茎尖嫁接脱除柑桔黄龙病病原体 [J]. 植物学保护学报，1987，14 (3): 184-188.

[16] 刘建雄，等．柑桔茎尖嫁接脱除裂皮病技术研究 [J]. 湖南农业科学，1988 (3): 30-32.

[17] Navarro L, Roistacher C N, Murashige T. Effect of size and source of shoot tips on psorosis-A and exocortis content of navel orange plants obtained by shoot-tip grafting in vitro [J]. 1976.

[18] Edriss M H, Burger D W. Micro-grafting shoot-tip culture of citrus on three trifoliolate rootstocks [J]. Scientia Horticulturae, 1984, 23 (3): 255-259.

[19] Fifaei R, Golein B, Taheri H, Tadjvar Y. Elimination of Citrus Tristeza Virus of Washington Navel Orange [Citrus sinensis (L.) Osbeck] Through Shoot-tip Grafting [J]. International Journal of Agriculture & Biology, 2007, 9 (1): 27-30.

[20] Lahoty P, Singh J, Bhatnagar P, Rajpurohit D, Singh B. In-vitro Multiplication of Nagpur Mandarin (Citrus reticulata Blanco) Through STG [J]. International Journal of Plant Research, 2013, 26 (2): 318-324.

[21] 徐雪荣，黎思娜，李映志．化州橘红茎尖微芽嫁接技术研究 [J]. 热带作物学报，2013，34 (7): 1237-1241.

[22] 罗君琴，李丽，聂振朋，等．柑橘试管内茎尖微芽嫁接技术试验初报 [J]. 浙江农业科学，2011 (5): 1031-1032.

[23] 李丽．柑橘茎尖微芽嫁接脱毒技术研究 [D]. 杭州：浙江大学，2014.

[24] 祁鹏志．柑橘茎尖微芽嫁接脱毒及黄龙病和衰退病分子检测的研究 [D]. 武汉：华中农业大学，2007.

[25] 陈如珠，李耿光，张兰英，等．提高柑桔茎尖微型嫁接成苗率研究 [J]. 广西植物，1991，11 (1): 63-66.

[26] Rangan TS, Murashige T, BittersWP. [J]. Hortscience, 1968, 3: 226-227.

[27] 马凤桐，T. Murashige. 柑桔属茎尖嫁接脱毒的研究 [J]. Journal of Integrative Plant Biology, 1989, 31 (7): 565-568.

[28] 宋瑞琳，吴如健，柯冲．茎尖嫁接脱除柑桔主要病原的研究 [J]. 植物病理学报，1999，29 (3): 275-279.

[29] 周常勇，蒋元晖，赵学源，何新华．脱除温州蜜柑萎缩病毒 (SDV) 的几种方法 [J]. 中国柑桔，1993，22 (2): 17-19.

侵染莱阳茌梨的苹果茎痘病毒的分子生物学和血清学特性研究*

李 柳，郑萌萌，王国平，洪 霓**
（华中农业大学植物科学技术学院，农业微生物学国家重点实验室，武汉 430070）

摘 要： 苹果茎痘病毒（*Apple stem pitting virus*，ASPV）是β线性病毒科（Betaflexiviridae）凹陷病毒属（*Foveavirus*）的代表种，在自然条件下主要侵染苹果（*Mulus* spp.）和梨（*Pyrus* spp.），在世界各苹果和梨产区均有分布。梨树对ASPV的侵染较敏感，常表现叶脉黄化或脉明症状，同时该病毒与其他病毒混合侵染可导致苹果及梨树的衰退。本研究从来源于中国山东省种植近百年的莱阳茌梨上检测到ASPV，通过设计引物进行RT-PCR扩增和扩增产物测序及序列拼接，获得了1个莱阳茌梨ASPV分离物（ASPV-LY）的其全基因组序列。序列比对的结果显示，分离物LY与NCBI上已报道的ASPV分离物具有相同的基因组结构，但全基因组序列相似性较低，为70.1%~78.2%，而与近些年报道的引起苹果绿皱果病的苹果绿皱果相关病毒（Apple green crinkle associated virus，AGCaV）分离物K10的序列相似性相对较高，为79.9%。*CP*基因及编码蛋白在核苷酸水平和氨基酸水平上与K10的相似性分别为78.1%和84.6%，而与其他ASPV分离物的核苷酸相似性为64.1%~69.5%，氨基酸相似性为67.8%~75.7%。该结果表明ASPV-LY为1个新的ASPV分离物。为明确该分离物的血清学特性，对ASPV-LY的CP进行原核表达，并采用3个ASPV分离物的抗体进行Western Blotting，结果显示，ASPV分离物间在血清学上存在差异，其中HN6-8抗体不能有效识别ASPV-LY的CP，该结果进一步表明ASPV-LY为在分子特性和血清学特性与其他ASPV分离物不同的1个新的分离物。

关键词： 苹果茎痘病毒；分子生物学；血清学

* 基金项目：国家农业部公益性行业计划（201203076-03）；梨现代农业技术产业体系（nycytx-29-08）
** 通信作者：洪霓；E-mail：whni@mail.hzau.edu.cn

台湾进境美人蕉上美人蕉黄斑驳病毒的检测与鉴定*

陈细红[1**]，蔡 伟[1]，高芳銮[2]，廖富荣[3]，沈建国[1***]

（1. 福建出入境检验检疫局检验检疫技术中心，福建省检验检疫技术研究重点实验室，福州 350001；2. 福建农林大学植物病毒研究所，福州 350002；3. 厦门出入境检验检疫局检验检疫技术中心，厦门 361012）

摘 要：美人蕉黄斑驳病毒（*Canna yellow mottle virus*，CaYMV）属于花椰菜花叶病毒科（CaulimoViridae）、杆状 DNA 病毒属（*Badnavirus*）成员，是美人蕉上为害较为严重的一种病毒，侵染美人蕉引起叶片褪绿斑、叶脉黄化、茎秆和花上出现条斑等症状。CaYMV 引起的美人蕉黄斑驳病防治非常困难，近年来在美国发生严重的地区被迫采取销毁病株的方式来防止其进一步扩散。为预防和控制该病毒的发生，建立快速、有效的检测方法，加强该病毒的早期检测具有重要意义。本研究以台湾进境的美人蕉植株为材料，提取 DNA 为模板，利用建立的 nested PCR 检测技术，从疑似病株上扩增出大小约 352bp 的目的片段。目的片段回收、纯化后，进行克隆测序。序列测定结果表明：目的片段序列全长为 352bp，与预先设计的片段大小完全相同，并与 GenBank 已报道的 CaYMV 病毒基因序列高度一致。上述结果表明：该批美人蕉植株上携带有 CaYMV。

关键词：美人蕉黄斑驳病毒；检测与鉴定

* 基金项目：质检公益性行业科研专项项目（201410076）；福州市科技计划项目（2013-N-54）

** 第一作者：陈细红，女，助理工程师，主要从事植物病害检测工作；E-mail：535095933@ qq. com

*** 通信作者：沈建国，研究员，主要从事植物病毒检测及其防治研究；E-mail：shenjg_ agri@ 163. com

Transcriptomic Changes in *Nicotiana benthamiana* Plants Inoculated with the Wild Type or an attenuated Mutant of *Tobacco Vein Banding Mosaic Virus**

Geng Chao[1,2], Wang Hongyan[2], Liu Jin[1], Yan Zhiyong[1,2], Tian Yanping[1,2], Yuan Xuefeng[1,2], Gao Rui[1], Li Xiangdong[1,2]**

(1. *Laboratory of Plant Virology*, *Department of Plant Pathology*, *College of Plant Protection*, *Shandong Agricultural University*, *Tai'an*, *Shandong*, 271018, *China*; 2. *Shandong Provincial Key laboratory for Agricultural Microbiology*, *Tai'an*, *Shandong*, 271018, *China*)

Abstract: *Tobacco vein banding mosaic virus* (TVBMV) is a potyvirus mainly infecting solanaceous crops. Helper component proteinase (HCpro) is an RNA silencing suppressor and the virulence determinant of potyviruses including TVBMV. The mutations of D198 to K and IQN motif to DEN in HCpro eliminated its RNA silencing suppression activity and attenuated the symptoms of TVBMV in *Nicotianabenthamiana* plants. Here, we used RNA-seq analysis to compare differential genes expression between the wild type (T-WT) TVBMV and an artificially attenuated mutant (T-HCm) carrying both mutations mentioned above at 1, 2 and 10 days post agroinfiltration (dpai). At 1 and 2 dpai, the *N. benthamiana* genes related to ribosome synthesis were up-regulated, whereas those related to lipid biosynthetic/metabolic process and responses to extracellular/externalstimuli were down-regulated in the inoculated leaves inoculated with T-WT or T-HCm. At 10 dpai, T-WT infection resulted in the repression of photosynthesis-related genes, which associated with the chlorosis symptom. T-WT and T-HCm differentially regulated RNA silencing pathway, suggesting the role of RNA silencing suppressor HCpro in virus pathogenesis. The salicylic acid and ethylene signaling pathway were induced but jasmonic acid signaling pathway were repressed after T-WT infection. The infection of T-WT and T-HCm differentially regulate the genes involved in the auxin signaling transduction, which associated with the stunting symptom caused by TVBMV. These results illustrate the dynamic nature of TVBMV－*N. benthamiana* interactionat the transcriptomic level.

* Funding: National Natural Science Foundation of China (NSFC; 31571984, 31501612)

** Corresponding author: Li Xiangdong; E-mail: xdongli@ sdau. edu. cn

Tobacco vein banding mosaic virus 6K2 Protein Hijacks NbPsbO1 for Virus Replication*

Geng Chao[1], Yan Zhiyong[1], Cheng Dejie[1], Liu Jin[1], Tian Yanping[1], Zhu Changxiang[1,2], Wang Hongyan[1], Li Xiangdong[1,2]**

(1. *Laboratory of Plant Virology*, *Department of Plant Pathology*, *College of Plant Protection*, *Shandong Agricultural University*, *Tai'an*, *Shandong*, 271018, *China*; 2. *Shandong Provincial Key laboratory for Agricultural Microbiology*, *Tai'an*, *Shandong*, 271018, *China*)

Abstract: Chloroplast-bound vesicles are key components in viral replication complexes (VRCs) of potyviruses. The potyviral VRCs are induced by the second 6ku protein (6K2) and contain at least viral RNA and nuclear inclusion protein b. To date, no chloroplast protein has been identified to interact with 6K2 and involve in potyvirus replication. In this paper, we showed that the Photosystem II oxygen evolution complex protein of *Nicotiana benthamiana* (NbPsbO1) was a chloroplast protein interacting with 6K2 of *Tobacco vein banding mosaic virus* (TVBMV; genus *Potyvirus*) and present in the VRCs. The first 6 ku protein (6K1) was recruited to VRCs by 6K2 but had no interaction with NbPSbO1. Knockdown of *NbPsbO1* gene expression in *N. benthamiana* plants through virus-induced gene silencing significantly decreased the accumulation levels of TVBMV and another potyvirus *Potato virus Y*, but not *Potato virus X* of genus *Potexvirus*. Amino acid substitutions in 6K2 that disrupted its interaction with NbPsbO1 also affected the replication of TVBMV. NbPsbP1 and NbPsbQ1, other two components of the Photosystem II oxygen evolution complex had no interaction with 6K2 and no effect on TVBMV replication. To conclude, 6K2 recruits 6K1 to VRCs and hijacks chloroplast protein NbPsbO1 to regulate potyvirus replication.

* Funding: National Natural Science Foundation of China (NSFC31201485, 31571984 and 31501612), Shandong Provincial Natural Science Foundation (ZR2011CM019)

** Corresponding author: Li Xiangdong; E-mail: xdongli@ sdau. edu. cn

甘蔗花叶病毒第 IV 组分离物 RT-PCR 检测体系的建立及应用*

闫志勇[1**]，程德杰[1**]，房　乐[1]，许小洁[1]，田延平[1]，贾　曦[2]，李向东[1,3***]

（1. 山东农业大学植物保护学院植物病毒学研究室，泰安　271018；
2. 山东省农业科学院，济南　250100；3. 山东省农业微生物重点实验室，泰安　271018）

摘　要： 甘蔗花叶病毒（*Sugarcane mosaic virus*，SCMV）是引起我国玉米矮化叶病的主要病毒。根据其基因组序列，SCMV 可以分为 4 个组，其中第 IV 组分离物属于新出现的强毒株系。为了快速检测 SCMV 尤其是 IV 组分离物的发生情况，笔者针对 SCMV I-IV 组及 IV 组分离物设计了 6 对引物，从中筛选出特异性最强的 2 对引物（I-IV-2F/I-IV-2R 和 IV-1F/IV-1R），进行 PCR 体系的优化。优化后的 PCR 体系为：退火温度 51℃，dNTP 终浓度为 0.1 mmol/L，引物终浓度为 0.20μM，*Taq*DNA 聚合酶终浓度为 0.015 U/μL。利用该体系，引物对 I-IV-2F/I-IV-2R 和 IV-1F/IV-1R 均能够从 50 ng 感病叶片或 0.05 ng 病毒 RNA 中检测出 SCMV。通过优化两对引物浓度比例建立了能同时检测 SCMV 所有分离物及第四组分离物的双重 PCR 体系。利用该体系从山东、河南及云南均检测出 SCMV 第 IV 组分离物的发生。本研究为 SCMV 尤其是 SCMV 第 IV 组分离物的快速检测提供了技术支持。

关键词： 甘蔗花叶病毒；第 IV 组；检测；反转录-聚合酶链式反应

Optimization and Application of RT-PCR Detection System for *Sugarcane mosaic virus* Group IV Isolates

Yan Zhiyong[1], Cheng Dejie[1], Fang Le[1], Xu Xiaojie[1],
Tian Yanping[1], Jia Xi[2], Li Xiangdong[1,3]

(1. *Laboratory of Plant Virology*, *College of Plant Protection*, *Shandong Agricultural University*, *Tai'an* 271018, *China*; 2. *Shandong Academy of Agricultural Sciences*, *Ji'nan* 250100, *China*; 3. *Shandong Provincial Key Laboratory of Agricultural Microbiology*, *Tai'an* 271018, *China*)

Abstract: *Sugarcane mosaic virus* (SCMV) is the prevalent virus inducing maize dwarf mosaic disease in China. According to their complete genome sequences, SCMV isolates aredivided into four groups, among which group IV is highly virulent. In order to rapidly detect the SCMV isolates, especially those of group IV, we designed 6 pairs of primers specific for SCMV groups I-IV and group

---

* 基金项目：山东省现代农业产业技术体系（SDAIT-02-09）；山东省农业重大应用创新项目

** 第一作者：闫志勇，男，山东冠县人，博士研究生，主要从事植物病毒学研究；E-mail：yzy_ 1990@163.com
程德杰，男，山东青州人，硕士研究生，主要从事植物病毒学研究；E-mail：sdauzbcdj@163.com

*** 通信作者：李向东，教授，主要从事植物病毒学研究；E-mail：xdongli@sdau.edu.cn

IV, respectively. Primer pairs I-IV-2F/I-IV-2R and IV-1F/IV-1Rwhich showedthe highest specificity were selected for optimization of RT-PCR detection system. The results of optimization were annealing temperature 51℃, the final concentrations of dNTP, primers and *Taq*DNA polymerase were 0. 1 mmol/L, 0. 2μM and 0. 015U/μL, respectively. With the optimized system and primer pairs I-IV-2F/I-IV-2R and IV-1F/IV-1, one can detect SCMV from 50 ng diseased leaves and 0. 05 ng viral RNA. A duplex PCR systemwas established by adjusting the concentration of two pairs of primers. Using this detect system, isolates of SCMV group IV were detected from Shandong, Henan and Yunnan provinces. This study provides technical support for the rapid detection of SCMV especially the fourth group.

Key words: *Sugarcane mosaic virus*; The fourth group; RT-PCR; Detection

甘蔗花叶病毒两个山东分离物的全基因组序列分析*

程德杰[1,2]**，闫志勇[1]**，黄显德[1]，田延平[1]，李向东[1,2]***
（1. 山东农业大学植物保护学院植物病毒研究室，泰安 271018；
2. 山东省农业微生物重点实验室，泰安 271018）

摘　要：甘蔗花叶病毒（*Sugarcane mosaic virus*，SCMV）是引起我国玉米矮花叶病的主要病毒。本文从山东泰安采集到两个表现矮花叶症状的玉米叶片样品（命名为 DWK1 和 DWK2），通过 RT-PCR 扩增全基因组片段并测定了其序列（GenBank 登录号分别为 KU171814 和 KU171815）。序列分析结果表明：DWK1 和 DWK2 基因组全长分别为 9 575个和 9 576个核苷酸（nucleotide，nt），开放阅读框均为 9 192nt，编码 3063 个氨基酸的多聚蛋白。DWK1 和 DWK2 的全基因组核苷酸一致率为 81. 7%，DWK1 与山西分离物 SX（AY569692）核苷酸一致率最高，为 90. 9%，DWK2 与河北分离物 BD8（JN021933）核苷酸一致率最高，达 99. 4%。二者在系统进化树中分别被聚类到 Ⅰ 组和 Ⅳ 组。重组分析发现，DWK1 是 HN（AF494510）、Guangdong（AJ310105）和 BD8 三个分离物的重组体。选择压力分析表明，SCMV 的 10 个蛋白的 d_N/d_S 值都小于 1，均处于负选择，但在 P1、P3 和 CP 中存在正选择位点。本研究结果可为甘蔗花叶病毒株系的监测及防控提供理论指导。

关键词：甘蔗花叶病毒；全基因组序列；重组分析；系统发育分析

* 基金项目：山东省现代农业产业技术体系（SDAIT-01-022-09）

** 第一作者：程德杰，男，硕士研究生，主要从事植物病毒学研究；E-mail：sdauzbcdj@ 163. com
闫志勇，男，博士研究生，主要从事植物病毒学研究；E-mail：yzy_ 1990@ 163. com
*** 通信作者：李向东，教授，主要研究方向为植物病毒学；E-mail：xdongli@ sdau. edu. cn

黄瓜绿斑驳花叶病毒弱毒突变体的筛选及交叉保护效果测定*

刘　锦**，许　帅**，闫志勇，冀树娴，黄显德，田延平，李向东***
（山东农业大学植物保护学院植物病毒研究室，泰安　271018）

摘　要：黄瓜绿斑驳花叶病毒（*Cucumber green mottle mosaic virus*，CGMMV）主要侵染葫芦科作物，给黄瓜、西瓜、甜瓜的生产造成严重损失。本研究将 CGMMV 济南分离物的全基因组克隆到带有 35S 启动子的双元载体 pCambia0390，获得含有 CGMMV 全基因组序列的质粒 pCGMMV，将其转入农杆菌后浸润本氏烟。结果表明：pCGMMV 能够成功侵染本氏烟且侵染效率为 100%。通过定点突变将 CGMMV 复制酶（RNA dependent RNA polymerase，RdRp）第 1 069 位的谷氨酸和衣壳蛋白（coat protein，CP）第 89 位的天冬氨酸分别突变变为丙氨酸均导致 CGMMV 致病力显著降低。交叉保护试验表明两个突变体对野生型 CGMMV 有非常好的交叉保护效果，保护效率分别为 91.7%和 100%。

关键词：黄瓜绿斑驳花叶病毒；侵染性克隆；弱毒突变体；交叉保护

---

* 基金项目：国家重点研发计划（SQ2017ZY060102）
** 第一作者：刘锦，男，博士研究生；E-mail：liujincc612@ 126. com
许帅，男，硕士
*** 通信作者：李向东，男，教授；E-mail：xdongli@ sdau. edu. cn

番木瓜环斑病毒西瓜株系侵染性克隆的构建及应用*

黄显德**，王　玉，程德杰，刘　锦，闫志勇，田延平，李向东***
（山东农业大学植物保护学院植物病理学系，泰安　271018）

摘　要： 利用cDNA末端快速扩增技术（RACE）和RT-PCR获得了番木瓜环斑病毒西瓜株系山东分离物（PRSV-SD）的全基因组序列。PRSV-SD基因组（GenBank登录号为MF085000）全长10 337个核苷酸（nucleotides，nt），5′-和3′-非翻译区分别为90nt和206nt，含有一个开放阅读框，编码3 346个氨基酸的多聚蛋白。PRSV-SD与GenBank中18个PRSV分离物的全基因组核苷酸一致率为82.05%~89.3%，多聚蛋白的氨基酸一致率为90.6%~94.7%。构建了可通过农杆菌浸润接种的侵染性克隆pCamPRSV，可侵染西瓜、甜瓜、西葫芦、黄瓜等葫芦科作物。通过在PRSV核内含体蛋白b（NIb）和衣壳蛋白（*CP*）基因之间插入绿色荧光蛋白（*GFP*）基因，得到pCamPRSV-GFP，在紫外灯下可观察到绿色荧光。

关键词： 番木瓜环斑病毒西瓜株系；全基因组序列；侵染性cDNA克隆

---

* 基金项目：国家重点研发计划（SQ2017ZY060102）
** 第一作者：黄显德，硕士研究生
*** 通信作者：李向东，男，教授，主要研究方向为植物病毒学；E-mail：xdongli@ sdau. edu. cn

山东省侵染马铃薯的马铃薯 Y 病毒株系鉴定*

张继武**，王姝雯，栾雅梦，冀树娴，黄显德，王　群，李向东，田延平***
（山东农业大学植物保护学院植物病理学系，泰安　271018）

摘　要：2015—2016 年在山东省马铃薯主产区采集马铃薯病毒病样品 584 份。ELISA 检测发现 515 个样品与马铃薯 Y 病毒（*Potato virus Y*，PVY）多克隆抗体呈现阳性反应，检出率为 88.18%。随机选取不同地区 133 个样品接种到普通烟后均能引起叶脉坏死症状。克隆了其中 42 个分离物的 P1、HC-pro 和 CP 三个基因，序列分析发现所有分离物在 *P*1 基因中均存在重组信号，系统进化分析显示 42 个分离物的 HC-pro 与 N 株系亲缘关系最近，41 个 CP 与 O 株系分为一组，1 个与 N 株系划分一组。通过扩增 7 个分离物全序列，重组和系统进化树结果显示其基因组类型属 PVY^{NTN-NW}（SYR-II）型。

关键词：马铃薯 Y 病毒；重组分析；系统进化树；复合侵染

---

* 基金项目：国家自然科学基金（31501612）
** 第一作者：张继武，男，山东淄博人，硕士研究生，主要研究方向为植物病毒学；E-mail：zhangjiwu9302@ 163. com
*** 通信作者：田延平，男，教授，主要研究方向为植物病毒学；E-mail：yanping. tian@ sdau. edu. cn

西瓜花叶病毒侵染性克隆的构建及其载体改造*

冀树娴**，王　璐，刘　锦，耿　超，李向东，田延平***
（山东农业大学植物保护学院植物病理学系，泰安　271018）

摘　要：西瓜花叶病毒（*Watermelon mosaic virus*，WMV）属于马铃薯 Y 病毒属（*Potyvirus*），通过蚜虫以非持久性方式传播，能够侵染葫芦科、藜科、豆科等 27 科 170 多种植物。WMV 侵染瓜类作物可引起花叶、果实畸形等症状，导致产量下降品质降低。本研究通过 RT-PCR 方法和 RACE 技术获得了来自西葫芦的 WMV 分离物全基因组序列，并成功将该分离物全基因组序列构建到带有 35S 启动子的双元载体 pCambia0390，获得 pCamWMV 全长 Cdna 克隆。获得的 pCam-WMV 可通过农杆菌浸润接种，并能够在西葫芦、黄瓜、西瓜等多种作物上引起典型花叶症状。将绿色荧光蛋白基因 *gfp* 克隆到 NIb 和 CP 编码区间，获得 pCamWMV-GFP，浸润西葫芦、黄瓜和西瓜等健康植株 10d 后，在紫外灯可观察到绿色荧光，说明 GFP 能够正常表达。

关键词：西瓜花叶病毒；侵染性克隆；GFP

* 基金项目：国家重点研发计划（SQ2017ZY060102）
** 第一作者：冀树娴，女，硕士研究生；E-mail：jishuxiansdau@163.com
*** 通信作者：田延平，男，教授，主要研究方向为植物病毒学；E-mail：Yanping.tian@sdau.edu.cn

宁夏银川设施大棚蔬菜病毒病检测[*]

申兆勐[**]，邓　杰，马占鸿[***]
（中国农业大学植物病理系，农业部植物病理学重点开放实验室，北京　100193）

摘　要：2017 年 5 月，笔者采用北京中检葆泰生物技术有限公司购买的 CMV，CGMMW 和 PVY 三种病毒检测试剂纸条，对宁夏银川地区设施大棚蔬菜病毒病进行了检测。方法是将采集的植物叶片放入样品袋中的网络衬里之间，在外部进行摩擦研磨样品，将试纸条绿色部分朝下插入样品袋中，3~5min 即可出结果。该试纸条上有两条线，质控线和测试线，质控线出现并显红色，说明测试正常，如果质控线不显色，说明测试无效。如果出现两条线，则说明测试为阳性，在样品中有病毒存在，如果只有一条线说明为阴性。检测结果表明：在银川军马场蔬菜大棚中调查，发现番茄大棚中有类似于病毒病症状的植株，但在黄瓜大棚中病害基本上都为黄瓜霜霉病，病毒病症状较少。分别在番茄大棚和黄瓜大棚里选了 5 个点，分别用 CMV、CGMMV、PVY 三种检测试剂纸条进行样品采集和试纸条检测，其中黄瓜样品中未检测到有病毒病的存在，在番茄大棚中，其中一个点的样品检测到有 CMV 存在。在银川园艺产业园的番茄和黄瓜大棚中调查，发现其中发病叶片很少，随机选取了 4 个大棚进行采样检测，每个大棚采集的样品分别用三种试纸条检测，均未检测到有病毒的存在。进一步对军马场在调查到有 CMV 病毒病存在的番茄大棚中进行更大面积的检测，发现很多植株上均存在有 CMV，选取大棚内 20 个点，每个点选 5 株进行病情指数调查，并采集病害样本带回市内做进行进一步的研究。

关键词：宁夏银川；设施大棚；蔬菜病毒病；检测

* 基金项目：宁夏回族自治区重点研发计划项目（2016BZ09）和国家重点研发计划项目（2016YFD0201302，2016YFD0300702）

** 第一作者：申兆勐，男，河南安阳人，硕士研究生，主要从事植物病害流行学研究；E-mail：771648345@ qq. com
*** 通信作者：马占鸿，教授，主要从事植物病害流行和宏观植物病理学研究；E-mail：mazh@ cau. edu. cn

小麦黄花叶病毒蛋白 P1 与 Rubisco 互作验证*

梁乐乐，刘丽娟，孙炳剑**，李洪连
（河南农业大学植物保护学院，郑州　450002）

摘　要：小麦黄花叶病毒（*Wheat yellow mosaic virus*，WYMV）是马铃薯 Y 病毒科（Potyviridae）大麦黄花叶病毒属（*Bymovirus*）成员之一，由土壤中禾谷多黏菌（*Polymyxa graminis*）介导传播。小麦感病后，一般可造成减产 10%~30%，重者减产 50% 以上甚至绝收。在我国江苏、河南、安徽、山东、陕西、湖北和四川等省均有分布，近年来由于感病品种的大面积种植，发病面积逐年增加，对小麦生产造成一定威胁。

P1 是小麦黄花叶病毒 RNA2 链上编码的病毒蛋白，以 P1 为诱饵蛋白，通过酵母双杂交系统筛选到其互作寄主蛋白 1，5-二磷酸核酮糖羧化酶/加氧酶 Rubisco（Ribulose bisphosphate carboxylase oxygenase）。Rubisco 是光合作用碳同化过程的关键酶，利用光反应中的同化力（ATP 和 NADPH），将 CO_2 转化为糖类等有机化合物，同时还是光呼吸代谢中的酶。作者将其大亚基 RbcL、小亚基 RbcS 扩增全长后与 P1 进行了互作验证，结果表明 P1 只与 RbcS 互作。通过转录组测序结果分析发现病毒侵染后的样品在光合作用碳固定 pathway 中核酮糖二磷酸羧化酶（ribulose-bisphosphate carboxylase large chain）有 10 个基因下调 22.8 倍，1 个基因上调 2.6 倍，推测是由于 P1 与 Rubisco 中 RucS 的互作，使该酶的基因表达量下调。而上游的 5-磷酸核酮糖异构酶（ribose 5-phosphate isomerase A）表达量下调 5.4 倍，下游的 3-磷酸甘油酸激酶（3-phosphoglycerate kinase）表达量上调 5.6 倍，推测是对这一变化的反馈。同时抑制光合作用碳同化使糖类等物质减少，使小麦生长不良，造成黄化、矮化等症状。

关键词：小麦黄花叶病毒；蛋白；互作

---

* 基金项目：NSFC-河南人才培养联合基金项目（U1304322）；国家公益性行业科研专项（201303021）；“十二五”国家科技支撑计划（2015BAD26B00）

** 通信作者：孙炳剑，副教授，主要从事植物病毒学研究；E-mail：sbj8624@ sina. com

水稻条纹花叶病毒在介体电光叶蝉体内的侵染

孙　想，贾东升，赵　萍，毛倩卓，吴　维，魏太云
（福建农林大学植物病毒研究所，福建省植物病毒学重点实验室，福建　350002）

摘　要：水稻条纹花叶病毒（Rice stripe mosaic virus，RSMV），属于弹状病毒科质型弹状病毒属的一个新种，由半翅目的电光叶蝉以持久增殖型方式传播。持久增殖型病毒均需在介体昆虫体内增殖扩散才能被有效传播。然而，由于 RSMV 是新发现的病毒，其在介体电光叶蝉内的侵染机制和传播特性尚不清楚。为此，本研究通过免疫荧光标记技术系统地分析了 RSMV 在电光叶蝉体内的侵染途径，明确 RSMV 首先侵染电光叶蝉滤室上皮细胞，增殖后即向细胞表面扩散，同时侵染前中肠。随后病毒可通过链接滤室表面和唾液腺的连接线扩散到唾液腺，同时处于叶蝉胸下神经中枢也会被病毒侵染。通过电镜切片观察发现，RSMV 病毒粒子在介体电光叶蝉脑组织和消化系统（唾液腺、滤室、前中肠）内大量聚集，且病毒粒子的形态存在一定的差异。其病毒粒子在昆虫体内主要存在有包膜和无包膜两种形态，其长度可达 50~600nm 不等。RSMV 病毒粒子在电光叶蝉唾液腺内腔规则排列，几乎铺满整个唾液腺内腔，且多为有包膜的病毒粒体，长度约为 200~300nm。而在脑组织和中肠内，病毒粒子倾向于在特异性膜结构上分布，且多为无包膜形态，病毒粒体长度差异较大，最长可达 600nm，最短仅有 50~60nm。综合以上研究，本课题初步明确了 RSMV 在介体电光叶蝉体内的侵染途径，为进一步研究 RSMV 的传播机制奠定了基础。

关键词：水稻条纹花叶病毒；电光叶蝉；侵染途径

水稻黄矮病毒在介体黑尾叶蝉中的侵染循回过程

王海涛，张 乾，张晓峰，王 娟，谢云杰，魏太云
（福建农林大学植物保护学院，闽台作物有害生物生态防控国家重点实验室，福州 350002）

摘 要：水稻黄矮病毒（*Rice yellow stunt virus*，RYSV）由介体黑尾叶蝉以持久增殖型的方式进行传播。目前关于 RYSV 在其介体黑尾叶蝉内的侵染和扩散过程尚不明确，为了探究 RYSV 在介体叶蝉内的侵染循回途径，本研究通过免疫荧光标记技术分析 RYSV 侵染黑尾叶蝉不同时间在不同组织和器官的定位。研究发现，在饲毒后第 2 天，极少量病毒积累在黑尾叶蝉的滤室上皮细胞，有 3%的黑尾叶蝉前中肠上皮细胞可以检测到病毒，表明病毒能够在滤室处突破中肠侵入屏障，实现初侵染进行复制增殖后有向临近组织扩散的趋势。在饲毒后第 4 天，有 13%的黑尾叶蝉前中肠上皮细胞可检测到病毒，在叶蝉滤室和中肠肌肉层均未检测到病毒。在滤室和前中肠被少量病毒侵染的情况下，可观察到被病毒侵染的神经纤维丝状结构。在饲毒后第 6 天，有 47%的黑尾叶蝉前中肠以及 30%的神经节可以检测到病毒，仅有 3%的黑尾叶蝉中中肠上皮细胞有病毒的分布，此时滤室上皮细胞的肌肉层、后中肠和唾液腺没有检测到病毒。在饲毒后第 8 天，有 13%的滤室肌肉层、47%的神经节以及 7%的黑尾叶蝉唾液腺可以检测到病毒，表明此时 RYSV 在黑尾叶蝉内突破唾液腺侵入屏障，进而侵染唾液腺。在饲毒后第 10 天，有 86%的中枢神经系统、67%的前中肠、60%复合神经节、23%的唾液腺、13%的滤室肌肉层、后中肠以及后肠可检测到病毒，表明 RYSV 对介体黑尾叶蝉神经系统具有倾向性。

以上结果表明 RYSV 在介体黑尾叶蝉滤室上皮细胞建立初侵染点，可能借助于神经组织并以逆神经轴突运输的方式扩散至中枢神经系统，进而侵染唾液腺，完成在介体叶蝉内的侵染循回过程。该研究为进一步解析 RYSV 与介体叶蝉的互作关系奠定了基础。

关键词：水稻黄矮病毒；黑尾叶蝉；侵染循回；免疫荧光技术

水稻黄矮病毒结构蛋白 P3 在其介体昆虫细胞中的功能研究*

王 娟**，张晓峰，谢云杰，王海涛，张 乾，魏太云***

（福建省植物病毒学重点实验室，福建农林大学植物病毒研究所，福州 350002）

摘 要：水稻黄矮病毒（Rice yellow stunt virus，RYSV）为弹状病毒科，植物弹状病毒组核型弹状病毒，可同时侵染水稻和介体昆虫叶蝉。由于植物细胞和动物细胞的结构不同，该类病毒在侵染两种不同寄主细胞过程中可能采取不同的侵染策略。前人的研究表明 RYSV 在侵染植物细胞过程中由其编码的结构蛋白 P3 介导通过细胞间连丝进行胞间运动。P3 蛋白被证明参与了病毒在植物细胞间的运动，为运动蛋白，但是其在病毒侵染介体昆虫黑尾叶蝉过程中的功能还不清楚。本研究通过在 RYSV 侵染的黑尾叶蝉细胞中进行免疫荧光定位，以及通过杆状病毒表达系统在 Sf9 昆虫细胞系中对其进行单独表达和与其他 RYSV 蛋白共表达和互作研究，结果发现，RYSV 编码的 P3 蛋白通过 N 蛋白介导进入细胞核，同时 P3 通过与 RYSV 编码的基质蛋白 M 互作，调节 M 蛋白的核质穿梭，完成病毒粒体的包装。另外在病毒侵染黑尾叶蝉细胞时，发现 N 蛋白、P3 蛋白、M 蛋白的表达顺序为 N>P3>M，即 P3 在其中的调节作用对病毒顺利完成侵染至关重要。综上所述，笔者的结果揭示了弹状病毒 RYSV 编码的 P3 蛋白在病毒侵染过程中的新功能，对该病毒的复制包装机制提供了新的模型，为开发新的抗病策略提供了理论依据。

关键词：水稻黄矮病毒；黑尾叶蝉；培养细胞；结构蛋白 P3

---

* 基金项目：国家杰出青年科学基金（31325023）；国家自然科学基金项目（31571979）

** 第一作者：王娟，硕士研究生，从事水稻病毒与介体昆虫互作机制研究；E-mail：wangjuan34567@163.com

*** 通信作者：魏太云，研究员；E-mail：weitaiyun@163.com

共生菌、卵黄原蛋白和水稻矮缩病毒三者互作介导 RDV 的经卵传播

吴　维，毛倩卓，贾东升，黄玲芝，李曼曼，陈红燕，魏太云
（福建农林大学植物病毒研究所，福建省植物病毒学重点实验室，福建　350002）

摘　要：水稻矮缩病毒（*Rice dwarf virus*，RDV）是水稻矮缩病的病原物，由介体昆虫黑尾叶蝉以持久增殖型方式进行传播，同时也通过经卵传播方式直接将病毒传至子代昆虫。同时叶蝉体内的初生共生菌 *Sulcia* 和 *Nasuia* 也可通过经卵方式进行垂直传播，现有研究表明在叶蝉体内初生共生菌和 RDV 入卵均发生在卵黄沉积期，此时卵巢大量摄入卵黄原蛋白；且共生菌和 RDV 均经由一类位于卵巢管柄处特异化的滤泡细胞-上皮鞘细胞侵入卵母细胞，而滤泡细胞可摄取血腔中的卵黄原蛋白提供给卵母细胞。这可能暗示共生菌和卵黄原蛋白在 RDV 入卵过程中起着一定作用。现有研究已证明 RDV 可通过与叶蝉初生共生菌 *Sulcia* 的外膜蛋白互作粘附在共生菌表面，伴随 *Sulcia* 入卵侵入卵母细胞；但是黑尾叶蝉的另一个初生共生菌 *Nasuia* 和卵黄原蛋白在 RDV 经卵传播过程中的作用仍不清楚。笔者通过免疫荧光观察发现：在卵巢上皮鞘处和卵巢内 *Nasuia*、卵黄原蛋白和 RDV 三者存在共定位，免疫电镜的结果也表明在上皮鞘和卵内共生菌 *Nasuia* 内部可以观察有 RDV 病毒粒子分布，且卵黄原蛋白胶体金颗粒会特异性的标记在 *Nasuia* 上。据此笔者初步判定 *Nasuia*、卵黄原蛋白可能介导 RDV 的经卵传播。通过酵母双杂交和 GST-Pull down 实验也表明 RDV 的主要外壳蛋白 P8 与 *Nasuia* 的孔蛋白以及卵黄原蛋白的第四个结构域存在互作。此外，黑尾叶蝉菌胞内会特异性的表达肽聚糖识别蛋白相关蛋白 *NC_ Prp*，干扰此基因表达会抑制 *Nasuia*，而不会对 *Sulcia* 造成明显的影响。因此笔者通过 RNAi 抑制 *NC_ Prp* 的表达，与对照注射 GFP 相比，叶蝉体内 *Nasuia* 量明显降低，而叶蝉体内的病毒量不受影响。但病毒的经卵传播效率受到明显的抑制，主要表现为病毒侵染卵巢比例降低，后代卵内病毒量和带毒率均明显降低。同时注射卵黄原蛋白抗体的抗体阻断实验表明：抑制卵巢对卵黄原蛋白的吸收，对 *Sulcia* 的入卵没有明显的影响，但会明显降低卵巢内 *Nasuia* 和 RDV 的含量，同时也会明显降低后代卵内 *Nasuia* 和 RDV 的侵染率以及卵内含量。根据笔者现有的结果笔者推断 RDV 还可以通过与 *Nasuia* 和卵黄原蛋白的互作，突破经卵传播屏障。

关键词：水稻矮缩病毒；经卵传播；*Nasuia*；卵黄原蛋白

水稻黄矮病毒非结构蛋白 P6 在介体昆虫侵染过程中的功能研究

谢云杰，张晓峰，王　娟，王海涛，魏太云
（福建省植物病毒学重点实验室，福建农林大学植物病毒研究所，福州　350002）

摘　要：水稻黄矮病毒（Rice yellow stunt virus，RYSV）是弹状病毒科，核弹状病毒属的成员，是负单链 RNA 病毒能够同时侵染水稻和介体叶蝉。该病毒编码的非结构蛋白 P6 在病毒侵染介体昆虫过程中的功能还不清楚。鉴于此，本研究通过免疫荧光技术在黑尾叶蝉培养细胞标记 P6 与各个蛋白之间的定位关系，结果发现 P6 分别能与病毒原质组分蛋白 N、P 在细胞核内发生共定位，并且利用酵母双杂交系统进一步验证了 P6 分别能与 N、P 发生明显互作。为了进一步明确 P6 在病毒复制增殖中发挥的作用，笔者利用了 RNAi 干扰技术，即在病毒侵染黑尾叶蝉培养细胞中干扰 P6，用共聚焦显微镜观察病毒侵染率变化情况，利用 RT-PCR 实验检测 N、P 以及基质蛋白 M 的相对表达量。结果表明，干扰 P6 后，病毒侵染率从 50%下降到 25%，N 与对照相比相对表达量下降约 37%，P 与对照组相比相对表达量下降约 32%，M 与对照组比相对表达量下降约 48%，总体上病毒量发生明显下降。因此，笔者推测水稻黄矮病毒在介体昆虫体内形成病毒原质时，P6 分别与 N、P 相互作用，利于病毒原质的形成，加快了病毒的复制，对病毒复制增殖起到了明显的促进作用。

关键词：水稻黄矮病毒；P6；黑尾叶蝉；互作

水稻条纹花叶病毒编码结构蛋白的抗体制备与检测

赵　萍，贾东升，郭剑光，魏太云
（福建农林大学植物病毒研究所，福建省植物病毒学重点实验室，福建　350002）

摘　要：水稻条纹花叶病毒（Rice stripe mosaic virus，RSMV），属于弹状病毒科质型弹状病毒属的一个新种，由电光叶蝉以持久增殖型方式传播。该病毒共编码 2 个非结构蛋白 P3 和 P6，以及 5 个结构蛋白：核衣壳蛋白（N）、糖蛋白（G）、核糖核酸聚合酶（L）、磷酸化蛋白（P）和基质蛋白（M）。由于是 2017 年首次报道的新病毒，目前对该病毒由介体昆虫传播的机制尚不清楚。为了便于研究 RSMV 在介体昆虫体内的侵染机制，本研究首先通过原核表达 N、P、M、G 蛋白，并免疫注射小鼠制备了 N、P、M、G 蛋白的多克隆抗体。通过 Western blot 检测，N、P、M、G 蛋白均可在 RSMV 侵染的昆虫体内检测到特异性蛋白条带。随后对抗血清进行纯化，并交联荧光素。通过免疫荧光标记检测 RSMV 侵染的电光叶蝉消化道组织，共聚焦显微镜观察发现 N、P 抗体可特异性标记电光叶蝉消化道内的病毒，表明 N 和 P 抗体可用于昆虫体内病毒的免疫荧光标记检测；而 M 和 G 抗体不能特异性标记 RSMV 编码的蛋白，相关抗体仍需改变方法进行重新制备。针对 N 和 P 抗体，笔者通过免疫荧光标记方法对实验室饲毒条件下电光叶蝉的带毒率进行检测，发现饲毒后 6 天的电光叶蝉带毒率高达 60%以上，RT-PCR 方法也验证了该检测结果。

综合以上研究结果，本研究制备的 N 和 P 抗体可用于检测病毒在介体昆虫体内的分布，并可用于昆虫带毒率的检测，为解析 RSMV 与介体昆虫间的互作机制和病害田间检测奠定了基础。

关键词：水稻条纹花叶病毒；抗体制备；Western blot 检测；免疫荧光标记检测

The interaction between *Turnip mosaic virus* encoded proteins and AtSWEET1 protein in *Arabidopsis thaliana*[*]

Sun Ying, Wang Yan, Zhang Xianghui, Zhang Yanhua,
Pan Hongyu, Liu Jinliang[**]

(*Department of Plant Protection*, *College of Plant Sciences*,
Jilin University, *Changchun* 130062, *China*)

Abstract: *Turnip mosaic virus* (TuMV) is an important species of the genus *Potyvirus* and has an exceptionally broad host range in terms of plant genera and families of any potyviruses. It occurs worldwide and causes great losses to agricultural production. The symptom formation of a plant viral disease results from molecular interactions between the virus and its host plant. SWEET (sugars will eventually be exported transporters) protein family is a new class of sugar transporters that play an important role in the interaction of host-pathogens. It has been reported that plants infected with fungi or bacteria will induce partial expression of the SWEET gene involved in pathogen-host interaction. However, the SWEET protein family has not been found to be involved in the interaction between virus and host.

In order to study whether *Arabidopsis thaliana* 17 *AtSWEET* genes were induced by virus infection. The expression of *AtSWEET* genes were detected to infect *Arabidopsis thaliana* by infectious cDNA clone of TuMV. The result showed that most of the *AtSWEET* genes were induced expression. It suggested that the *AtSWEET* gene family is involved in the interaction of TuMV and *Arabidopsis thaliana*.

A yeast two hybrid library of *Arabidopsis thaliana* cDNAs was screened using the P3 protein as bait. AtSWEET1 in *Arabidopsis thaliana* was identified to interact with TuMV P3 protein. It in deeply confirmed that AtSWEET1 interacted with TuMV-encoded P3, HC-Pro, VPg and NIa-Pro protein by yeast two-hybrid assay and Bimolecular fluorescence complementation assay (BiFC). It suggested that SWEET protein family play an important role in the interaction between the virus and the host. The role of AtSWEET1 in TuMV infection was under investigation.

* Funding: Natural Science Foundation of China (31201485)

** Corresponding author: Liu Jinliang, E-mail: jlliu@jlu.edu.cn

第四部分 细 菌

Identification of Wheat Blue Dwarf Phytoplasma Effectors Targeting Plant Proliferation and Defence Responses

Wang N.[1], Li Y.[2], Chen W.[2], Yang H. Z.[1], Zhang P. H.[1], Wu Y. F.[1]

(1. *State Key Laboratory of Crop Stress Biology for Arid Areas*, *College of Plant Protection*, *Northwest A&F University*, *Yangling* 712100, *Shaanxi*, *China*; 2. *Oil Crops Research Institute of Chinese Academy of Agricultural Sciences*, *Key Laboratory of Biology and Genetics Improvement of Oil Crops*, *Ministry of Agriculture*, *Wuhan* 430062, *Hubei*, *China*)

Abstract: Identification of effectors from pathogenic microbes is one of the most important themes for elucidating pathogenic mechanism. Wheat blue dwarf (WBD) phytoplasma causes dwarfism, witches' broom and yellow leaf tips in wheat plants, and results in severe yield loss in northwest China. In this study, 37 candidate effector proteins were expressed in *Nicotiana benthamiana* plants. Plants expressing the SAP11-like protein SWP1 showed typical witches' broom. Interestingly, SWP11 induced cell death and triggered defence responses, such as H_2O_2 accumulation and callose deposition. qRT-PCR analysis showed that two marker genes of hypersensitive response (HIN1 and HSR203J), the pivotal defence regulatory gene NPR1 and three pathogenesis-related genes PR1, PR2 and PR3 were significantly up-regulated in *N. benthamiana* leaves expressing SWP11. In addition, SWP12 and SWP21 (TENGU-like) were shown to suppress SWP11-, BAX-and/or INF1-induced cell death. These results indicated that SWP21 has distinct role in virulence compared with its homologs in other phytoplasmas, and that WBD phytoplasma possesses effectors that may target plant proliferation and defence responses. The ability of these effectors to trigger or suppress plant immunity provides new insights into the phytoplasma-plant interaction.

Key words: SAP11-like; Witches' broom; Hypersensitive response; TENGU-like

西瓜噬酸菌菌毛基因 *Aave*-2726 的功能研究*

王万玉**，杨丙烨，童亚萍，胡方平，蔡学清***
（福建农林大学植物保护学院，福州 35002）

摘 要：西瓜细菌性果斑病是以西瓜、甜瓜等为主的葫芦科作物上一种严重的世界性病害，主要是通过种子传播。据报道，植物病原细菌的Ⅳ型菌毛与其定殖、表面黏附、遗传物质摄取、生物膜形成和毒力因子的注入等有关。NCBI 上关于西瓜噬酸菌菌毛基因 *Aave*-2726 的注释为 hypothetical protein，即功能尚不明确，本研究通过定点突变和互补，并测定了生物学特性。结果表明菌毛基因 *Aave*-2726 缺失后病菌的游动性减弱、生物膜形成能力降低，互补菌株的游动性和生物膜形成能力恢复；缺失菌株在 NA 培养基上的菌落形态发生变化，晕圈宽度变窄，互补后菌落的晕圈形成能力部分恢复；缺失菌株对甜瓜和西瓜幼苗的致病性降低，互补后对甜瓜和西瓜幼苗的致病力部分恢复；但是缺失菌株的群体感应和过敏性反应没有变化。试验结果证明菌毛基因 *Aave*-2726 与西瓜噬酸菌的致病性、游动性和生物膜的形成有关。

关键词：西瓜噬酸菌；菌毛基因；生物学特性

---

* 基金项目：福建省自然科学基金项目（2014J01084）；福建农林大学科技创新专项基金（CXZX2016133，KF2015058）；农业部行业专项项目（201003066-2）

** 第一作者：王万玉，男，研究方向：植物保护

*** 通信作者：蔡学清，女，博士，教授，研究方向：植物细菌及植病生防；E-mail：caixq90@163.com

西瓜噬酸菌游动性减弱突变体的筛选及其功能探讨*

杨丙烨**，林志坚，胡方平，蔡学清***

（福建农林大学植物保护学院，福州　35002）

摘　要：由西瓜噬酸菌（*Acidovorax citrulli*）引起的西瓜细菌性果斑病（Bacterial Fruit Broth，BFB），是一种发生范围广，毁灭性强的种传病害，也是我国农业植物检疫性有害生物之一。本研究通过游动性测定对所构建的西瓜嗜酸菌突变体文库进行筛选，获得 1 株游动性减弱的突变株，并对突变体进行亚克隆，结果显示插入位点位于鞭毛蛋白基因 *fliS* 内。另外，通过 PCR 扩增、酶切、转化等步骤构建互补载体，采用三亲杂交的方法构建 *fliS* 基因的互补菌株。对野生菌株、突变菌株和互补菌株的过敏反应、群体感应、菌膜、游动性、致病性、生长速率、鞭毛的合成等生物学功能测定。结果显示：突变菌株的过敏反应和对西、甜瓜的致病性均丧失、游动性减弱、菌膜形成能力增强、生长速率减慢；而互补后游动性、生长速率恢复至野生菌状态，过敏反应、致病性未恢复；电镜下观察，鞭毛蛋白基因 *fliS* 突变菌株的鞭毛明显变短，互补后鞭毛形成能力恢复到野生菌株状态。试验结果证明鞭毛蛋白基因 *fliS* 与西瓜噬酸菌游动性、生长速率和鞭毛合成有关。

关键词：西瓜噬酸菌；游动性；*FliS*；致病性

---

* 基金项目：福建省自然科学基金项目（2014J01084）；福建农林大学科技创新专项基金（CXZX2016133，KF2015058）；农业部行业专项项目（201003066-2）

** 第一作者：杨丙烨，女，硕士研究生，研究方向：植物病理学

*** 通信作者：蔡学清，女，博士，教授，研究方向：植物细菌及植病生防；E-mail：caixq90@163.com

广东南瓜青枯病病原鉴定*

佘小漫[1,2**]，何自福[1***]，汤亚飞[1]，蓝国兵[1]，于　琳[1]
（1. 广东省农业科学院植物保护研究所，广州　510640；
2. 广东省植物保护新技术重点实验室，广州　510640）

摘　要：茄科雷尔氏菌 *Ralstonia solanacearum*（Smith）Yabuuchi *et al.* 是世界上最重要的植物病原细菌之一，广泛分布于热带、亚热带及温带地区。该病原细菌的寄主范围很广，可侵染 50 个科 200 多种植物。中国南瓜（*Cucurbita moschata* Duch.，俗称南瓜）和印度南瓜（*Cucurbita maxima* Duch.，俗称笋瓜）是南瓜属作物中的主要栽培种，现在我国南北各地均广泛种植。2016 年 5 月，在广东的印度南瓜种植地发生青枯病，田间病株率约 23%。印度南瓜病株叶片萎蔫、失去光泽且呈灰绿色，病株维管束变褐，最后整株枯萎。在含 1% TZC 的 LB 琼脂平板上，30℃培养 48h 后，可从病株茎基部组织中分离到菌落形态较一致的细菌分离物，菌落呈近圆形或梭形，隆起，中间粉红色，周围乳白色。人工接种致病性测定显示，该分离菌株能侵染印度南瓜植株并引起与田间症状相同的青枯病，说明其为引起印度南瓜青枯病的病原菌。进一步细菌学试验及 16S rDNA 序列比较，鉴定该病原菌为茄科雷尔氏菌。碳水化合物利用试验结果表明：16 个菌株均能利用甘露醇、山梨醇和甜醇等 3 种醇，不能利用乳糖、麦芽糖、纤维二糖 3 种糖，属于生化变种 IV，8 个菌株均能利用甘露醇、山梨醇和甜醇、乳糖、麦芽糖、纤维二糖 3 种糖，属于生化变种 III。此外，演化型复合 PCR 鉴定结果显示，24 个菌株均能扩增出 280bp 和 144bp 特异条带。因此，引起印度南瓜青枯病的病原鉴定为茄科雷尔氏菌，属于生化变种 III 和 IV、演化型 I 型（亚洲组）。

关键词：印度南瓜青枯病；茄科雷尔氏菌；病原鉴定

---

* 基金项目：广东省科技计划项目（2015B020203002，2015A020209057）；广东省农业科学院院长基金项目（201513）

** 第一作者：佘小漫，副研究员，硕士，研究方面植物细菌；E-mail：lizer126@ 126. com

*** 通信作者：何自福，研究员；E-mail：hezf@ gdppri. com

First Report of *Serratia plymuthica* Causing Ginseng Root Rot in Northeastern China*

Li Weiguang**, Han Chao***

(*College of Plant Protection*, *Shandong Agricultural University*, *Taian*, *Shandong* 271018, *China*)

Abstract: Specific bacterial disease symptoms were observed on ginseng root in almost all regions in Dandong, Liaoning Province in China. As we all know, ginseng is one of the major cash crops grown in Northeastern China. Over the past few years, soil-borne diseases of ginseng caused very serious problems, particularly the root rot which made the control and management difficult. Slowing plant growth and apex putrescence are the typical symptoms for root rot whichis mostly causedby pathogenic fungi. However, bacterial soft of ginseng root is rarely reported. For the purpose of identification of agents causing bacterialroot rot, pathogen was isolated from the infected root and purified on nutrient agar. Its pathogenicity was confirmed using pathogenicity test on ginseng root. These bacteria were identified biochemically and molecularly as *Serratia plymuthica* using molecular techniques−16S rRNA sequence analysis. To our best knowledge, this is the first report of ginseng root rotcaused by pathogenic bacteria as *Serratia plymuthica*.

Key words: *Serratia plymuthica*; Bacterial disease; Ginseng; Root rot

---

* Funding: Young Innovators Awards Reach

** First author: Li Weiguang, the research direction for microbial resource utilization; E-mail: sdsyli@163. com

*** Corresponding author: Han Chao, lecturer, the research direction for microbial resource utilization; E-mail: hanch87@163. com

基于锁式探针快速检测玉米细菌性枯萎病菌和内州萎蔫病菌研究*

李志锋[1**]，冯建军[2,3**]，吴绍精[2,3]，王忠文[1]，程颖慧[2,3]

（1. 广西大学农学院，南宁　530004；2. 深圳出入境检验检疫局动植物检验检疫技术中心，深圳　518045；3. 深圳市检验检疫科学研究院，深圳　518010）

摘　要：玉米细菌性枯萎病菌（*Pantoea stewartii* subsp. *stewartii*，PSST）和玉米内州萎蔫病菌（*Clavibacter michiganensis* subsp. *nebraskensis*，CMN）是两种主要为害玉米维管束的毁灭性细菌。随着我国进口玉米种子逐年增加，这两种细菌传入我国的风险越来越高，并且被列入 2007 年颁布的《中华人民共和国进境植物检疫性有害生物名录》。因此，针对玉米产品上这两种细菌建立一种特异性强、灵敏度高的二重检测方法，既是快速通关的需要，也是严格检疫和保护我国玉米安全生产的需要。本研究根据 PPST 的 *Cps*D 基因和 CMN 的 *ITS* 基因序列分别设计了 PPST-PLP 和 CMN-PLP 锁式探针，建立二重滚环扩增体系，并利用生物素标记的下游引物进行扩增，将扩增产物与偶联上微球的捕获探针进行特异性杂交，最后利用液相悬浮芯片仪进行检测。结果表明：第一，16 株供试单一或混合菌株样品中，该二重检测体系可在同一反应同时检测出 CMN 与 PSST 协同混合 DNA 模板，也可检测出单一靶标菌 DNA 模板，而近似种或其他属供试菌株检测结果呈阴性；第二，通过对菌株 CMN 和 PSST 混合配制的 10^2-10^8CFU/mL 的不同梯度菌悬液进行二重检测，该体系能够检测到菌悬液浓度阈值为 10^3CFU/mL，同时相互之间无交叉反应；第三，该二重检测体系 3 次重复试验的平均荧光值变异系数均小于 10%，表明具有良好的重复性；第四，通过系列梯度浓度的纯菌悬液混合健康种子浸泡液检测，在所有处理样品中利用二重检测方法可同时检测到该混合浸泡液体系中含有 1×10^5CFU/mL 的 CMN 和 1×10^4CFU/mL 的 PSST，而对应的常规 PCR 检测结果均为阴性。因此，本研究所设计的特异的锁式探针建立的玉米细菌性枯萎病菌和玉米内州萎蔫病菌的二重检测方法是一种特异性强，灵敏度高的检测新方法，可为进境玉米检疫提供新的技术保障。

关键词：玉米细菌性枯萎病菌；玉米内州萎蔫病菌；锁式探针；滚环扩增；液相芯片

---

* 基金项目：深圳市海外高层次人才创新创业专项资金项目（KQC201109050077A）和深圳局科技项目（SZ2011002）

** 通信作者：冯建军；E-mail：sccfjj@126.com

广西蔗区检测发现由白条黄单胞菌引起的检疫性病害甘蔗白条病*

李文凤**，单红丽，张荣跃，仓晓燕，王晓燕，尹　炯，罗志明，黄应昆***
（云南省农业科学院甘蔗研究所/云南省甘蔗遗传改良重点实验室，开远　661699）

摘　要：甘蔗白条病（Sugarcane leaf scald disease）由白条黄单胞菌 *Xanthomonas albilineans*（Ashby）Dowson 引起，是中国进境植物重要的检疫对象之一。2016 年在中国广西北海、来宾、百色蔗区发现疑似甘蔗白条病蔗株。为了明确病原，从这 3 个蔗区共采集了 62 份疑似病样，提取基因组 DNA，利用白条黄单胞菌特异性引物 XAF1/XAR1 对 DNA 样品进行 PCR 扩增。结果显示，62 份 DNA 样品都能够获得约 600 bp 的特异性片段；随机选取 21 个 PCR 产物进行测序分析，21 条片段大小均为 608 bp，序列完全一致（GenBank 登录号：KY315183－KY315203）。BLAST 检索结果表明所得序列是导致甘蔗白条病的白条黄单胞菌 *raxB*1 基因核苷酸序列，其与 GenBank 中登录的白条黄单胞菌 *raxB*1 基因核苷酸序列（GenBank 登录号：FP565176）一致性为 100%，并在系统进化树中处于同一分支。这是引起检疫性病害甘蔗白条病的白条黄单胞菌在我国大陆的首次报道。田间调查结果显示：桂糖 46 号、桂糖 06-2081 高度感病，病株率为 18%～50%，严重田块高达 100%；福农 41 号中度感病，病株率为 5%～15%；桂糖 42 号、柳城 05-136、桂糖 49 号、云蔗 08-1095、粤甘 47 号、闽糖 06-1405、柳城 07-150、海蔗 22 号、桂糖 08-1180、福农 09-7111、福农 09-2201、德蔗 07-36、新台糖 22 号 13 个试验示范品种未见发病，在田间表现抗病；感病品种染病严重蔗株全叶枯萎，茎的节部长出许多侧芽，侧芽叶片具白色条纹，造成大幅度减产减糖。本研究在广西蔗区检测发现的甘蔗白条病是一种极易随种苗传播的危险性病害，其对甘蔗产业的发展有潜在威胁。建议有关部门采取相应的预防与管理措施，从源头上控制其扩散蔓延，确保甘蔗安全生产和蔗糖产业可持续发展。

关键词：广西蔗区；甘蔗白条病；PCR 检测；白条黄单胞菌；序列分析

---

* 基金项目：国家现代农业产业技术体系建设专项资金资助（CARS-20-2-2）；云南省现代农业产业技术体系建设专项
** 第一作者：李文凤，女，研究员，主要从事甘蔗病害研究；E-mail：ynlwf@163.com
*** 通信作者：黄应昆，研究员，主要从事甘蔗病害防控研究；E-mail：huangyk64@163.com

坡柳（*Dodonaea viscosa* L.）丛枝植原体新病原的分子鉴定*

笈小龙**，吴育鹏，谢慧敏，余乃通***，王健华，张秀春，刘志昕
（中国热带农业科学院热带生物技术研究所，农业部热带作物生物学与遗传资源利用重点实验室，海口　571101）

摘　要：从四川攀枝花杧果园中表现为丛枝、小叶和黄花等症状的坡柳植株发病样品中，利用植原体 16S rDNA 基因的通用引物 R16mF2/R16mR1 和 R16F2n/R16R2，对发病植株总 DNA 进行巢式 PCR 检测，同时设计植原体抗原膜蛋白基因（*AntMP*）的保守引物 AntMP-F/AntMP-R 进行 PCR 验证。结果显示，坡柳样品巢式 PCR 的第一轮、第二轮 DNA 条带大小分别为 1 400bp 和 1 200bp，经 NCBI 序列相似性比较均为植原体 16S rDNA 序列，GenBank 登录号为 KT957205 和 KT957206；PCR 结果显示抗原膜蛋白基因大小约 600bp，与目标基因大小一致。将测得的 16S rDNA 基因序列与 GenBank 数据库中登录的 16SrⅠ～XV 组植原体 16S rDNA 序列进行同源性比对，构建系统进化树，结果显示，四川坡柳丛枝植原体（DOVI-SC）属于 16SrⅠ组（即翠菊黄花组），与已报道的 5 个 16SrⅠ组（AY101386，AY566302，AY389822，L33760 和 KP662119）同属一个组。利用 iPhyClassfier 在线分析软件对获得的 2 条植原体序列进行虚拟 RFLP 分析，结果显示 KT957205，KT957206 与 16SrI-B 亚组植原体（GenBank accession：NC_ 005303）相似度分别是 1.0、0.97，归属于 16SrⅠ-B 亚组。本研究对引起四川坡柳丛枝的植原体病害进行检测，获取了该植原体的分子生物学信息，确定其分类地位和同源关系，为坡柳感染植原体的早期诊断和快速检测，以及防控措施的制定提供科学线索。

关键词：坡柳；植原体；分子鉴定；系统进化树

---

* 基金项目：国家公益性行业（农业）科研专项（201403075）；海南省热带果树生物学重点实验室开放课题（KFZX2017002）

** 第一作者：笈小龙，男，硕士研究生，研究方向：植物病毒学

*** 通信作者：余乃通；E-mail：yunaitong@ 163. com

Origin and Evolution of the Kiwi fruit Canker Pandemic*

Honour C. McCann[1 **], Li Li[2**], Liu Yifei[3], Li Dawei[2], Pan Hui[2], Zhong Caihong[2], Erik H. A. Rikkerink[4], MatthewD. Templeton[4,5], Christina Straub[1], Elena Colombi[1], Paul B. Rainey[1,6,7 ***], Huang Hongwen[2,3***]

(1. *New Zealand Institute for Advanced Study*, *Massey University*, *Auckland*, *New Zealand*; 2. *Key Laboratory of Plant Germplasm Enhancement and Specialty Agriculture*, *Wuhan Botanical Garden*, *Chinese Academy of Sciences*, *Wuhan*, *China*; 3. *Key Laboratory of Plant Resources Conservation and Sustainable Utilization*, *South China Botanical Garden*, *Chinese Academy of Sciences*, *Guangzhou*, *China*; 4. *The New Zealand Institute for Plant and Food Research Limited*, *Auckland*, *New Zealand*; 5. *School of Biological Sciences*, *University of Auckland*, *New Zealand*; 6. *Department of Microbial Population Biology*, *Max Planck Institute for Evolutionary Biology*, *Plön*, *Germany*; 7. *ÉcoleSupérieure de Physique et de ChimieIndustrielles de la Ville de Paris* (*ESPCI ParisTech*), *CNRS UMR* 8231 *PSL Research University*, *Paris*, *France*)

Abstract: Recurring epidemics of kiwifruit (*Actinidia* spp.) bleeding canker disease are caused by *Pseudomonas syringae* pv. *actinidiae* (*Psa*). In order to strengthen understanding of population structure, phylogeography, and evolutionary dynamics, we isolated *Pseudomonas* from cultivated and wild kiwifruit across six provinces in China. Based on the analysis of 80 sequenced Psa genomes, we show that China is the origin of the pandemic lineage but that strain diversity in China is confined to just a single clade. In contrast, Korea and Japan harbor strains from multiple clades. Distinct independent transmission events marked introduction of the pandemic lineage into New Zealand, Chile, Europe, Korea, and Japan. Despite high similarity within the core genome and minimal impact of within-clade recombination, we observed extensive variation even within the single clade from which the global pandemic arose.

Key words: Pathogen evolution; Genomic epidemiology; Bacterial plant pathogen; Plant-microbe interactions; Disease emergence

These authors contributed equally to this work.

Senior authors.

Corresponding authors.

---

* Funding: This work was funded by grants from the New Zealand Ministry for Business, Innovation and Employment (C11X1205), Canada Natural Sciences and Engineering Research Council (NSERC PDF), Chinese Academy of Sciences President's International Fellowship Initiative (Grant no. 2015PB063), China Scholarship Council (Grant no. 201504910013), National Natural Science Foundation of China (Grant no. 31572092), Science and Technology Service Network Initiative Foundation of The Chinese Academy of Sciences (Grant no. KFJ-EW-STS-076), Protection and Utilization of Crop Germplasm Resources Foundation of Ministry of Agriculture (Grant no. 2015NWB027)

** First author: Li Li, assistant professor; E-mail: lili@ wbgcas. cn

*** Corresponding author: Huang Hongwen; E-mail: huanghw@ scbg. ac. cn

探究柑橘黄龙病菌与植原体在长春花上的关系

陈俊甫，郑　正，邓晓玲*
（华南农业大学柑橘黄龙病研究室，广州　510642）

摘　要：本研究利用长春花作为实验材料研究柑橘黄龙病菌［*Candidatus* Liberibacter asiaticus”（CLas）］和植原体两者之间的关系，设计了 3 个实验组来进行观察。

A 组：实验组（感染 CLas 的长春花×感染植原体的长春花接穗）+对照组（感染 CLas 的长春花×健康的长春花接穗）。

B 组：实验组（感染植原体的长春花×感染 CLas 的长春花接穗）+对照组（感染植原体的长春花×健康的长春花接穗）。

C 组（对照组）：C1：健康的的长春花×感染 CLas 的长春花接穗，C2：健康的长春花×感染植原体的长春花接穗。

通过取样嫁接口以下的叶片观察和进行 PCR 检测，结果显示：A 组接穗在嫁接 5~10d 之后萎蔫枯死，且实验组材料未检测到植原体病菌的存在，感染 CLas 的长春花自嫁接后 2~3 周后枯死；B 组实验组的 CLas 接穗能存活 2~3 周，30%接穗成功感染含有植原体的长春花且在植株上的缓慢扩散至全株，3 个月左右植株枯死，70%接穗成功感染嫁接部分的枝条而后枝条枯死，植株其他部位没有检测到 CLas 病菌且一直存活，出现花变叶等症状，对照组的健康接穗存活且 3 周之后检测到植原体；C 组的 C1 组中自嫁接成功感染 CLas 病菌的长春花在 1 个月左右枯死，C2 组嫁接成功感染植原体病菌的长春花存活至今且出现花变叶等症状。结论：该实验显示两种病原菌可能没有直接的互作关系。但 B 组实验组与 C 组实验结果表明在存在植原体病菌的同时感染 CLas 病菌能延长长春花植株的存活时间，是否存在植原体病菌能增强长春花的树势还有待验证。

接下来的工作包括：优化 A 组实验的材料、条件；设计 D 组实验：用“V”形嫁接法同时嫁接分别含有 CLas 和植原体病菌的长春花接穗；进行 CLas、植原体以及两者同时存在的样品的基因组序列分析以便更深入的了解两者间的关系。

关键词：柑橘黄龙病菌；植原体；长春花

* 通信作者：邓晓玲；E-mail：xldeng@ scau. edu. cn

Distribution and Genetic Diversity of "*Candidatus* Liberibacter asiaticus" in citrus shoots in the field

Chen Yanling, Tang Rui, Zheng Zheng, Deng Xiaoling*, Xu Meirong*

(*Citrus Huanglongbing ResearchLaboratory, South China Agricultural University, Guangzhou*, 510642, *China*)

Abstract: Huanglongbing (HLB), a devastating disease affecting citrus, is associated with the bacteria *Candidatus* Liberibacter asiaticus (CLas). CLas is unevenly distributed in the citrus phloem and causes a range of symptoms on the leaves and fruit. However, no data has been shown the genetic diversity of CLas in individual trees or citrus branches. This study indicated that CLas titers in fruit and leaf samples within one branch were varied. The highest titers were measured in symptomatic fruit. CLas populations were also more diverse throughout a single tree than within samples collected from a single shoot. 4 (7.55%), 7 (13.21%), and 11 (20.75%) of the 53 groups of Chinese samples collected from different shoots of individual branches showed different polyacrylamide gel electrophoresis (PAGE) band types when amplified by three primer sets respectively. Comparatively, the Florida strains had relative simple PAGE band profile when amplified with all of the three primer sets. This study first identified the multi-infection severity of CLas in citrus shoots in the field, which will guide the sampling and survey methods.

Key words: *Candidatus* Liberibacter asiaticus; Huanglongbing; Detection; Distribution

* Corresponding author: xldeng@ scau. edu. cn; meirongxu@ scau. edu. cn

沙田柚耐黄龙病的机理研究

戴泽翰，吴丰年，郑　正，许美容，邓晓玲*
（华南农业大学柑橘黄龙病研究室，广州　510642）

摘　要：柑橘黄龙病（Citrus Huanglongbing）由候选的韧皮部杆菌 *Candidatus* Liberibacter spp. 引起，该病害给全世界柑橘生产造成严重威胁。目前尚未发现完全抗黄龙病的柑橘栽培品种，据田间调查，柚类植株感染黄龙病后表现出一定耐病性状，但目前对于柚类的耐病机理尚未有系统的研究。随着首个柚类柑橘 *Citrus grandis* 全基因组的公布，加上在目前尚未有对黄龙病的抗病基因标记的背景下，探究柚类柑橘对黄龙病的耐病机理对于挖掘耐病种质，了解黄龙病的发病机制有着重要意义。本研究分析和比较了感染黄龙病的沙田柚在组织结构和转录水平上的差异。结果表明：发病沙田柚的维管形成层具有比感病品种具有更旺盛的分化能力，能够向外分化出更多的次生韧皮部细胞，且单个光合细胞合成和容纳淀粉的能力更强。通过转录组分析发现了 1 105个差异表达基因，关于叶绿体、类囊体膜、过氧化物酶及物质转运等 13 个通路的基因被明显富集，这些基因可能在沙田柚应答黄龙病胁迫下起着关键作用。

关键词：黄龙病；沙田柚；耐病性；转录组

* 通信作者：邓晓玲；E-mail：xldeng@ scau. edu. cn

手持式近红外光谱仪判别砂糖橘黄龙病的研究

王 娟，黄家权，黄洪霞，邓晓玲*，许美容*
（华南农业大学柑橘黄龙病研究室，广州 510624）

摘 要： 柑橘黄龙病（Citrus Huanglongbing，HLB）是柑橘上的一种毁灭性病害，严重制约了广东乃至世界的柑橘产业的发展。我国和世界大多病区的黄龙病的病原为 *Candidatus* Libribacter asiaticus（CLas）。探究快速、高效的柑橘黄龙病诊断和鉴定的方法，对柑橘黄龙病的流行的预测预警及该病害的及时防控具有重要的生产意义。本研究以 500 份砂糖橘叶片（包含 206 个阳性样品及 294 个阴性样品）为材料，分别使用广州讯动网络科技有限公司的手持式近红外光谱仪收集和分析黄龙病阳性和阴性样品、不同症状类型（斑驳、花叶、黄化、各种缺素症状及无显著症状）、不同叶片部位（叶基部、中部和尾部）、不同成熟度（老叶、成熟叶和新叶）、不同叶面（正面和反面）的近红外光谱，以每个样品的 Clas 实时荧光定量 PCR 测定结果为参照，建立样本标准数据库模型。结果表明：病叶的最佳扫描的叶面为叶背面（正面和背面的阳性符合率分别为 68.9%和 80.0%），而健康叶片的最佳扫描的叶面为正面（正面和背面的阴性符合率分别为 71.8%和 58.8%）；柑橘叶片沿主脉的基部、中部、尾部 3 个部位的最佳扫描部位是基部（3 个部位的检测符合率分别为 68.2%、66.5%和 65.3%）；目前，500 个样品的黄龙病检测综合符合率为 78.6%（其中阳性符合率为 73.6%，假阳性率为 10.7%，阴性符合率为 82.5%，假阴性率为 10.0%）。进一步分析表明主要的假阳性来源于烂根造成的叶脉黄化叶片、缺锌缺镁状叶片以及幼嫩的叶片；主要的假阴性叶片来源于轻微斑驳黄化和无明显症状的叶片；健康叶片样品和感病叶片样品的主要差异体现在 1 420~1 460nm 的谱峰上。本研究成果为快速检测柑橘黄龙病提供了一定的理论依据，为黄龙病的早期诊断提供了新的途径和思路。

关键词： 柑橘黄龙病；快速检测；近红外光谱

* 通信作者：邓晓玲；E-mail：xldeng@ scau. edu. cn；
许美容；E-mail：meirongxu@ scau. edu. cn

Factors influencing the vector-pathogen interaction in Asian Citrus Psyllid (*Diaphorina citri* Kuwayama) and "*Candidatus* Liberibacter asiaticus"

Wu Fengnian, Huang Jiaquan, Xu Meirong,
Cen Yijing, Deng Xiaoling*
(*Citrus Huanglongbing ResearchLaboratory, South China Agricultural University, Guangzhou*, 510642, *China*)

Abstract: *Candidatus* Liberibacter asiaticus (CLas) is an unculturable alpha-proteobacteria associated with Huanglongbing (HLB), currently the most serious citrus disease worldwide. CLas is primarily transmitted by Asian citrus psyllid (ACP, *Diaphorina citri* Kuwayama) in a persistent-circulative manner, new knowledge of the CLas acquisition and circulative regularity by ACP is necessary forimproving strategies for HLB control. In this study, CLas proportion and concentration of infected ACPs were checked based on five factors (insect life stage, temperature, CLas concentration of hosts, sex, and host species). The results indicated nymph stage, favorable temperature, and high CLas concentration of host plants could increase the acquisition efficiency of CLas by ACP. We further found out the circulative regularity of CLas in ACP after 3 d acquisition access period (AAP, i. e. the period of latency necessary for the bacteria to enter the salivary gland) on the different life stages were significantly different. CLas would not multiply after 3 d AAP in adults however multiplies in their hemolymphafter 15 d, whilstin the nymphs multiplication in the salivary gland occurs after 12 d post infection. Alowproportion (ca. 10.00%) of the infected adults had CLas in the salivary glands after 18 d post infection. The results of this research can serve as a template to explain the mechanism of CLas acquisition by ACP.

Key words: *Candidatus* Liberibacter asiaticus; *Diaphorina citri*; CLas acquisition; Circulative regularity; Huanglongbing

---

* Corresponding author: Deng Xiaoling; E-mail: xldeng@ scau. edu. cn

烟草角斑病菌分离纯化及生防菌剂抑制作用的研究*

程　璐**，吴元华**，夏　博***
（沈阳农业大学植物保护学院，沈阳　110866）

摘　要：烟草角斑病（tobacco angular leaf spot）是严重危害烟草叶部的细菌性病害。本研究在辽宁省阜新地区采集疑似角斑病样品进行分离和鉴定，分离后的病菌在肉汁胨琼脂（NA）培养基上进行恒温培养，菌落最初半透明，渐变为灰白色，中间不透明，边缘透明，圆形，稍凸起，表面平滑，有光泽；通过光学显微镜下对菌体形态学观察表明，菌体杆状，（0.5～0.6）μm×（1.5～2.1）μm，极生鞭毛3～6根；革兰氏染色法进行染色表明，病菌显示革兰氏染色阴性。提取病菌DNA进行16S rDNA的分子生物学方法对菌株进行测定，经与NCBI的GenBank数据库对比，确认该病菌为丁香假单胞菌（*Pseudomonas syringae*）。

收集5%中生菌素粉剂、1.8%嘧肽·多抗水剂、枯草芽孢杆菌等10余种生物杀菌剂，采用含菌平板抑菌圈测定法对该菌进行抑菌效率测定。结果表明：5%中生菌素粉剂和1.8%嘧肽·多抗水剂效果对该菌抑菌效果最好，其他次之。中生菌素稀释600倍液和900倍液的抑菌率分别达95.99%和92.89%；嘧肽·多抗稀释800倍液和1 000倍液的抑菌率分别达96.98%和93.25%。试验结果为该病原菌的田间药剂防治奠定了一定基础。

关键词：烟草；角斑病菌；分离鉴定；抑菌测定

---

* 基金项目：国家烟草重大专项［中烟办（2016）259号］，合同号：110201601026（LS-06）

** 第一作者：程璐，女，在读硕士，研究方向：病害生物防治；E-mail：605349616@qq.com
吴元华，男，教授，博士生导师，研究方向：植物病毒学和生物农药；E-mail：wuyh09@vip.sina.com

*** 通信作者：夏博，男，讲师，主要从事寄生性种子植物防控及其与寄主互作机制的研究；E-mail：xiabo0522@.163.com

柑橘黄龙病菌亚洲种微滴式数字 PCR 检测体系的建立*

钟 晰**，刘雪禄，王雪峰***
（西南大学柑桔研究所/中国农业科学院柑桔研究所，国家柑桔
工程技术研究中心，重庆 400712）

摘 要：柑橘黄龙病（Citrus Huanglongbing，HLB）是严重影响柑橘生产的检疫性细菌病害，该病害传播流行快，为害区域广，目前尚无有效的治疗方法，砍除病树、防控木虱和种植无病苗是当前防控的主要措施。我国发生的黄龙病由韧皮部杆菌亚洲种（*Candidatus* Liberibacter asiaticus）引起，该病原目前尚不能培养，因此当前的检测主要依靠常规 PCR 和定量 PCR 方法。这些检测手段存在灵敏性、一致性等方面的不足，制约了其在病害早期快速诊断中的应用。微滴式数字 PCR（Droplet digital PCR，ddPCR）是近年兴起的快速、灵敏检测新技术，本研究对反应中的退火温度、引物浓度和探针浓度进行了优化，建立了柑橘黄龙病菌亚洲种的数字 PCR 检测方法，同时比较了 qPCR 和 ddPCR 方法的线性范围、灵敏度并进行相关性分析。研究结果表明 ddPCR 反应中的最佳退火温度为 54℃，引物浓度为 1 000 nmol/L，探针浓度为 500nmol/L。ddPCR 的线性范围较 qPCR 小，从 $10 \sim 10^5$ 拷贝/μL 而 qPCR 线性范围从 $10^2 \sim 10^8$ 拷贝/μL。ddPCR 的灵敏度较 qPCR 高，达到 8.8 拷贝/20μL，qPCR 和 ddPCR 检测结果呈现显著的相关性，相关系数 $r = 0.99$，$P < 0.001$。应用建立的 ddPCR 检测方法对定量 PCR 检测中 Ct 值（Cycle threshold）较高而无法判断是否为阳性的柑橘叶片 DNA 样品进行了检测，结果 ddPCR 能准确读出样品中目标 DNA 的分子数，表现良好的低丰度检测稳定性。

关键词：柑橘黄龙病；早期诊断；数字 PCR

---

* 基金项目：国家自然科学基金（31671992）

** 第一作者：钟晰，硕士研究生；E-mail：767911863@qq.com

*** 通信作者：王雪峰；E-mail：wangxuefeng@cric.cn

野油菜黄单胞菌中4-羟基苯甲酸降解途径及其在侵染过程中的功能

陈　博，何亚文*

（上海交通大学生命科学技术学院，微生物代谢国家重点实验室，上海　200240）

摘　要： 野油菜黄单胞菌（*Xanthomonas campestris* pv. *campestris*，*Xcc*）是一类专性需氧的革兰氏阴性细菌，主要侵染十字花科植物，包括卷心菜、花椰菜、甘蓝、萝卜等，引起植物黑腐病。*Xcc* 主要通过叶片排水孔或伤口侵染，到达维管束后进行繁殖和扩散，通过维管系统传播而引起宿主感染。作为一种全球性病害，*Xcc* 侵染导致农经作物严重减产，造成重大经济损失。因此，研究 *Xcc* 与寄主植物相互作用的分子机理，在此基础上研发新型绿色环保的防治措施，对黑腐病的控制以及农业、生态和经济的可持续性发展有重要意义。

4-羟基苯甲酸（简称4-HBA）是在自然界中广泛存在的酚类化合物，除少量由微生物产生并作为呼吸链辅酶Q的前体外，自然界游离的4-HBA主要由植物产生。通过生物信息学比对、基因敲除、基因回补等手段，笔者在 *Xcc* 中鉴定出一系列与4-HBA降解和转运相关的基因。研究结果显示：*Xcc* 首先通过4-HBA-3-单加氧酶（PobA）将4-HBA转变成原儿茶酸，然后经原儿茶酸开环途径将4-HBA逐步降解为琥珀酰CoA和乙酰CoA，最终进入三羧酸循环。转录水平检测及EMSA等实验表明 *pobA* 基因的表达受到4-HBA及AraC家族转录因子PobR的调控。UPLC-TOF-MS证实萝卜和卷心菜中含有4-HBA等酚类化合物，致病性检测显示4-HBA降解缺陷菌株在萝卜叶片上的致病力减弱，表明4-HBA降解对 *Xcc* 侵染具有重要意义。除此之外，笔者发现几种酚类化合物能够干扰 *Xcc* 的4-HBA降解能力，表明宿主可能有其他机制应对 *Xcc* 的侵染。因此本研究阐述了 *Xcc* 中4-HBA的降解途径及调控方式，推测这条途径在植物与 *Xcc* 相互作用中具有生物学意义，旨在为探究 *Xcc* 的致病机理和防治策略提供理论依据。

关键词： 野油菜黄单胞菌；4-HBA；降解途径

---

* 通信作者：何亚文；E-mail：yawenhe@ sjtu. edu. cn

广西部分地区植物青枯病菌演化型的鉴定*

陈媛媛**，韦云云，张耀文，黎起秦，林　纬，袁高庆***
（广西大学农学院，南宁　530005）

摘　要：由 *Ralstonia solanacearum* 引起的植物青枯病是一种难以防治的土传病害，该病在广西普遍发生且为害严重，对当地农业和林业生产造成较大经济损失。本研究根据广西各地种植作物特色，从中选择有代表性的县市，广泛采集植物青枯病标样，进行病菌分离纯化。采用演化型分类框架所设计的复合 PCR 引物对广西不同地区和寄主来源的代表性青枯菌菌株进行演化型鉴定，以明确广西不同地区和寄主来源的青枯菌菌株的演化型。2015—2016 年，从广西 13 个地级市选择了有代表性的 30 个县市，广泛采集植物青枯病典型标样，分离纯化获得 126 个青枯菌菌株，寄主范围涉及茄科、豆科、葫芦科等 8 个科 17 种植物。对其中 98 个代表菌株的演化型鉴定结果表明：97 个菌株可以同时扩增得到片段大小为 144bp 和 280bp 的 2 条特异性片段；1 个菌株（采集于广西大学农科教学实习基地的马铃薯病株）扩增出大小分别为 280bp 和 372bp 的两个片段。其中大小 280bp 片段为青枯雷尔氏菌特异性扩增条带，144bp 和 372bp 分别为演化型Ⅰ和演化型 II 的特异性扩增条带；说明广西这 13 个地级市的植物青枯病菌主要属于青枯雷尔氏菌的演化型Ⅰ（亚洲组）。新发现了南瓜嫁接毛节瓜青枯病，按照柯赫氏法则进行了致病性验证，该病菌形态、培养性状和生理生化特性与雷尔氏菌的相符；利用 16S rRNA 通用引物和青枯雷尔氏菌的特异性引物对代表菌株 Cq01 进行 PCR 扩增，扩增产物分别为 1 400bp 和 280bp，与 NCBI 中多株 *R. solanacearum* 的同源性达 99%，且该菌株可扩增得到 144bp 的演化型Ⅰ特异条带。该结果表明：引起嫁接毛节瓜青枯病的病原菌为青枯雷尔氏菌 *R. solanacearum*，且属于演化型Ⅰ。研究结果为深入研究广西植物青枯病菌种内遗传多样性打下基础。

关键词：青枯病菌；演化型；鉴定

---

* 基金项目：广西自然科学基金（2015GXNSFAA139076）

** 第一作者：陈媛媛，硕士研究生，研究方向为植物病原生物学；E-mail：724919508@ qq. com
*** 通信作者：袁高庆，副教授，研究方向为植物病原生物学；E-mail：ygqtdc@ sina. com

广西香蕉细菌性软腐病病原鉴定*

杜婵娟[1]，潘连富[1]，付 岗[1**]，叶云峰[2]，杨 迪[1]，张 晋[1]
（1. 广西农业科学院微生物研究所，南宁 530007；
2. 广西农业科学院园艺研究所，南宁 530007）

摘 要：香蕉是我国南亚热带地区的特色作物，也是广西的重要经济作物。近年来，细菌性病害在广西蕉园的发生呈上升趋势。2014 年起，陆续在玉林发现粉蕉感染软腐病。苗期发病，在球茎或球茎与假茎交界处产生褐色斑点，病斑随后向周边扩展，球茎很快腐烂发臭；假茎形成海绵状软腐，内部维管束变褐色。病株生长点坏死，生长迟缓，心叶萎缩或黄化，叶片逐渐变黄枯萎。成株期植株，极易推倒或被风吹倒。该病有时也可感染果实，导致个别果实由内而外变褐腐烂。

为了明确该病病原，从发病蕉园采集染病蕉头和蕉果，采用稀释分离法从病组织中分离获得细菌 GR-1 菌株，柯赫氏法则确定其致病性。该菌株经单菌落纯化后，进行形态学观察、生理生化测试和 16S rDNA 序列测定。结果表明：该菌在 NA 培养基上生长，菌落乳白，略带灰色。菌体杆状，周生鞭毛，革兰氏染色阴性。生理生化测试结果与菊欧文氏菌（*Erwinia chrysanthemi*）一致。采用菊欧文氏菌特异性引物扩增该菌 DNA，得到对应大小目的产物。GR-1 菌株 16S rDNA 序列（GenBank 登录号：MF041857）与菊欧文氏菌（GenBank 登录号：HM590189）的序列同源性为 100%。根据以上结果将引起广西香蕉细菌性软腐病的病原菌鉴定为菊欧文氏菌。

关键词：香蕉；细菌性软腐病；*Erwinia chrysanthemi*；鉴定

---

* 基金项目：国家自然科学基金（31560006）；广西农业科学院科技发展基金（桂农科 2016JZ15，桂农科 2017JM33，2015YT79）

** 通信作者：付岗，博士，副研究员，主要从事植物病害及其生物防治研究；E-mail：fug110@ gxaas. net

患黄龙病和溃疡病的柑橘韧皮部组织内微生物多样性分析*

贾纪春[1,2]，姜道宏[1,2]，谢甲涛[1]，程家森[1]，陈　桃[1]，付艳苹[1**]

（1. 湖北省作物病害监测和安全控制重点实验室（华中农业大学），武汉　430070；
2. 农业微生物学国家重点实验室（华中农业大学），武汉　430070）

摘　要：柑橘黄龙病是由韧皮部杆菌属类细菌（*Candidatus* Liberibacter）引起的柑橘毁灭性细菌病害。柑橘溃疡病是由薄壁菌门黄单胞杆菌属地毯草黄单胞杆菌柑橘致病变种（*Xanthomonas axonopodis* pv. *citri*）引起的柑橘重要细菌性病害。研究柑橘植株内生微生物和黄龙病菌、溃疡病菌之间的相互关系，有助于从微生态角度防控黄龙病菌和溃疡病菌。本研究利用高通量测序技术对患有黄龙病菌或溃疡病及健康的柑橘植株韧皮部内细菌和真菌多样性进行了测序分析。对 OTU 进行统计分析发现，韧皮部组织内生细菌群落结构在属水平上，患黄龙病植株与健康植株内群落存在显著差异，而患溃疡病植株与健康植株内微生物群落差异不显著，但三者都是拟杆菌属（*Bacteroides*）最为丰富。对韧皮部组织内生真菌进行丰度聚类分析，在属水平上，患黄龙病植株、溃疡病柑橘植株与健康植株特异性积累的真菌均存在显著性差异。健康植株中主要以 *Medicopsis*、*Periconia*、曲霉属（*Aspergillus*）、核盘菌属（*Sclerotinia*）、镰刀菌属（*Fusarium*）内的真菌为主，而患黄龙病植株中主要以毛双孢属（*Lasiodiplodia*）、德福霉属（*Devriesia*）和 *Sarcopodium* 内的真菌为主，患溃疡病植株中主要是以赤霉属（*Gibberella*）和 *Bacidina* 内的真菌为主。全样本菌群结构分析结果表明柑橘黄龙病菌和溃疡病菌对柑橘韧皮部内生微生物均有不同程度的影响，健康植株韧皮部内生微生物的种类和积累量均显著多于患病植株。

关键词：黄龙病；溃疡病；微生物；多样性分析

---

* 第一作者：贾纪春，男，硕士研究生，主要从事分子植物病理学相关研究；E-mail：jiajichun@ outlook. com

** 通信作者：付艳苹，教授，主要从事分子植物病理学及生物防治的相关研究；E-mail：yanpingfu@ mail. hzau. edu. cn

玉米种子中洋葱腐烂病菌的分离和鉴定

陈　青[1]，陈红运[1]，方志鹏[1]，谢毅璇[3]，廖富荣[1]，易建平[2]
（1. 厦门出入境检验检疫局，厦门　361026；2. 上海出入境检验检疫局，
上海　2001352；3. 厦门市农业局植保站，厦门　361012）

摘　要：洋葱腐烂病是葱属植物生产的毁灭性病害之一，其病原菌洋葱腐烂病菌（*Burkholderia gladioli* pv. *alliicola*）是我国禁止进境的植物检疫性有害生物，主要为害洋葱、郁金香和葱属等经济植物的鳞茎和球茎。2016年厦门检验检疫局在隔离种植台湾地区旅客携带物（玉米种子）时观察到玉米叶片出现软腐症状，从症状组织处分离到一株细菌分离物3676-1，对该分离物进行了烟草过敏性反应测试、致病性测试、Biolog测试和序列分析。试验结果表明：分离物3676-1能引起烟草过敏性坏死反应；人工接种洋葱鳞片和玉米叶片均能引起明显病变症状；Biolog测试结果为唐菖蒲伯克霍尔德菌 *B. gladioli*；16S rDNA序列与GenBank中 *B. gladioli*（CP009323）、*B. gladioli* pv. *alliicola*（GU936679）和 *B. gladioli* pv. *agaricicola*（GU936678）的序列相似性均为100%（755/755bp），与 *B. gladioli* pv. *gladioli*（GU479033）的序列相似性为99.74%（753/755bp），序列差异为2bp；16S-23S序列与GenBank中 *B. gladioli* pv. *alliicola*（EF552069，D87082，KU507040-44）的序列相似性为99.84%~100%，序列差异为0-1bp；与 *B. gladioli* pv. *agaricicola*（EF552068，KU507039）的序列相似性为96.81%~97.24%，序列差异为20bp；与 *B. gladioli* pv. *gladioli*（EF552070）的序列相似性为97.10%，序列差异为19bp；*gyrB* 基因序列与GenBank中 *B. gladioli* pv. *alliicola*（KF857488，KF857489，KF857494，KF857496）的序列相似性为99.55%~9.77%，序列差异为1~2bp；与 *B. gladioli* pv. *agaricicola*（AB220902）的序列相似性为99.77%，序列差异为2bp。根据试验结果将分离物3676-1鉴定为洋葱腐烂病菌 *Burkholderia gladioli* pv. *alliicola*。

关键词：玉米；洋葱腐烂病菌；鉴定

柑橘黄龙病菌在寄主根部的消长规律研究

李 敏，鲍敏丽，黄家权，吴丰年，邓晓玲*
（华南农业大学柑橘黄龙病研究室，广州 510642）

摘 要：柑橘黄龙病是目前柑橘生产上最严重的病害，分布广泛。该病为系统性病害，且病原菌在植株体内分布不均匀。以往的观察和研究黄龙病所取的植株材料多为地上部位，而对根部的研究较少。黄龙病菌侵染到根部后，对根系造成严重破坏，进而影响植株生长。本研究跟踪黄龙病菌在植株根部的周年动态，以研究该病害在根部的发病规律。2016 年秋季于广东省惠州市柑橘果园中筛选出 27 株感病植株（19 株砂糖橘和 8 株红心蜜柚），每月采集特定枝条的叶片及对应方位地下须根，运用实时荧光定量 PCR 测定样品中黄龙病菌的数量。结果表明不同季节条件下病原菌在两种柑橘植株的显症叶片中的数量均无显著性差异，但叶片中的病原菌浓度高于根系中的病原菌浓度；而根系中的病原菌浓度则随季节呈显著变化：秋冬季病菌含量显著高于春季；黄龙病菌浓度在不同品种的根系中也有显著差异：红心蜜柚根部菌含量高于砂糖橘。该结果将有利于明确黄龙病菌对寄主的致病机理，为黄龙病防治提供理论依据。

关键词：柑橘黄龙病菌；根系；消长规律

---

* 通信作者：邓晓玲；E-mail：xldeng@ scau. edu. cn

利用酵母双杂交系统研究柑橘溃疡菌 c-di-GMP 信号系统功能和作用机制

江美倩，陈小云，徐领会
（华南农业大学植物病理学系，广东省群体微生物创新团队，广州　510642）

摘　要： 柑橘溃疡病是由 *Xanthomonas axonopodis* pv. *citri*（*Xac*）地毯草黄单胞柑橘致病变种引起的细菌性植物病害，属于黄单胞杆菌属，是严重为害柑橘生产的主要病害之一。主要为害柑橘的枝、叶、果实，最直接的病害症状表现为出现黄色晕圈，两面隆起及火山口状的圆形病斑。果实出现病斑会严重影响柑橘的品质，病害严重时会导致落果、落叶、枝条枯萎从而导致柑橘减产。环鸟苷二磷酸（cyclic diguanylate，c-di-GMP）是细菌内广泛存在的一类全新的第二信使分子，参与调控细菌多种生物学功能并影响其致病性，如生物膜形成、运动性、胞外多糖和胞外酶的分泌，毒性因子产生以及三型分泌系统相关基因的表达等。具有鸟苷酸环化酶（diguanylate cyclase，DGC）活性的 GGDEF 结构域蛋白以及具有磷酸二酯酶（phosphodiesterases，PDE）活性的 EAL 或 HD-GYP 结构域蛋白分别负责 c-di-GMP 的合成与降解。本研究将在蛋白-蛋白互作水平，采用酵母双杂交系统，探讨 c-di-GMP 信号系统在 *Xac* 致病过程的致病机制。笔者已经构建了 *Xac* 基因组酵母双杂交系统表达文库，为筛选互作蛋白奠定了基础。笔者将含有 GGDEF 结构域的 XAC1570 构建在诱饵表达载体 PGBKT7，进行了酵母双杂交文库的筛选，在缺陷型培养基 SD-Trp-Leu-His-Ade 筛选出了一些与 XAC1570 相互作用的蛋白，其互作的真实性及可能的作用机制有待进一步深入研究。本研究为能更加精确认识 c-di-GMP 的调控机理和其下游的受体蛋白提供了新的思路。

关键词： 柑橘溃疡病；环鸟苷二磷酸；酵母双杂交

马铃薯疮痂病菌不同致病种的分子鉴别*

杨德洁**，邱雪迎，关欢欢，赵伟全***，刘大群***
（河北农业大学植物保护学院，河北省农作物病虫害生物
防治工程技术研究中心，保定 071000）

摘　要：马铃薯疮痂病（Potato common scab）是由植物病原链霉菌引起的一种重要经济性病害，不但影响马铃薯的外观和品质，而且会降低薯块的商品价值。马铃薯疮痂病的病原菌较为复杂，目前国内外已报道十多种疮痂病原链霉菌可引起马铃薯疮痂病，对这些病原菌目前还缺乏有效的分子鉴别方法。笔者在前期工作中从我国不同地区已经采集并纯化到大量不同的病原菌菌株，本研究对 4 个不同致病种的分子区别方法进行了探索。根据 NCBI 的 GenBank 数据库中已登记的疮痂病菌特有毒素合成基因簇中的基因及 16S rRNA 序列及 16S~23S ITS 区域序列设计不同致病种的特异性引物，对选择的致病种 *Streptomyces scabies*（典型菌株 CPS-1）、*Streptomyces diastatochromogeness*（典型菌株 Sd-2）、*Streptomyces turgidiscabies*（典型菌株 St-2）、*Streptomyces acidiscabies*（典型菌株 CPS-3）等 4 个典型菌株的基因组进行扩增。挑选 Tm 值相近且扩增产物长度不同，特异性和稳定性较高的的引物，经温度梯度筛选和优化反应体系后，获得了 4 对不同致病种的检测引物 D1/D2、T1/T2、A1/A2、S3/S4，扩增产物大小分别为 1 280 bp、723bp、468bp、252bp。利用 4 对引物和引物组合可以对以上 4 种致病菌进行快速分子鉴定。为验证检测方法的实用性，本研究分别对含有马铃薯疮痂病菌 *S. scabies*、*S. turgidiscabies* 及 *S. acidiscabies* 的 3 个发病地块的土壤样品进行了检测，经土壤 DNA 小量提取试剂盒提取土壤总基因组 DNA，同时用所得 4 对引物扩增检测，扩增产物的电泳结果表明：3 个土壤样品均获得了相应的特异性条带，说明建立的鉴别方法可以对疮痂病菌不同致病种的土壤样品进行有效区分。本研究建立了一种马铃薯疮痂病菌不同致病种快速鉴别的方法，该方法可对疮痂病田土壤进行快速病原定种检测，有助于提高生产中疮痂病菌种类鉴定的检测效率，也为马铃薯疮痂病的病原检测和预测预报提供了新的途径。

关键词：马铃薯疮痂病菌；致病种；分子鉴别

---

* 基金项目：河北省自然科学基金（C2014204109）；河北农业大学科研发展基金计划项目
** 第一作者：杨德洁，硕士研究生；E-mail：yangdjie@foxmail.com
*** 通信作者：赵伟全，教授，博士生导师；E-mail：zhaowquan@126.com
刘大群，教授，博士生导师；E-mail：ldq@hebau.edu.cn

柑橘溃疡病菌磷酸二酯酶及含 PilZ 结构域 c-di-GMP 受体的功能和作用机制研究

史 瑜，刘 琼，徐领会
（华南农业大学植物病理学系，广东省群体微生物创新团队，广州 510642）

摘 要：柑橘溃疡病菌（*Xanthomonas axonopodis* pv. *citri*，*Xac*）是地毯草黄单胞柑橘致病变种引起的细菌性病害，是当前全球危害柑橘生产重要病害之一。细菌胞内 c-di-GMP 信号分子是广泛存在的第二信使，调控细菌生物膜、运动性和毒性等多种生物学功能。c-di-GMP 分子是在鸟苷酸环化酶（diguanylate cyclase，DGCs）的作用下由两个 GTP 分子生成，又由磷酸二酯酶（phosphodiesterase，PDEs）分解为 pGpG，最终得到两分子的 GMP。在结构上，通常 DGCs 含有保守的 GGDEF 结构域；而 PDEs 包含 EAL 或 HD-GYP 结构域。含有 EAL 和 HD-GYP 结构域的磷酸二酯酶，能够把 c-di-GMP 信号线性化成 PGPG 最终水解成 GMP，从而终止 c-di-GMP 信号的合成。柑橘溃疡病菌编码 4 个蛋白 XAC1184、XAC2152、XAC3958 和 XAC2868 具有 EAL 结构域及编码 3 个蛋白 XAC0350、XAC2493 和 XAC1877 具有 HD-GYP 结构域。本研究主要利用基因敲除来获得一些 c-di-GMP 信号系统功能缺失的突变体，比如 ΔXAC1971、ΔXAC3402、ΔXAC1184、ΔXAC2152、ΔXAC3958、ΔXAC0350、ΔXAC2493；生物学功能试验发现 PilZ 结构域的基因参与调控运动性、毒性和 EPS 的产生。酵母双杂交系统筛选到含一系列与 PilZ 结构域蛋白 XAC3402 互作的候选蛋白，为了进一步探究 c-di-GMP 信号系统对柑橘溃疡病菌的致病调控分子机理打下基础。这些互作蛋白的功能和作用机制还有待进一步实验证明。

关键词：柑橘溃疡病菌；PilZ 结构域；运动性；毒性；蛋白-蛋白互作

纽荷尔脐橙和温州蜜柑内生细菌富集方法的比较*

吴思梦**，刘 冰***，蒋军喜，周 英，鄢明峰
（江西农业大学农学院，南昌 330045）

摘 要：柑橘溃疡病是一种重要的检疫性病害，该病对柑橘品种的选择性较强。在探索柑橘内生细菌在寄主品种抗溃疡病的作用中，用分子生物学方法研究植物内生细菌菌群结构时，植物叶绿体 DNA 通常会产生较大的干扰。为了解决这个问题，本研究以温州蜜柑和纽荷尔脐橙的叶片和果实为材料，样品组织表面消毒后经过溶液 A（离析酶 R-10+纤维素酶 R-10+0.25 mol/L蔗糖）、溶液 B（离析酶 R-10+纤维素酶 R-10+12.8%甘露醇+0.12%MES+0.36%$CaCl_2 \cdot 2H_2O$+0.011%NaH_2PO_4）和对照溶液 C（0.25mol/L 蔗糖）静置后，轻微震荡离心管后置于 37℃培养箱中培养 3h。200×g 离心 5min，重复离心 3 次，收集上清液，16500×g 离心 20min，弃上清液，用全式金 EasyPure Bacteria Genomic DNA Kit 提取宏基因组 DNA，并应用 Illumina HiSeq 平台测序，分析对比样品中的细菌菌群组成。分析结果表明：脐橙叶片经过溶液 A 处理，植物叶绿体 DNA 占内生细菌基因组 DNA 的比例更少；脐橙果实、蜜柑果实和叶片经过溶液 B 处理后，植物叶绿体 DNA 占内生细菌基因组 DNA 的比例更少；相比于 C 处理，样品经过 A 或 B 处理能增加 OTU 分类单元数量，获得内生细菌的更多种类。果实内生细菌主要分布于变形菌门，脐橙和蜜柑果实经 A、B 两种方法获得的变形菌门含量均明显高于 C；叶片内生细菌主要分布于变形菌门、厚壁菌门，脐橙叶片经三种方法获得的变形菌门和厚壁菌门总含量没有显著性差异，而蜜柑叶片经 A、B 两种方法获得的总含量均明显高于 C。由此可以得出，用溶液 A 或 B 对柑橘组织进行预处理，其内生细菌能够更好地被富集，有效减少柑橘叶绿体 DNA 的干扰。本实验结果可为柑橘及其他植物内生细菌的研究提供参考依据。

关键词：温州蜜柑；纽荷尔脐橙；内生细菌；富集

* 基金项目：国家自然科学基金项目（31460139）；江西省科技支撑计划项目（20121BBF60024）
** 第一作者：吴思梦，女，硕士研究生，主要从事植物病理学研究；E-mail：1367894992@qq.com
*** 通信作者：刘冰，副教授，博士，主要从事植物病理学研究；E-mail：lbzjm0418@126.com

基于高通量测序分析白三叶草种带细菌多样性*

况卫刚[1**]，高文娜[2]，罗来鑫[3]，李健强[3***]

（1. 江西农业大学农学院，南昌　330045；2. 北京出入境检验检疫局检验检疫技术中心，北京　100026；3. 中国农业大学植物保护学院，北京　100193）

摘　要：我国每年进口大量草坪草和牧草种子，草种携带的部分病原菌对种子发芽和种苗活力具有一定影响。种子寄藏病菌是植物病害的重要初侵染来源，也是病害远距离传播的主要途径。本研究利用 IlluminaMiSeq 测序平台，通过对 16S rDNA V3-V4 高变区进行测序，对 2015 年来源于丹麦（DB）、美国（MB）、阿根廷（TB）和新西兰（XB）的 4 个批次白三叶草（*Trifolium repens*）种子携带的细菌多样性进行了分析。结果表明：来源不同的 4 个白三叶草种子携带的细菌差异较大。通过对原始序列进行质控，共得到 484 715条高质量 Reads 进行后续分析。在 97%相似性水平，4 个样品分别获得 341（DB）、340（MB）、382（TB）、297（XB）个 OTUs，其中共有 74 个相同的 OTUs，占总 OTUs 的 9.9%，来自阿根廷的种子样品 TB 细菌丰富度最高。在门水平上，MB、TB 和 XB 携带细菌主要属于 Proteobacteria，DB 主要属于 Bacteroidetes。在属水平上，DB 携带细菌主要属于 *Streptophyta*（16.7%）和 *Prevotella*（11.9%）。MB、TB、和 XB 携带细菌主要的属为 *Sphingomonas*，所占的比例分别为 46.9%、55.08%、47.2%。本研究比较分析了不同来源国家的 4 个白三叶草种子携带的细菌多样性，为我国进境口岸草种病原细菌检验检疫提供了理论参考，为预防种传病原细菌的传播和为害提供了检测技术支撑，同时为挖掘潜在的有益内生细菌资源提供了借鉴。

关键词：白三叶草；细菌多样性；高通量测序；16S rDNA

* 基金项目：江西农业大学博士启动金

** 第一作者：况卫刚，博士，讲师，主要从事种子病理及杀菌剂药理学研究；E-mail：kwing23@126.com

*** 通信作者：李健强，博士，教授，主要从事种子病理及杀菌剂药理学研究；E-mail：lijq231@cau.edu.cn

湖北恩施马铃薯疮痂病的致病菌鉴定*

王　甄[1,2,3**]，沈艳芬[1,2,3***]，肖春芳[1,2,3]，高剑华[1,2,3]，
张远学[1,2,3]，叶兴枝[1,2,3]，李大春[1,2,3]
（1. 湖北恩施中国南方马铃薯研究中心，恩施　445000；2. 恩施土家族苗族自治州
农业科学院，恩施　445000；3. 湖北省农业科技创新中心鄂西综合试验站，恩施　445000）

摘　要：随着马铃薯种植面积的不断扩大，马铃薯病害也日渐凸显，其中马铃薯疮痂病发生呈加重趋势，该病不仅影响马铃薯种薯和商品薯的外观和销售价格，造成减产，此外，对种薯生产也构成了严重的威胁，在马铃薯生产上造成了很大的经济损失。马铃薯疮痂病是一种世界性病害，各种植区均有不同程度的发生。由于种薯调运等原因，该病在西南马铃薯主产区发生也日趋严重。通过调查发现，部分疮痂病害发生严重的田块发病率高达 100%，使块茎表面“体无完肤”，失去了商品价值，对农户造成了巨大经济损失。当种薯带病或土壤带菌时，会导致整株的马铃薯新薯块发病，加上连作，病害逐年加重。

马铃薯疮痂病是由病原链霉菌（*Streptomyces* sp.）引起的一种土传病害，目前已经有十几种链霉菌被发现可引起病害的发生。通过 16S rDNA 序列分析，聚类结果发现中国马铃薯疮痂病菌存在遗传多样性，分布较复杂，不同地区的病原菌菌株间差异较大，中国马铃薯疮痂病菌具有明显的遗传多样性。

本实验室通过调查发现，湖北恩施地区马铃薯疮痂病也有发生，且发病有加重的趋势。通过在恩施马铃薯产地采集病薯，进行病原菌的纯化分离和分子鉴定发现，恩施地区马铃薯疮痂病菌的致病种类主要是 *S. scabies*、*S. acidiscabies*、*S. turgidiscabies* 三种，其中危害面积比较广的是 *S. scabies*，*S. acidiscabies* 目前仅发现一例。马铃薯作为恩施地区的主要粮食作物之一，是老白姓餐桌上备受喜爱的食物，随着马铃薯主粮化的推进，马铃薯种植产业会更加稳定发展，因此马铃薯病害越来越受到重视。研究马铃薯疮痂病致病菌的多样性，为深入分析疮痂病菌的致病机制奠定了基础。

关键词：恩施；马铃薯疮痂病；致病菌

---

* 基金项目：湖北省技术创新专项（鄂西民族专项 2016AKB052）；2016 中央引导地方科技资金；现代农业产业技术体系专项资金资助（CARS-10）

** 第一作者：王甄，女，硕士，农艺师，主要从事水稻病毒和马铃薯病害防治与遗传育种研究

*** 通信作者：沈艳芬，研究员，从事马铃薯遗传育种及病害防治研究；E-mail：13872728746@163.com

核桃细菌性疫病菌多糖、色素及其竞争能力研究*

邹路路[1**]，李美凤[2]，魏 蜜[1]，毛雅慧[1]，傅本重[1***]
（1. 湖北工程学院生命科学技术学院，特色果蔬质量安全控制湖北省重点实验室，孝感 432000；2. 四川大学轻纺与食品学院，成都 610064）

摘 要： 树生黄单胞菌核桃致病变种（*Xanthomonas arboricola* pv. *juglandis*，*Xaj*）是引起核桃细菌性疫病（walnut bacterial blight）的致病菌，它引起全世界范围内核桃生产上最严重的病害。病原菌 *Xaj* 除了引起核桃细菌性黑斑病之外，还可以引起核桃幼果褐色顶端坏死（brown apical necrosis，BAN）及树干纵渗溃疡（vertical oozing canker，VOC）。本研究比较了来源于北京地区核桃病原菌 *Xaj*1（中国农业大学周涛提供）在 12 种不同种培养基上的生长状况，确定最适生长培养基。通过苯酚硫酸法测定了 *Xaj*1 菌株在 CKTM、King'B、LB 和 NA 等 10 种培养基条件下的多糖（Exopolysaccharides）产生量，并用紫外分光光度计法测定病菌在 NAG、PDA 和 YPGA 等 11 种培养基条件下的色素产生量。用平板菌落计数法，测定了病原菌与 3 株核桃内生菌 Y7、J13 和 G5 的竞争能力。实验结果表明：病原菌 *Xaj*1 在这 12 种培养基上都能生长，在 NYG、King'B 等培养基上生长较好，而在 mTMB 和 BS 培养基上生长状况较差。在 LB、PD 和 YPGA 等培养基中产生多糖较多，OD_{490}大于 2.0，在 NA 上产生多糖最少，$OD_{490}=0.148$。色素的产生量以 LB 培养基上最高，YPGA 上最少。细菌竞争性试验结果表明：混合培养后，病原菌 *Xaj*1 比单独培养的菌落数要多，在两者都增加的情况下，病原菌增加的量更大，因此，核桃病原菌 *Xaj*1 在培养基上比其内生菌的生长竞争能力更强。此外，对病原菌的致病范围进行了离体接种实验。研究结果为进一步理解病原菌的致病性，以及病原菌与寄主和内生菌的互作奠定了一定的基础。

关键词： 核桃；细菌性疫病菌；多糖；色素；竞争性

---

* 基金项目：湖北省教育厅重点项目（D20162701）

** 第一作者：邹路路，在读硕士研究生，植物病理学；E-mail：911162909@ qq. com

*** 通信作者：傅本重，副教授，研究方向：植物病原细菌；E-mail：benzhongf@ yahoo. com

核桃细菌疫病菌趋化、生物膜及基因组分析*

朱洁倩[1**]，陈倩倩[2]，陈　铭[1]，杨瑞青[1]，毛雅慧[1]，魏　蜜[1]，傅本重[1***]

（1. 湖北工程学院生命科学技术学院，特色果蔬质量安全控制湖北省重点实验室，孝感　432000；2. 华南农业大学园艺学院，广州　510642）

摘　要：本研究测试了核桃细菌性疫病菌 *Xaj*1（*Xanthomonas arboricola* pv. *juglandis*，*Xaj*）对 4 种趋化物 NaCl、L-谷氨酸、HCl 和脲素的趋化作用，用改进的毛细管法和改进的滴落法进行实验。结果表明：脲素对核桃内生菌 Y6 的趋化性反应有抑制作用，NaCl、L-谷氨酸及 HCl 对核桃病原菌 XCC1 的趋化性反应有促进作用。

通过结晶紫染色法测定了 25 株核桃内生菌和病原菌 Xaj1（中国农业大学周涛提供）体外形成生物膜的能力。结果发现 26 菌株都不同程度的形成了生物膜，但形成能力各有差异。其中根部内生菌生物膜形成能力都很强，而菌株 G1 最强，茎部内生菌中菌株 J3 生物膜形成能力最强，叶部内生菌中菌株 Y5 生物膜形成能力最强，*Xaj*1 的生物膜形成能力较强。

作者对来源于湖北省丹江口市核桃幼果褐色顶端坏死的病原进行分离，经过 16S rDNA、黄单胞菌属和树生黄单胞菌核桃致病变种（*Xanthomonas arboricola* pv. *juglandis*，*Xaj*）特异引物 Xan-F/R 和 Xaj1F/R 进行 PCR 检测，以及致病性接种试验，得到一株强致病性的病原菌 Xaj-DW3F3。对 DW3F3 进行全基因组测序和序列分析，共得到 6 853 478 条 reads 和 2 056 043 400bp。基因组大小为 5. 14Mbp，G+C 含量为 65. 36%。编码蛋白基因 4 381个，其中最小基因长度为 114bp，最大基因长度为 13 317bp，平均基因长度为 988. 75bp。总的编码基因碱基 4 331 709bp，编码率为 84. 11%。此外，编码 rRNA 和 rRNA 基因分别为 53 个和 7 个。对全基因组进行了 KEGG 注释，共有 1 277个基因共注释上 205 个 Pathway；对预测基因进行 COG 功能分类预测，共有 4 065个 Gene 被注释上 20 种 COG 分类。对我国核桃重大病原菌的全基因组测序和分析的完成，为更好的理解和研究核桃与病原菌 Xaj 的相互作用奠定了基础。

关键词：核桃；疫病菌；*Xanthomonas arboricola* pv. *juglandis*，*Xaj*；趋化性；生物膜；基因组

---

* 基金项目：湖北省教育厅重点项目（D20162701）

** 第一作者：朱洁倩，硕士研究生，植物病理学；E-mail：zhu_ jieqian@ sina. com

*** 通信作者：傅本重，副教授，研究方向：植物病原细菌；E-mail：benzhongf@ yahoo. com

河南濮阳丝棉木丛枝病原菌的检测与鉴定*

王圣洁[1**]，张文鑫[1**]，李　永[1]，林彩丽[1]，汪来发[1]，朴春根[1]，谢守江[2]，田国忠[1]
（1. 中国林业科学研究院森林生态环境与保护研究所国家林业局森林保护学重点实验室，北京　100091；2. 河南省濮阳市林科院，濮阳　457100）

摘　要：丝棉木（*Euonymus maackii* Rupr），属卫矛科植物，由于其适应能力强，观赏价值高，已逐渐成为园林绿化中落叶小乔木的主流树种。从河南濮阳采集到表现丛枝症状的感病丝棉木植株，主要表型症状为：叶片变小、叶部黄化卷缩、节间缩短、枝条顶部丛生严重。本研究旨在确定丝棉木丛枝的病原。以丝棉木丛枝叶片总 DNA 为模板，利用植原体通用引物对 P1/P1 和 FDtufF1/FDtufR1 进行 16S rDNA 和 *tuf* 基因的扩增和测序，获得了大小为 1.6kb 的 16S rDNA 片段和 932bp *tuf* 基因片段，用所测 16S rDNA 基因序列和 *tuf* 基因序列与其他已知种类植原体序列进行系统进化分析。丝棉木丛枝植原体的这两个基因序列都分别和枣疯病的两个序列在一枝系上，同属植原体 16S rV-B 组。再将获得的 16S rDNA 基因序列输入到 *i*PhyClassifier 中进行虚拟 RFLP 分析，结果显示丝棉木丛枝样品与枣疯样品具有完全相同的酶切图谱。以本实验室建立的 16S rVb-LAMP 诊断方法，同样获得了阳性的结果。本文首次对丝棉木丛枝病害进行了分子诊断与鉴定，确立了其植原体的分类地位。本文在增加植原体侵染寄主种类、丰富植原体多样性的同时，也为丝棉木丛枝植原体病害防治提供了第一手资料，对丝棉木丛枝植原体的检测、预防、治疗都有积极作用，也为丝棉木栽培管理和病害防治提出了更高的要求。在植原体检测方面，应用了新的检测手段 LAMP，为植原体检测增加了可靠性。关于丝棉木丛枝植原体和枣疯病植原体的关系，笔者注意到了丝棉木和枣树栽培管理上有一定关联，但这种关联是否有构成植原体传播的因果关系尚待验证，后续研究将从植原体跨物种传播、植原体分类、致病分子机理、遗传本质等方面进行更深入的研究。

关键词：丝棉木丛枝；植原体；16S rDNA 序列；*tuf* 基因序列；系统进化树；RFLP 分析

---

* 基金项目：国家自然科学基金项目（31370644）

** 第一作者：王圣洁，男，博士研究生，主要方向：分子植物病理；E-mail：wsjguoyang@126.com
张文鑫，男，博士研究生，主要方向：分子植物病理；E-mail：61960467@qq.com

青枯劳尔氏菌 GMI1000 胞外多糖合成缺陷突变体筛选及生物学特征研究

神方芳
（华南农业大学，广州 510642）

摘　要：青枯劳尔氏菌（*Ralstonia solanacearum*），又称青枯菌，起初命名为青枯假单胞杆菌（*Pseudomonas solanacearum*），革兰氏阴性菌，是一种土传植物细菌性青枯病的病原菌。青枯菌主要为害烟草、番茄、茄子以及辣椒等植物，发病植株出现萎蔫状，取出植株可见其地下组织软化，呈褐黄色，并散发臭味；断其茎部，可以挤出乳白色液体菌脓。青枯菌环境适应能力强，寄主范围广，严重影响作物产量，对农业生产为害巨大。

本实验利用青枯劳尔氏菌 GMI1000 构建 Tn5 插入突变体库，在 50 000多个可隆中筛选得到 56 个胞外多糖明显减少的突变体，利用 Hitail PCR 的方法获得 Tn5 转座子插入的位点基因，通过纤维素酶活测定，生物膜检测、运动性检测和接种实验分析突变体的表型与致病性，为进一步快速鉴定青枯菌致病力强弱的方法以及研究青枯菌胞外多糖的致病性和调控机理奠定了基础。

关键词：青枯菌；Tn5 突变体库；胞外多糖；致病性

马铃薯黑胫病菌致病相关蛋白 PGL 的预测与功能验证*

胡连霞**，杨志辉，赵冬梅，张 岱，潘 阳，朱杰华***
（河北农业大学植物保护学院，保定 071001）

摘 要：马铃薯黑胫病由 *Pectobacterium atrosepticum* 引起，是我国各马铃薯产区普遍发生和为害严重的细菌病害。当前对病菌致病基因功能的研究是植物病理学研究热点和焦点，而马铃薯黑胫病菌 JG10-08 菌株的全基因测定为该菌致病相关基因功能的研究奠定了基础。本研究利用生物信息学预测了果胶酸裂解酶 endo-PGLI 和 exo-PGLA 的结构与功能，并通过分子生物技术对其致病功能进行了验证。该研究的主要结果如下。

（1）预测了果胶酸裂解酶 endo-PGLI 和 exo-PGLA 的结构与功能：在分析蛋白序列及基本性质的基础上，利用 ANTHEPROT5.0 软件分析 endo-PGLI 和 exo-PGLA 蛋白二级结构，均主要由 α-螺旋和 β-折叠组成。通过 SWISS-MODEL 蛋白同源模建数据库构建蛋白三级结构，结果显示 endo-PGLI 蛋白的三级结构为马鞍状，而 exo-PGLA 蛋白的三级结构为扁筒状。利用 NCBI 中的 CDD 数据库比对分析保守结构域和功能位点，确定了 endo-PGLI 氨基酸序列 13-568 位为周质的果胶酸裂解酶家族 Pectate_ lyase_ 2 保守结构域，exo-PGLA 氨基酸序列 478～646 位为果胶酸裂解酶家族 β-折叠保守结构域。因此，推测 endo-PGLI 和 exo-PGLA 均具有降解植物细胞壁果胶成分的致病功能。

（2）通过基因敲除与回复验证了果胶酸裂解酶 endo-pglI 和 exo-pglA 的致病功能：利用同源重组双交换原理，敲除了马铃薯黑胫病菌 *P. atrosepticum* 野生型菌株 JG10-08 中目的基因 endo-pglI 和 exo-pglA，获得了两株突变株 ΔJG10-08$_{endo\text{-}pglI}$ 和 ΔJG10-08$_{exo\text{-}pgl\,A}$。将目的基因 endo-pglI 和 exo-pglA 回复到突变株中，得到回复株 $\Delta\Delta$JG10-08$_{endo\text{-}pglI}$ 和 $\Delta\Delta$JG10-08$_{exo\text{-}pglA}$。用注射法分别将野生型菌株 JG10-08 和突变株 ΔJG10-08$_{endo\text{-}pglI}$ 和 ΔJG10-08$_{exo\text{-}pglA}$ 接种到马铃薯植株茎基部，7 d 后突变株 ΔJG10-08$_{endo\text{-}pglI}$ 和 ΔJG10-08$_{exo\text{-}pglA}$ 接种处周围有轻微的发病症状，其病情指数分别为 19.26 和 20.74；而接种野生型菌株 JG10-08 的植株茎基部发病严重，接种处出现腐烂症状，其病情指数高达 50.37。利用浸泡法分别将野生型菌株 JG10-08 和突变株 ΔJG10-08$_{endo\text{-}pglI}$ 和 ΔJG10-08$_{exo\text{-}pglA}$ 接种马铃薯块茎，7d 后，突变株 ΔJG10-08$_{endo\text{-}pglI}$ 和 ΔJG10-08$_{exo\text{-}pglA}$ 接种的块茎表面出现了轻微的褐色斑点，病情指数分别为 16.30 和 18.52。野生型菌株 JG10-08 接种的马铃薯块茎已经开始腐烂，病情指数为 47.41。回复株和野生型菌株相比，对马铃薯植株和块茎的致病性，无显著性差异。这表明 endo-pglI 和 exo-pglA 两个基因在黑胫病菌致病性中发挥着重要作用。

关键词：果胶裂解酶；致病基因；功能预测；功能验证

* 基金项目：现代农业产业技术体系建设专项资金资助（CARS-10-P12）

** 第一作者：胡连霞，博士生，主要从事马铃薯黑胫病研究

*** 通信作者：朱杰华，教授，主要从事马铃薯病害和分子植物病理学研究；E-mail：zhujiehua356@ 163. com

野生酸枣内生细菌分离鉴定及抑菌活性初探*

刘慧芹[1**]，刘　东[2]，王远宏[1]，韩巨才[2]，刘慧平[2***]
（1. 天津农学院园艺园林学院，天津　300384；2. 山西农业大学农学院，太谷　030801）

摘　要：为了广泛地开发植物内生菌的资源，本文从野生酸枣中进行生防菌的分离和筛选。通过组织分离法，从野生酸枣的根、茎、叶和果实中分离得到 76 株内生细菌，通过平板对峙法，以 8 种病原菌为指示菌，测定了所得菌株的活性。利用生物学和分子学方法对菌株 SZG-23 进行了鉴定。通过对所有菌株进行初筛和复筛，最终得到 8 株颉颃作用较强的菌株，其中菌株SZG-23 对梨黑斑病菌和黄瓜枯萎病菌的抑制作用最强，达到 90.3%和 72.4%。另外，SZG-23 菌株发酵液对其他几种植物病原菌生长均有一定抑制作用，其中对梨黑斑病菌、番茄灰霉病菌作用超过 90%。其突变菌株能在酸枣幼苗中定殖和移动。通过形态观察、生理生化特征测定及 16S rDNA 同源性分析鉴定为解淀粉芽孢杆菌（*Bacillus amyloliquefaciens*），16S rDNA 序列在 GenBank 中注册号为 KF48366。

关键词：野生酸枣；内生细菌；筛选；鉴定；抑菌活性

* 基金项目：山西省财政支持农业科技成果转化资金（SCZZNCGZH201304，SCZZNCGZH201403）；天津市蔬菜现代农业产业技术体系（ITTVRS2017011）

** 第一作者：刘慧芹，博士，副教授，主要从事植物病原细菌致病性研究；E-mail：wjxlhq@ 126. com
*** 通信作者：刘慧平，博士，教授，博士生导师，主要从事植物病害生物防治；E-mail：sxndlhp@ 163. com

钙离子介导的丁香假单胞菌细菌素作用机理研究[*]

李俊州[**]，周丽颖，范　军[***]
（中国农业大学植物病理系，北京　100193）

摘　要： 丁香假单胞菌番茄致病变种（*Pseudomonas syringae* pv. *tomato*）DC3000 菌株（DC3000）作为模式病原菌与拟南芥的互作体系已被广泛用于研究细菌与植物互作的机理。通过转座子突变和高通量细菌生长测定体系笔者筛选得到一株对植物提取物敏感并且丧失致病力的 DC3000 突变体菌株。进一步分析发现突变体中因转座子插入导致一个隐含的 *saxE* 基因发生转录缺陷，造成细菌在植物提取液中生长受抑制且丧失在寄主植物中的定殖能力。围绕这一发现，综合运用遗传学，植物化学以及分子生物学等多种手段，笔者首先发现钙离子是植物和提取物中抑制 saxE 缺陷型菌株（Δ*saxE*）生长的物质基础，同时发现细菌侵染过程中其所处环境中游离态钙离子的动态变化与病程相关。其次，通过筛选回复突变体鉴定得到 DC3000 中细菌素编码基因是钙离子抑制 Δ*saxE* 生长的必需基因。序列相似性分析表明该细菌素类似于大肠杆菌中的细菌素 M（colicin M），进一步研究发现存在钙离子条件下，该细菌素可能依赖 sec 系统被分泌至周质，这一特征在 colicin M 及 colicin M like 细菌素中首次被发现，为这类细菌素的释放机制提供了思路。本研究揭示了钙离子参与的一种新的植物抑菌机制，同时也为未来农作物抗病性改良提供参考。

关键词： 丁香假单胞菌；细菌素；大肠菌素 M；钙离子

---

* 基金项目：国家自然科学基金（No. 31272006）

** 第一作者：李俊州，男，博士研究生，研究方向为植物病理学
*** 通信作者：范军，男，教授，研究方向为植物病理学；E-mail：jfan@ cau. edu. cn

Molecular Variability and Phylogenetic Relationship Among *Acidovorax avenae* Strains Causing Red Stripe of Sugarcane in China

Li Xiaoyan[1*], Sun Huidong[1*], Wang Jinda[1], Fu Huaying[1], Huang Meiting[1], Zhang Qingqi[2], Gao Sanji[1**]

(1. *National Engineering Research Center for Sugarcane*, *Fujian Agriculture and Forestry University*, *Fuzhou*, *Fujian* 350002, *China*; 2. *College of Crop Science*, *Fujian Agriculture and Forestry University*, *Fuzhou*, *Fujian* 350002, *China*)

Abstract: Sugarcane red stripe is a bacterial disease that has worldwide prevalence and is caused by *Acidovorax avenae* subsp. *avenae* (*Aaa*). In this study, we collected 108 sugarcane leaf samples from nine different sugarcane planting regions in China between 2013 and 2016. The causal *A. avenae* bacteria were successfully isolated in the red stripe leaf samples. PCR assays were carried out on DNA isolated from all collected samples using novel primers targeted to the *A. avenae* 16S-23S rDNA spacer regions. Positiveresults were obtained for 96.4% (81/84) of symptomatic (red stripe) and 83.3% (20/24) of asymptomatic leaves. All the amplified fragments were cloned and sequenced, which revealed that the Chinese *A. avenae* strains hada significant degree of length variation (436~454 base pairs, bp) and sequence identity (89.2% ~ 100%). The 142 *A. avenae* sequences originated from this study were clustered into six phylogenetic groups (H, HE, E1-1, E1-2, E2-1, E2-2) and had sequence identity that ranged between 89.2% and 99.5%. *In silico* and polyacrylamide gel electrophoresis analysis of PCR products digested with *Hind* III and *Eco* RI revealed that these sequences showed various restriction fragment length polymorphism (RFLP) profiles that were closely associated according to phylogenetic grouping. Furthermore, phylogeny analysis based on these Chinese *A. avenae* sequences, 22 corresponding sequences from two *Acidovorax* species and two other *A. avenae* subspecies, revealed that all the tested strains in the *A. avenae* species fell into clade I, including the six distinct Chinese *A. avenae* groups mentioned above, *A. avenae* subsp. *cattleyae* (*Aaca*), and *A. avenae* subsp. *citrulli* (*Aac*). *A. konjaci* and *A. facilis* species appeared in clades II and III, respectively. This is the first report of PCR-based identification and molecular characterization of *A. avenae* infecting sugarcane plants in China.

Key Words: *Saccharum* spp.; Red stripe disease; *Acidovorax avenae*; PCR detection; Genetic variation

* First authors: These authors contributed equally to this work

** Corresponding author: Gao Sanji; E-mail: gaosanji@yahoo.com

Screening of Specific Primers for Detection of *Acidovorax citrulli* from Cucurbits Seed by Bio-PCR*

Kan Yumin, Li Yuwen, Jiang Na, Li Jianqiang, Luo Laixin**

(*Department of Plant Pathology*, *China Agricultural University/Beijing Key Laboratory of Seed Disease Testing and Control* (*BKL-SDTC*), *Beijing* 100193, *China*)

Abstract: Bacterial fruit blotch (BFB), caused by *Acidovorax citrulli*, is one of the most destructive seed-borne diseases of watermelon as well as other cucurbit crops. This quarantine seed-borne disease has been reported in most watermelon and melon producing regions worldwide. Since the infected seeds are the most important primary inoculums source of BFB, seed health testing (SHT) and then eliminating the contaminated seeds is one of the most effective methods to control the disease. Bio-PCR is widely used in SHT including the detection of *A. citrulli*. As *A. citrulli* have abundant genetic diversity and many other closely related species, specific and sensitive primers is very important in the seed health testing. To screen the best primers for detection of *A. citrulli* by Bio-PCR, in this study, 33 bacterial strains including 17 *Acidovorax citrulli*, 10 strains of other species in *Acidovorax* and 6 more strains of other bacterial genera were chose to value the reported 7 pairs of primers reported at present. The results indicated thatthe primers pairSEQID4^{m}/SEQID5 had the best specificity that could distinguish all the tested strains of *A. citrulli* from other species and genera. Then the sensitivity of this primer was tested with Bio-PCR by the use of ASCM and EBBA, two semi-selective media for *A. citrulli*. Itindicated that the detection limit of pure bacterial culture of *A. citrulli* was 10^2CFU/mL. While the detection limit in the watermelon seeds by Bio-PCR was 0. 01CFU/g of *A. citrulli*in 96 g seeds.

Key words: *Acidorvorax citrulli*; Bio-PCR; Primer; Specificity; Sensitivity

---

* Funding: This research was supported by the Special Fund for Agro-scientific Research in the Public Interest (No. 201003066)

** Corresponding author: Luo Laixin; E-mail: luolaixin@ cau. edu. cn

低温寡营养条件对番茄溃疡病菌存活状态的影响

谈 青*，韩思宁，白凯红，陈 星，罗来鑫，李健强，蒋 娜**
（中国农业大学植物病理学系/种子病害检验与防控北京市重点实验室；
农业部作物有害生物监测与绿色防控重点实验室，北京 100193）

摘 要：番茄溃疡病是一种系统性侵染的病害，在番茄生产中造成了严重的经济损失。笔者实验室前期研究表明番茄溃疡病菌 *Clavibacter michiganensis* subsp. *michiganensis*（简写为“Cmm”）在寡营养状态下经铜离子或温和酸性条件诱导可进入有活力但不可培养（Viable but Non-Culturable，简称 VBNC）状态。VBNC 状态被普遍认为是不产芽孢细菌的一种抗逆机制。为探明 Cmm 在田间是否能以 VBNC 状态越冬，本研究采用 4℃寡营养条件对 Cmm 进行诱导，利用实验室前期建立 PMA-qPCR 结合平板培养计数方法，分别检测 Cmm 的活性及可培养性。将 10^7 CFU/mL 的 Cmm 菌悬液分别置于 4℃或 28℃条件下培养，在不同时间取样检测。在检测的 149d 内，4℃、28℃下总细菌数均维持在 10^7 cell/mL 至 10^8 cell/mL 之间；有活性菌体数呈下降趋势，4℃和 28℃处理均比初始浓度下降约 1.5～2.0 个数量级，约为 10^5～10^6 cell/mL；至第 149d，28℃下 Cmm 可培养菌体数下降至 1 CFU/mL 以下，4℃下可培养细胞数为 10^2 CFU/mL。由有活性细菌数减去可培养细菌数即得到 VBNC 状态菌体数量。结果显示，在第 149d，与最初细菌添加量相比，4℃、28℃下分别有 72.69%和 78.76%的 Cmm 进入 VBNC 状态。研究结果表明在此长期诱导条件下，低温 4℃能诱导 Cmm 进入 VBNC 状态，但与 28℃条件相比，4℃下保持细菌可培养状态的时间更长。据此推测 Cmm 在田间极有可能以 VBNC 状态越冬，成为下一生长季可能初侵染来源。后期将对田间土壤或病残体样品中是否存在 VBNC 状态的 Cmm 进行检测。

关键词：番茄溃疡病菌；VBNC 状态；低温；PMA-qPCR

---

* 第一作者：谈青，女，硕士研究生，主要从事种传细菌病害研究；E-mail：15321543395@163.com
** 通信作者：蒋娜，实验师，主要从事种子病理学研究；E-mail：jn2009@139.com

细菌性条斑病菌 non-TAL 效应蛋白 Xop101 抑制水稻免疫的功能研究*

吴文章**，王燕平**，刘丽娟**，崔福浩**，李 倩，
汪澂扬，周 爽，王善之，孙文献***
（中国农业大学植物保护学院，农业部作物有害生物
监测与绿色防控重点实验室，北京 100193）

摘 要：细菌性条斑病菌（*Xanthomonas oryzae* pv. *oryzicola*, *Xoc*）引起的细菌性条斑病是我国南方稻区的主要病害之一。水稻条斑病菌分泌到植物细胞内的转录激活子样（transcription activator-like, TAL）和非转录激活子样（non-TAL）效应蛋白对病原菌的致病性至关重要，但目前对单个 non-TAL 效应蛋白发挥毒性功能的分子机制了解十分匮乏。

前期研究发现，功能未知的条斑病菌 non-TAL 效应蛋白 Xop101 在水稻中的异源表达能够抑制水稻的基础免疫以及对细条病菌的抗性。酵母双杂交、免疫共沉淀、双分子荧光互补等实验结果表明，Xop101 能与水稻中的 E3 泛素连接酶 OsET1 互作，并且 Xop101 与 OsET1 E3 连接酶酶活关键突变体的互作增强。OsET1 作为 E3 泛素连接酶能够发生自身泛素化，并且在体外 OsET1 能够将 Xop101 泛素化。进一步研究表明，Xop101 能够在烟草叶片中强烈促进 OsET1 的降解，而 MG132 能显著抑制 Xop101 介导的 OsET1 降解，说明 OsET1 的降解依赖于 26S 蛋白酶体途径。在以上结果的基础上，笔者将解析 Xop101 与 OsET1 互作的分子基础，阐明 Xop101 可能的生化功能及其对 OsET1 功能的影响，最终明确 OsET1 调控水稻免疫的分子机制。研究结果将为深入理解水稻条斑病菌的致病机理以及水稻抗病机理提供新的思路。

关键词：细菌性条斑病；水稻；Xop101；先天免疫

---

* 基金项目：国家自然科学基金（31671991）；农业部公益性行业专项（201303015）；“高等学校学科创新引智计划”（B13006）

** 第一作者：吴文章，硕士研究生，研究方向：水稻与病原细菌的分子互作；E-mai：524792791@ qq. com
王燕平，主要从事植物病理学研究；E-mail：wang_ yanping123@ 163. com
刘丽娟，主要从事植物病理学研究；E-mail：melody_ alpha@ 163. com
崔福浩，副教授，主要从事水稻与病原细菌、真菌互作；E-mail：cuifuhao@ 163. com
*** 通信作者：孙文献，教授，主要从事水稻与病原细菌、真菌的互作；E-mail：wxs@ cau. edu. cn

第五部分　线　虫

不同施肥条件下番茄根际土壤线虫种类鉴定及各营养类群丰度比较*

丁晓帆**，杞世发，毛　颖，龙　慧
（海南大学热带农林学院，海口　570228）

摘　要：为了研究不同施肥条件对海南儋州番茄根际土壤线虫群落组成的影响，试验设不施底肥（CK）、有机底肥（羊粪商品有机肥，OF）、无机底肥（康普无机化肥 NPK12－12－17，NPK）3 个处理，分别在整地施用底肥前及番茄不同生长期间分 6 次采集土壤样本，用浅盘分离法分离线虫，完成线虫系统形态学观测及数量统计。结果表明：共鉴定出 2 纲 8 目 22 科 41 属，其中食细菌线虫（Bacterivores）8 科 13 属，食真菌线虫（Fungivores）3 科 4 属，植食性线虫（Plant-parasites）6 科 9 属，杂食/捕食性线虫（Omnivores-predators）7 科 15 属，并且肾属 *Rotylenchulus* Linford & Olivera，1940 为该番茄田的绝对优势属（丰度 88%以上）。各线虫属及营养类群线虫数量统计分析表明：不同施肥条件改变番茄根际土壤线虫的群落组成，施用有机、无机底肥均明显增加线虫属总数及食细菌线虫的丰度，但均明显降低杂食/捕食线虫的丰度；施用有机底肥使食真菌线虫丰度上升，但无机底肥使食真菌线虫丰度明显下降；施用有机、无机底肥均不能明显改变植食性线虫在番茄根际土壤线虫中的占比，这是否与肾属在该番茄田中占绝对优势有关，仍需进一步研究。

关键词：土壤线虫；种类鉴定；丰度；施肥条件；番茄

* 基金项目：2016 年度海南省自然科学基金项目（20163040）；2016 年度海南省高等学校教育教学改革研究项目（Hnjg2016ZD-2）；海南大学教育教学研究项目（hdjy1605）

** 通信作者：丁晓帆，女，副教授，硕士；从事植物线虫学研究；E-mail：dingxiaofan526@163.com

Ultrastructure, Morphological and Molecular Characterization of New and New Recoreded Species of *Hemicriconemoides* from China*

Maria Munawar**, Tian Zhongling, Eda Marie Barsalote,
Cai Ruihang, Li Xiaolin, Qu Nan, Zheng Jingwu***
(*Laboratory of Plant Nematology, Institute of Biotechnology, College of Agriculture & Biotechnology, Zhejiang University, Hangzhou* 310058, *China*)

Abstract: Sheathoid nematodes of the genus *Hemicriconemoides* are migratory root-ectoparasites of many plants including various agricultural crops and fruit trees. They are generally found inhabiting tropical and subtropical areas of the world and have been reported from Africa, America, Asia, Australia and southern Europe, presently genus consists of 52 valid species. In this study we provide morphological and molecular characterization of one new and one new recorded species *viz. Hemicriconemoides* n. sp. and *H. kanayanesis* isolated from *Metasequoia glyptostroboides* and *Camelliae uruku* in Hangzhou, Zhejiang, respectively.

The new species is most similar to *H. californianus* and considered as a cryptic species. Morphologically the new species can be characterized by cuticular sheath well separated at anterior region, cuticle smooth with retrorse annuli, lip region slightly set off and roughly square in outline. Stylet 73~83μm long, excretory pore 4~5 annuli posterior to pharyngeal bulb, vulva posteriorly located 88%~91% of the total body, tail conical with bluntly pointed terminus. *H. kanayaensis* can be characterized by its exclusive lip morphology (set off, marked with constriction), cuticular sheath attached at anterior and vulval region, stylet 66~78μm long, excretory pore located 4~5 annuli posterior to pharyngeal bulb, vulva posteriorly located 88%~90% of the body, tail elongated widely conoid, ended up in a rounded terminus. Scanning electron microscopic observations and molecular characterization based on COI of mtDNA, 18S, D2-D3 28S and ITS of rDNA sequences for both species also provided. To date sixteen *Hemicriconemoides* species have been reported from China, where *H. kanayaensis* is the 17th recorded species.

Our results suggested that the concomitant use of morphological and molecular characterization help to remove the taxonomic opaqueness and paucity of knowledge regarding species identity. This will help to improve its relevance for posterity and will further elucidate the deeper levels of phylogenetic relationships, among species of genus *Hemicriconemoides*.

Key words: *Hemicriconemoides*; Morphology; Ultrastructure; rDNA; China

---

* 基金项目：国家自然科学基金（No. 31371921）
** 第一作者：Maria Munawar，博士后，主要从事植物线虫分类学研究
*** 通信作者：郑经武，教授，主要从事植物线虫分类及线虫病害治理方面的研究；E-mail：jwzheng@ zju. edu. cn

大豆孢囊线虫生防真菌 HZ-9 的鉴定及其防效评价*

田忠玲**，Barsalote Eda Marie，蔡瑞航，李晓琳，郑经武***
（浙江大学生物技术研究所，杭州　310058）

摘　要：大豆孢囊线虫（*Heterodera glycines*）属于定居性内寄线虫，其引起的病害是典型的土传性病害。在针对该线虫的综合治理措施中，生物防治是主要的防治对策，随着人们意识到化学防治造成有害生物的抗药性及对环境的污染，有关植物寄生线虫生物防治成为植物线虫防治的热点之一。国内外在大豆孢囊线虫生防菌的筛选及其商品化应用方面做了大量工作。本文报道筛选获得的防治大豆孢囊线虫有潜力的生防菌株 HZ-9 鉴定及其防效评价的结果。

基于形态学和分子特征将菌株 HZ-9 鉴定为钩状木霉（*Trichoderma hamatum*）。测定了菌株 HZ-9 孢子悬浮液（浓度为 10^8CFU/mL）对大豆孢囊线虫的的毒性，发现该菌株对大豆孢囊线虫 J2 的活性较强，致死率达到 83.3%；此外发现该菌株孢子悬浮液对农林业生产上其他一些重要的线虫——南方根结线虫（*Meloidogyne incognita*）、甘薯茎线虫（*Ditylenchus destructor*）、水稻干尖线虫（*Aphelechoides besseyi*）和松材线虫（*Bursaphelenchus xylophilus*）等都有一定活性，其中对松材线虫毒性较强，致死率达到 85.6%，对南方根结线虫的致死率达到 58.3%。菌株温室盆栽防治大豆孢囊线虫的试验表明其对大豆孢囊线虫具有较好的防治效果，随着施用孢子悬浮液浓度的增加，其孢囊抑制率呈升高的趋势，侵入根内的线虫数量呈递减的趋势。当施用孢子浓度达到 10^9CFU/mL 时，菌株对孢囊抑制率达到 87.5%，根内侵入线虫数量与对照相比减少 81.3%。此外还发现该菌株对大豆的生长具有一定的促进作用。

关键词：大豆孢囊线虫；生防真菌；鉴定；钩状木霉；防效

---

* 基金项目：公益性行业（农业）科研专项（201503114）

** 第一作者：田忠玲，博士研究生，主要从事植物线虫病害研究

*** 通信作者：郑经武，博士，教授，主要从事植物线虫分类及线虫病害治理方面的研究；E-mail：jwzheng@ zju. edu. cn

Characterization and diagnostics of stubby root nematode *Trichodorus cedarus* Yokoo, 1964 (Dorylaimida: Trichodoridae) from Zhejiang Province, China[*]

Barsalote Eda Marie[**], Munawar Maria, Tian Zhongling, Cai Ruihang, Li Xiaolin, Qu Nan, Zheng Jingwu[***]

(*Laboratory of Plant Nematology*, *Institute of Biotechnology*, *College of Agriculture & Biotechnology*, *Zhejiang University*, *Hangzhou* 310058, *China*)

Abstract: The study was conducted to characterize morphological and molecular profiles of seven isolates of *Trichodorus cedarus* recovered from deciduous trees and ornamentals in Zhejiang Province, Eastern China. Identification of species was derived from polytomous key using prime character codes: E, G, H and subgroup codes: J, K, L, M for male populations while prime character codes: D, C, L, K and subgroup codes: G, H, J for female populations. *T. cedarus* is characterized by a short body (0.63~0.90mm), onchiostyle 55 ~ 69μm and secretory-excretory pore (S-E) situated along the anterior 2/3 of pharyngeal bulb. In females, *T. cedarus* is distinguished by possessing didelphic genital system, transverse slit vulva located 52%~60% along the body, vaginal refractive pieces approximately 1~1.9μm size, vagina pear-shaped withtriangular to rounded triangular vaginal sclerotization. In males, *T. cedarus* is characterized by having 3 ventromedian cervical papillae (CP) located between onchiostyle base and S-E pore, presence of 3 ventromedian precloacal supplements with posterior supplement (SP1) at level of retracted spicule, large sperms with sausage-shaped nucleus, spicule short (38.1~45.4μm) appearing slightly curved or nearly straight, manubrium offset without capitular extension and shaft more or less cylindrical with continuous fine ornamentations except at extremities. The codes for identification of species are: D1-C1-L1-K2-A12-B23-E3-F1-G2-H1-I2-J1-M1-N1-O1-P1-Q2-R1 in females and F34-D3-P2-A21-B23-C21-E0-G1-H3-I1-J21-K32-L2-M2-N1-O5 in males. Analysis based on partial 18S, ITS1 and D2-D3 segments of 28S obtained a fragment length of 1 726~1 743bp, 1 221~1 225bp and 796~799bp, respectively. Blastn results revealed 99% sequence identity to *T. cedarus* ZJ8 (HM106502). Similarly, phylogenetic reconstruction showed cluster of present isolates to *T. cedarus* putative species from Zhejiang with 99% clade support. This paper takes into account the morphological and molecular comparisons among intraspecies of *T. cedarus*, the most prevalent trichodorid nematodes in Zhejiang, China.

Key words: Stubby root nematode; Morphology; rDNA; China

---

* 基金项目：国家自然科学基金（No. 31371921）

** 第一作者：Barsalote Eda Marie，博士生（菲律宾籍），主要从事植物线虫分类学研究

*** 通信作者：郑经武，教授，主要从事植物线虫分类及线虫病害治理方面的研究；E-mail：jwzheng@ zju. edu. cn

水稻干尖线虫接种方法的比较研究*

谢家廉**，杨　芳，彭云良，姬红丽***
（四川省农业科学院植物保护研究所，西南农作物有害生物治理重点实验室，成都　610066）

摘　要：水稻干尖线虫（*Aphelenchoides besseyi* Christie）是引起世界各稻米产区上水稻干尖病的重要水稻病原线虫。该线虫侵染水稻后为害地上部分，引起水稻叶尖干枯、"小穗头"等症状，造成水稻产量及品质下降。水稻干尖线虫主要取食水稻的地上部分，侵染后期进入小穗取食胚、子房和雄蕊等器官并迅速繁殖，造成谷粒带虫。为了更好地在实验室和田间研究水稻干尖线虫的侵染机制和筛选抗性品种，寻找接种成功率高并且简单方便的接种方法是十分必要的。本研究初步探索了不同的干尖线虫接种法的接种效率和成功率。研究发现，水漫法接种线虫在接种成功率和线虫数量上都是最优的，接种成功率在95%以上，接种效率达到11.11%。浸种法和苗期喷雾法接种成功率较低，均在40%左右，接种成功率分别约为2.82%和1.28%，但这两种接种方法操作简单方便，可在田间通过与其他方法联合使用以达到理想的接种效果。

关键词：水稻干尖线虫；接种方法；水漫法；浸种法；喷雾法

* 基金项目：四川省国际合作项目（2015HH0041）

** 第一作者：谢家廉，女，博士，助理研究员，研究方向：植物线虫学；E-mail：xiejialian1985@163.com
*** 通信作者：姬红丽，女，博士，研究员，研究方向：植物线虫学；E-mail：Hongli.Ji@Jihongli.com

不同药剂浸种对水稻干尖线虫的毒力测定*

李少军**，谢家廉，杨　芳，彭云良，姬红丽***
（四川省农业科学院植物保护研究所，西南农作物有害生物治理重点实验室，成都　610066）

摘　要：水稻干尖线虫（*Aphelenchoides besseyi* Christie）是全世界分布最广泛的植物病原线虫之一，侵染水稻后能引起水稻叶尖干枯、“小穗头”等症状，造成水稻产量下降，严重时减产达 40%。该病的初侵染源为种子带虫，其最有效的防治方法是播种前种子进行药剂处理。为明确不同农药药剂浸种对水稻干尖线虫的防治效果，探索其最佳使用剂量，为田间推广使用提供科学依据，笔者进行了 4 种药剂浸种对干尖线虫的室内毒力测定试验，在 25℃下浸种 48h 后，调查各药剂对干尖线虫的毒力效果。结果显示，3. 2% 阿维菌素 EC 防效最好，其 10 000倍稀释液浸种对水稻干尖线虫致死率在 90%以上，校正防效为 91. 27%，10. 5%噻唑膦 EC 的防效次之，其 500 倍液的防效为 79. 9%，而 25g/L 的溴氰菊酯 EC 以及 90%杀虫单 SP 的各个稀释液防效均在 65%以下。

关键词：水稻干尖线虫；药剂；浸种；防效

---

* 基金项目：四川省国际合作项目（2015HH0041）
** 第一作者：李少军，男，硕士，研究实习员，研究方向：植物病理；E-mail：finch1985@ 126. com
*** 通信作者：姬红丽，女，博士，研究员，研究方向：植物线虫学；E-mail：Hongli. Ji@ Jihongli. com

Effect of abamectin on the cereal cyst nematode (CCN, *Heterodera avenae*)[*]

Ji Xiaoxue[1**], Wang Hongyan[1], Wang Dong[2], Qiao Kang[1***]

(1. *College of Plant Protection, Shandong Agricultural University, Tai'an*, 271018, *China*;
2. *College of Agronomy, Shandong Agricultural University, Tai'an* 271018, *China*)

Abstract: The cereal cyst nematode (CCN, *Heterodera avenae*) is a major pest in wheat and until now there is no pesticide registered to control this pest in China. Development of effective methods of controlling CCN is urgently needed. Abamectin is a biological pesticide that has a high nematicide activity. However, the efficacy of abamectin soil application to control CCN in wheat and its effect on yield in China remains unknown. Therefore, laboratory, greenhouse and field tests were carried out to evaluate the potential of abamectin soil applications for CCN control and improvement of wheat yield. Laboratory tests showed that abamectin exhibited knockdown toxicity to CCN, with LC_{50} and LC_{90} values 9.8mg/L and 59.4mg/L. Greenhouse experiment and field trials showed that soil applications of abamectin provided significant CCN control and higher straw dry weights and wheat grain yields. There was an 8.5% to 19.3% yield increase from the various abamectin treatments compared with the control. The results of this study demonstrated that abamectin exhibited a high nematicidal activity to *H. avenae* and adequate performance to enhance wheat crop yields.

Key words: Abamectin; *Heterodera avenae*; Wheat

* 基金项目：国家自然科学基金项目（31601661）；公益性行业（农业）科研专项（201503130）

** 第一作者：姬小雪，女，实验师，主要从事农药毒理与有害生物抗药性研究；E-mail：xxfzim@ 163. com

*** 通信作者：乔康，男，副教授，博士，主要从事土壤消毒技术和农药毒理学研究；E-mail：qiaokang11-11@ 163. com; qiaokang@ sdau. edu. cn

蔬菜根结线虫生防菌株的分离与筛选*

张　洁**，杨丽荣，孙润红，夏明聪，武　超，薛保国***
（河南省农业科学院植物保护研究所，郑州　450002）

摘　要：随着保护地蔬菜种植面积的扩大和连茬种植等因素的影响，蔬菜根结线虫病呈逐年加重的趋势，并上升为蔬菜生产上的主要病害，严重威胁着我国蔬菜的品质，因此生产上亟需一种安全有效的防治方法。本研究利用选择性培养基从多地采得的番茄根结线虫的根结及卵囊上分离到 56 株真菌菌株，122 株细菌菌株以及 25 株放线菌菌株。通过对二龄线虫的致死作用、寄生能力测定以及盆栽防效测定，最终筛选出 2 株生防真菌（NF-22 和 NF-45）、3 株生防细菌（NB-12、NB-58 和 NB-101）以及 2 株生防放线菌（NS-06 和 NS-24）进行后续深入研究。其中，生防真菌 NF-45 对线虫的寄生率达到 82%，盆栽防效达到 65%；生防细菌 NB58 的发酵液对二龄线虫的致死率达到 98%，盆栽防效达到 60%；生防放线菌 NS-06 的发酵液对二龄线虫的致死率达到 95%，盆栽防效达到 58%。由此可以看出，以上菌株都具有很大的开发利用潜力。

关键词：蔬菜；根结线虫；生防菌株

---

* 基金项目：国家科技支撑计划课题（2012BAD19B04）；农业部“948”引进技术项目（2014-Z63）；小麦主要病虫害绿色防控产品的研发与产业化（141100111100）

** 第一作者：张洁，女，助理研究员，博士，主要从事植物寄生线虫综合防治研究；E-mail：zhangjie656@ 126. com
*** 通信作者：薛保国，男，研究员，博士，博士生导师，主要从事分子微生物学研究；E-mail：13613714411@ 163. com

黄淮四省小麦根腐线虫不同种群的遗传多样性分析*

杜　鹃，张　含，逯麒森，赵　贝，李　宇，王　珂，李洪连**
（河南农业大学植物保护学院，郑州　450002）

摘　要：根腐线虫又称短体线虫（*Pratylenchus* spp.），是一类重要的迁移性内寄生线虫，寄主范围广泛，在亚洲、澳洲、美洲和欧洲等地区均有发生，对全球作物的危害程度仅次于根结线虫和胞囊线虫。该类线虫引起的小麦根腐线虫病是一种世界性病害，在一些地区对小麦生产构成了较大威胁。为了明确黄淮麦区根腐线虫的区域分布及种类遗传多样性，本次研究主要对黄淮麦区四省（河北、山东、山西、安徽）部分地区的小麦根腐线虫进行了初步的调查。主要研究结果如下：

对河北邢台市邢台县、山西运城市平陆县等8个种群进行单条线虫DNA的提取，并对PCR扩增的rDNA区包括18S区以及ITS区进行扩增。利用DNAMAN软件对8个小麦根腐线虫种群的18SrDNA序列进行了分析，结果表明供试8个种群与GenBank中其他短体线虫群体分别具有92%~99%（EU669927）、92%~96%（EU130812）的同一性，说明小麦根腐线虫不同群体的18SrDNA序列具有较高的保守性。根据小麦根腐线虫供试8个不同种群和GenBank中其他多个短体线虫群体18SrDNA基因的核苷酸序列，利用MEGA 5.0软件构建系统进化树，结果表明：所有18个小麦根腐线虫种群主要聚为两支，其中第一支可以分为两个亚支，第一亚支包含群体PraN1、PraN2和PraN4，群体PraN3、PraN5-PraN8和其他的4个种群聚为第二亚支；种群*P. hippeastri*（KJ001716）、*P. scribneri*（EU669958）、*P. loosi*（AB90529）、*P. loosi*（AB905296）、*P. loosi*（AB905286）和*P. speijeri*（KF974683）6个种群组成了第二支。供试小麦根腐线虫8个种群ITS rDNA的核苷酸序列在GenBank中与其他小麦根腐线虫具有91%~99%的同一性。ITS rDNA核苷酸序列构建的进化树与18SrDNA系统进化树相似，所有19个种群主要聚为两支，Pra2和Pra3两个群体聚为第一支，后者又可以分为两个亚支，其中第一个亚支由供试的Pra4、Pra6群体和GenBank中的其他4个群体*P. scribneri*（JX046933）、*P. scribneri*（JX046934）、*P. scribneri*（KT873860）和*P. scribneri*（JX046932）组成，然而Pra1、Pra5群体和GenBank中其他7个群体亲缘关系较近，共同构成第二亚支。初步研究结果表明：黄淮麦区小麦根腐线虫不同种群之间存在显著的遗传变异和丰富的遗传多样性。

关键词：小麦根腐线虫；种群；遗传多样性

---

* 基金项目：国家公益性行业科研专项（201503112）；国家自然科学基金（31601619）

** 通信作者：李洪连，教授，主要从事植物土传病害研究；E-mail：honglianli@ sina. com

Morphology and molecular analysis of one new species of *Trophurus* (Nematoda: *Telotylenchinae*) from the soil associated with *Cinnamomum camphora* in China*

Wang Ke, Du Juan, Wang Zhenyue, Li Yu, Li Honglian**

(*Department of Plant Pathology*, *Henan Agricultural University*, *Zhengzhou* 450002)

Abstract: The subfamily Telotylenchinae represents a large group of plant-parasitic nematodes. These nematodes are obligate root ectoparasites of a large variety of plants. The female reproductive system in most genera of Telotylenchinae is didelphic, except in *Trophurus* Loof, 1956. *Trophurus* is one small genus in the subfamily Telotylenchinae, and there are only 15 *Trophurus* species have been described in the world. During nematode surveys in Huanghui area of China, one plant nematode population was collected from the soil associated with *Cinnamomum camphora* in Anhui Province, China. The nematode population belongs to the genus *Trophurus*, and is regarded as a new species. The new species can be separated from morphologically similar species by the body length, stylet length, knobs type, ornamentation in the lateral field, post-vulval uterine sac length, with or without post-rectal intestinal sac, tail shape, ornamentation on the tail terminus in the female; and spicules length in the male. The internal transcribed spacer sequences of ribosomal DNA (ITS rDNA), and partial 18S ribosomal DNA (18S rDNA) from the new species will be amplified and sequenced. Phylogenetic relationships of the new species and other species will be further studied.

Key words: Nematode; Morphology; Molecular analysis; New species

---

* Funding: This study was funded by the National Natural Science Foundation of China (31601619) and Special Fund for Agro-scientific Research in the Public Interest (201503112)

** Corresponding author: Li Honglian; E-mail: honglianli@ sina. com

第六部分
抗病性

30个甘蔗新品种（系）抗褐锈病基因 *Bru*1 分子检测*

李文凤**，张荣跃，单红丽，王晓燕，尹 炯，罗志明，仓晓燕，黄应昆***
（云南省农业科学院甘蔗研究所/云南省甘蔗遗传改良重点实验室，开远 661699）

摘 要：甘蔗褐锈病是目前中国蔗区发生最普遍，危害最严重的病害之一，选育和种植抗病品种是防治该病最为经济有效的措施。为明确近年国家甘蔗体系育成的新品种（系）对甘蔗褐锈病的抗性，确定其应用潜力，为合理布局和推广使用这些新品种（系）提供依据。本研究采用已报道的抗褐锈病基因 *Bru*1 稳定的分子标记 R12H16 和 9O20-F4，于 2016 年对中国近年选育的 30 个甘蔗新品种（系）进行抗褐锈病基因 *Bru*1 分子检测，明确了新品种（系）中抗锈病基因 *Bru*1 分布状况，检测筛选出福农 09-4095、云蔗 08-1609、福农 07-3206、云瑞 10-701、云瑞 10-187、粤甘 48 号、粤甘 50 号、桂糖 08-1533、桂糖 06-1492、福农 09-12206、德蔗 09-78、粤甘 47 号、闽糖 06-1405、桂糖 08-1180、桂糖 06-2081 等 15 个新品种（系）含抗褐锈病基因 *Bru*1，具有抗褐锈病的应用潜力，为有效防控甘蔗褐锈病提供了科学依据和优良抗性新品种（系）。

关键词：甘蔗；优良新品种（系）；抗褐锈病基因 *Bru*1；分子检测

---

* 基金项目：国家现代农业产业技术体系建设专项（CARS-20-2-2）；云南省现代农业产业技术体系建设专项

** 第一作者：李文凤，女，研究员，主要从事甘蔗病害研究；E-mail：ynlwf@ 163. com
*** 通信作者：黄应昆，研究员，从事甘蔗病害防控研究；E-mail：huangyk64@ 163. com

甘蔗新品种（系）花叶病病原分子检测及自然抗性评价*

李文凤**，单红丽，张荣跃，王晓燕，尹　炯，仓晓燕，罗志明，黄应昆***
（云南省农业科学院甘蔗研究所/云南省甘蔗遗传改良重点实验室，开远　661699）

摘　要：甘蔗花叶病是一种重要的世界性甘蔗病害，近年来已成为中国蔗区发生最普遍，危害最严重的病害之一，选育和种植抗病品种是防治该病最为经济有效的措施。为明确近年国家甘蔗体系育成的新品种（系）花叶病田间自然发病情况，分析评价其对花叶病的自然抗病性，确定其应用潜力，为合理布局和推广使用这些新品种（系）提供依据。本研究采用 5 级分级调查法（1 级高抗发病率 0.00%、2 级抗病发病率 0.01%～10.00%、3 级中抗发病率 10.01%～33.00%、4 级感病发病率 33.01%～66.00%、5 级高感发病率 66.01%～100%），对 2016 年国家甘蔗体系开远综合站繁殖示范的 33 个甘蔗新品种（系）花叶病田间自然发病情况进行系统调查，同时每个新品种（系）采集 10 份样品共 330 份，利用 RT-PCR 法对其病原（SCMV、SrMV、SCSMV）进行检测。田间自然发病调查结果表明：33 个新品种（系）中，表现 1 级高抗到 3 级中抗的有 27 个，占 81.82%，4 级感病到 5 级高感的有 6 个，占 18.18%。分子检测结果显示，引起新品种（系）花叶病病原有 SCSMV、SrMV 两种，所有品种中均检测到 SCSMV，SCSMV 检出率高达 83.94%，SrMV 检出率为 13.33%，且存在 2 种病原复合侵染（复合侵染率 7.58%），SCSMV 已成为云南甘蔗花叶病主要病原。研究结果明确了各新品种（系）对甘蔗花叶病的自然抗性，揭示了引起新品种（系）花叶病病原有 SCSMV 和 SrMV，SCSMV 为最主要致病病原，为生产用种选择和有效防控甘蔗花叶病提供了科学依据。

关键词：甘蔗；新品种（系）；花叶病；病原检测；自然抗病性

---

* 基金项目：国家现代农业产业技术体系建设专项（CARS-20-2-2）；云南省现代农业产业技术体系建设专项

** 第一作者：李文凤，女，研究员，主要从事甘蔗病害研究；E-mail：ynlwf@163.com

*** 通信作者：黄应昆，研究员，从事甘蔗病害防控研究；E-mail：huangyk64@163.com

Tryptophan decarboxylaseplays a positive role in the resistance to cereal cyst nematode in *Aegilops varabilis* No. 1*

Huang Qiulan[1,2], Li Lin[3], Zheng Minghui[3], Long Hai[1],
Deng Guangbing[1], Zhang Haili[1]**, Yu Maoqun[1]**

(1. *Chengdu Instituteof Biology*, *Chinese Academy of Sciences*, *Chengdu* 610041, *China*;
2. *College of Life Science*, *Sichuan University*, *Chengdu*, 610064, *China*;
3. *School of Basic Medicine*, *Zunyi Medical University*, *Zunyi*, 563003, *China*)

Abstract: Cereal cyst nematode (CCN) is a vital rootinfected pathogen in wheat. The yield losses caused by CCN reached 10% to 40%. *Aegilops varabilis NO.* 1 is relative of wheat and resistant to CCN and root-knot nematode (RKN). RNA-seq analysis of the interaction between *Aegilops varabilis NO.* 1 and second juvenile stage of CCN indicates alkaloid metabolism plays an important role in the defense responses to early infection. *Tryptophan decarboxylase* (*TDC*) gene is the key gene of the metabolic pathways. Silencing of *TDC*1 and pretreatment with its activity inhibitor both increased the infection number of CCN J2 in the root of *Aegilops varabilis NO.* 1. Furthermore, *AeVTDC*1 transgenic tobacco (*Nicotiana tabacum*) overexpression *AeVTDC*1 was obtained and their resistance to root-knot nematode was tested. Resistance assay showed that the number of root-knot in transgenic plants was significant less than that in wild-type plants. The downstream second metabolites of alkaloid metabolism, tryptamine and serotonin, their contents in roots were also measured, and UPLC-MS analysis revealed that the tryptamine in *AeVTDC*1 transgenic tobacco was much more than that in wild-type plants, whereas the serotonin was less than that in wild-type plants. These preliminary results indicated that *AeVTDC*1 plays a positive role in resistance tocereal cyst nematode and root knot nematodes in *Aegilops varabilis No.* 1.

Key words: *AeVTDC*1; Cereal cyst nematodes; Root-knot nematode; *Aegilops varabilis NO.* 1

---

* Funding: This work was supported by the National Science Foundation of China (31470097)

** Corresponding authors: Yu Maoqun, E-mail: yumq@ cib. ac. cn
Zhang Haili, E-mail: zhanghl@ cib. ac. cn

71 个甘蔗新品种（系）双抗 SCSMV 和 SrMV 鉴定与评价*

李文凤**，单红丽，张荣跃，王晓燕，尹　炯，罗志明，仓晓燕，黄应昆***

（云南省农业科学院甘蔗研究所/云南省甘蔗遗传改良重点实验室，开远　661699）

摘　要：甘蔗花叶病是一种重要的世界性甘蔗病害，目前已成为我国蔗区（云南、广西、广东、海南、福建、四川和浙江等蔗区）发生最普遍，危害最严重的病害之一，利用抗病品种是控制该病害最经济、安全、有效的方法。本研究以我国蔗区甘蔗花叶病的 2 种主要病原甘蔗条纹花叶病毒分离物（SCSMV-JP1，GenBank 登录号 JF488064）和高粱花叶病毒分离物（SrMV-HH，GenBank 登录号 DQ530434）为接种毒源，采用人工切茎接种和 RT-PCR 检测相结合方法，于 2015—2016 年 2 次对中国近年选育的 71 个甘蔗新品种（系）进行了双抗 SCSMV 和 SrMV 鉴定与评价。结果表明：71 个甘蔗新品种（系）中，对 SCSMV 表现 1 级高抗到 3 级中抗的有 18 个，占 25.35%，4 级感病到 5 级高感的有 53 个，占 74.65%；对 SrMV 表现 1 级高抗到 3 级中抗的有 19 个，占 26.76%，4 级感病到 5 级高感的有 52 个，占 73.24%。综合分析结果显示，粤糖 55 号、福农 30 号、云蔗 03-258、云蔗 05-51、粤糖 34 号、云蔗 06-80、粤糖 40 号、闽糖 01-77、桂糖 02-467、粤糖 96-86 等 10 个新品种（系）双抗 SCSMV 和 SrMV 2 种病毒，占 14.08%。研究结果明确了 71 个甘蔗新品种（系）对甘蔗花叶病 2 种主要致病病原的抗性，筛选出双抗 SCSMV 和 SrMV 的甘蔗新品种（系）10 个，为生产用种选择提供了科学依据。

关键词：甘蔗；新品种（系）；甘蔗条纹花叶病毒（*Sugarcane streak mosaic virus*，SCSMV）；高粱花叶病毒（*Sorghum mosaic virus*，SrMV）；抗病性鉴定

---

* 基金项目：国家现代农业产业技术体系建设专项（CARS-20-2-2）；云南省现代农业产业技术体系建设专项

** 第一作者：李文凤，女，研究员，主要从事甘蔗病害研究；E-mail：ynlwf@163.com

*** 通信作者：黄应昆，研究员，从事甘蔗病害防控研究；E-mail：huangyk64@163.com

四川小麦品系对主要病害抗性及变异监测*

夏先全**，肖万婷，叶慧丽，章振羽，魏会廷，徐　志
（四川省农业科学院植物保护研究所，成都　610000）

摘　要：条锈病、白粉病和赤霉病是四川乃至全国小麦生产上的重要病害，常因其危害而造成严重减产、品质变劣。同时，四川也是全国小麦条锈病重要菌源基地和小种变异区之一。为持续稳定控制小麦条锈病、减缓白粉病和赤霉病等为害，2016年笔者在四川成都、绵阳、雅安、南充、宜宾和广安等不同生态区设立多个小麦病害鉴定监测圃，收集了各育种单位（公司）的小麦材料、高代品系以及主要生产品种1 663份，进行小麦抗条锈病、白粉病和赤霉病性鉴定评价以及主要病害发生动态和条锈小种变异监测。

受2016年春节前后两次降温和寒潮的影响，秋苗和越冬菌源降低，省内小麦条锈病总体发生较轻，同时由于不同麦区条锈菌地方优势小种存在差异，导致个别品种在不同病圃出现抗性差异，参试小麦品系中条锈病高抗、中抗和感病的比例分别占到45.5%、29.4%和25.1%；由于开春后的气候条件适合白粉病的侵染、繁殖和流行，小麦白粉病的发生为害略重，参试品系中高抗、中抗和感白粉病比例分别为18.4%、25.6%和55.9%；由于小麦抽穗扬花期连续小雨较多，对禾谷类镰刀菌的繁殖有利，导致多数地市小麦赤霉病发生普遍，参试品种中有8份田间表现高抗赤霉病，需进一步鉴定验证，其余中抗、中感和高感赤霉病的品系分别占比为13.2%、61.3%和25%。

2015年四川省小麦生长季节条锈病生理小种组成与之前年份相比出现明显变化，条中32和条中33小种出现频率下降特别明显，而贵农22致病类群的出现频率上升极快，占总数的46.97%。其中条中34（贵农22-9）已经上升为优势小种，其对19个鉴别寄主中除中四和*Triticum. spelta* album之外的其他品种均具有毒性，标志着条锈病菌毒性进化的一个新水平。由于条锈小种在转主寄主小檗上的有性繁殖和变异，加之生产上部分感病哺育小麦品系大面积种植，是导致类似条中34等的原稀有小种逐步上升为优势菌系的主要原因。为了避免把本身自然界中比例极低的毒性菌源人为助推成优势菌系，造成生产上的风险蔓延，在抗锈选育、病害试验等科研活动中，提出了尽量减少人为扩繁和扩散各类新菌系突变体，特别是不要在自然界中人工接种各类稀有菌系突变体的建议。

关键词：小麦；条锈病；抗性鉴定；生理小种；变异监测

* 基金项目：四川省农作物育种攻关项目；国家现代农业产业技术体系四川麦类作物创新团队

** 第一作者：夏先全，副研究员，主要从事小麦品种抗病性评价及病害防治研究；E-mail：xiaxianquan@163.com

TaSBT1 参与小麦抗条锈病防卫反应的初步研究

陈发晶，余　洋，毕朝位，杨宇衡*
（西南大学植物保护学院，重庆　400715）

摘　要：由条形柄锈菌（*Puccinia striiformis* f. sp. *tritici*，*Pst*）引起的小麦条锈病是全球小麦生产上危害程度最大、发生范围最广的病害之一。笔者从前期的在小麦——*Pst* 非亲和互作蛋白质组数据中筛选出 1 个小麦类枯草杆菌蛋白酶 TaSBT1 在互作初期被特异诱导表达，但其具体的功能和作用方式尚不清楚。本研究首先在小麦品种“水源 11”中克隆得到了 *TaSBT*1 的完整 ORF，随后通过 qRT-PCR 技术对其进行小麦与条锈菌的互作表达谱分析，结果显示 *TaSBT*1 在非亲和组合中的 12~24h 有强烈的上调表达，而在亲和组合中表达量无显著变化，表明其可能参与小麦对条锈菌的抗病诱导过程。利用 BSMV 介导的基因瞬时沉默技术将 *TaSBT*1 基因沉默后，表型结果显示基因沉默植株对非亲和生理小种 CYR20 的抗性显著下降。同时采用瞬时表达技术在烟草叶片中过表达 *TaSBT*1 基因，可以诱发植物细胞死亡。此外，利用酵母信号肽筛选系统成功验证了其编码蛋白信号肽的分泌功能。以上结果表明 *TaSBT*1 在小麦抗条锈过程中具有重要作用。本研究为深入研究 TaSBT1 对小麦抗锈机制和持续全面防治小麦条锈菌具有重要的理论意义。

关键词：小麦条锈病；*TaSBT*1；qRT-PCR；基因沉默；瞬时表达；信号肽

* 通信作者：杨宇衡，主要从事植物病理学研究；E-mail：yyh023@swu.edu.cn

花生 *HyPGIP* 基因的克隆及荧光定量表达分析*

王麒然[1**]，王 琰[2]，王志奎[3]，张茹琴[1]，迟玉成[4]，夏淑春[1]，鄢洪海[1***]
（1. 青岛农业大学农学与植物保护学院，青岛 266109；2. 山东省乳山市农业局，青岛 264500；3. 青岛市即墨市农业局，青岛 266200；4. 山东省花生研究所，青岛 266101）

摘 要：山东省是我国主要花生产区之一，2012 年山东花生的种植面积约为 78.71 万 hm^2，花生的产量则达到了 348.7 万 t。而花生病害的侵染是影响花生产量主要因素之一，近来由立枯丝核菌（*Rhizoctonia solani*，AG-1-IA）侵染引起的花生叶腐病日益加重，成为一些地区花生的主要病害。据王麒然等报道花生叶腐病病菌分泌细胞壁降解酶，尤其是产生高活性的多聚半乳糖醛酸酶（PG）是该病菌的重要致病机制。然而，寄主植物为了抵御病原物的为害，在其进化过程中也往往会形成相应的防御机制，为此，笔者探寻了花生体内多聚半乳糖醛酸酶抑制蛋白（PGIP）特性，基因表达与寄主抗病性的关系，意在为寻找花生叶腐病防治新策略提供理论依据。

通过测定花生植株接种叶腐病病菌（*Rhizoctonia solani*，AG-1-IA）后植株内 PG 的活性变化及 PGIP 蛋白的生成量变化，来研究 PG 在花生叶腐病菌致病中的作用及其 PGIP 蛋白在寄主抗病中的作用；根据 GenBank 数据库的 5 个花生 *PGIP* 基因的 mRNA 序列的 CDs 区（无内含子）设计引物，通过总 RNA 提取，反转录到 cDNA，PCR 扩增、目的条带回收纯化后连接 T 载体、并转化 *E. coli* DH 5α，克隆获得花生多聚半乳糖醛酸酶抑制蛋白基因 *HyPGIP*，经测序和 ORF Finder 分析 *HyPGIP* 基因序列，预测 HyPGIP 蛋白的基本性质和生物学功能；并通过在花生植株上接种叶腐病菌诱导 *HyPGIP* 基因表达，来分析 HyPGIP 蛋白与花生的抗病性关系。结果表明：抗病品种花育 25 接种叶腐病菌后叶片的病斑比感病品种白沙 1016 小，PG 活性开始阶段升高，但很快下降，和白沙 1016 比 PG 活性显著弱。而生成的 PGIP 蛋白量要比白沙 1016 显著多，说明 PG 活性与花生叶腐病菌致病性关系密切，而 PGIP 蛋白与寄主的抗病性紧密相关。经基因克隆和测序获得 *HyPGIP* 基因全长序列为 1 427bp，该基因序列中含有一个 1 029bp 的完整 ORF，编码 342 个氨基酸，无内含子，终止密码子为 TGA，命名为 *HyPGIP* 基因，该 ORF 序列与已报道的 5 个花生 PGIP 序列的 ORF 区同源性都很高；结构预测发现该序列存在 8 个 LRR 结构域，存在信号肽，疏水性较强，定位于细胞壁上；经寄主接种病菌处理后，花生植株中的 *PGIP* 基因表达量明显增加，并随接种时间延长而不断升高，且在 36~48h 时间段内表达量升高迅速，然后生成量增加缓慢。RT-PCR 检测表明在正常健康植株中，*PGIP* 基因表达量甚少，但经接种叶腐病菌处理后增加明显，且在 36~48h 时间段内还出现一个快速增加过程，之后生成量缓慢减少，说明 *HyPGIP* 基因的表达受病原菌侵染影响，增加过程也不是匀速进行。

关键词：花生；基因；克隆；荧光定量表达

---

* 基金项目：山东省自然科学基金项目（ZR2011CL005）；山东省“泰山学者”建设工程专项经费资助（BS2009NY040）

** 第一作者：王麒然，男，硕士研究生，主要从事分子植物病理学研究；E-mail：814532720@ qq. com

*** 通信作者：鄢洪海，男，教授，博士，主要从事植物病理学与玉米病害研究；E-mail：hhyan@ qau. edu. cn

胶孢炭疽菌激活杧果防御酶基因的差异表达研究*

苏初连**，康　浩，梅志栋，刘晓妹，蒲金基，张　贺***
（中国热带农业科学院环境与植物保护研究所，海口　571101）

摘　要：本文通过实时荧光定量 PCR 的方法，运用相对定量方法发现，胶孢炭疽菌分生孢子侵入杧果嫩叶和成熟果过程中，不同程度地激活活性氧相关的 3 个基因（*MiSOD*，*MiPOD* 和 *MiPPO*）和抗病相关的 4 个基因（*MiCHI*，*MiARE*，*MiPR*1 和 *MiPAL*）等 7 个防御酶基因的活性。7 个防御酶基因在接种成熟果皮 12h 时达到活性的最大值，随后基因活性呈下降趋势，但仍显著高于 0h 接种时期的活性；在接种嫩叶时，7 个防御酶基因活性呈递增趋势，72h 达最大值，说明了寄主植物在抵御病原菌侵染的过程中不同器官、不同发育阶段，同一类基因的表达差异情况；揭示了寄主植物在受病原菌侵染时会激活自身抗病防御酶基因高效表达，提高酶的活性，从而抵抗病原菌的侵染，提高其抗病性。本文通过研究杧果抗病相关防御酶基因的表达为提高其抗病性奠定了基础。

关键词：胶孢炭疽菌；防御酶基因；差异表达

注：本摘要的全文已被《中国南方果树》接收。

* 基金项目：国家自然科学基金（31460455）；中央级公益性科研院所基本科研业务费专项-中国热带农业科学院院级创新团队项目（项目编号：17CXTD-09）

** 第一作者：苏初连，女，在读硕士研究生，研究方向：微生物学；E-mail：1373945693@ qq. com

*** 通信作者：张贺，男，硕士，助理研究员，研究方向：热带果树病理学；E-mail：atzzhef@ 163. com

“金艳”猕猴桃 *XTH7* 基因克隆与序列分析*

贺　哲**，王园秀，刘　冰，黄春辉，蒋军喜***
（江西农业大学农学院，南昌　330045）

摘　要：木葡聚糖内糖基转移酶/水解酶（XTH）能够调控细胞壁的松弛和伸展，对细胞壁的重构具有重要作用。由葡萄座腔菌（*Botryosphaeria dothidea*）引起的猕猴桃果腐病已严重为害猕猴桃果实的产量和品质，为了寻找猕猴桃抗性品种“金艳”抗果腐病的相关基因，本研究在猕猴桃果腐病发病盛期，根据猕猴桃果腐病发病特性，设置 24h，72h，144h 三个时间点，对“金艳”猕猴桃果实人工接种葡萄座腔菌，诱导其抗病性基因表达，然后通过 RNA-Seq 的方法获得所有转录本，筛选差异基因。经对比发现 *XTH7* 基因在设置的 3 个时间点内和空白对照相比均持续上调。利用 RT-PCR 和 cDNA 末端快速扩增（RACE）技术从“金艳”猕猴桃中克隆出 *XTH7* 基因全长序列，该序列与 GenBank 中来自美味猕猴桃的对应序列具有 99%的同源性。克隆的 cDNA 全长为 1 278bp，其中开放阅读框长度为 885bp，编码 294 个氨基酸。预测相对分子质量为 33. 17ka，等电点为 8. 35，属于亲水性蛋白。本研究结果为进一步研究“金艳”猕猴桃 *XTH* 基因的结构与生物学功能以及其抗病机制奠定了基础。

关键词：“金艳”猕猴桃；木葡聚糖内糖基转移酶/水解酶（XTH）；RACE

---

*　基金项目：国家自然科学基金（31460452）；江西省科技计划项目（20141BBF60019）

**　第一作者：贺哲，硕士研究生，研究方向为分子植物病理学；E-mail：hezhe1993@ 163. com

***　通信作者：蒋军喜，教授，博士，主要从事植物病害综合治理研究；E-mail：jxau2011@ 126. com

水稻感病和抗病品种对纹枯病菌抗性的蛋白组学分析*

圣　聪**，马洪雨，乔露露，赵弘巍，牛冬冬
（南京农业大学植物保护学院植物病理系，南京　210095）

摘　要：水稻纹枯病是由立枯丝核菌侵染引起的水稻三大病害之一，造成水稻减产 10%～50%，但目前对水稻抵抗纹枯病的抗病分子机制了解较少。本研究利用同位素标记相对和绝对定量（isobaric tags for relative and absolute quantification，*iTRAQ*）方法分别对抗、感水稻纹枯病品种进行蛋白组比较分析。笔者共鉴定到 6 560个蛋白，其中与对照相比，在误差小于 2，蛋白变化倍数大于 1.5 倍或者小于 0.6 倍的条件下，共有 755 个蛋白的表达水平在纹枯病菌侵染后显著升高或降低。进一步分析发现这些差异蛋白主要包括乙醛酸和二羧酸代谢，甘氨酸，丝氨酸和苏氨酸代谢、非饱和脂肪酸合成和糖酵解/糖异生调节等途径（$P<0.01$）。此外，笔者通过 qRT-PCR 检测了部分蛋白在 mRNA 水平的表达情况，检测结果表明这些基因表达水平和蛋白测序结果基本一致。此外，笔者通过农杆菌注射的方法在本氏烟中体外检测其对纹枯病菌的抗病功能。以上研究工作为进一步阐述水稻对纹枯病菌的抗病机制奠定了基础。

关键词：水稻；立枯丝核菌；蛋白组；抗病机制

* 通信作者：牛冬冬，副教授，主要从事水稻-病原物互作的分子机制研究；E-mail：ddniu@ njau. edu. cn
** 第一作者：圣聪，在读博士生，研究方向为水稻纹枯病的抗病机制

QTL Mapping of Blast Resistance Genes of Two Elite Rice Varieties in the USA

Chen Xinglong[1], Jia Yunlin[2*], Jia Melisa H[2],
Shannon Pinoon[2], Hu Donghong[3], Wang Xueyan[2], Wu Boming[1]
(1. *Department of Plant Pathology, China Agricultural University, Beijing, China* 1001932, *Department of Agriculture-Agricultural Research Service (USDA-ARS), Dale Bumpers National Rice Research Center (DB NRRC), Stuttgart, AR* 721603, *Crop Science Division, Taiwan Agricultural Research Institute, Council of Agriculture (COA), Taichung, Taiwan*)

Abstract: 'Cybonnet' and 'Saber' are two blast resistant tropical japonica rice varietiesthat have been effectively deployed in the Southern USA. In the present study, a recombinant inbred line (RILs) population from these two parents with 251 individuals was analyzed to identify blast resistance genes. Tenfield isolates of the races IB1, IB17, IB49 and IE1k (TM2) of *Magnaporthe oryzae* collected from 1988 to 2015 in the Southern USA were used to inoculate the parents and RILsunder greenhouse conditions. The disease reactions were evaluated with two disease rating systems-the percentage of lesion area and visual rating. Both the parental lines were resistant to all the isolates except IE1K, Cybonnet was susceptible with mean rating 6 (0~9). A linkage map with 186 single nucleotide polymorphism (SNP) markers was constructed and used to locate quantitative trait loci (QTL), and a total of 7 resistant QTLs were mapped on chromosome 2, 8, 9, 10, 11 and 12. Two of them performed extremely strong, with 18 and 20 phenotype traits confirmed them separately. Of them, qBLAST12 (Q1) located on chromosome 12 was mapped around SNP marker FNP_ 2 or id12004228 with a LOD value ranging from 19. 84~51. 19. This QTL accounted for 25. 63% to 53. 19% of disease reactions, and was located at 10. 60~10. 86 megabases (MB), almost the same place of the major blast R gene Pita. The other QTL, qBLAST2 (Q2) was located at the last SNP marker id2015934 of chromosome 2 with a LOD value ranging from 14. 49 to 73. 46. The proportion of phenotypic variance it explained (PVE) was 16. 46% ~ 74. 3%, and its physical position was located around 35. 03 MB, very near to the major blast R gene *Pib*. The other 5 minor QTLs accounted for 2. 14%~4. 49% of disease reactions and 2 of them were new rice blast resistance QTLs. The results also showed that Q1 (qBLAST12) itself can explain for 37. 14% (25. 63%~53. 19%) and 26. 45% (18. 54%~74. 3%) disease reactions (except IE1K) while Q2 (qBLAST2) can account for 26. 67% (13. 80%~69. 40%) and 24. 92% (17. 11%~58. 02%) when using the 0-9 rating scale and the percentage of lesion area, respectively. More importantly, while Q1 plus Q2, they can still make a function means of 14. 27% (5. 86%~20. 47%) and 17. 53%(12. 04%~20. 70%) (ExceptIE1K) for explaining disease response. And while Q1 plus Q2, they can still make a function to decreases the disease score by-0. 90 (-1. 10~0. 58) and-4. 70 (-6. 55~2. 22) (except IE1K) under the two investigating methods.

* Corresponding author: Jia Yunlin; E-mail: Yulin. jia@ ars. usda. gov

不同抗性油菜品种接种黑胫病菌后 PPO 活性及同工酶变化研究

宋培玲[1]，吴 晶[1,2]，燕孟娇[1]，皇甫海燕[1]，郝丽芬[1]，皇甫九如[1]，贾晓清[1]，李子钦[1]
（1. 内蒙古农牧业科学院植物保护研究所，呼和浩特 010031；
2. 内蒙古大学生命科学学院，呼和浩特 010020）

摘 要：采用比色法和聚丙烯酰胺凝胶电泳法，对油菜感染黑胫病菌 *Leptosphaeria biglobosa* 后叶内多酚氧化酶（PPO）活性及同工酶进行比较研究。接种后，油菜抗病品种的 PPO 活性呈现先上升后下降的单峰曲线变化；感病品种的 PPO 活性呈现不变，上升、下降，再微升的变化趋势。抗、感两个品种接种处理的 PPO 活性均高于对照处理的活性。黑胫病菌可以诱导油菜抗、感 2 个品种体内的 PPO 活性发生变化，且接种后抗、感 2 个品种的 PPO 活性差异极显著（$P<0.01$）。即 PPO 活性与油菜品种抗病性呈正相关。接种后，抗、感两个品种的 PPO 同工酶谱带数量及染色深度明显不同，与感病品种及对照无菌水处理相比，抗病品种分别于接种后的 48h、72h、96h 增加了相对迁移率 Rf 为 0.46 的 P_2酶带，且抗病品种新增 PPO 酶带的染色深度随着接种时间的延长而增强；即黑胫病菌可诱导抗性品种产生 Rf 为 0.46 的 PPO 同工酶，Rf 为 0.46 的 PPO 同工酶谱带的有无与油菜品种对黑胫病菌的抗性相关。

关键词：油菜；黑胫病菌；多酚氧化酶；酶活；同工酶

8个成株抗叶锈基因在河北省的可利用性研究*

郭惠杰**，闫红飞，郝焕焕，张天野，孟庆芳***，刘大群***
（河北农业大学植物保护学院/国家北方山区农业工程技术研究中心/河北省农作物病虫害生物防治工程技术研究中心，保定　071000）

摘　要：小麦叶锈病是由小麦叶锈菌（*Puccinia triticina*）引起的一种世界性病害，在全世界小麦种植区广泛分布。2015年小麦叶锈病在我国黄淮海麦区发生总体偏重，给小麦的生产带来严重威胁。本实验利用2016年采自河北省的105株小麦叶锈菌株，对成株抗叶锈基因*Lr*12、*Lr*13、*Lr22a*、*Lr22b*、*Lr*34、*Lr*35、*Lr*37和*LrT*3进行成株期鉴定。结果发现，85.7%和86.7%的菌株分别在*Lr22b*和*LrT*3上表现感病，说明这两个基因在河北省利用价值不大；57.1%和53.3%的菌株分别在*Lr22a*和*Lr*37上表现感病，说明这两个基因在河北省可与其他基因联合应用；13.3%和7.6%的菌株在*Lr*12和*Lr*13上表现感病，说明这两个基因在河北省具有有效性；21.9%的菌株在*Lr*34表现潜育期长，在接种后12~14d未出现孢子堆或仅有零星孢子堆；55.2%的菌株表现孢子堆小或少，说明*Lr*34在河北省可以利用；没有发现对*Lr*35有毒性的菌株。

关键词：小麦叶锈病；成株抗叶锈基因；河北省；有效性

---

* 基金项目：国家重点基础研究发展计划（2013CB127700）；河北自然基金项目（C2015204105）

** 第一作者：郭惠杰，在读硕士生研究生，研究方向为植物病理学；E-mail：1558213489@qq.com

*** 通信作者：孟庆芳，副教授，主要从事植物病害生物防治与分子植物病理学研究；E-mail：qingfangmeng500@126.com
刘大群，教授，主要从事植物病害生物防治与分子植物病理学研究

Over-expression of Pokeweed Antiviral Protein Enhances Plant Resistance Against *Tobacco mosaic virus* Infection in *Nicotiana benthamiana*[*]

Zhou Yangkai, Che Yanping, Xu Yujiao, Zhu Feng[**]

(*College of Horticulture and Plant Protection, Yangzhou University, Yangzhou, Jiangsu 225009, China*)

Abstract: Pokeweed antiviral protein (PAP) is members of the type I ribosome inactivating proteins (RIPs) that are isolated from the extracts of pokeweed plant (*Phytolacca americana*). In this report, in order to investigate the role of PAP in plant systemic resistance response against virus infection in *Nicotiana benthamiana*, *N. benthamiana* leaves that overexpressed *PAP* gene (35*S*: *PAP*) was generated through *Agrobacterium*-mediated transient expression. The reduction levels of virus accumulation were detected over a 15-day time course in 35*S*: *PAP*-overexpressing *N. benthamiana* leaves compared with only promoter expressed plants. The results suggest that overexpression of PAP in *N. benthamiana* plants yielded a significant reduction in TMV tagged with the green fluorescent protein (GFP) fluorescence in the noninoculated upper leaves. Furthermore, quantitative real-time PCR (qRT-PCR) analysis showed that the accumulation levels of TMV were significantly reduced in the systemic leaves of 35*S*: *PAP*-overexpressing *N. benthamiana* plants compared with the levels observed in control (35*S*: 00) plants. Overall, our results suggest that overexpression of *PAP* enhances plant systemic resistance against TMV infection in *N. benthamiana*.

* Funding: This work was supported by the National Natural Science Foundation of China (31500209), Natural Science Foundation of the Higher Education Institutions of Jiangsu Province of China (15KJB210007) and Natural Science Foundation of Yangzhou (YZ2015106)

** Corresponding author: Feng Zhu; E-mail: zhufeng@ yzu. edu. cn

Tobacco Alpha-expansin EXPA4 Involves in *Nicotiana benthamiana* Defense Against *Tobacco mosaic virus*

Chen Lijuan, Zou Wenshan, Wu Guo, Lin Honghui, Xi Dehui*

(*Ministry of Education Key Laboratory for Bio-Resource and Eco-Environment, College of Life Science, Sichuan University, Chengdu* 610064, *China*)

Abstract: Expansins are cell-wall-loosening protein, known for their endogenous function in cell wall extensibility during plant growth. Expansin genes are classified into four subfamilies: α-and β-expansins (*EXPA* and *EXPB*, respectively), expansin-like A (*EXLA*) and expansin-like B (*EXLB*). Effects of expansins on plant growth, developmental processes and environment stress responses were well studied. By contrast, relatively limited information is available on expansins regulation of plant response to pathogen attack. Only several studies reported that plants response to pathogen may involve changes in the expression levels of expansin genes. For instance, Cantu *et al.* identified that simultaneous suppression of polygalacturonase (*LePG*) and *LeExp*1 dramatically reduced the susceptibility of ripening fruit to necrotrophic fungi *Botrytis cinerea*. The absence or down-regulation of *AtEXLA*2 leads to increased resistance of *Arabidopsis* to *B. cinerea* in a CORONATINE INSENSITIVE 1 (COI1) -dependent manner. By contrast, the roles of expansin genes in plant defense response to virus pathogen have not been studied in any detail.

In this study, virus-induced gene silencing (VIGS) combined with agrobacterium-mediated transient overexpression method were conducted to investigate the role of *Nicotiana tabacum* alpha-expansin 4 (*EXPA*4) in modulating *Tobacco mosaic virus* (TMV-GFP) resistance in *Nicotiana benthamiana*. Results indicated that silencing of *EXPA*4 reduced the sensitivity of *N. benthamiana* to TMV-GFP, and overexpression of *EXPA*4 accelerated the virus reproduction on inoculated plants. In addition, our data suggested that the effects of *EXPA*4 on tobacco defense against virus attack were caused by affecting the antioxidative metabolism and phytohormone-mediated immunity responses in tobacco. The changes of *EXPA*4 expression has more effects on SA and ET not JA in *N. benthamiana* induced by TMV-GFP. The expressions of SA and ET-mediated defense genes were also significantly affected by *EXPA*4, indicating that the effects of *EXPA*4 on the susceptibility of *N. benthamiana* to TMV-GFP have relation to the changes of the defense pathway mediated by SA or ET. Furthermore, VIGS approach combining with exogenous phytohormone treatments suggested that *EXPA*4 have different responses to different phytohormones. Taken together, these results presented here is the first report of expansins involved in tobacco defense against virus attack.

Key words: Expansins; *Tobacco Mosaic Virus*; *Nicotiana benthamiana*; Plant defense

---

* Corresponding author: Xi Dehui; E-mail: xidh@scu.edu.cn

31 个小麦品种（系）抗叶锈性鉴定*

安 哲**，张 涛，李建嫄，刘 鹏，张瑞丰，杨文香***，刘大群***
（河北农业大学植物保护学院/国家北方山区农业工程技术研究中心/河北省农作物病虫害生物防治工程技术研究中心/植物病理系分子植物病理学实验室，保定 071001）

摘 要：为探测 31 个小麦育种材料的抗叶锈基因，选用 12 个有鉴别力的中国小麦叶锈菌菌系在苗期对测试品系进行抗叶锈基因推导分析，并对这些材料进行成株抗叶锈性鉴定与慢叶锈性筛选。结果显示，藁麦 8911、B466、949-14-3、小偃 166、陕优 225、陕 515、A111、ZB3、92R137、陕农 981、ZB5、ZB9、9925、西农 335 这 14 个品种中可能含有 *Lr*1、*Lr*2*a*、*Lr*11、*Lr*17、*Lr*18、*Lr*25、*Lr*30、*Lr*3*bg* 与 *Lr*50 这 9 个已知苗期抗叶锈基因的一个或多个，另外周麦 18、贵 15（异白）、兴育 7 号、区 23、GF-19、ZA4、陕 538、L6001-2、4163、浚麦 35 这 10 个品种含有与供试的已知基因不同的抗性基因，周优 102、豫教 0388、L201、贵农 775、113、贵农 13、方穗小偃 22 这 7 个品种中不含有对 12 个供试锈菌苗期抗病的基因。田间抗叶锈性鉴定发现周优 102、藁麦 8911、贵 15 异白、L201、贵农 775、陕优 225、113、4163、ZB3、92R137、贵农 13、陕农 981、陕 538、ZB5、ZB9、方穗小偃 22、9925、L6001-2、西农 335 为慢叶锈性品系材料，可用于小麦抗病育种。

关键词：小麦叶锈病；叶锈菌；慢锈性；基因推导

---

* 基金项目：国家重点基础研究发展计划（2013CB127700）；河北省现代农业产业体系小麦产业创新团队建设项目（HBCT2013010204）；河北省高等学校科学技术研究项目（Z2014065）

** 第一作者：安哲，硕士研究生，研究方向为分子植物病理学；E-mail：645766371@ qq. com
*** 通信作者：杨文香，教授，研究方向为分子植物病理学；E-mail：wenxiangyang2003@ 163. com
刘大群，教授；E-mail：ldq@ hebau. edu. cn

黑龙江省主栽马铃薯品种及重要育种材料对北方根结线虫抗性评价*

毛彦芝[1**]，李春杰[2]，胡岩峰[2]，华　苹[2]，尤　佳[2]，
王鑫鹏[2]，刘喜才[1]，白彦菊[1]，王从丽[2***]
（1. 黑龙江省农业科学院克山分院，克山　161606；2. 中国科学院东北地理与农业生态研究所，黑土区农业生态重点实验室，哈尔滨　150081）

摘　要： 根结线虫（*Meloidogyne* spp.）是一种寄主范围广泛，为害马铃薯的重要的植物寄生线虫。为明确黑龙江省主栽马铃薯品种和重要育种资源对北方根结线虫病害的抗感反应，本研究选用了黑龙省主栽马铃薯品种和部分育种材料共 28 份进行温室盆栽试验，随机区组设计，5 次重复。结果表明：北方根结线虫在这 28 份材料上都能繁殖，且繁殖系数（RF）均大于 1（1.53~36.93）。白头翁、会-2 和红参的繁殖系数分别为 2.41、2.02 和 1.53，与青薯 3 号（36.93）和 608Kennebec（37.9）差异显著（$P<0.05$）；线虫在参试的马铃薯植株根上繁殖的卵块数量范围为每克鲜重 26.6~217.4 个，根据卵块数量分级方法，白头翁（3 级）为抗病材料；植株上的虫瘿指数（Galling Index）范围为 1.5~6.8，根据评价虫瘿指数方法，白头翁（1.8）、会-2（1.5）和红参（2.2）为抗病材料。综合块茎评价结果在参试的 28 份材料中白头翁是较抗北方根结线虫的材料，会-2 和红参的根部对北方根结线虫具有一定的耐病性，可作为培育抗病品种资源，感病材料青薯 3 号和 608Kennebec 可作为筛选材料的感病对照。

关键词： 马铃薯；主栽品种及育种材料；北方根结线虫；抗性评价

---

* 基金项目：黑龙江省农业科技创新工程院级科研项目（Q001）；中国科学院百人计划
** 第一作者：毛彦芝；E-mail：kshpotato@163.com
*** 通信作者：王从丽；E-mail：wangcongli@iga.ac.cn

东乡野生稻对稻曲病抗性研究*

胡建坤**，黄　蓉，华菊玲，李湘民，黄瑞荣***
（江西省农业科学院植物保护研究所，南昌　330200）

摘　要：东乡野生稻（*Oryzae sativa* f. *spontanea* Roschev.，OSFSR）是迄今为止我国乃至全球分布最北的普通野生稻，因其具有遗传多样性及蕴藏着耐寒、耐旱、耐瘠、广亲和性、胞质雄性不育、抗病虫等有利基因，对当代水稻的研究有着举足轻重的作用。东乡野生稻抗水稻细菌性条斑病和白叶枯病已有报道，但对稻曲病的抗性至今不为人所知。为此，作者通过人工接种方法对东乡野生稻抗稻曲病进行了研究。

将东乡野生稻南昌异地圃保存的 3 个居群 9 个群落的 200 个单株繁育后代移栽到稻曲病常发稻田，单本栽培，株行距 60cm×100cm，常规管理。以两优培九和两优 036 为稻曲病接种感病对照品种。采用马铃薯蔗糖液体培养基摇培稻曲病菌。以稻曲病菌菌丝片段-孢子混合液（孢子液浓度为 5×10^5个/mL）为接种体，在东乡野生稻稻剑叶与倒二叶距离为 2~3cm 时，采用人工注射接种法，接种东乡野生稻 3 个居群 9 个群落的单株繁殖材料，共接种 88 份。

鉴定结果表明不同单株繁殖材料对稻曲病抗性差异明显，存在抗病、中抗、中感和感病等 4 种类型，无免疫、高抗或高感材料。参鉴材料中抗病和中抗材料分别占 46. 6%和 29. 5%，以抗病类型居多，占参鉴材料的 76. 1%；感病和中感材料分别占 10. 3%和 13. 6%，感病类型占参鉴试材料的 23. 9%。结果表明：东乡野生稻群体对稻曲病抗性较强，具有抗稻曲病特质，有望在其中挖掘到抗稻曲病优异种质，具有进一步研究开发利用价值。

东乡野生稻不同群落的单株后代在南昌异地圃自然条件下生长，观察多年尚未发现稻曲病株。东乡野生稻的农艺性状与栽培稻有显著差异，其抗稻曲病机理目前尚不清楚，因此，研究东乡野生稻抗稻曲病机制与抗性生理具有重要意义。

关键词：东乡野生稻；稻曲病；抗性研究

---

* 基金项目：江西省科技支撑计划（2012BBF60004）；江西省农业科学院创新基金项目（2014CQN007）
** 第一作者：胡建坤，助理研究员，硕士，微生物学；E-mail：hjk0206@ 126. com
*** 通信作者：黄瑞荣，研究员；E-mail：huangruirong073@ 163. com

高品质樱桃番茄对晚疫病的抗性评价*

叶思涵[1**]，董　鹏[2]，孙现超[1***]

（1. 西南大学植物保护学院，重庆　400715；2. 重庆市农业技术推广总站，重庆　400020）

摘　要：樱桃番茄，又名珍珠番茄，是一种以鲜食为主的新型蔬果，外观优美，食味甜脆、口感诱人，以其果似樱桃、色泽鲜红、营养丰富而深受广大消费者所喜爱。随着人们生活水平的迅速提高，市场对高品质樱桃番茄的需求量急剧增加，栽培樱桃番茄的经济效益十分可观。

设施条件下樱桃番茄的产量和品质都较大田有很大提升，因而在重庆多地均有种植。但樱桃番茄是一种抗病能力较差的茄科植物，其中晚疫病是危害设施樱桃番茄较为严重的一种病害，病原物是致病疫霉（*Phytophthora infestans*），传染性强，通常会造成樱桃番茄减产10%～50%，严重时甚至使植株坏死而导致绝收。该病在20～30℃和空气相对湿度为80%以上时容易发生，通常从植株下部叶片开始发病，产生圆形或者不规则形边缘不明显的暗绿色或褐色水渍状病斑，温湿度适宜时叶片病健交界处会出现稀疏白霉，随后病叶腐烂，空气干燥时病部青白色，叶片变得脆而易破。果实发病多在青果期，近果柄处有深褐色的云纹状硬斑，潮湿时病果表面可产生稀疏白霉。根据国家番茄品种试验调查标准，笔者对重庆市潼南区国家农业科技园区的大棚春植高品质樱桃番茄无土基质栽培区的晚疫病发生情况进行了观测与调查。园区目前主栽品种主要有6种，分别为冬美、黑宝石、千禧、佳西娜、珍海299、QQ黄钻。以上品种从播种到第一穗果成熟均为100d左右，根据往年晚疫病发生情况，5月中旬至6月常为发病高峰期，今年的樱桃番茄于3月中旬播种，5月上旬移栽，5月下旬开花，进入易感病生育期，此时正值南方的梅雨季节，晚疫病于今年5月中旬普遍发生。各品种病情调查采用对角线五点取样法进行取样，每点10株，共50株。调查结果显示，各品种平均发病率，珍海299最高，发病率为86%，其次是千禧和黑宝石，分别为68%和60%，此外佳西娜和QQ黄钻均为16%，冬美最低为12%；各品种病情指数，千禧最高，病情指数为68，其次是珍海299，为48，此外黑宝石为31.33，佳西娜和QQ黄钻均为16，冬美最低为12。根据番茄晚疫病群体或品种抗病性划分标准：I（免疫）：0；HR（高抗）：0～10.0；R（抗）：10.1～30.0；MR（中抗）：30.1～50.0；S（感）：50.1～70.0；HS（高感）：≥70.1可知，供试的6个樱桃番茄品种中，属于抗晚疫病的品种有3种，分别为冬美、佳西娜和QQ黄钻，其中冬美抗晚疫病抗性较好；属于中抗晚疫病的品种有2种，分别为黑宝石和珍海299，且黑宝石抗性优于珍海299；属于感晚疫病的品种仅1种，即千禧。

---

* 基金项目：重庆市社会事业与民生保障创新专项（cstc2015shms-ztzx80011，cstc2015shms-ztzx80012）；中央高校基本科研业务费专项资金资助项目（XDJK2016A009，XDJK2017C015）

** 第一作者：叶思涵，硕士研究生，从事植物病理学研究

*** 通信作者：孙现超，研究员，博士生导师，主要从事植物病毒学及植物病害生物防治研究；E-mail：sunxianchao@163.com

剑麻防御素基因的克隆与表达分析*

黄　兴[1**]，汪　涵[2]，梁艳琼[1]，习金根[1]，郑金龙[1]，
贺春萍[1]，吴伟怀[1]，李　锐[1]，易克贤[1***]
（1. 中国热带农业科学院环境与植物保护研究所，农业部热带农林有害生物入侵检测与控制重点开放实验室，海南省热带农业有害生物检测监控重点实验室，海口　571101；2. 海南大学热带农林学院，海口　570228）

摘　要：植物防御素是植物中广泛存在的抗菌活性物质，在真菌、细菌和昆虫入侵植物过程中对其分泌物均有抑制作用。本研究根据NCBI数据库中检索到的多种植物中的防御素基因序列设计简并引物，从剑麻中克隆获得1段防御素基因片段，后采用侧翼序列克隆方法获得其完整编码序列。其开放阅读框具有237bp，编码78个氨基酸，含有8个保守Cys残基，具备Knot1功能域。系统进化分析表明AsPDF1与芦笋、油棕防御素基因亲缘关系较近，可能具有蛋白酶抑制活性。荧光定量分析结果显示，该防御素基因在剑麻斑马纹病病原菌疫霉入侵过程中表达水平显著提高，表明该基因具有一定的抑菌活性。

关键词：剑麻；防御素基因；表达

* 基金项目：国家麻类产业技术体系建设项目（CARS-19-E17）；国家自然科学基金项目（31371679）

** 第一作者：黄兴，博士，助理研究员，研究方向：分子抗性育种；E-mail：huangxingcatas@126.com

*** 通信作者：易克贤，博士，研究员，研究方向：分子抗性育种；E-mail：yikexian@126.com

几种植物源活性物质的抑菌、抗病毒效果及其机理研究*

张　旺[1**]，薛　杨[1]，李　斌[2]，张　攀[1]，陈德鑫[3]，孙现超[1***]

（1. 西南大学植物保护学院，重庆　400716；2. 中国烟草总公司四川省公司，成都　610000；3. 中国农业科学院烟草研究所，青岛　266101）

摘　要：寻找植物源活性物质是天然产物农药研发的重要途径。本文结合相关研究成果和报道，广泛搜集可能具有抗病毒或抑菌活性的植物源物质，研究其对烟草赤星病菌［*Alternaria alternata*（Freis）Keissler］和烟草花叶病毒（*Tobacco mosaic virus*，TMV）的抑制效果，并初步探讨其作用方式和机制。首先，以烟草花叶病毒为靶标，对收集到的15种植物源物质进行抗病毒活性测定，经过初筛、复筛并结合防御酶活性情况，筛选出熊果酸与4-甲氧基香豆素2种能够降低病毒初侵染含量的活性物质。研究其作用方式与机制，结果显示，熊果酸与4-甲氧基香豆素能够不同程度提高烟草中SOD、PPO、POD、CAT防御酶活性，降低烟叶中过氧化物含量；在病毒侵染前期降低植物细胞的膜脂过氧化程度，减轻细胞膜受损；并且能够提高PR1、NPR1和PR2等防卫反应相关基因表达量，促进病程相关蛋白的合成，从而降低病毒初侵染含量，具有保护和延缓病程的作用。体外作用研究发现，2种活性物质与TMV混合1 h后，病毒粒体发生了明显的聚集现象，而溶剂处理后的病毒分散均匀无聚集，其中熊果酸处理后的聚集程度较4-甲氧基香豆素更为明显。将处理后的病毒摩擦接种至本氏烟，荧光的斑点数量和强度都有所降低，说明2种物质能在体外直接作用病毒粒体，降低病毒的侵染活性或体内复制水平，可能对病毒具有一定钝化作用。其次，利用平板抑菌法，以烟草赤星病菌为靶标，测定15种植物源物质的抑菌活性，发现丁香油与丁香油酚具有较好的抑菌作用。丁香油含量为4.13%时能够强效抑制赤星病菌孢子的萌发，丁香油含量为0.52%时显著降低菌丝的生长速度，减少菌丝的生长量。进一步研究作用机理发现，丁香油的作用位点之一是细胞膜，其能破坏赤星病菌的细胞膜结构，增加质膜通透性，使细胞内含物外渗，并提高膜脂过氧化程度，增加丙二醛含量，致使菌丝细胞死亡，抑制病原菌的生长和繁殖。

关键词：活性物质；抑菌；抗病毒

---

* 基金项目：国家自然科学基金（31670148）；中央高校基本科研业务费专项资金资助项目（XDJK2016A009，XDJK2017C015）；中国烟草总公司四川省公司科技项目（SCYC201703）

** 第一作者：张旺，硕士研究生，从事植物病理学研究

*** 通信作者：孙现超，研究员，博士生导师，主要从事植物病毒学及植物病害生物防治研究；E-mail：sunxianchao@163.com

两种草莓抗连作障碍突变体的田间抗性评价*

贾　薇**，苏　媛[1]，罗　敏[1]，齐永志[1]***，甄文超[2]***
（1. 河北农业大学植物保护学院，保定　071001；2. 河北农业大学农学院，保定　071001）

摘　要：连作障碍严重制约了草莓生产的可持续发展。以两种草莓抗连作障碍突变体‘DA-133’与‘TZ-125’为试材，通过田间试验研究连作条件下两种突变体与各自野生型（达赛莱克特，DA和童子一号，TZ）的生长发育状况及土壤微生物变化特征，旨在明确两种草莓突变体的抗连作障碍能力。结果表明：①果实收获期，除单株叶片数外，TZ-125各生长指标均优于野生型，其茎粗、地上部鲜重、地上部干重、根系鲜重和根系干重分别是TZ的1.22倍、1.19倍、1.20倍、2.08倍和1.90倍。TZ-125实际产量比TZ提高17.4%，而DA-133实际产量也比DA提高13.8%。222kg/hm^{2}氯化苦用量熏蒸地块，两突变体各生长指标及产量均与300kg/hm^{2}氯化苦用量（生产田常规熏蒸剂量）熏蒸地块其各自野生型无显著差异；②在促成栽培地块，果实膨大期，300kg/hm^{2}氯化苦用量熏蒸处理DA和TZ根际土壤可培养真菌数量分别比DA-133和TZ-125高67%和40%。从现蕾期开始，222kg/hm^{2}氯化苦用量熏蒸地块DA-133与TZ-125根据土壤与根表真菌总量均分别与DA和TZ 300kg/hm^{2}氯化苦用量熏蒸地块差异不明显。除定植期外，各时期300kg/hm^{2}氯化苦用量熏蒸DA-133和TZ-125地块根际土壤可培养细菌数量均明显高于野生型相同处理。且222kg/hm^{2}氯化苦用量熏蒸处理DA-133和TZ-125根际土壤细菌总量均与DA、TZ 300kg/hm^{2}氯化苦用量熏蒸处理相当。不同氯化苦用量处理草莓根际土壤和根表可培养放线菌动态变化与细菌变化基本相同。

关键词：草莓；抗连作障碍突变体；生长发育；可培养土壤微生物；抗性评价

---

* 基金项目：国家自然科学基金项目（31140064）
** 第一作者：贾薇，硕士研究生，主要从事植物病理生态学研究；E-mail：18833206551@163.com
*** 通信作者：齐永志，博士，讲师，主要从事植物病理生态学研究；E-mail：qiyongzhi1981@163.com
甄文超，博士，教授，主要从事农业生态学与植物生态病理学研究；E-mail：wenchao@hebau.edu.cn

Differences in the Production of the Pathogenesis-related Protein NP24 in Fruits*

Tong Zhipeng**, Wang Zehao, Liang Yue***

(*College of Plant Protection*, *Shenyang Agricultural University*, *Shenyang* 110866)

Abstract: Pathogenesis-related (PR) proteins are produced in response to biotic and abiotic stresses, and are important for plant defenses that mediate resistance against environmental stresses. The PR-5 protein has been identified as a thaumatin-like protein based onthe fact its sequence is highly similar to that of a sweet-tasting thaumatin from *Thaumatococcus daniellii*. A previous study revealed that NP24, which belongs to the same family as PR-5, is specifically produced in tomato roots. In the current study, NP24 was identified in fresh tomato fruits using SDS-PAGE and MALDI-TOF-MS analyses. An examination of the fruit structures indicated that NP24 was most abundant in the exocarp. Moreover, a phylogenetic analysis involving proteins with an amino acid sequence similar to that of tomato NP24 suggested the existence of five families among the analyzed plants. These families consisted of subclades or included only one taxon [i. e., *Solanaceae* (with 100% bootstrap similarity), *Vitaceae* (single taxon), *Actinidiaceae* (single taxon), *Rutaceae* (single taxon), and *Sapindaceae* (single taxon)]. The family *Rosaceae* appears to be polyphyletic, with three species being identified in this study. Therefore, NP24may be widely distributed in fruits from *Solanaceae* and other plant species. Meanwhile, NP24 was detected in pepper (*Capsicum annuum*), potato (*Solanum tuberosum*), black nightshade (*Solanum nigrum*), blueberry (*Vaccinium corymbosum*), persimmon (*Diospyros kaki*), and cherry (*Cerasus pseudocerasus*), but not in Goji (*Lycium barbarum*), ginkgo (*Ginkgo biloba*), and grape (*Vitis vinifera*) based on SDS-PAGE results (i. e., presence or absence of a band at the expected size). In summary, the NP24 is localized to specific tissues of tomatoes and other fruits. The presented results may provide novel insights into the PR-5 protein and post-harvest disease management.

* 基金项目：沈阳农业大学引进人才科研基金（20153040）

** 第一作者：佟志鹏，硕士研究生，研究方向为植物病理学；E-mail：tongzhipeng0727@163.com

*** 通信作者：梁月，教授，博士生导师；E-mail：liangyuet@126.com

十字花科蔬菜品种对根肿菌 4 号和 9 号小种的抗性分析*

黄　蓉**，胡建坤，黄瑞荣***，叶敬秋

（江西省农业科学院植物保护研究所，南昌　330200；江西省婺源县溪头乡农业综合服务站，婺源　333214）

摘　要：鞭毛菌亚门芸苔根肿菌侵染导致的十字花科蔬菜根肿病是世界性的植物土传真菌病害。为满足江西省十字花科蔬菜生产对抗根肿病品种的需求，降低根肿病的发生及为害，本试验根据江西省内根肿病菌的生理小种类群分布，分别在根肿病发病地区婺源和南昌设立 4 号生理小种和 9 号生理小种的田间病圃，对 223 个小白菜、大白菜、花菜、萝卜、油菜品种的抗病性进行田间抗性鉴定。结果表明：56 个小白菜品种，除紫冠一号表现为抗病外，其余品种对根肿菌 4 号小种均表现为感病；而对根肿菌 9 号小种的抗性表现则存在差异，其中上海黑叶五月慢表现高抗，上海七宝青等 6 个品种表现抗病。42 个大白菜品种，除北京桔红心对根肿菌 4 号小种表现为抗病外，其余品种均为感病；而对根肿菌 9 号小种的抗性则表现为北京大牛心等 7 个品种免疫，京春绿等 4 个品种高抗，京研快菜 4 号等 3 个品种抗病。49 个花菜和 6 个萝卜品种的整体抗性水平都较强，对 9 号生理小种全部表现为免疫；对 4 号生理小种，除 16 个花菜品种和 2 个萝卜品种表现为感病，其余均表现抗病或高抗。根肿菌 9 号生理小种对 70 个油菜品种全部表现为免疫；而 4 号生理小种对油菜的抗性表现则存在差异，5K83 等 3 个品种表现为抗病，其余品种均为感病。以上结果为科学评判十字花科作物不同种类及同种蔬菜不同品种对根肿病的抗性水平，指导根肿病区调整品种布局，以及优化十字花科蔬菜种植结构奠定了基础。

关键词：根肿菌；抗性鉴定；品种布局

---

* 基金项目：国家公益性行业（农业）科研专项（201003029）；江西省科技支撑计划项目（20151BBF60068）；江西省科研院所基础设施配套项目（20151BBA13034）；江西现代农业科研协同创新专项（JXXTCX2015005-004）；江西省现代农业蔬菜产业体系病虫害防治岗位专项（JXARS）；江西省重点研究计划项目（20161ACF60015）；南昌市科技支撑计划项目（洪财企［2012］80 号）

** 第一作者：黄蓉，副研究员，植物病害防治；E-mail：huangrong8229@163.com

*** 通信作者：黄瑞荣，研究员，植物病害防治；E-mail：huangruirong073@163.com

水稻白叶枯病抗性资源筛选及遗传分析

杨　军，郭　涛，王海凤，薛芳，张士永
（山东省水稻研究所，济南　250100）

摘　要：白叶枯病是严重为害水稻生产的重要病害之一。通过发掘抗性基因培育抗病品种是控制白叶枯病的有效途径。为获得抗白叶枯病种质资源，利用强致病菌株 PXO99 对 331 个水稻品种（系）进行剪叶法人工接种鉴定，并对鉴定出的抗病材料进行基因分析。结果表明：水稻材料 Y1623 对 PXO99 表现抗病，其他 330 个品种（系）对 PXO99 高度感病；通过分子标记对 Y1623 分析确定其不含有已知的抗 PXO99 基因，推测 Y1623 含有新的抗病基因，是潜在的抗病资源。本研究结果为白叶枯病抗病育种工作提供了借鉴。

关键词：白叶枯病；PXO99；资源筛选；遗传分析

水稻转录因子 WRKY45 的重组表达、纯化和晶体生长*

程先坤，赵彦翔，蒋青山，彭友良，刘俊峰**
（中国农业大学植物保护学院植物病理学系，北京 100193）

摘 要：WRKY 家族是高等植物中十个最大的转录因子家族之一，它们在绿色植物中广泛存在。水稻 *WRKY* 基因的功能广泛地涉及了由病原细菌和真菌等引起的病害反应及水杨酸（SA）和茉莉酸（JA）相关的信号通路，在水稻抗病虫害及抗逆过程中发挥着重要的作用。因此，研究 WRKY 转录因子的作用机理具有理论和实践意义。OsWRKY45 是水稻中第 III 组 WRKY 转录因子，拥有 1 个典型的 WRKY 结构域和 C2HC 型锌指结构。其中 WRKYGQK 保守序列的和锌指结构在这一相互作用中起到重要的作用。研究表明尽管 W-box（T/CTGACC/T）的核心序列一致，但是邻近的侧翼序列对 WRKY 蛋白核酸位点的偏好性也有影响。虽然有单独 WRKY 晶体结构的信息，相互作用的信息尚不完整等本研究利用原核表达系统，将编码 OsWRKY45 全长和截断体基因分别构建到不同的表达载体，并将构建好的载体分别转化到大肠杆菌 BL21（DE3）、Rosetta-gami（DE3）等表达菌株表达。通过小量诱导、试纯化筛选到合适的载体、表达菌株和纯化条件。在此基础上，通过亲和层析、离子交换和分子筛得到稳定、均一的蛋白，符合晶体生长。同时，利用凝胶迁移实验（Electrophoretic mobility shift assays，EMSA）分析了与 WRKY45 特异性结合的 DNA 序列，并将筛选到特异结合的 DNA 与 WRKY45 蛋白样品混合制备蛋白质和 DNA 的复合物，利用气相扩散法对复合物进行晶体生长，获得了复合物晶体，并收集了衍射数据，正在进行晶体结构解析和 WRKY45 与 W-box 相互作用原理以及在 DNA 结合序列上偏好性的分析，以解释 W-box 侧翼序列对 WRKY 蛋白核酸位点的偏好性的影响。

关键词：稻瘟病；OsWRKY45；W-box；重组蛋白

---

* 第一作者：程先坤，博士研究生，植物病理学专业；E-mail：cxk20130808@126.com
** 通信作者：刘俊峰，教授；E-mail：jliu@cau.edu.cn

Relationships among Deadwood Symptom Grades, Salicylic Acid Content and Diameter at Breast Height of *Malus sieversii* in China*

Yu Shaoshuai**, Zhao Wenxia***,
Yao Yanxia, Huai Wenxia, Xiao Wenfa

(*Key Laboratory of Forest Protection of State Forestry Administration, Research Institute of Forest Ecology, Environment and Protection, Chinese Academy of Forestry, Beijing* 100091, *China*)

Abstract: *Malus sieversii* distributed in Xinjiang, China is of serious exsciccation and its living condition is severely threatened. Relationships among deadwood symptom grades, salicylic acid (SA) content and diameter at breast height (DBH) of *M. sieversii* plants collected in the same sampling site and habitat conditions was analyzed to indicate phenotypic features of *M. sieversii* that maybe related to the symptoms of exsciccation. According to the deadwood symptom, *M. sieversii* under the same sampling area and habitat was divided into 6 grades. The difference of salicylic acid composition and content in different symptom grades of *M. sieversii* was investigated by ultrasonic extraction, HPLC and statistics methods. The SA content in the different individuals was in the range of 3.61~20.25mg/100g, with significant difference ($P<0.05$). The SA content in samples of different deadwood symptom grades was in order of Ⅴ grade > 0 grade > Ⅳ grade > Ⅲ grade > Ⅰ grade ($P<0.05$). The variation coefficient of DBH of *M. sieversii* was in order of Ⅴ grade > 0 grade > Ⅱ grade > Ⅳ grade > Ⅲ grade > Ⅰ grade, basically same as the trend of the change of SA content. There was a negative correlation between SA content and DBH of *M. sieversii* ($P>0.05$), which was similar to that of the samples with symptom of deadwood rate <40% ($P>0.05$). While there was positive correlation between SA content and DBH of the samples with symptom of deadwood rate >40% ($P>0.05$). There was a certain correlation among deadwood symptom grades, SA content and DBH of *M. sieversii*. The results achieved in the study would provide some reference clues for ecological conservation and restoration of wild fruit forest in Tianshan Mountains and screening and cultivation of high-quality varieties of cultivated apple.

Key words: Wild apple; Deadwood rate; Correlation analysis; Stress resistance

---

* Funding: National Key R&D Plan (2016YFC0501503)

** First author: Yu Shaoshuai, PhD, engaged in plant pathology; E-mail: hzuyss@163.com

*** Corresponding author: Zhao Wenxia, Prof, engaged in forest pest quarantine & forest pathology; E-mail: zhaowenxia@caf.ac.cn

小麦病程相关蛋白 TaLr19TLP1 的 Gateway 克隆载体的构建

王　菲，梁　芳，张艳俊，张家瑞，王海燕，刘大群
（河北农业大学植物保护学院，河北省农作物植物病虫害生物防治工程技术研究中心，
国家北方山区农业工程技术研究中心，保定　071000）

摘　要：病程相关蛋白 *PR*5，属类甜蛋白家族（Thaumatin-like proteins，TLPs），前期笔者对该基因结构域进行预测，综合结果分析该基因有 95.1%的可能性为胞外蛋白，含有信号肽。现笔者将利用 Gateway 技术构建小麦病程相关蛋白 *TaLr*19*TLP*1（不含信号肽）和 *TaLr*19*TLP*1-*SP*（含信号肽）的克隆载体。依据 Gateway 中定向克隆 TOPO 技术的要求设计引物，将目的基因连接到 pENTR™/D-TOPO（入门载体）上，并对终载体 pEarleyGate 103（PEG103）进行酶切改造，使目的基因在 LR 酶的作用下，连接到 PEG103 上，形成重组载体 TaLr19TLP1-PEG103 和 TaLr19TLP1-SP-PEG103，将二者转化到农杆菌 GV3101 中，通过烟草瞬时表达技术，使用激光共聚焦显微镜进行亚细胞定位的观察。亚细胞定位结果显示，*TaLr*19*TLP*1 定位于细胞膜，而 *TaLr*19*TLP*1-*SP* 通过质壁分离实验验证，该基因定位于胞外。Western blot 证明，笔者构建的融合蛋白在烟草叶片中可以表达。本实验中使用的 Gateway 技术不需进行 BP 反应，只需要一套体系和一支 LR 酶即可完成，打破了传统的酶切连接转化的方法构建载体，更加方便快捷。

关键词：小麦；Gateway 克隆技术；亚细胞定位技术；Western blot

异源表达几种核糖体失活蛋白及对TMV的抑制作用研究*

魏周玲**，蒲运丹，薛　杨，叶思涵，张永至，吴根土，青　玲，孙现超***
（西南大学植物保护学院，重庆　400715）

摘　要：植物病毒病使主要的经济作物受到严重的危害，仅次于真菌病害，广泛存在于世界范围内。近年的研究表明，将核糖体失活蛋白（ribosome-inactivating proteins，RIPs）基因转化植物，转基因植物会具有抗植物病毒能力。因此，本文根据 Gen Bank 中已报道的商陆抗病毒蛋白基因、丝瓜核糖体失活蛋白基因和苦瓜素基因全序列，设计并合成扩增 *PAP-c* 和 *α-MC*、*MAP*30、*Luffin-a* 基因全长引物，通过 RT-PCR 及基因克隆方法，克隆得到基因 *α-MC*、*MAP*30、*Luffin-a* 和 *PAP-c*，分别为 861bp、861bp、831bp 和 711bp。Wolf PSORT 预测显示 RIPs 主要定位于细胞质膜上。用共聚焦荧光显微镜下观察发现，标记 GFP 的 RIPs 均定位在本氏烟叶片表皮细胞质膜上，与 Wolf PSORT 预测的 RIPs 定位结果相一致。异源表达 4 种 RIPs 的植物细胞无明显毒性，表达部位细胞完整。在本氏烟中异源表达 RIPs 后，再接种 TMV-GFP，在紫外灯下观察发现 4 种 RIPs 处理后的本氏烟在接种 TMV-GFP 48h 后没有出现绿色荧光，而对照组出现荧光。72h 后处理组出现零星荧光，但对照组的绿色荧光开始扩散，连续观察，处理组几乎没有变化，接种 TMV-GFP 6d 后，发现处理组的绿色荧光几乎没有扩大的趋势，而对照组的绿色荧光已扩散至心叶；ELISA 检测表明，在接种 TMV-GFP 6d 后的叶片中，对照组与健康植物的 OD_{492} 比值几乎已达到处理组的 10 倍以上；实时荧光定量 PCR 检测 TMV RNA 的含量，结果显示对照组 TMV RNA 表达量是处理组的 149 倍左右，表明 4 个 RIPs 对 TMV 复制和移动都有明显抑制；实时荧光定量 PCR 结果分析显示，*NPR*1 基因在只注射 TMV、单独分别表达 α-MC、MAP30、Luffin-a 以及分别表达 α-MC、MAP30、Luffin-a 后注射 TMV 的本氏烟中都被诱导表达，但后者的表达量是前两个处理的约 1.5~3 倍，在只分别接种 α-MC、MAP30、Luffin-a 和分别表达 α-MC、MAP30、Luffin-a 后注射 TMV 的本氏烟中都检测到 *PR*1、*PR*2，但后者的表达量显著高出前者约 5~7 倍，表明异源表达 RIPs 对植物病毒的抗性作用可诱导植物中防卫相关基因 *NPR*1、*PR*1、*PR*2 的表达，从而引起更强的防御反应。异源表达 RIPs 显著抑制 TMV，能够激活植物防卫反应，且对植物细胞无明显毒性。研究结果为利用异源表达 RIPs 方法开发控制植物病毒新产品提供了参考依据。

关键词：植物病毒；异源表达；TMV；抑制

* 基金项目：国家自然科学基金（31670148）；重庆市社会事业与民生保障创新专项（cstc2015shms-ztzx80011，cstc2015shms-ztzx80012）；中央高校基本科研业务费专项资金资助项目（XDJK2016A009，XDJK2017C015）；中国烟草总公司四川省公司科技项目（SCYC201703）

** 第一作者：魏周玲，女，硕士研究生，从事植物病理学研究

*** 通信作者：孙现超，博士，研究员，博士生导师，主要从事植物病毒学及植物病害生物防治研究；E-mail：sunxianchao@163.com

云芝多糖诱导免疫本氏烟抗烟草花叶病毒的转录组研究*

谢咸升[1,2]**，陈雅寒[2]，安德荣[2]***，陈　丽[1]，张红娟[1]
（1. 山西省农业科学院小麦研究所，临汾　041000；2. 旱区作物逆境生物学国家重点实验室/西北农林科技大学植物保护学院，杨凌　712100）

摘　要：为解释云芝多糖抗植物病毒活性，本研究选择云芝多糖为药剂研究对象，本氏烟为寄主植物，最具代表性的烟草花叶病毒（TMV）为防治对象，设置喷清水对照（T01）、喷云芝多糖（T02）、接 TMV 前 24h 喷施云芝多糖（T03）、接 TMV（T04）、接 TMV 后 24h 喷施云芝多糖（T05）5 个处理，3 次重复，处理 72h 后混样提取 RNA 进行转录组测序。试验结果表明：本氏烟各处理样品共得到 20.56Gb Clean Data，Q_{30}碱基百分比均大于 88.54%，GC 含量 44.04%。Clean Reads 与本氏烟参考基因组比对效率从 75.68%到 85.92%，Uniq Mapped Ratio 与参考基因组比对效率从 71.21%到 79.18%，新发掘出 4 万多个 SNP 位点，转换（62.5%）多于颠换（37.5%），多位于基因区，杂合型约占 58%。新预测约 31 000个可变剪接事件，其中喷施云芝多糖（T02）与接 TMV 前 24h 喷云芝多糖（T03）相比差异不显著，但比清水对照（T01）多 845 个；接 TMV（T04）与接 TMV 后 24h 喷施云芝多糖（T05）相比差异不显著，但比清水对照（T01）多 3 128个。共优化 22 783个本氏烟基因结构，发掘出 2 837个新基因，其中 1 522个获得注释。通过转录本定量分析，共获得 10 个差异基因表达集。与清水对照（T01）相比，接 TMV 后 24h 喷施云芝多糖（T05）、接 TMV 前 24h 喷施云芝多糖（T03）、接 TMV（T04）、喷云芝多糖（T02）处理差异基因分别为 5 032种、3 124种、1 549种和 341 种，说明云芝多糖在病毒胁迫时会调动本氏烟大量基因转录，免疫调节作用明显。通过对差异表达基因功能注释和富集，云芝多糖在寄主与病毒互作过程中具有明显的免疫调控活性，主要涉及核酶活性、光合作用和能量代谢、病毒诱导的基因沉默等，诱导抗坏血酸、苯丙素生物合成、芪类化合物，二芳基庚烷和姜酚生物合成、植物-病原菌相互作用、植物激素信号转导、基因沉默、参与基因沉默的 miRNA 合成等途径。预测了差异表达基因蛋白互作网络相关基因调控途径。该研究结果对研究云芝多糖诱导植物抗病毒及药剂、植物、病毒之间的互作有指导意义。

关键词：云芝多糖；诱导免疫；烟草花叶病毒；转录组

---

* 基金项目：国家重点研发计划（SQ2017ZY060102）；山西省重点研发计划（指南）项目（201603D321030）；山西省农业科学院博士研究基金（YBSJJ1705）

** 第一作者：谢咸升，男，副研究员，博士，主要从事植物保护研究；E-mail：xxshlf@ 163. com

*** 通信作者：安德荣，男，教授，博士生导师，主要从事微生物资源利用及植物病毒学研究；E-mail：anderong323@ 163. com

复合生物菌剂对苹果再植病害的防控效果研究*

赵　璐**，刘　胜，王树桐***，曹克强***

（河北农业大学植物保护学院，保定　071001）

摘　要：苹果再植病害广泛发生在世界各苹果产区，严重制约苹果产业可持续发展。利用溴甲烷、氯化苦等进行土壤熏蒸是防治该病害的主要手段。但化学熏蒸不但对土壤环境造成污染，溴甲烷还会对臭氧层造成损坏。利用生物防治再植病害作为化学熏蒸的替代方法近年来得到了广泛关注和研究。本研究以市场上较为普遍应用的木美土里复合微生物菌剂作为试验材料，开展了苹果再植病害的田间防控试验，以株高、干径、叶绿素含量和产量等苹果树生长指标作为评价指标。结果表明：木美土里复合微生物菌肥对一年生再植果树植株生长有促进作用，一年生再植苹果树，定植时施用木美土里生物菌肥并结合根宝贝灌根，其植株的干茎增长率是对照的 1.5 倍、株高增长率是对照的 1.4 倍、叶片叶绿素含量高于对照植株，最高达对照的 1.16 倍，分枝数、平均分枝长度等生长指标均优于对照再植植株和正茬植株。通过对再植病害防效的连续观察，木美土里生物菌肥对再植后多年植株的生长有显著促进作用，定植时施用木美土里生物菌肥并结合根宝贝灌根，其生长 2~4 年的果树植株的干茎平均是对照的 1.29 倍、株高平均是对照的 1.2 倍、分枝长度平均是对照的 1.25 倍，树龄为 5 年的再植果树，单株产量是对照的 1.86 倍，显著优于有机肥处理的对照植株，防效显著。以上研究结果表明：木美土里复合微生物菌肥在田间对再植苹果树具有促生长、增产量的作用，可以有效缓解苹果再植病害症状。本研究为推广苹果再植病害的生物防治提供了理论基础。

关键词：复合微生物菌剂；苹果再植病害；生物防治

---

* 基金项目：国家苹果产业技术体系（CARS-28）和河北省自然科学基金（c2016204140）

** 作者简介：赵璐，在读硕士研究生，研究方向：植物病害流行与综合防治；E-mail：1391492647@qq.com

*** 通信作者：王树桐，博士，教授，从事植物病害流行与综合防治研究；E-mail：bdstwang@163.com

曹克强，博士，教授，从事植物病害流行与综合防治研究；E-mail：ckq@hebau.edu.cn

国外引进马铃薯资源对早疫病的田间抗性评价*

娄树宝**，孙旭红，田国奎，王海艳，李凤云，李成军，王立春***
（黑龙江省农业科学院克山分院/农业部马铃薯生物学与遗传育种重点实验室，克山 161606）

摘　要：通过 4 年对 57 份国外引进的马铃薯资源早疫病田间抗性鉴定结果表明：国外抗早疫病资源比较丰富，筛选出 2 份高抗材料，筛选出 391180.6、395109.29、399049.22、399050.3、зарево、Early Gem 等 6 份抗性较好且稳定的材料，为马铃薯抗早疫病育种提供了优良的亲本。

关键词：马铃薯；旱疫病；抗性评价

---

* 基金项目：农业部“948”项目（2015-Z52）；黑龙江省经济作物现代农业产业技术协同创新体系；国家科技支撑计划（2012BAD02B05）；黑龙江省应用技术研究与开发（GA15B102-03）；齐齐哈尔市农业公关项目（NYGG-201504）

** 第一作者：娄树宝，男，助理研究员，主要从事马铃薯遗传育种工作

*** 通信作者：王立春，副研究员，主要从事马铃薯遗传育种工作

油菜黑胫病抗病相关基因 *WRKY*70 和 *LRR-RLK* 的序列分析及功能验证*

郝丽芬[1,2]**，燕孟娇[1]，皇甫海燕[1]，宋培玲[1]，贾晓清[1]，皇甫九茹[1]，李子钦[1]***

（1. 内蒙古农牧业科学院，呼和浩特　010031；

2. 内蒙古农业大学生命科学学院，呼和浩特　010018）

摘　要：油菜黑胫病是油菜生产上重要的病害，病原菌为小球腔菌属真菌，具有广泛的寄主范围，但主要为害十字花科植物。目前，我国有 14 个省发现油菜黑胫病的弱侵染型菌株 *Leptosphaeria biglobosa*，发病呈现逐年加重的趋势，并且在局部地区造成严重危害。研究发现，油菜响应病原菌侵染，体内会产生一系列生理生化反应，而且伴随有植物病原菌互作基因的特异表达，本研究在弱侵染型菌株 nm-1 侵染油菜青杂 5 号的转录组测序基础上，通过差异基因分析和 GO、KEGG 功能分析，筛选获得抗病相关的两个基因 *BnWRKY*70（BnaA04g02560D）和 *BnLRR-RLK*（BnaC02g30150D）。WRKY70 转录因子可以结合到顺式作用元件上，调节基因表达的强度，响应病原菌的胁迫参与植物防御反应；LRR-RLK 是典型的植物跨膜类受体激酶，可以直接作为受体或与受体形成复合体参与免疫反应，包括参与病原分子模式识别系统。

本研究对 *BnWRKY*70 和 *BnLRR-RLK* 进行生物信息学序列分析和功能验证，结果发现，*BnWRKY*70 基因 CDS 全长 858bp，编码 285 个氨基酸，含有一个 WRKY 保守结构域，属于Ⅲ类 WRKY 转录因子；*BnLRR-RLK* 基因 CDS 全长 2775bp，编码 924 个氨基酸，含有蛋白激酶和丝氨酸苏氨酸蛋白激酶活性位点。利用病毒诱导的基因沉默（VIGS）技术，构建 TRV 载体诱导的油菜基因沉默体系。沉默后 *BnWRKY*70 和 *BnLRR-RLK* 基因的表达量显著下降，表型观察发现基因沉默油菜株系叶片比未沉默株系病斑面积更大，更易感病；病情指数的统计结果显示，沉默株系的相对病情指数显著高于对照，说明 *BnWRKY*70 和 *BnLRR-RLK* 基因表达增加了油菜的抗病性。本研究为加速鉴定油菜抗病通路中的重要基因提供了数据信息。

关键词：油菜；黑胫病；*BnWRKY*70 和 *BnLRR-RLK*；病毒诱导的基因沉默

---

* 基金项目：内蒙古科技厅计划项目（201505018）

** 第一作者：郝丽芬，博士，助理研究员，主要从事分子病理学研究；E-mail：haolifen616@163.com

*** 通信作者：李子钦，研究员，主要从事油料作物病害研究；E-mail：ziqinli88@yahoo.com

RNA-seq analysis reveals genes associated with *NPR*1-mediated acquired resistance in barley

Gao Jing[1], Bi Weishuai[1], Li Huanpeng[1], Jorge Dubcovsky[2,4], Yu Xiumei[3], Yang Wenxiang[1], Liu Daqun[1], Wang Xiaodong[1]*

(1. *College of Plant Protection*, *Biological Control Center for Plant Diseases and Plant Pests of Hebei*, *Agriculture University of Hebei*, *Baoding*, *Hebei* 071000, *China*.
2. *Department of Plant Science*, *University of California*, *Davis*, *CA* 95616, *USA*.
3. *College of Life Science*, *Agriculture University of Hebei*, *Baoding*, *Hebei* 071000, *China*.
4. *Howard Hughes Medical Institute* (*HHMI*), *Chevy Chase*, *MD* 20815, *USA*.)

Abstract: In Arabidopsis, systemic acquired resistance (SAR) is established beyond the initial infection of pathogen or directly induced by salicylic acid (SA) treatment. NPR1 protein is considered as the master regulator of the SAR in both SA signal sensing and transduction. Our previous study showed that *NPR*1 homologs in wheat and barley regulate expression of genes encoding pathogenesis-related (PR) proteinsduring acquired resistance (AR) triggered by *Pseudomonas syringae* pv. *tomato* DC3000. However, genes associated with this *NPR*1-mediated AR are largely unknown. Here in this study, RNA-seq analysis was used to reveal differentially expressed genes during AR among barley transgenic lines over-expressing wheat *NPR*1 (*wNPR*1-OE), transgenic lines suppressing barley *NPR*1 (*HvNPR*1-Kd) and wild-type plants. Several *PR* genes and SA-sensitive *BCI* genes were designated as downstream genes of *NPR*1 during AR. Expression of few transcription factors, especially WRKYs, showed intensive association with *NPR*1 expression. A profile of genes associated with *NPR*1-mediated AR in barleywas drafted.

* Corresponding author: Wang Xiaodong (E-mail: zhbwxd@ hebau. edu. cn) and Daqun Liu (E-mail: liudaqun@ caas. cn)

水稻 bZIP 转录因子 APIP5 的重组表达、纯化与晶体生长

程希兰*，张 鑫，彭友良，齐晓萱，刘俊峰**
（中国农业大学植物保护学院植物病理学系，北京 100193）

摘 要：水稻是世界上重要粮食作物之一，由稻瘟病菌引起的稻瘟病严重影响水稻的品质。水稻中 bZIP 转录因子 APIP5 是细胞死亡的负调控因子，稻瘟病菌效应蛋白 AvrPiz-t 能够通过抑制 APIP5 转录活性并降低蛋白水平，从而激发效应蛋白介导的细胞死亡。本研究试图纯化样品，获得晶体，为解析 bZIP 转录因子 APIP5 晶体结构，以阐明其在水稻抗病性中的分子机理提供结构基础。

首先，APIP5N 端（1-93AA）是与 AvrPiz-t 互作的必须区段，因此作者构建了 APIP5 全长和 APIP5（1-93AA）、APIP5（1-148AA）、APIP5（1-164AA）等截短体的不同表达载体，并将全长和不同截断体构建好的表达载体分别热激到大肠杆菌 BL21（DE3）、Rosetta-gami（DE3）等不同的表达菌株中，经 IPTG 诱导，筛选最佳的表达条件。通过亲和层析、离子交换层析、凝胶过滤层析等纯化方法对 APIP5 全长极其截断体蛋白进行进一步纯化，最终获得符合晶体生长要求的蛋白样品，利用质谱技术鉴定所制备的样品为目的蛋白。目前作者正在利用气相扩散法对纯化好的蛋白样品进行大量晶体生长条件的筛选，以期获得蛋白晶体；另外，利用 ITC 等互作方法验证 APIP5 以及截短体与 AvrPiz-t 体外互作方式，并制备二者复合物，为获得复合物的结构以解释效应蛋白与寄主互作机理奠定基础。

关键词：稻瘟病菌；APIP5；原核表达；蛋白纯化

* 第一作者：程希兰，硕士研究生，植物保护专业；E-mail：chengxilan1026@163.com
** 通信作者：刘俊峰，教授；E-mail：jliu@cau.edu.cn

苹果 MdASMT1 的克隆与原核表达*

吴成成**，练 森，李保华，王彩霞***
（青岛农业大学植物保护学院，山东省植物病虫害综合防控重点实验室，青岛 266109）

摘 要：褪黑素能够清除自由基且具有抗氧化作用，在植物中具有广泛的生理功能。不同植物、不同组织内褪黑素含量存在较大差异，一些逆境胁迫条件均可诱导褪黑素的产生。外源褪黑素也可影响植物内褪黑素的含量，进而减缓病原菌侵染、高温、盐胁迫等对植物造成的损伤。植物体内褪黑素的合成需要 4 步连续的酶促反应，N-乙酰基-5-羟色胺-甲基转移酶（N-acetylserotonin methyltransferase，ASMT）是褪黑素合成途径中的最后一个催化酶，*ASMT* 基因过表达，可以提高褪黑素的含量，增强植物的抗旱性，但该基因是否参与植物的抗病作用尚无报道。

本研究以“富士”和“嘎啦”（*Malus domestica* Borkh）苹果叶片为材料，获得 *ASMT* 的 cDNA 序列，命名为 *MdASMT*1。该基因开放阅读框（ORF）为 1 077bp，编码 358 个氨基酸。生物信息学分析结果表明 *MdASMT*1 编码的蛋白理论分子量约为 39.7 ku，等电点为 5.42（pI），没有信号肽，为非分泌型蛋白；有 34 个 O-糖基化位点；催化位点及 S-腺苷-L-蛋氨酸结合位点分别为 His[263]、Glu[225]。多序列比对及系统进化树分析表明，*MdASMT*1 编码的蛋白与珠美海棠 *MzASMT*1 氨基酸序列同源性最高为 99.2%，其次为葡萄和可可树，氨基酸序列相似性分别为 66.4%和 52.4%。原核表达产物经 SDS-PAGE 分析，MDASMT1 融合蛋白分子量约 58 ku，与预测分子量大小一致，该蛋白主要存在于上清中，为可溶性蛋白；1 mmol/L IPTG，25℃，诱导 8h 可实现该蛋白的高效表达；Western blot 检测得到了特异性条带，进一步证实 *MdASMT*1 编码的蛋白得到成功表达，为进一步研究 *MdASMT*1 基因的功能奠定了基础。

关键词：苹果；*MdASMT*1；克隆；原核表达

---

* 基金项目：现代农业产业技术体系建设专项资金（CARS-28）；国家自然科学基金（31272001）

** 第一作者：吴成成，女，硕士研究生，研究方向果树病理学；E-mail：15610567060@ qq. com

*** 通信作者：王彩霞，女，教授，主要从事果树病害研究；E-mail：cxwang@ qau. edu. cn

A Receptor Like Kinase Gene SBRR1 Positively Regulates Resistance to Sheath Blight in Rice

Feng Zhiming, Zhang Yafang, Wang Yu, Zhang Huimin, Cao Wenlei, Chen Zongxiang, Pan Xuebiao and Zuo Shimin*

(*Key Laboratory of Crop Genetics and Physiology/Co-Innovation Center for Modern Production Technology of Grain Crops, Key Laboratory of Plant Functional Genomics of the Ministry of Education, Yangzhou University, Yangzhou* 225009, *China*)

Abstract: Sheath blight (SB), caused by the necrotrophic fungal pathogen *Rhizoctonia solani* Kühn, poses a great threat to the rice grain yield as one of the most serious rice diseases. Receptor like kinases play important roles in plant defense responses to disease. However, receptor like kinases associated with rice SB have not been reported. We analyzed the differential gene expression profiles of the rice variety YSBR1 with high resistance to SB before and after inoculation with *R. solani*, and isolated a rice receptor like kinase gene *SBRR*1 (*sheath blight-related RLK gene* 1) induced by *R. solani*. *SBRR*1 is preferentially expressed in leaf sheaths and leaves, which is consistent with the main infective site of *R. solani*, implying its role in regulating rice resistance to SB. The *SBRR*1 loss-of-function mutant *sbrr*1 is more susceptible to SB compared with its wild type. Moreover, knockdown of *SBRR*1 expression by RNAi reduces the resistance of rice to SB. These results demonstrated that *SBRR*1 positively regulates resistance to SB. We further identified an SBRR1-interaction protein SIP1 (SBRR1 interaction protein 1) that encodes an ankyrin repeat protein via yeast two hybrid screening. *SIP*1 is also induced by *R. solani* and preferentially expressed in leaf sheaths and leaves. Our results enrich the knowledge of the molecular mechanism of resistance to rice SB, which will provide the theories basis for establishing a new strategy to prevent and control this disease.

Key words: Rice sheath blight; Receptor like kinase; *SBRR*1; Resistance mechanism

* Corresponding author: Zuo Shimin; E-mail: smzuo@yzu.edu.cn

Rapid Identification of Stripe Rust Resistant Gene in a Space-induced Wheat Mutant Using Specific Length Amplified Fragment (SLAF) Sequencing*

Zhang Xing**, Lu Chen, Cai Sun, Peng Zhang, Ma Dongfang***

(*College of Agriculture*, *Yangtze University*, *Jingzhou*, *Hubei* 434025, *China*)

Abstract: Stripe rust, caused by *Puccinia striiformis* f. sp. *tritici*, is one of the most devastating diseases on wheat. Growing resistant cultivars is the preferred strategy to control the disease. The space-induced wheat mutant R39 performed adult-plant resistance to stripe rust. Genetic analysis revealed single recessive gene, designated as *YrR*39, is responsible for the resistance of R39 to stripe rust race CYR33. To finely map *YrR*39, bulked segregant analysis (BSA) combined with SLAF sequencing (SLAF-seq) strategy was applied and finally *YrR*39 was located into the candidate regions on chromosome 4B with a size of 17. 39 Mb. Totally 126 genes were annotated in the regions and 21 genes with annotations associated to resistance were selected for further qRT-PCR analysis. As a result, the candidate gene *Traes_ 4BS_ C868349E*1 (annotated as F-box/LRR-repeat protein) was obviously up-regulated after 12, 24, 48, 96 hours post infected by stripe rust, suggesting that it was the key candidate gene for controlling the stripe rust resistance in wheat. The current study demonstrated the BSA combined with SLAF-seq for SNP discovery is an efficient approach to map complex genomic region for isolation of functional gene.

Key words: Bulked segregant analysis; SLAF-seq; Stripe rust

---

* 基金项目：国家自然基金（31501620）；植物病虫害生物学国家重点实验室开放课题（SKLOF201707）

** 第一作者：张兴，硕士，主要从事小麦抗条锈病研究；E-mail：zxchinacn@ 163. com

*** 通信作者：马东方，博士，副教授，主要从事小麦真菌病害综合防治研究；E-mail：madongfang1984@ 163. com

A single Amino-acid Substitution of Polygalacturonase Inhabiting Protein (OsPGIP2) Allows Rice to Acquire Higher Resistance to Sheath Blight

Chen Xijun[1], Li Lili[1], Zhang Lina[1], Xu Bin[1],
Chen Zhongxiang[2], Zuo Shimin[2], Xu Jingyou[1]

(1. *Horticulture and Plant Protection College*, *Yangzhou University*, *Yangzhou* 225009, *China*; 2. *Key Laboratory of Crop Genetics and Physiology of Jiangsu Province/Key Laboratory of Plant Functional Genomics of the Ministry of Education*, *College of Agriculture*, *Yangzhou University*, *Yangzhou* 225009, *China*)

Abstract: A possible strategy to control plant diseases is improvement of natural plant defense mechanisms against the infection of pathogens. As one kind of leucine-rich repeat proteins, polygalacturonase inhibiting proteins (PGIPs) have been found involved in plant defense against the pathogens by inhibiting the activity of fungal endo-polygalacturonases (endo-PGs). To dig more resistant gene, a single amino acid of the OsPGIP2, which has no effect on the endo-PGs activity of *Rhizoctonia solani*, was substituted and it allowed OsPGIP2 to acquire the inhibiting ability. Overexpression of *OsPGIP*2 significantly increased rice resistance to rice sheath blight and counteracted the tissue degradation caused by RsPGs. Furthermore, *OsPGIP*2 overexpression did not affect rice agronomic traits or yield components. All these indicated that *OsPGIP*2 was an important potential gene that could be applied to rice sheath blight resistance breeding.

Key words: Single amino acid substitution; Polygalacturonase inhibiting protein; Sheath blight resistance; *Oryzae sativa*

A Sheath Blight Resistance QTL $qSB\text{-}11^{LE}$ Encodes a Receptor Like Protein Involved in Rice Early Immune Response to *Rhizoctonia solani*

Xue Xiang, Wang Yu, Feng Zhiming, Zhang Huimin, Shan Wenfeng, Cao Wenlei, Zhang Yafang, Chen Zongxiang, Pan Xuebiao and Zuo Shimin*

(*Key Laboratory of Crop Genetics and Physiology/Co-Innovation Center for Modern Production Technology of Grain Crops, Key Laboratory of Plant Functional Genomics of the Ministry of Education, Yangzhou University, Yangzhou* 225009, *China*)

Abstract: Sheath blight (SB) is one of the three most serious diseases, which leads to severe grain loss annually. Rice resistance to SB disease is a typical of quantitative trait controlled by polygenes or quantitative trait loci (QTLs). Lots of QTLs for SB resistance have been reported while none of them was isolated so far, which significantly hinders the elucidation of resistant mechanism. Previously, we have mapped an SB resistance QTL on chromosome 11 into a 78kb region, in which the resistant allele (here after called $qSB\text{-}11^{LE}$) is from Lemont (LE). Here, through a series of experiments like RNA interference, overexpression and complementation tests, we confirmed that a receptor-like protein lack of intracellular domain in the fine mapping region is $qSB\text{-}11^{LE}$. RT-PCR data showed that $qSB\text{-}11^{LE}$ transcription was significantly induced by SB pathogen *Rhizoctonia solani* but not by rice blight pathogen *XOO*. The $qSB\text{-}11^{LE}$ is highly expressed in leaf sheath at the booting stage. *Subcellular localization* assay indicated that $qSB\text{-}11^{LE}$ protein localizes in both *nucleus and* membrane. Through the treatments of known pathogen associated molecular patterns (PAMPs), chitin and flg22, we found that knock down of $qSB\text{-}11^{LE}$ significantly affected rice innate immune response to chitin. Some known receptor/co-receptor like kinase proteins were found no interaction with $qSB\text{-}11^{LE}$ protein in *vitro*. Upon the infection of SB fungus, the synthesis of hormone ethylene (ET) was apparently induced in wild plants, while it was only weakly increased in $qSB\text{-}11^{LE}$ RNAi plants as well as in susceptible near isogenic line (NIL). Spray of ET was able to recover the resistance phenotype of $qSB\text{-}11^{LE}$ RNAi and NIL plants. Collectively, we conclude that $qSB\text{-}11^{LE}$ is involved in rice early immune response to SB fungus and regulates ET synthesis upon SB pathogen infection to activate ET-dependent defense response.

Key words: Rice sheath blight; QTL; $qSB - 11^{LE}$; Receptor like protein; Early immune response; Ethylene

* Corresponding author: Zuo Shimin; E-mail: smzuo@ yzu. edu. cn

拟南芥中有复杂的遗传机制调控 NLP 引起的细胞死亡*

陈俊斌**，房雅丽，范　军***
（中国农业大学植物病理系，北京　100193）

摘　要：NLP（Necrosis and ethylene inducing peptide like protein）是微生物中广泛存在的可以在双子叶植物上引起细胞死亡和诱导乙烯生成的一类蛋白。虽然目前已经确定 NLP 为一类微生物相关分子模式，但是对于其在双子叶植物上引起细胞死亡的机制尚不清楚。本研究利用拟南芥为材料，研究拟南芥中介导 NLP 引起细胞死亡的遗传因子。经过筛选 200 多个拟南芥生态型后发现与 Col-0 相比，El-0、Gie-0 和 Stw-0 三个生态型对 NLP 引起的细胞死亡较不敏感。利用 NLP 处理 Stw-0 X Col-0 F2、El-0 X Col-0 F2 和 Col-0XGie-0 重组自交系群体，通过测量植物细胞电解质泄漏水平，发现 NLP 引起的细胞死亡呈现数量性状（QTL）变异。QTL 分析发现 6 个位点参与调控 NLP 在拟南芥上引起的细胞死亡。其中，1 号染色体上有 2 个 QTL 位点（QG1. 1 和 QE1. 1），2 号及 4 号染色体各有 1 个 QTL 位点（QE2. 1 和 QS4. 1），5 号染色体有两个 QTL 位点（QG5. 1 和 QS5. 1）。5 号染色体上的两个 QTL 发生部分重叠，其余各个 QTL 位置均不相同。因此，拟南芥中存在复杂的遗传机制来控制 NLP 引起的细胞死亡。

关键词：NLP；QTL；细胞死亡

* 基金项目：国家自然基金（No. 31571946）
** 第一作者：陈俊斌，男，博士研究生，主要研究方向植物病理学
*** 通信作者：范军，男，教授；E-mail：jfan@ cau. edu. cn

聊城地区不同玉米品种对灰斑病菌的抗性分析*

杨明明，王桂清**，马　迪
（聊城大学农学院，聊城　252059）

摘　要：为了明确聊城地区主栽玉米品种对灰斑病的抗感性差异，2016 年 9 月采取田间自然诱发鉴定方法，分析了 70 个主栽玉米品种对灰斑病的抗性。结果表明：聊城地区玉米灰斑病发生很轻，发病率 1%左右；不同玉米品种对灰斑病抗性差异明显，其中抗病品种居多，为 90%，易感品种仅占 10%。该研究为抗灰斑病品种的选育和推广奠定了基础。

关键词：聊城市；玉米品种；玉米灰斑病；抗性鉴定

Resistance Analysis of Maincorn Varieties to Gray Leaf Spot in Liaocheng Area

Yang Mingming，Wang Guiqing，Ma Di
(*College of Agronomy*，*Liaocheng University*，*Liaocheng* 252059，*Shandong*，*China*)

Abstract：Using the method of natural inducement in field，the resistance of 70 main corn varieties to gray leaf spot was identified in Liaocheng area in September 2016. The results showed that the gray leaf spot was very light，the incidence rate was about 1%. Different corn varieties were significantly different in resistance to gray leaf spot. Among them，resistant varieties were mostly，was 90%，susceptible varieties only 10%. The study made clear the difference of resistance to gray leaf spot of main corn varieties in Liaocheng area. It laid a foundation for the promoting and breeding of resistant varieties.

Key words：Shandong province Liaocheng area；Corn varieties；Gray leaf spot；Resistance identification

玉米灰斑病又称尾孢菌叶斑病，是玉米生产中的一种严重叶部病害，特别是 20 世纪 90 年代后，灰斑病发生地域扩大，在全球的分布更加广泛，对生产的影响也变得更为明显。在我国该病的致病病原主要有两种，北方地区为玉蜀黍尾孢（*Cercospora zeae-maydis*），云南和湖北等地为玉米尾孢（*C. zeina*）[1-2]。我国南北方玉米灰斑病致病菌种类不同，推广应用品种的抗性程度存在差异，新型抗病品种的选育已成为控制玉米灰斑病为害的最根本最重要的措施。

* 基金项目：山东省中青年科学家科研奖励基金（2006BS06014）；山东省自然科学基金联合专项（ZR2012CL17）；2017 年度聊城大学大学生科技文化创新基金项目（26312178821）

** 通信作者：王桂清，女，博士，教授，主要从事植物保护的教学与科研工作；E-mail：wangguiqing@ lcu. edu. cn

1 材料与方法

1.1 调查时间与地点

2016年9月于山东省聊城市汇德丰种业有限公司玉米品种展示基地进行发病程度调查，该展示圃地势肥力、管理措施等均与自然农田均一，但整个生育期不施用农药，耕作制度为小麦-玉米两熟制。

1.2 供试玉米品种

供调查玉米品种共70个，为已通过审定的聊城地区主推品种，且均为夏玉米区推广品种。

1.3 田间病情调查

采用田间自然诱发鉴定方法，在展示圃内调查玉米品种在自然条件下的发病水平。每个玉米品种小区（10m×5m）调查10行，采用五点取样法，每点10株，共50株，调查穗位叶灰斑病的发生情况，确定其占叶片面积的比率，记录病情级别，计算发病率、病情指数等。根据病情指数（DI）划分品种的抗性类别：DI≤10为高抗（HR），10<DI≤30为抗病（R），30<DI≤50为中抗（MR），50<DI≤70为感病（S），70<DI为高感（HS）。玉米灰斑病分级标准[3]为：0级，免疫（I），穗位叶无病斑；1级，高抗（HR），穗位叶病斑面积占整叶面积0~5%；3级，抗病（R），穗位叶病斑面积占整叶面积6%~10%；5级，中抗（MR），穗位叶病斑面积占整叶面积11%~35%，7级，感病（S），穗位叶病斑面积占整叶面积36%~70%；9级，高感（HS），穗位叶病斑面积占整叶面积70%~100%。

$$\text{发病率}(\%) = \frac{\text{发病株数}}{\text{调查总株数}} \times 100$$

$$\text{病情指数}\ DI = \frac{\sum(\text{各级病株数} \times \text{各级代表值})}{\text{调查总株数} \times \text{最高级代表值}} \times 100$$

1.4 数据分析

利用EXCEL对数据进行统计和分析。

2 结果与分析

2.1 不同玉米品种抗病性

70个玉米品种的抗玉米灰斑病结果如表1。聊城地区玉米灰斑病发生较轻，仅振杰2号的发病率为13.3%，其余品种的发病率均低于1%。

表1 70个玉米品种的抗玉米灰斑病结果

品种	母本	父本	发病率（%）	发病指数	抗性评价	品种	母本	父本	发病率（%）	发病指数	抗性评价
振杰1号	聊112	Lx9801	0	0	I	泰玉7号	泰08	泰03	0.46	9.34	HR
振杰2号	京89	Lx9801	13.3	2.66	HR	泰玉11	泰玉08	泰玉09	0.3	6.60	HR
聊玉18	835-2	3087	0.13	2.66	HR	泰玉14	泰039系	泰13	0.13	2.66	HR
聊玉19	L219	聊85-308	0.47	16	R	泰玉D4	DⅡ	泰15系	0.2	4.32	HR
聊玉20	L113	聊308	0.2	6	HR	天泰10	PC57	PCH42	0.93	28.04	R
聊玉93-1	3189	聊85-308	0.2	6	HR	天泰14	Pc9	PcH47	0.73	16	R
丰聊008	聊508	昌7-2	0.47	9.34	HR	天泰16	Pc67	PcH45	0.26	5.34	HR

（续表）

品种	母本	父本	发病率（%）	发病指数	抗性评价	品种	母本	父本	发病率（%）	发病指数	抗性评价
承玉10	承系24	853	0.13	2.66	HR	天塔5号	HOO-108	825	1	76	HS
济丰96	8233	昌7-2	0.4	8	HR	枣2044	3054431	96122	0.26	5.34	HR
LN1	L496	L172	0.53	10.66	R	金穗一号	协青早A	R978	0.3	6.66	HR
LN3	L686	9401-1	0	0	I	金穗18	96-4	K002	0.6	14.66	R
LN14	L189	H21	0.67	1.34	HR	郁青一号	郁858	丹598	0.46	9.34	HR
金海5号	JH78-2	JH3372-1	1	25.34	HR	五岳97-1	TM3	TM100	0.46	9.34	HR
金海601	JH2631	武314-8	0.27	7.06	HR	谷育178	301	2302	0.73	14.66	R
金海604	P138	JH3372	0	0	I	淄玉2号	3178	黄801	0.6	16	R
金海702	JH673	JH3372	0.47	9.34	HR	淄玉9号	Lz1578	H21	0.66	1.34	HR
汶农玉6号	LZ288	H21	0.13	2.66	HR	郑单958	郑58	昌7-2	0.87	22.66	R
登海3号	DH08	P138	0.73	15.34	R	浚单20	9058	浚92-8	0.26	5.34	HR
登海11	DH65232	DH40	0.8	40.34	MR	先行2号	X9508-1	Lx9801	0.26	26.66	R
登海3688	DH158	DH13	1	34.66	MR	先行3号	X9601-2	Lx9801	0.2	4	HR
登海3622	DH158	DH323	1	48	MR	先行5号	X1902	Lx9801	0.67	14.66	R
鲁单6003	Lx2394	Lx2394	0.67	1.34	HR	农大108	黄C	178	1	70.66	HS
鲁单9002	郑58	Lx9801	0.3	9.34	HR	M9	F229	昌7-2	1	69.34	S
鲁单9006	Lx00-6	Lx9801	0.87	25.34	R	鑫丰1号	XF0138	SX211	0.3	8	HR
鲁单999	5318	Lx9801	0.4	14.66	R	鑫丰6号	SX053	SX2	0.2	5.34	HR
鲁单984	478	Lx9801	0.3	9.34	HR	圣瑞16	SRY8	SRY16	1	57.34	S
鲁单661	齐319	308	0.67	1.34	HR	先玉335	PH6WC	PH4CV	1	28	R
鲁单6018	118	P138	0.46	13.34	R	莘州158	莘5812	莘1101	0.2	4	HR
鲁单981	齐319	Lx9801	0.2	4	HR	东单60	A801	C101	0.8	32	MR
鲁单8013	Qx901	H53	0.93	50.34	S	连胜15	9648	686	0.47	10.66	R
鲁单8009	沈137	Lx9801	0.53	16	R	中科4号	CT019	Lx9801	0.13	2.66	HR
鲁单6028	Lx9223	Lx2394	1	30.66	MR	联创3号	CT08	CT609	0.73	14.66	R
鲁种99118	PL95	LZ01	1	30.66	MR	费玉2号	费04	费06	0.93	41.34	MR
鲁原种22号	鲁原476	8112C9-331	0.8	22.66	R	费玉3号	费玉03	费玉04	0.86	52	S
泰玉2号	齐319	3841	0.4	8	HR	费玉4号	费玉03	费玉05	0.93	50.66	S

根据表2统计了不同抗性类型品种所占的比例，如下图。70个玉米品种对灰斑病的病情指数为0~76，品种间差异显著，说明不同玉米品种对灰斑病的抗感程度不同，其中表现为抗性

（包括高抗、抗病和中抗）的品种多达 63 个，占 90%，说明聊城地区主栽玉米品种对灰斑病多表现为抗病。振杰 1 号、LN3、金海 604 等 3 个品种的病情指数为 0，振杰 2 号、聊玉 18、聊玉 20 等 34 个品种的病情指数小于 10，即高抗品种共 37 个，占 52. 86%，在夏玉米种植中可作为抗灰斑病优良品种进行推广；M9、圣瑞 16、鲁单 8013、费玉 3 号和费玉 4 号 5 个品种的病情指数大于 50，品种天塔 5 号和农大 108 的病情指数大于 70，即感病和高感品种共 7 个，占 10%，在玉米灰斑病高发地区进行玉米栽植时要有的放失。

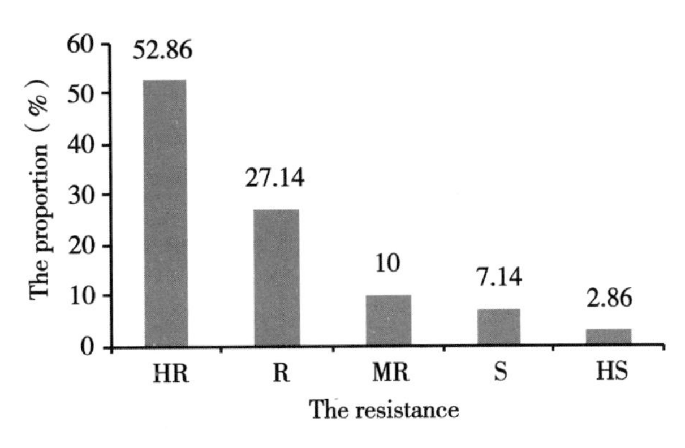

图　不同抗性类型品种所占的比例

2. 2　玉米亲本抗病性

姚清鉴[4]对选育的新品系进行抗性来源分析得出：表现高抗的品系，其亲本中一般至少有一个高抗材料或两个亲本均为抗病材料；杂交种的抗感病性趋向母本，母本的抗感病性对杂交种的影响较大。据此作出如下推断：

丰聊 008 和 M9 两个品种的父本皆为昌 7－2，而丰聊 008 为高抗品种，而 M9 为感病品种，从而推测丰聊 008 的母本聊 508 对玉米灰斑病表现为抗病，M9 的母本 F229 对玉米灰斑病表现为感病。

品种鲁单 661、鲁单 981 和泰玉 2 号对玉米灰斑病均表现为高抗，三者的母本均为齐 319，说明齐 319 对玉米灰斑病应为抗性；品种费玉 3 号和费玉 4 号对玉米灰斑病均表现为感病，二者的母本均为费玉 03，说明费玉 03 较感玉米灰斑病。

品种振杰 1 号、振杰 2 号、鲁单 9002、鲁单 9006、鲁单 999、鲁单 984、鲁单 981、鲁单 8009、先行 2 号、先行 3 号、先行 5 号和中科 4 号对灰斑病均表现为抗性和高抗，其父本均为 Lx9801，可以推测 Lx9801 抗玉米灰斑病。

关于亲本对灰斑病的抗性，有待于田间直接鉴定结果予以证明。今后，在组配玉米杂交组合时，应注意选用抗病性强且病情指数低的自交系作母本，从而获得抗性较强的杂交种。

3　结论与讨论

灰斑病的发生受气候条件影响较为明显，尤以湿度最为关键。玉米抽穗扬花至灌浆期，气温较低、降雨偏多、雾日较多，则加重病害发生[5]，故温暖、湿润的山区和沿海地区病害易于发生。2016 年聊城地区 7～9 月份有雨的天数仅 15d 左右，却多为阵雨，雨量较小，即湿度较小，不利于灰斑病的发生，因此，2016 年聊城地区灰斑病发病率不足 1%。

对玉米品系进行灰斑病抗性鉴定的常用的方法包括病害发生率（DI）法和病害严重度（DS）法[6]。玉米杂交品种和自交系对玉米灰斑病表现的抗性差异较大，凡具有热带血缘的自交系都是抗病的[7]，杂交组合的抗病性介于双亲抗性之间，多数趋向于抗病强的亲本[8]。Roane 等[9]在美国弗吉尼亚州评价了 49 个杂交种，发现几乎所有品种均有相同程度的灰斑病发生，因而断定缺乏抗性；Hilty[10]评价几个杂交种和一些自交种，发现仅在一个自交种上（T222）表现高抗，高抗的品种 T222 表现型与其原父本 PA875 相似。不同地区主栽品种不同，其抗灰斑病程度不同，本研究表明，聊城地区主栽玉米品种对灰斑病多表现为抗病，占比 90%；陕西省商洛地区的抗病品种包括陕单 2001、正大 999、奥利 10 号、安森 7 号、安玉 2166 等[11]；云南地区玉米杂交种抗灰斑病占鉴定总数的 1/3，其中高抗（HR）的品种有屯玉 7 号、海禾 1 号和海禾 2 号共 3 个，占鉴定材料总数的 3. 75%[12]。

玉米灰斑病已经成为我国北方和西南地区玉米上继玉米大、小斑病之后的又一种重要叶部病害，危害严重，必须加强对该病病原物的发生规律以及抗病品种培育等系统、深入研究，以有效控制玉米灰斑病的扩延和危害。

参考文献

[1] 刘庆奎，秦子惠，张小利，等．中国玉米灰斑病病原菌的鉴定及其基本特征研究［J］．中国农业科学，2013，46（19）：4044-4057.

[2] 张小飞，李晓，崔丽娜，等．西南地区玉米灰斑病病原种类分子鉴定［J］．西南农业学报，2014，27（3）：1079-1081.

[3] 吴景芝，刘世建，沙本才，等．利用 F_2 代建立玉米灰斑病的分级标准［J］．植物保护学报，2009，36（5）：479-480.

[4] 姚清鉴．玉米新选育品系的抗病性表型鉴定及抗性来源初步分析［D］．北京：中国农业科学院，2013.

[5] 林伟锋，文家富，郑小惠，等．商洛市玉米灰斑病的重发原因与防控对策［J］．陕西农业科学，2016，62（1）：70-72.

[6] 王桂清，陈捷．玉米灰斑病抗病性研究进展［J］．沈阳农业大学学报，2000，31（5）：418-422.

[7] 吕国忠，张益先，梁景逸，等．玉米灰斑病发生流行规律及品种抗病性［J］．植物病理学报，2003，33（5）：462-467.

[8] 吴继昌，马丽君，王作英．玉米抗尾孢菌叶斑病鉴定与抗病材料利用［J］．辽宁农业科学，1997，（5）：25-28.

[9] Roane C W, *et al*. Observations on gray leaf spot of maize in Virginia［J］. Plant Disease, 1974, 58: 456-459.

[10] Hilty J W, *et al*. Response of maize hybrids inbred lines to gray leaf spot disease and the effects on yield in Tennessee［J］. Plant Disease, 1979, 63（6）: 515-518.

[11] 林伟锋，文家富，郑小惠，等．商洛市玉米灰斑病的重发原因与防控对策［J］．陕西农业科学 2016，62（1）：70-72.

[12] 胡正川，段进章，张朝富，等．玉米杂交种对玉米灰斑病抗性的初步研究［J］．云南农业科技，2013（1）：17-19.

Identification and Characterization of Plant Cell Death-Inducing effector proteins from *Rhizoctonia solani*[*]

Li Shuai[**], Fang Rui, Jia Shentong, Wei Songhong[***]
(*College of Plant Protection*, *Shenyang Agricultural University*, *Liaoning* 110866, *China*)

Abstract: Rice sheath blight, caused by *Rhizoctonia solani*, is one of the most important fungus diseases of rice. However, little has been known about molecular mechanisms underlying the function (s) of effectors in *R. solani* virulence pathogenicity so far. Identification and characterization of novel effector proteins are crucial for understanding pathogen virulence and host-plant defense mechanisms. In previous studies, genome of *R. solani* AG1 IA reveals many potential genes coding effector. Bioinformatics analysis indicated that some effectors contain structure which may contribute to pathogenicity. Here, we performed functional studies on putative effectors in *R. solani* and demonstrated that 3 of 40 putative effectors caused necrosis or necrosis-like phenotypes in *Nicotiana benthamiana*. Among them, 1 protein signal peptide play important role in triggering cell death in *N. benthamiana*. These results will provide useful information for insight into pathogenic factors for *R. solani*.

Key words: *Rhizoctonia solani*; Effector; Plant Immunity; Pathogenicity

* Funding: Rice Industry Technology System of Modern Agriculture (CARS-01) and Liaoning Rice Research System (Liao Agri-Science <2013> No. 271)

** First author: Li Shuai, Ph. D.; E-mail: lishuai@ syau. edu. cn

*** Corresponding author: Wei Songhong, professor; E-mail: songhongw125@ 163. com

十字花科蔬菜黑腐病菌 VBNC 状态下抗逆相关基因的表达

白凯红*，阚玉敏，陈　星，蒋　娜，李健强，罗来鑫**

（中国农业大学植物病理学系，种子病害检验与防控北京市重点实验室，北京　100193）

摘　要：十字花科蔬菜黑腐病（Black rot of crucifers）是一种重要的种传细菌性病害，在我国主要为害甘蓝、白菜、花椰菜等十字花科蔬菜，该病是由野油菜黄单胞杆菌野油菜致病变种（*Xanthommonas campestris* pv. *campestris*，简称 Xcc）引起。病原细菌菌体呈短杆菌，大小为（0.4~0.5）μm×（0.7~3.0）μm，革兰氏染色呈阴性。已有文献报道，Xcc 在营养缺乏、Cu^{2+}诱导下能够进入有活力但不可培养（Viable but nonculturable，VBNC）的状态，而 VBNC 状态被认为是多种革兰氏阴性细菌和某些不产芽孢的革兰氏阳性细菌对环境的一种抗逆反应。为探明细菌初始浓度对诱导后 VBNC 状态菌体数量的影响，本试验选用 Xcc 8004 菌株，以对数生长期的菌体为起点，采用两种初始菌量为 10^8 CFU/mL 的 50μmol/L Cu^{2+}体系（即初始诱导浓度分别为 OD_{600}=0.18 与 OD_{600}=0.45）进行诱导，结果表明这两种初始浓度诱导分别在 24h 与 48h 所有有活力的细菌进入 VBNC 状态，而 VBNC 状态菌量分别为 3.3×10^7cell/mL 与 1.5×10^7cell/mL，因此后续试验选用初始诱导浓度为 OD_{600}=0.18 的起始浓度。在此诱导体系下，选取 0min、5min、12h、1d、2d、10d 6 个时间点收集菌体；选取文献报道中与抗逆相关的 *rpoS*（Master regulator of the general stress response in *E. coli*）、*relA*（（p）ppGpp synthetase）、*spoT*（（p）ppGpp hydrolase/synthetase）、σ^{54}（Control nitrogen use in *E. coli*）、σ^{H}（Responsible for the expression of heat shock promoters in *E. coli*）、*clpX*（ATP-dependent Clp protease ATP-binding subunit ClpX）、HP（Activate the expression of a regulon of hydrogen peroxide-inducible genes，e. g. *ahpC et al.*）等 7 个基因，以 16SrRNA、σ^{70}为内参基因，结果表明 *relA*、*spoT* 在试验设置中各时间段均显著上调表达；*clpX*、HP 在 Cu^{2+}诱导初期（5min）时显著上调表达，后续阶段无显著变化；*rpoS* 在初期下调表达，2d 显著上调表达，随后无明显变化；σ^{54}、σ^{H}在诱导 5 min 到 10d 的各个阶段均无显著变化。试验表明，*relA*、*spoT*、*ropS*、*clpX*、HP 在 Xcc 抗逆反应中发挥一定功能，还需要后续深入研究。

关键词：十字花科蔬菜黑腐病菌；VBNC；抗逆反应；基因表达量

* 第一作者：白凯红，博士研究生；E-mail：bkh0616@126.com

** 通信作者：罗来鑫，副教授，主要从事种传病害研究；E-mail：luolaixin@cau.edu.cn

六个苦瓜品种对枯萎病的抗病性鉴定试验*

王永阳**，田叶韩，李雪玲，何邦令***，高克祥***
（山东农业大学植物保护学院，山东省蔬菜病虫生物学重点实验室，泰安　271018）

摘　要：苦瓜枯萎病是由尖孢镰刀菌苦瓜专化型（*Fusarium oxysporum* f. sp. *momordicae*）引起的一种土传病害，连作会加重病害的危害，严重时甚至会造成绝产，极大地制约了苦瓜产业的发展。目前，生产中常用的化学防治措施对苦瓜枯萎病防治效果不佳，对该病害最经济有效的防治措施之一是选择种植抗枯萎病品种。

本试验采用病土培育法，制备含有 1×10^4 个孢子/g 的病土种植苦瓜。对来自南方的 6 个苦瓜品种（编号分别为 0822、1420、1510、1512、1522、1534）进行了枯萎病的抗病性鉴定，同时测定了苦瓜株高、茎粗等生长指标，评估各苦瓜品种对枯萎病的抗病性及其在北方露地环境下的适应性。试验测定了苦瓜苗移栽 45 d 和 75 d 时各品种的发病率和病情指数，选择高感品种 9208 作为对照，评估比较各苦瓜品种的抗病水平。结果显示，各鉴定苦瓜品种的发病率和病情指数均优于对照高感品种，其中品种 1512 表现最好，发病率和病情指数最低，75 d 时分别为 13. 04%和 11. 59，抗病效果达到了 87. 68%。品种 1534、1510 也表现良好，抗病效果均达到 80%以上。由此可知，1512、1534、1510 三个苦瓜品种对枯萎病抗病效果较好，均达到高抗水平，适合作为抗枯萎病品种在北方推广应用。

关键词：苦瓜；品种；枯萎病；抗病性

* 基金项目：公益性行业（农业）科研专项经费项目（201503110-12）

** 第一作者：王永阳，硕士研究生，研究方向：植物病害生物防治；E-mail：1425076736@ qq. com

*** 通信作者：何邦令，副教授，硕士生导师，研究方向：植物病虫害防治；E-mail：hebangling@ 126. com
高克祥，博士，教授，博士生导师，研究方向：植物病害生物防治；E-mail：kxgao63@ 163. com

Identification of *Phytophthora sojae*-resistance Soybean Germplasm from Huang-Huai-Hai Region and Northeast of China

Yang Jin, Ye Wenwu, Ren Linrong, Yao Yan, Wang Yan,
Dong Suomeng, Zheng Xiaobo, Wang Yuanchao
(*Plant Pathology Department*, *Nanjing Agricultural University*, *Nanjing*, *China*, 210095)

Abstract: *Phytophthora sojae*-caused root and stem rotis a destructive disease on soybean. Utilization of resistance cultivars is the most effective method to manipulate this soybean disease. In this study, we collected 223 soybean cultivars (lines) from Huang-Huai-Hai region and Northeast of China, and identified the response of these soybeans to 8 *P. sojae* strains which represent 8 major virulence types of China, respectively, using an etiolated hypocotyl-slit inoculation method. Results showed that 19 cultivars (lines) resistant to all 8 *P. sojae* strains and 53 resistant to 7 isolates, indicating that there are still some useful *P. sojae*-resistance soybean germplasm. The 223 soybean cultivars (lines) produced 36 reaction types, and 30 of them were new types that possibly carried new resistance genes. These results will benefit the soybean production and breeding program.

Key words: *Phytophthora sojae*; Resistance; Reaction types; Resistance genes

灵芝多糖诱导棉花抗枯萎病的效应*

王红艳**，赵 鸣，薛 超，隋 洁
（山东棉花研究中心，济南 250100）

摘 要：诱导抗性是利用植物内在抗性机制来防治病害的有效途径，符合我国农药减量施用和农产品安全的战略需求。灵芝多糖（*Ganoderma lucidum* polysaccharide，GLP）是灵芝提取物中的关键药效成分，其作为一种寄主免疫增强剂已广泛应用于医药领域，然而目前在植物病害方面对于灵芝多糖进行防治的研究甚少。棉花是我国种植面积最大的经济作物，也是重要的战略储备资源。棉花枯萎病是棉花生产上的重要病害，在国内各大棉区均有发生。

本课题组前期研究表明灵芝多糖对棉花枯萎病和立枯病均有一定的抑菌活性，并可以诱导棉花产生系统抗病性。本研究通过温室盆栽防治试验研究了灵芝多糖通过不同浓度、不同诱导方式处理对棉株生长的影响及对棉花抗病性的作用。25~100mg/L 灵芝多糖浸种处理后，对棉花种子发芽率、幼苗株高、植株质量均表现出一定的促进作用。其中 100mg/L 灵芝多糖处理的棉花种子发芽率最高，达 89.5%，与清水对照 80.5%的发芽率相比有显著提高；且该处理对棉花幼苗的株高和地下部分鲜重的促进作用也最大，分别较清水对照增加了 10.1%和 32.2%。棉株生长至两叶一心期时，100mg/L 灵芝多糖采用灌根、喷雾及茎部注射 3 种诱导方式处理棉苗，2d 后挑战接种棉花枯萎病菌孢子悬浮液。接种后 2~6d，采用灌根和茎部注射方式处理的系统叶片中植物抗病防御相关酶 SOD、POD、PAL 活性以及根系活性显著高于清水对照，喷雾处理的系统叶片酶活及根系活性略高于清水对照。棉株两叶一心期，采用 100mg/L 灵芝多糖灌根处理 2d 后测定棉株体内部分抗病相关基因的表达量。结果表明：灵芝多糖处理可以诱导棉株苯丙烷类代谢途径的关键基因 *PAL*、*4CL* 以及抗病相关基因 *basic chitinase*、*β*-1，3-*glucanase* 相对表达量的提高。

综上，灵芝多糖可以诱导棉花产生对枯萎病的系统抗病性，为将灵芝多糖开发为新型生物农药提供了理论基础。今后需要进一步研究灵芝多糖诱导抗病的信号转导途径。

关键词：灵芝多糖；棉花；枯萎病；诱导抗性

* 基金项目：山东省自然科学基金（ZR2015CQ024）
** 第一作者：王红艳，女，助理研究员，博士，从事植物病害生物防治、生物农药研究；E-mail：sdauwhy@ 163. com

不同玉米品种对斯克里布纳短体线虫的抗性鉴定*

逯麒森，杜　鹃，周志远，李　宇，王　珂，王振跃，李洪连**
（河南农业大学植物保护学院，郑州　450002）

短体线虫（*Pratylenchus* spp.）是一类重要的植物病原线虫。该线虫主要在植物根部的皮层组织取食，引起组织坏死，影响植物地上部的生长而造成作物产量的损失。据研究报道，短体线虫可以为害玉米，造成玉米产量和质量的下降。在国内，高学彪首次报道斯克里布纳短体线虫对玉米具有较强致病力，并指出该线虫会造成病株根系稀少，根部发生坏死病斑，地上部瘦弱矮小。Waudo 等的研究结果表明玉米自交系 B68Ht 对斯克里布纳短体线虫具有较好的抗性。Smolik 等的研究结果表明斯克里布纳短体线虫在玉米自交系 A619Ht、SD45、84742、84763 都有较高的繁殖力。关于不同玉米推广品种对该线虫的抗性报道相对较少，有待进一步加强相关研究。

本研究对郑州郊区玉米根际采集到的根腐线虫进行了分离，并利用形态和分子生物学方法对其进行了种类鉴定，确定了该病原线虫为斯克里布纳短体线虫（*P. scribneri*）。在对该线虫利用胡萝卜愈伤组织进行单雌扩繁培养的基础上，利用盆栽接种实验检测了豫安 3 号、鲁单 818、豫单 112、秋润 100 等 22 个推广玉米品种对该线虫的抗性。接种 60 d 后的调查结果表明：斯克里布纳短体线虫能侵染所有供试玉米品种的根部，并对玉米的生长造成一定的影响，导致玉米株高降低，地上鲜重减轻，根重减轻，但对玉米的叶片数和根长的影响不大；供试的 22 个玉米品种中无高抗品种，只有 MC703、秋润 100 两个品种平均病情指数分别为 46.7 和 50.0，表现为中抗，占供试品种的 9.1%；亿科 209 等 8 个品种平均病情指数在 50.1～75.0，表现为中感，占供试品种的 36.4%；其余 12 个品种平均病情指数均在 75.0 以上，表现为高感，占供试品种的 54.5%，发病最重的品种为陕科 6 号，平均病情指数达到 100。鉴定结果表明：目前生产上主推的玉米品种对斯克里布纳短体线虫的抗性普遍较差，应引起重视。

---

* 基金项目：河南省玉米产业体系植保岗位科学家科研专项（S2015-02-G05）和国家自然科学基金（31601619）

** 通信作者：李洪连，教授，主要从事植物土传病害研究；E-mail：honglianli@ sina. com

水稻抗稻瘟病基因 *Piyj* 的精细定位*

周　爽[1]，李腾蛟[1]，王　丽[1]，方安菲[1]，赵文生[1]，
李振宇[2]，张士永[3]，彭友良[1]，孙文献[1]
（1. 中国农业大学植物保护学院，北京　100193；2. 辽宁省盐碱地利用研究所，
盘锦　124010；3. 山东省农业科学院水稻所，济南　250100）

摘　要：对辽宁省审定的高产、优质、抗病性强的水稻品种盐粳 456 进行抗瘟性分析，推测其含有至少 2 个抗瘟基因（其中 1 个为新抗瘟基因，暂时命名为 *Piyj*。通过稻瘟菌小种 11-856-57-1，11-856-9-2 和 11-856-20-1 对盐粳 456 进行划伤接菌鉴定，利用简单序列重复标记（Simple Sequence Repeat，SSR）对盐粳 456 与感病亲本丽江新团黑谷的 F_2 后代进行抗瘟基因的连锁分析，将新抗瘟基因定位在水稻第 12 号染色体～6.2 Mbp 物理距离内。随后，通过对盐粳 456 进行全基因组测序，拼接获得了盐粳 456 基因组序列，将其与丽江新团黑谷基因序列进行对比，获得了在新抗瘟基因初定位的区间内两品种间的插入与缺失片段（Insertion and Deletion，InDel），并据此设计了 27 对 InDel 分子标记。接着，运用多态性 SSR 和 InDel 分子标记筛选约 14 000个 F_2 单株，得到 60 个在初定位区间内发生基因重组的个体。通过对重组个体及其后代抗感表型进行了鉴定，结合重组个体的基因型，将抗瘟基因 *Piyj* 定位在分子标记 InDel4 与 InDel25 间的～580kb 的物理距离内。比对水稻品种日本晴与盐粳 456 的基因序列，在该区域中有 2 个基因编码具有核苷酸结合位点（NBS）和富含亮氨酸重复（LRR）结构域的蛋白。目前，对这两个基因的功能鉴定正在进行中。

关键词：稻瘟病；抗瘟基因；图位克隆；分子标记

* 基金项目：“十二五” 国家科技支撑计划（21032267）

Genetic Dissection of *Arabidopsis* Genes Governing Immune Gene Expression (*Aggie*) *

Cui Fuhao[1,2**], Xu Guangyuan[2], Zhou Jinggeng[3],
Sun Wenxian[1], He Ping[3***], Shan Libo[2]

(1. *Department of Plant Pathology, China Agricultural University, Beijing* 100193, *China*;
2. *Department of Plant Pathology & Microbiology, and Institute for Plant Genomics & Biotechnology, Texas A&M University, College Station, TX* 77843, *USA*;
3. *Department of Biochemistry & Biophysics, and Institute for Plant Genomics & Biotechnology, Texas A&M University, College Station, TX* 77843, *USA*)

Abstract: Plants and animals possess pattern recognition receptors (PRRs) to instantaneously detect the presence and determine the nature of infection by recognizing microbe-associated molecular patterns (MAMPs). MAMP perception elicits profound transcriptional reprogramming in hosts to fend off pathogen invasion. The signaling networks orchestrating immune gene expression remain elusive. We have developed a sensitive genetic screen based on the transcriptional changes upon MAMP perception in *Arabidopsis* transgenic plants carrying an early and specific immune responsive gene *FRK*1 promoter fused with a luciferase (LUC) reporter. A series of mutants with altered *pFRK*1:: *LUC* activity were identified and named as *Arabidopsis* genes governing immune gene expression (*aggie*). The *aggie*101 mutant showed the reduced *pFRK*1:: *LUC* induction in response to multiple MAMP treatments. Real-time RT-PCR analysis confirmed that the endogenous *FRK*1 gene activation by MAMPs was also compromised in the *aggie*101 mutant. The activation of evolutionarily conserved MAP kinase (MAPK) cascades constitutes an early and convergent step downstream of multiple PRR signaling. MAMP perception also triggers a rapid burst of reactive oxygen species (ROS). Significantly, MAPK activation and ROS production are reduced in *aggie*101 mutant, suggesting that the mutation occurs at an immediately early step after MAMP perception. Map-based cloning coupled with next-generation sequencing revealed that the *aggie*101 mutant encodes a kinase-inactive BAK1, a signaling partner/co-receptor of multiple PRRs and the BRI1 receptor for plant growth hormone brassinosteroids (BRs). Consistently, the *aggie*101/*bak*1-16 mutant exhibited compromised responses to BR treatment. Furthermore, the *aggie*101/*bak*1-16 mutant exhibited seedling lethality when combined with the mutation of its closest homolog BKK1/SERK4. The data indicate that the BAK1 kinase activity is required for its multiple functions in plant immunity, development and cell death control. We are deploying the next generation sequencing coupled with map-based cloning to uncover the gene identities of other *aggie* mutants. The isolation and characterization of these *Aggie* genes will help understand host immune signaling and provide genetic resources to improve crop resistance.

Key words: Arabidopsis; Innate immunity; Signaling transduction; Mutant screen

---

* Funding: National Natural Science Foundation of China (Grant No. 31671991)

** First author: Cui Fuhao, Associate Professor; E-mail: cuifuhao@ 163. com

*** Corresponding authors: He Ping, Professor, E-mail: pinghe@ tamu. edu; Shan Libo, Professor, E-mail: lshan@ tamu. edu

水稻 PTI 关键调控组分高效遗传筛选系统的建立*

齐 婷**，吴文章**，柳建英，张 亢，孙文献，崔福浩***

（中国农业大学植物保护学院，农业部作物有害生物监测与
绿色防控重点实验室，北京 100193）

摘 要：水稻是我国最重要的粮食作物之一，而白叶枯病、稻瘟病、稻曲病等细菌性和真菌性病害对水稻产量和品质造成了严重威胁，提高水稻抗病性已成为保障我国水稻生产安全的重要途径。双子叶模式植物拟南芥的抗病机制已得到深入研究，但对单子叶模式植物水稻抗病途径的了解仍旧十分匮乏。前期研究发现，细菌鞭毛蛋白、肽聚糖、真菌几丁质等微生物相关分子模式（Microbe-associated Molecular Patterns，MAMPs）能够显著诱导水稻抗病相关基因 *OsPBZ*1 的表达。因此，笔者构建了 *pOsPBZ*1：*luciferase* 转基因水稻，发现转基因水稻愈伤中 *pOsPBZ*1：*LUC* 的活性能被细菌鞭毛蛋白 flg22 强烈诱导。获得 *pOsPBZ*1：*LUC* 转基因水稻纯合株系后，笔者计划在此基础上创建水稻 PTI 关键调控组分 EMS 突变体库，快速筛选 MAMPs 处理后水稻 PTI 信号显著改变的突变体，并检测候选突变体对水稻白叶枯菌、稻瘟菌、稻曲菌等重要病原菌的抗性，对抗病性显著改变的突变体进行图位克隆与功能鉴定。水稻 PTI 关键调控组分高效遗传筛选系统的建立，不仅能加速解析水稻 PTI 信号传导的分子调控机制，而且能为提高水稻抗病性提供新的靶标和思路。

关键词：水稻抗病机制；PTI；突变体库；信号传导

---

* 基金项目：国家自然科学基金（31671991）

** 第一作者：齐婷，硕士研究生，研究方向：水稻与病原细菌的分子互作；E-mail：328715042@ qq. com
吴文章，硕士研究生，研究方向：水稻与病原细菌的分子互作；E-mail：524792791@ qq. com

*** 通信作者：崔福浩，副教授，主要从事水稻与病原细菌、真菌互作；E-mail：cuifuhao@ 163. com

小麦抗源兴资 9104 抗白粉病 QTL 定位*

黄　硕，王琪琳[2]，穆京妹[1]，曾庆东[2]，吴建辉[2]，黄丽丽[2]，康振生[2**]，韩德俊[1**]
（1. 旱区作物逆境生物学国家重点实验室；2. 旱区作物逆境
生物学国家重点实验室，杨凌　712100）

摘　要：小麦白粉病严重威胁小麦安全生产，发掘小麦抗白粉病对于培育新品种，实现病害持久控制具有重要意义。本研究以小麦抗源兴资 9104 与 Avocet S 杂交创制的 177 个重组自交系（RILs）为材料，利用 Illuminaiselect 90K SNP 芯片进行全基因组扫描，利用 Carthagene（V1.3）软件构建遗传连锁图谱；分别在湖北武汉和陕西杨凌进行两年两点成株期抗白粉病表型鉴定，以定位小麦抗白粉病 QTLs。结果表明：小麦抗感双亲间有 12 496个 SNP 多态性位点，获得包含 6 335个 SNP 和 19 个 SSR 标记的遗传连锁图；获得覆盖小麦 21 条染色体的 36 个连锁群，总图距为 3 148.9cM，其中 1B 染色体上的标记在短臂区段极度压缩，与之前报道过的兴资 9104 存在 1RS/1BL 易位相符。QTL Icimapping（V4.0）软件完备区间作图法，以最严重度（MDS）为指标进行 QTL 定位，鉴定出抗白粉病 QTL*Qpmxznwafu*－*3BS* 解释的表型变异平均为 17%；*Qpmxznwafu*-*5BL* 解释的表型变异平均为%。本研究为进一步开发以上 QTL 的分子标记进行分子标记辅助选择育种和研究小麦成株期抗白粉病遗传机制奠定了基础。

关键词：小麦；白粉病；QTL；SNP

---

* 基金项目：国家基础研究“973”计划（2013CB127700）；国家自然科学基金（31371924）

** 通信作者：康振生，教授，主要从事小麦与条锈菌互作研究；E-mail：kangzs@nwsuaf.edu.cn
韩德俊，教授，主要从事小麦抗条锈病遗传育种研究；E-mail：handj@nwsuaf.edu.cn

利用HIGS技术创制短柄草抗赤霉病材料*

谭成龙，马 微，郭 军**
（西北农林科技大学植物保护学院，旱区作物逆境生物学国家重点实验室，杨凌 712100）

摘 要： 小麦赤霉病（Fusarium head blight）是由禾谷镰刀菌（*Fusarium graminearum*）引起的一种真菌病害，是世界范围内的一种毁灭性病害。小麦受到赤霉菌侵染后不仅产量下降，而且会产生多种毒素，这些毒素对人和动物都会产生毒害作用。培育抗性品种是针对小麦病害最为经济环保的防控策略。然而，在目前的研究中，应对赤霉病抗病品种相对较少。因此，鉴定赤霉病的致病相关基因和创制抗赤霉病的种质资源显得尤为重要。

近年来，寄主诱导的基因沉默（Host-Induced Gene Silencing，HIGS）技术已被广泛的应用于基础理论研究及培育新的种质资源。在本实验中，以禾谷镰刀菌的分泌蛋白和激酶基因作为沉默的靶标基因。构建沉默表达载体后通过农杆菌转化的方法转入短柄草中，以期筛选出沉默后抗病性明显增强的基因，为后续创制小麦抗赤霉病种质资源奠定基础。本实验构建了FGSG_ 00677、FGSG_ 08731、FGSG_ 13463、CYP51、FGSG_ 09660、FGSG_ 05775、FGSG_ 12300、FGSG_ 11564、FGSG_ 10056共9个基因的沉默表达载体。通过农杆菌介导的方法将构建好的沉默表达载体转入短柄草中。FGSG_ 00677、FGSG_ 08731、FGSG_ 13464、CYP51四个基因获得了稳定的转化植株。通过PCR检测和GUS染色检测，FGSG_ 00677获得阳性株系7株；FGSG_ 08731获得阳性株系3株；FGSG_ 13464获得阳性株系5株；CYP51获得阳性株系3株。将FGSG_ 00677、FGSG_ 08731、FGSG_ 13464、CYP51 T2代接种禾谷镰刀菌PH-1后，野生型短柄草小穗发病明显，而转基因植株几乎没有发病，且PH-1在转基因短柄草小穗中的侵染率显著低于在野生型短柄草小穗中的侵染率，说明转基因短柄草植株对赤霉病具有一定的抗病性。组织学观察显示侵染野生型植株的禾谷镰刀菌菌丝网较为密集，而侵染转基因植株中的菌丝网较为稀疏，有的甚至没有菌丝网，只有稀疏的菌丝，菌丝膨大畸变。说明在转基因植株中，禾谷镰刀菌的侵染力有所下降。从而进一步证明了转基因短柄草植株对赤霉病的抗病性。本实验通过HIGS技术鉴定赤霉病的致病相关基因，创制短柄草抗赤霉病新材料，从而为创制小麦抗赤霉病种质资源提供基因资源和理论基础。

关键词： 小麦赤霉病；寄主诱导的基因沉默；农杆菌介导；短柄草

---

* 基金项目：国家重点基础研究发展计划（No. 2013CB127700）；国家自然科学基金资助项目（No. 31371889；No. 31171795）

** 通信作者：郭军；E-mail：guojunwgq@ nwsuaf. edu. cn

参与小麦抗条锈菌基因 *Yr10* 过敏性坏死反应 *TaMYB29* 的功能研究

王晓静[1]，王亚茹[1]，黄丽丽[2]，康振生[2]
（1. 西北农林科技大学生命学院，旱区作物逆境国家重点实验室，杨凌　712100；
2. 西北农林科技大学植保学院，旱区作物逆境国家重点实验室，杨凌　712100）

摘　要：通过实时荧光定量 PCR 技术、亚细胞定位技术、酵母单杂分析、瞬时过表达和病毒诱导的基因沉默分析 TaMYB29 的表达和功能。结果表明：克隆 TaMYB29 基因全长 786bp，编码 262 个氨基酸。定量分析表明 TaMYB29 在条锈菌与寄主互作过程中的表达谱分析在前期表达升高，并且在非亲和反应中表达显著升高。TaMYB29 受外源激素和非生物胁迫干旱诱导表达上调。小麦原生质体和烟草瞬时表达 TaMYB29 定位于细胞核。酵母自激活实验说明 TaMYB29 具有较弱的自激活活性。VIGS 瞬时沉默 TaMYB29，接菌含有 Yr10 的小麦后坏死斑减少。PVX 介导的病毒表达系统过表达 TaMYB29，能引起烟草叶片的 PCD。说明 TaMYB29 能够促进植物细胞的坏死，在 Yr10 引起的过敏性坏死反应中发挥作用。

关键词：条锈菌；小麦条锈菌基因 Yr10；TaMYB29；过敏性坏死反应

小麦类钙调磷酸酶亚基 B 蛋白基因 *TaCBL4* 的功能分析*

薛庆贺，刘　芃，康振生**，郭　军**
（旱区作物逆境生物学国家重点实验室，西北农林科技大学植物保护学院，杨陵　712100）

摘　要：类钙调磷酸酶亚基 B 蛋白（calcineurin B-1ike protein，CBL）作为一类钙离子结合蛋白，通过与一类蛋白激酶（CBL-interacting protein kinase，ClPK）结合，在钙信号依赖的生理生化过程中发挥作用。*CBL* 基因在植物非生物胁迫上的重要作用已经得到广泛的证实，然而关于其在生物胁迫上的作用知之甚少。本研究通过对小麦基因组中 CBL 家族基因的鉴定并利用 RT-PCR，在条锈菌诱导的小麦叶片中克隆得到小麦 CBL 家族中的一个基因，命名为 *TaCBL4*。并利用 qRT-PCR 技术、亚细胞定位技术及病毒介导的基因沉默技术（Virus-Induced Gene silence，VIGS）分析了其功能特性。结果显示，*TaCBL4* 可编码 219-aa 的氨基酸序列，包含 4 个保守的 EF 手结构。与水稻、拟南芥的 CBL 蛋白具有高度相似性，其中与水稻 OsCBL4 相似性达 88.28%。亚细胞定位实验表明 TaCBL4 的分布在整个小麦细胞中，即为细胞质、细胞膜和细胞核。定量分析表明，在小麦与条锈菌亲和与非亲和互作的过程中，*TaCBL4* 均受到条锈菌的诱导表达。通过 VIGS 技术沉默 *TaCBL4* 后发现小麦的抗病性明显减弱，沉默植株接种条锈菌非亲和小种 CYR23 后，活性氧积累明显受到抑制。同时，病程相关蛋白基因 *TaPR*1、*TaPR*2 表达也受到了抑制。综上结果，*TaCBL4* 可能通过调控活性氧积累参与了小麦对条锈菌的抗病过程。

关键词：小麦；小麦条锈菌；CBL；亚细胞定位；qRT-PCR；VIGS

---

* 基金项目：国家重点基础研究发展计划（No. 2013CB127700）；国家自然科学基金资助项目（No. 31371889；No. 31171795）

** 通信作者：康振生；E-mail：kangzs@ nwsuaf. edu. cn
郭军；E-mail：guojunwgq@ nwsuaf. edu. cn

Blufensin1 负调节小麦条锈病

张新梅[1]，裴晨铃[1]，李　兴[1]，冯　浩[2]，康振生[2]
（1. 西北农林科技大学生命科学学院，旱区作物逆境生物学国家重点实验室，杨凌　712100；
2. 西北农林科技大学植物保护学院，旱区作物逆境生物学国家重点实验室，杨凌　712100）

摘　要：由条形柄锈菌（*Puccinia striiformis* f. sp. *tritici*）引起的小麦条锈病是一种世界性范围内普遍流行的气传病害，严重威胁小麦生产。CYR31、CYR32 和 CYR33 等强致病性小种的出现和发展，使得我国小麦再度处于条锈病大流行的威胁中，研究小麦抗条锈菌的分子机制进而培育抗性品种是目前急需解决的重大问题。随着 Mlo 基因在大麦白粉病防治方面的应用，在育种过程中突变感病基因（Sgene）作为一种新型的抗病育种策略。本研究根据实验室构建的条锈菌 CYR31 与小麦亲和互作的 cDNA 文库，分析获得差异表达序列，结合荧光定量 PCR 的实验结果筛选到候选小麦感病基因 *Blufensin*1 基因 EST 序列，通过电子克隆已获得该基因全长序列。此外，通过 BIFC 技术，发现 TaBln1 与钙调蛋白互作。通过病毒介导的基因沉默技术（VIGS）发现沉默 *Blufensin*1 基因后，小麦抵抗条锈病能力增强，表明 *Blufensin*1 基因在小麦与条锈菌互作中起到负调节的作用。研究结果为解析小麦与条锈菌互作中的感病机制以及通过沉默感病基因为创制持久、广谱抗病品种奠定基础。

关键词：小麦；条锈菌；感病基因

第七部分
病害防治

Sensitivity of *Cochliobolus heterostrophus* to DC and QoI Fungicides, and Their Control Efficacy Against Southern Corn Leaf Blight in Fujian Province*

Dai Yuli**, Gan Lin, Ruan Hongchun, Shi Niuniu, Du Yixin, Chen Furu, Yang Xiujuan***

(*Fujian Key Laboratory for Monitoring and Integrated Management of Crop Pests, Institute of Plant Protection, Fujian Academy of Agricultural Sciences, Fujian Fuzhou* 350013)

Abstract: To determine the sensitivity of *Cochliobolus heterostrophus* to DC and QoI fungicides, and evaluate their control efficacy against southern corn leaf blight (SCLB), a total of 73 *C. heterostrophus* isolates collected from seven regions in Fujian Province were tested to determine the sensitivity to iprodione and pyraclostrobin, the representative fungicides for DC and QoI families respectively, by the method of measuring the mycelial growth on the fungicide-amended media. The results indicated that the tested isolates were very sensitivity to iprodione and pyraclostrobin, and the EC_{50} of these isolates for iprodione and pyraclostrobin were in the range of 0.168 4~2.990 2μg/mL and 0.017 4~6.583 4μg/mL, with the mean value of 1.003 3±0.532 7μg/mL and 1.437 0±1.490 3μg/mL, respectively. The frequency curves of iprodione and pyraclostrobin were continuous and unimodal, and followed the normal distribution. Hence, the mean EC_{50} value of 1.003 3±0.532 7 μg/mL and 1.437 0±1.490 3 μg/mL could be severed as baseline sensitivities of *C. heterostrophus* to iprodione and pyraclostrobin, respectively. Results of field trials showed that two sprays of 50% iprodione SC at 500μg a. i. /mL and 25% pyraclostrobin EC at 250μg a. i. /mL exhibited greater control efficacy for the control of SCLB, with the control efficacities ranging of 61.51% ~ 70.69% and 69.54% ~ 78.98%, respectively. These efficacities were the equal of spraying of 25% propiconazole EC at 250μg a. i. /mL, but high than the efficacities of reference fungicides chlorothalonil and mancozeb. These resultsare instructive for guidance on reasonable selection of fungicides for effective management of SCLB, and guidance on resistance risk assessment and monitoring of *C. heterostrophus* in the future.

Key words: *Cochliobolus heterostrophus*; Baseline sensitivity; Iprodione; Pyraclostrobin; Control efficacy

---

* 基金项目：福建省农业科学院博士启动基金项目（2015BS-4）；福建省自然科学基金项目（2016J05073）；福建省农业科学院青年科技英才百人计划项目（YC2016-4）；福建省属公益类科研院所专项（2014R1024-5）

** 第一作者：代玉立，安徽霍邱人，助理研究员，博士，研究方向：真菌学及植物真菌病害；E-mail：dai841225@126.com

*** 通信作者：杨秀娟，福建建瓯人，研究员，研究方向：植物病理学；E-mail：yxjzb@126.com

Inhibitive Effect of Chinese Leek Extract on the Incidence of Two Important Apple Disease: Apple Ring Rot and Apple Valsa Canker*

Huang Yonghong[1**], Zuo Cunwu[2], Zhang Weina[2]

(1. *College of horticulture*, *Qingdao agricultural university*, *Qingdao*, *Shandong*, 266109, *P. R. China*, 2. *College of horticulture*, *Gansu agricultural university*, *Lanzhou*, *Gansu*, 730070, *P. R. China*)

Abstract: Apple ring rot and Apple valsa canker, caused by fungus *Botryosphaeria berengeriana* and *Valsa mali*, are two major disease of apple in China, and are difficult to control using conventional methods such as planting resistant varieties or application of fungicides. Here, we describe an effective method for controlling the disease using Chinese leek (*Allium tuberosum*). In a confrontation experiment, Chinese leek extract (CLE) significantly suppressed the growth of the two fungus. Growth of *V. mali* colonies after a 5-day treatment with 1 : 250 (1 g Chinese leek/250 mL) and 1 : 100 preparations of CLE was decreased by 73. 2% and 97. 13%, respectively, compared with that of the untreated control. Furthermore, 100% inhibition was obtained using a 1 : 20 preparation of CLE. Growth of *B. berengeriana* colonies after a 5-day treatment with 1 : 250 (1 g Chinese leek/250 mL) and 1 : 100 preparations of CLE was decreased by 56. 75% and 95. 38%, respectively, compared with that of the untreated control. Furthermore, 100% inhibition was obtained using a 1 : 20 preparation of CLE. CLE clearly disrupted the normal morphology of *V. mali* and *B. berengeriana* mycelium, and significantly suppressed the incidence of the two diseases. Forapple valsa canker, Compared with the untreated controls, the rate of spread of disease spots decreased by 96. 4% and 94. 75%, respectively, on the shoots and fruits of apples treated with a 1 : 20 preparation of CLE. Forapple ring rot, compared with the untreated controls, the rate of spread of disease spots decreased by 100% and 100%, respectively, on the shoots and fruits of apples treated with a 1 : 20 preparation of CLE. The findings of this study indicate that CLE and its main active components significantly suppress the growth of *V. mali* and reduce the incidence of apple valsa canker caused by this fungus.

Key words: *Allium tuberosum*; Apple ring rot; Apple valsa canke; Biocontrol

---

* Funding: This research was supported by theNational Natural Science Foundation of China (Grant No. 31272151, 31471864) and the Qingdao Agricultural University High-level Personnel Start-up Fund (6631115024)

** Corresponding author: Huang Yonghong; E-mail: gstshh@ 126. com

低毒真菌病毒在植物病害生物防治中的作用研究*

刘　忱**，张美玲，舒灿伟，周而勋***
（华南农业大学农学院植物病理学系，广州　510642）

摘　要：真菌病毒（Mycovirus 或 Fungal virus）是一类侵染真菌并能在其中复制的病毒，广泛存在于各大真菌类群中。多数真菌病毒侵染寄主真菌后并不会引起寄主真菌出现明显的症状，只有少部分能对其造成影响。其中，有一些真菌病毒在减弱寄主致病力方面具有显著的效果，这些病毒统称为低毒真菌病毒，因其具有弱毒特性而被认为是一种重要的生防菌资源。本文主要从以下几个方面对低毒真菌病毒在植物病害生物防治中的作用研究进行了较为全面的综述，旨在为该类病毒的生防应用研究提供借鉴：①低毒真菌病毒的发现：自第一例低毒真菌病毒——栗疫病菌（*Cryphonectria parasitica*）CHV1 被发现以来，在其他植物病原真菌中均陆续发现了低毒真菌病毒的存在。如核盘菌 DNA 病毒 SsHADV-1、灰霉线粒体病毒 BcDRV、立枯丝核菌双分体病毒 RsPV2 等。②低毒真菌病毒的生物防治应用：最经典的例子当属栗疫病菌低毒病毒 CHV1 在防治栗疫病上的应用，该病毒在 20 世纪 90 年代欧洲栗疫病流行期对栗疫病的生防效果取得了骄人的成效，挽救了大批濒临灭绝的欧洲栗树林；此外，华中农业大学姜道宏教授课题组从油菜菌核病分离的 DNA 病毒 SsHADV-1 能在田地中自然传播，具有重要生防应用前景。③低毒真菌病毒与寄主真菌之间的互作：主要列举栗疫病菌— CHVI 互作系统、核盘菌— SsDRV 互作系统、镰刀菌— FgV1 互作系统以及灰霉菌— BVX/BVF 互作系统这四个互作系统的研究进展。其中，栗疫病菌— CHVI 互作系统是目前唯一已经建立反向遗传学系统的互作系统，因而研究得相对较透彻，是低毒病毒与寄主真菌互作的模式系统。④低毒真菌病毒的生防优势：与其他生物防治方法相比，低毒真菌病毒在生防上的应用优势表现在：低毒真菌病毒一旦侵染到其他强毒力的寄主真菌，能快速发挥作用；低毒菌株和毒力菌株共享相似的生态位，低毒菌株不仅不会对寄主植物造成任何伤害，反而很可能会产生病原菌致病相关分子模式（pathogen-associated molecular patterns，PAMPs）或效应子（effector）而被寄主植物识别，从而诱导寄主植物产生防御反应以抵御毒力菌株的侵染；低毒真菌病毒进入农田生态系统后，既可以长久的定殖，又可以进行传播。⑤低毒真菌病毒应用中存在的问题：低毒真菌病毒虽然具有重要的生防潜力，但其存在极大的应用困境，主要原因是真菌菌株间的营养体不亲和性和真菌病毒体外传播途径不明朗。但具有强侵染力的低毒病毒能克服菌株间营养体不亲和性的缺陷，挖掘此类病毒成为解决低毒真菌病毒当前生防应用困境的主要手段。

关键词：低毒真菌病毒；植物病害；生物防治；dsRNA 真菌病毒

---

* 基金项目：国家自然科学基金项目（31470247）

** 第一作者：刘忱，硕士研究生，研究方向：植物病原真菌及真菌病害；E-mail：415576700@ qq. com

*** 通信作者：周而勋，教授，博导，研究方向：植物病原真菌及真菌病害；E-mail：exzhou@ scau. edu. cn

Evolution of Dimethomorph and Azoxystrobin Fungicides Resistance in Potato Pathogen *Phytophthora infestans*

Waheed Abdul*, Shen Linlin, Zou Wenchao,
Xie Jiahui, Wang Tian, Zhan Jiasui**
(*Institute of Plant Virology, Fujian Agriculture and Forestry University/Key Laboratory of Plant Virology of Fujian Province, Fuzhou* 350002)

Abstract: Fungicide resistance is very important problem currently in agriculture fields. In-vitro research was conducted for the evolution of fungicide resistance to *Phytophthora infestans*. Two fungicides dimethomorph and azoxystrobin were used for training of two different genotypes of isolates for evolutionary process. Measured the EC_{50} for both fungicides, 50% inhibition zone of dimethmorph and azoxystrobin fungicides were 0.1μg/mL and 0.2μg/mL respectively. After measured the 50% inhibition zone, different concentrations of fungicides amended in rye agar media for growth of pathogens. Training of pathogen started from wild types parental isolates. Gradually One year passage of original parental and mutant isolates from low concentration of fungicides to high concentrations. Parental isolates directly transferred to every concentration of fungicides but mutant isolates continuously transferred. Completely suppressed the growth of Parental isolates at 0.9μg/mL of azoxystrobin and 1μg/mL of dimethmorph but mutant isolates still grown upto1.5μg/mL of both fungicides, its mean adaption of pathogen to fungicides due to continuously using of fungicides on same pathogens. Sensitivity test was done by selection of 3 concentrations for each fungicide for all mutant and parental isolates. Almost all the mutant and parental isolates were sensitive to selective concentrations of fungicides. There was no evidence of pathogen resistance to both fungicides on the base of phenotypic growth.

Key words: *Phytophthora infestans*; Dimethomorph; Azoxystrobin; Evolution; Resistant

---

* First author: Waheed Abdul, Male, Master student, research direction for potato late blight; E-mail: waheed90539@gmail.com
** Corresponding author: Zhan Jiasui, Professor, research interests for population genetics; E-mail: Jiasui.zhan@fafu.edu.cn

Pleiotropy and Temperature-mediated Evolution of Plant Pathogen Adaptation to a Non-specific Fungicide in Agricultural Ecosystem*

He Menghan[1,2**], Li Dongling[1,2], Zhu Wen[1,2], Wu Ejiao[1,2], Yang Lina[1,2], Waheed Aboul[1,2], Zhan Jiasui[2,3***]

(1. *State Key Laboratory of Ecological Pest Control for Fujian and Taiwan Crops*, *Fujian Agriculture and Forestry University*, *Fuzhou*, 350002, *China*; 2. *Fujian Key Laboratory of Plant Virology*, *Institute of Plant Virology*, *Fujian Agriculture and Forestry University*, *Fuzhou*, 350002, *China*; 3. *Key Lab for Biopesticide and Chemical Biology*, *Ministry of Education*, *Fujian Agriculture and Forestry University*, *Fuzhou*, 350002, *China*)

Abstract: The spread of antimicrobial resistance and climate change represent two major phenomena that are exerting a disastrous impact on natural and social issues but investigation of the interaction between these phenomena in an evolutionary context is limited. In this study, a statistical genetic approach was used to investigate the evolution of antimicrobial resistance in agricultural ecosystem, and its association with air temperature. We found no resistance to mancozeb, a non-specific fungicide widely used in agriculture for more than half a century, in 215 *Alternaria alternata* isolates sampled from geographic locations along a climatic gradient and cropping system representing diverse ecotypes in China, consistent with low resistance risk of non-specific fungicides. Genetic variance accounts for 35% of phenotypic variation while genotype-environment interaction is negligible, suggesting that heritability plays a more important role in the evolution of resistance to non-specific fungicides in plant pathogensthan phenotypic plasticity. We also found that resistance to mancozeb in agricultural ecosystem is under constraining selection and significantly associated with local air temperature, possibly resulting from a pleiotropic effect of resistance with thermal and other ecological adaptations. The implication of these results for Antimicrobial management in the context of climate change is discussed.

Key words: *Alternaria alternata*; Negative pleiotropy; Climate change; Epistasis; Evolution of antimicrobial resistance; Population genetics

---

* Funding: The study was supported by national natural science foundation of China (grant No. 31371901 and U1405213)

** First author: He Menghan, female, PhD student, research direction for potato early blight; E-mail: hemenghan55@163.com

*** Corresponding author: Zhan Jiasui, Professor, research interests for population genetics; E-mail: Jiasui.zhan@fafu.edu.cn

两株生防链霉菌对烟草主要病原菌的颉颃作用研究

李继业[1]，彭立娟[2]，Paolo Cortesi[3]，Marco Saracchi[3]，Cristina Pizzatti[3]，陈孝玉龙[2]*
(1. 贵州大学农学院，贵阳 550002；2. 贵州大学烟草学院，贵阳 550002；3. 意大利米兰大学食品、环境与营养学院，米兰，意大利)

摘　要：链霉菌（*Streptomyces*）是土壤中含量丰富的一类有菌丝体的细菌，可以产生多种有益的次生代谢产物，使之具有成为生物农药和生物肥料的潜力。*S. exfoliatus* FT05W 和 *S. cyaneus* ZEA17I 是核盘菌、镰刀菌等土传病原菌的高效颉颃链霉菌，在田间能够良好地防控生菜上由核盘菌引起的菌核病。贵州省是中国第二大烟草生产省份，其烟草种植长期受到几种病原菌的高度威胁。在本研究中，通过平板对峙实验，笔者探索了 *S. exfoliatus* FT05W 和 *S. cyaneus* ZEA17I 对贵州省烟草上 3 种主要病原菌的颉颃作用。结果显示，*S. exfoliatus* FT05W 和 *S. cyaneus* ZEA17I 对烟草黑胫病病原，寄生疫霉菌烟草致病型（*Phytophtora parasitica var. nicotianae* Tucker）的抑制率分别达到了 68.82%和 76.18%。另外，*S. cyaneus* ZEA17I 对烟草赤星病病原——致病链格孢菌（*Alternaria alternata*）的抑制率达到了 35.5%，而 *S. exfoliatus* FT05W 对致病链格孢菌则无明显的颉颃作用。最后，两株链霉菌对于烟草青枯病病原，青枯劳尔氏菌（*Ralstonia solanacearum*）并无直接颉颃作用，不能抑制该病原菌的生长。研究结果表明：*S. exfoliatus* FT05W 和 *S. cyaneus* ZEA17I 对于贵州省烟草上主要的 3 种病原菌的颉颃作用差异较大，但对于烟草黑胫病的防治具有良好的应用潜力，其对烟草的定殖（EGFP 标记）及黑胫病的田间防治效率正在积极探索之中。

关键词：链霉菌；生物防治；平板对峙；烟草病原菌

* 通信作者：陈孝玉龙；E-mail：chenxiaoyulong@ sina. cn

多雨湿润蔗区甘蔗重要病害发生流行动态与防控策略*

李文凤**，尹　炯，单红丽，张荣跃，仓晓燕，王晓燕，罗志明，黄应昆***
（云南省农业科学院甘蔗研究所/云南省甘蔗遗传改良重点实验室，开远　661699）

摘　要：近年我国蔗区雨季来得早且持续时间长，阴雨天多、日照少、温凉高湿，为甘蔗病害大发生流行创造了极其有利条件，感病品种遇上适宜气候条件（多雨高湿）导致甘蔗梢腐病、锈病、褐条病等多种重要病害并发流行、大面积为害成灾，并存在日趋严重趋势。本文在调查研究基础上，紧扣寄主植物、病原物和环境条件等植物病害暴发流行3要素，分析探讨了多雨湿润蔗区甘蔗重要病害的发生动态，并提出了相应的防控策略措施。据调查，以梢腐病发生最广、为害最重，粤糖93-159、新台糖1号、川糖79-15、粤糖00-236、闽糖69-421、福农91-21等高感品种连片种植区域常大面积爆发流行，高温高湿、通风不良、偏施氮肥等极易发生，感病品种发病严重，常使大量蔗茎枯死，造成减产减糖，据小样点分析评估，病株率平均为81.1%、严重的100%，甘蔗产量损失率平均为38.42%、最多的48.5%，甘蔗糖分平均降低3.14%、最多的降低4.21%，重力纯度平均降低4.15%、最多的降低8.14%；锈病主要发生于云南、广西湿润蔗区，雨多、露水重、湿度大极易流行，规模化连片种植粤糖60号、德蔗03-83、柳城03-1137、桂糖46号等高感品种和多雨高湿是锈病暴发流行主要诱因；褐条病多发生于土壤瘦瘠或缺磷少钾和雨多（长期阴雨天气）、湿度大的蔗区，长期连片种植福农91-21、桂糖02-761、粤糖93-159、粤糖00-236、新台糖25号、闽糖69-421等高感品种最易爆发流行。7—10月多雨高湿季，既是甘蔗伸长大拔节奠定产量关键时期，也是多种甘蔗重要病害并发为害高峰期。为防止甘蔗病害大面积暴发流行、为害成灾，确保甘蔗后期安全生长，提出注重合理布局和选用抗病品种，大力推广使用脱毒健康种苗，因时、因地制宜强化田间管理，加强病情监测和发病初期选用50%多菌灵可湿性粉剂、75%百菌清可湿性粉剂或65%代森锌可湿性粉剂500~600倍液+0.2%~0.3%磷酸二氢钾液（300~500倍液）及时喷药防治，可有效控制病害大面积发生流行。

关键词：多雨湿润蔗区；甘蔗病害；发生动态；防控策略；技术措施

* 基金项目：国家现代农业产业技术体系建设专项资金资助（CARS-20-2-2）；云南省现代农业产业技术体系建设专项资金资助

** 第一作者：李文凤，女，云南石屏人，研究员，主要从事甘蔗病害研究；E-mail：ynlwf@163.com

*** 通信作者：黄应昆，研究员，从事甘蔗病害防控研究；E-mail：huangyk64@163.com

地衣芽孢杆菌 W10 抗菌蛋白的研究*

陈丽丽**，赵文静，朱　薇，谈　彬，纪兆林***，董京萍，童蕴慧，徐敬友
（扬州大学园艺与植物保护学院，扬州　225009）

摘　要：芽孢杆菌（*Bacillus* spp.）是植物病害生防中研究和应用较多的一类生防细菌，生防机制主要有营养和空间位点竞争、产生多种抑菌物质（如脂肽类抗生素、降解胞壁酶类、细菌素、抗菌蛋白、多肽类化合物等）、溶菌作用、诱导植物抗病性和促进植物生长等方面。地衣芽孢杆菌（*Bacillus. licheniformis*）W10 是笔者实验室从植物根际筛选获得的一株生防细菌，对多种植物病原真菌均有较好的抑制作用，而且具有定殖和诱导抗病能力，田间对番茄灰霉病、油菜菌核病和桃枝枯病等防效与化学农药腐霉利、多菌灵相当，另外该菌对植物还有一定的促生作用。该菌生防机制主要是产生一种抗菌蛋白。本文在此基础上对其抗菌蛋白进一步研究。W10 培养上清滤液通过硫酸铵沉淀、透析袋透析获得 W10 粗蛋白，再经柱层析（HiPrep 16/60 Sephacryl S-100 High Resolution，AKTA purifier）获得纯化的抗菌蛋白，经 MALDI-TOF/MS-MS 质谱分析和 Mascot 搜索串级质谱数据鉴定为丝氨酸蛋白酶。据此设计引物，从 W10 gDNA 中扩增获得基因序列，经测序含有一个开放阅读框，编码基因 1 347bp。经生物信息学分析，该基因编码 448 个氨基酸残基，分子量 48 794.16Da，等电点 6.04，是一种亲水性蛋白，在二级和三级结构中无规则卷曲占主要部分，α-螺旋结构和β-折叠结构所占比例相似，含有 Peptidase_ S8 的保守结构域。这为高效生防工程菌的构建和 W10 的进一步应用奠定了理论基础。

关键词：地衣芽孢杆菌（*Bacillus. licheniformis*）；抗菌蛋白；丝氨酸蛋白酶

---

* 基金项目：国家自然科学基金（31101475）；国家现代农业产业技术体系建设专项（CARS-31-2-02）；江苏省农业科技自主创新资金项目［CX（14）2015，CX（15）1020］；江苏省无锡市农业科技支撑项目（CLE01N1410）

** 第一作者：陈丽丽，硕士，研究方向：分子植物病理学研究；E-mail：1432465479@ qq. com

*** 通信作者：纪兆林，副教授，研究方向：植物病害生物防治及分子植病研究；E-mail：zhlji@ yzu. edu. cn

小麦纹枯病颉颃细菌的筛选鉴定及其对小麦幼苗的促生作用研究*

季　鹏**，庞　欢，盛　典，韩　超***
（山东农业大学植物保护学院，泰安　271018）

摘　要：为了探寻对小麦纹枯病有明显防效的生物菌剂，本研究通过采集山东泰安、莱芜、济南和诸城等地小麦连作田根际土壤，利用平板稀释法分离得到68株细菌，以小麦纹枯病致病菌立枯丝核菌为指示菌，通过平板对峙法，筛选获得了4株具较强颉颃效果的细菌，分别为TA28、TA31、Z-5和Z-3。经形态学观察、生理生化特性检测和16S rDNA基因序列分析，鉴定此4株颉颃菌分别为甲基营养型芽孢杆菌（*Bacillis methylotrophicus*）、枯草芽孢杆菌（*Bacillus subtilis*）、解淀粉芽孢杆菌（*Bacillus amyloliquefaciens*）和贝莱斯芽孢杆菌（*Bacillus velezensis*）。抑菌谱测定结果表明：4株颉颃细菌对腐霉菌、镰刀菌、致病疫霉、轮枝菌等多种土传病害病原真菌也具有良好的颉颃效果。温室盆栽试验结果表明，利过细菌菌液灌根不仅对小麦纹枯病发生具有明显的预防作用，并且对小麦根系发育具有促进作用。本研究结果表明，筛选出的4株细菌是立枯丝核菌等农作物常见土传病害病原菌潜在的优良生防菌株，作为生防菌剂具有潜在的开发应用价值。

关键词：生防细菌；小麦纹枯病；土传病害；促生作用

---

* 基金项目：山东农业大学泉林黄腐酸肥料工程实验室开放研究基金项目；山东农业大学青年创新基金

** 第一作者：季鹏，男，山东邹城人，研究方向：微生物资源及病害防治研究；E-mail：sdauzbjp@163.com

*** 通信作者：韩超，博士研究生，讲师，研究方向：微生物资源及病害防治研究；E-mail：hanch87@163.com

2016—2017 年河南省小麦条锈病流行的原因分析及防控措施*

于思勤，彭　红，徐永伟
（河南省植物保护植物检疫站，郑州　450002）

小麦条锈病由条形柄锈菌（*Puccinia striiformis*）侵染引起，是一种随气流远距离传播、流行速度快、暴发性强、危害损失严重的病害。主要为害叶片，其次是叶鞘、茎秆和穗部。河南省小麦条锈病秋苗期就能发病，发生盛期在 4 月中旬至 5 月上旬，以豫南麦区发生严重。2016 年 12 月 14 日唐河县在全国最先发现小麦条锈病发病中心，比常年早 3 个月以上。2017 年 3 月豫南麦区进入普发期，4 月黄河以南麦区普遍流行，截止 5 月 27 日全省发生面积 2 321万亩，是 2000 年以来发生最严重的年份。

（1）发生特点。①发病时间早、范围广。2016 年 12 月 14 日唐河县发现小麦条锈病发病中心 9 个，中心面积 0.2～1m^2，严重度 10%～80%，小麦品种为衡观 35。南阳市淅川县、驻马店市正阳县也在 12 月份发现条锈病发病中心，均为多年所罕见。5 月 23 日调查，全省 18 个地市 118 个县见病，全省平均病叶率 5.1%，最高 100%，发病田平均严重度 14.7%，最高 100%。②菌源量大、侵染为害时间长。冬前多地出现发病中心，春季迅速扩展蔓延，发生程度严重，4 月上旬豫南地区小麦条锈病进入发病盛期，一直持续到 5 月中旬，较常年延长 15d 以上。③小麦品种之间发病程度差异明显。由于小麦条锈菌存在许多生理小种，优势生理小种不断变化，致病性增强，感病品种发生严重。

（2）流行原因分析。①菌源充足。研究结果表明，小麦条锈菌小麦条锈病菌能够在河南省的豫北、豫西海拔 1 400m 以上地区越夏，越夏的主要寄主是自生麦苗；受超强“厄尔尼诺”天气影响，2016 年越夏菌源量大，加上外来菌源丰富，导致秋季条锈病发病早，范围广。②气候条件有利。2016 年河南省秋冬季气温偏高，土壤墒情较好，有利于条锈菌的侵染和越冬繁殖；2017 年 3—5 月全省降水量偏多，平均气温偏高，有利于条锈病的扩展蔓延。③主导品种综合抗性较差。麦播期推迟，田间湿度大，麦苗长势弱，抗病性差，有利于病原菌侵染危害，发病程度严重。

（3）防治措施。①种植抗耐病品种。抗病性鉴定和发病程度调查结果表明，周麦 22、矮抗 58、郑麦 366、百农 207、西农 979、豫农 416、周麦 18、周麦 31、周麦 32、豫麦 158、豫麦49-198、中育 9307、洛麦 31、新麦 18、新麦 20 等对小麦条锈病抗性较好，可以因地制宜选择种植，合理布局抗条锈品种，防止抗性丧失。②加强栽培防病措施。在小麦播种前，消灭条锈病菌越夏的豫西、豫北山区自生麦苗，减少初侵染菌源；提倡施用腐熟有机肥，增施磷钾肥，氮磷钾合理搭配施用，增强小麦抗病能力。③科学用药。全面实行三唑类药剂拌种，从源头控制苗期条锈病发生，降低春季条锈病流行速度。在条锈病点片发生期，严格封锁控制发病中心，发现一点控制一片，发现一片控制全田；当条锈病病叶率达 1%时，喷洒 20%三唑酮乳油 1 000倍液、12.5%烯唑醇可湿性粉剂 1 500倍液、43%戊唑醇或 25%丙环唑乳油 2 000倍液，做到普治与挑治相结合，注意喷匀喷透，严重发生地块间隔 7d 再喷一次。

* 基金项目：河南省小麦产业技术体系（S2015-01-G08）；公益性行业科技专项（201303030）

解淀粉芽孢杆菌 PG12 对水稻细菌性条斑病菌的颉颃效果研究*

杨　旸**，段雍明，隋书婷，李　燕，王　琦***
（中国农业大学植物病理学系，农业部植物病理学重点开放实验室，北京　100193）

摘　要：芽孢杆菌（*Bacillus* spp.）在自然环境中广泛存在，也是植物体内常见的内生细菌，具有适应性广、作用机制多样、抗逆性强、环境友好等优点，被认为是一种重要的生防因子。解淀粉芽孢杆菌（*Bacillus amyloliquefaciens*）PG12 是一株防治苹果轮纹病的生防菌，在苹果采前或采后施用，均可在一定储藏期内减轻苹果采后病害的发生。PG12 可明显抑制苹果轮纹病菌菌丝并导致菌丝形态异常。进一步研究表明，PG12 对苹果轮纹病具有较好防效的主要原因是产生性质稳定的脂肽类抗生素。目前关于 PG12 对细菌病害的广谱防效尚无相关报道。现阶段对于细菌性病害的防治主要依赖于化学农药，随着人民生活水平的提高，对农药残留问题愈加重视，因此，开发高效、无毒、无残留的生物防治微生物制剂迫在眉睫。

黄单胞菌稻生致病变种（*Xanthomonas oryzae* pv. *oryzicola*）引起的水稻细菌性条斑病是水稻上重要的细菌性病害，迄今水稻条斑病在我国南方某些稻区发病严重，逐渐成为水稻上仅次于白叶枯病的细菌性病害。目前，该病害一旦发生，主要依赖于化学农药进行治理。

为研究解淀粉芽孢杆菌 PG12 对细菌病害的生物防治潜力，本研究将水稻细菌性条斑病菌与解淀粉芽孢杆菌 PG12 在 NA 培养基平板上共培养，研究其颉颃能力。结果表明，解淀粉芽孢杆菌 PG12 对该病原菌有一定的颉颃效果。解淀粉芽孢杆菌 PG12 基因组中含有脂肽类抗生素伊枯草素、丰原素和表面活性素的关键合成基因，这可能是 PG12 对水稻细菌性条斑病菌有颉颃作用的原因。PG12 颉颃上述重要致病菌的分子或生化机制以及对上述病害的田间防效还有待进一步研究。

关键词：生物防治；解淀粉芽孢杆菌；细菌病害

---

* 基金项目：中国农业大学基本科研业务专项资金项目；国家自然科学基金项目（31672074）
** 第一作者：杨旸，女，山东聊城人，博士，中国农业大学植物病理学系；E-mail：katherine523@163.com
*** 通信作者：王琦，教授，主要从事植物病害生物防治与微生态方向研究；E-mail：wangqi@cau.edu.cn

放线菌 JY-22 发酵液对烟草赤星病菌的抑制机理*

邓永杰[1**]，王　丽[1]，薛　杨[1]，李　斌[2]，张　攀[1]，陈德鑫[3]，孙现超[1***]

（1. 西南大学植物保护学院，重庆　400716；2. 中国烟草总公司四川省公司，成都　610000；3. 中国农业科学院烟草研究所，青岛　266101）

摘　要：自 2007 年笔者实验室对缙云山土壤颉颃放线菌资源进行调查研究，筛选获得一株颉颃效果显著的放线菌株 JY-22，鉴定为链霉菌属吸水链霉菌 *Streptomyces hygroscopicus*。实验室前期工作优化了该菌发酵条件，并明确 JY-22 菌株的发酵液对温度与紫外具有很好的稳定性，在 80℃水浴 3h 后，紫外照射 30min 后均不会对菌株 JY-22 发酵滤液抑菌活性造成影响。对真菌病原菌抑菌谱广，抑菌率达 41.9%~75.7%。因此，有必要进一步研究 JY-22 抑菌机理和代谢活性产物特性，本文以烟草赤星病菌（*Alternaria alternata*）为靶菌，初步明确放线菌 JY-22 的活性代谢产物抑菌机理及其活性产物的分离纯化工艺为进一步研究提供依据。将 10 块培养 4d 的烟草赤星病菌菌饼加入 100mL PD 培养基中，摇床培养 36 h 后加入无菌发酵滤液，使其浓度分别为 10%、5%、2.5%、1%、0%，24 h 后采用电导率法测对细胞膜渗透性影响，96 h 后采用菌丝湿重法和紫外吸收法测对菌丝生长，可溶性蛋白，麦角甾醇及丙二醛含量的影响。结果表明，JY-22 对烟草赤星病菌具有颉颃作用，菌丝尖端出现膨大畸形、菌丝内原生质体外渗、细胞质凝聚出现囊状膨大物；JY-22 无菌发酵滤液处理后，10%无菌滤液对菌丝抑制率达到 75%，电导率增大电解质外漏，丙二醛含量显著升高，麦角甾醇与可溶性蛋白含量降低。表明放线菌 JY-22 对烟草赤星病菌具有显著颉颃作用，主要通过抑制麦角甾醇合成，引发细胞膜脂质过氧化，使细胞膜受损，导致菌丝生长受损。细胞膜是 JY-22 主要作用位点之一。

关键词：放线菌；JY-22；烟草赤星病菌；菌丝抑制

---

* 基金项目：国家自然科学基金（31670148）；中央高校基本科研业务费专项资金资助项目（XDJK2016A009，XDJK2017C015）；中国烟草总公司四川省公司科技项目（SCYC201703）

** 第一作者：邓永杰，男，硕士研究生，从事植物病理学研究

*** 通信作者：孙现超，博士，研究员，博士生导师，主要从事植物病毒学及植物病害生物防治研究；E-mail：sunxianchao@163. com

PhoR/PhoP 双组分对枯草芽孢杆菌 NCD-2 菌株中 surfactin 合成的影响*

董丽红[1,2]**，郭庆港[2]，王培培[2]，李社增[2]，鹿秀云[2]，张晓云[2]，赵卫松[2]，马　平***

（1. 河北农业大学植物保护学院，保定　071000；2. 河北省农林科学院植物保护研究所/河北省农业有害生物综合防治工程技术研究中心/农业部华北北部作物有害生物综合治理重点实验室，保定　071000）

摘　要：Surfactin 是由芽孢杆菌类细菌产生的一种脂肽类物质，具有溶血活性和排油能力，并且与细菌生物膜形成和根际定殖能力相关。PhoR/PhoP 双组分系统是枯草芽孢杆菌中普遍存在的一种调控系统。在低磷环境下，位于细胞膜上的组氨酸蛋白激酶 PhoR 自磷酸化后将自身获取的磷酸基团转移到位于细胞质中的调控因子 PhoP 上，磷酸化的调控因子 PhoP 通过与下游目的基因的启动子区域相结合进而调控目的基因的表达。笔者分析了 PhoR/PhoP 双组分对生防枯草芽孢杆菌 NCD-2 菌株中 surfactin 的合成的影响。结果发现，在低磷环境下，同 NCD-2 野生型菌株相比，PhoR 和 PhoP 突变子显著降低了 NCD-2 菌株的溶血活性；NCD-2 菌株的排油能力很强，而 PhoR 和 PhoP 突变子均丧失排油能力；NCD-2 野生型菌株的生物膜形成能力分别是 PhoR 和 PhoP 突变子的 2.0 倍和 2.2 倍。通过 AKTA Purifier 10 fast protein liquid chromatography（FPLC）色谱分析技术证明 NCD-2 菌株可以产生 surfactin，并发现在低磷条件中，野生型菌株 NCD-2 所产生的 surfactin 分别是 PhoR 和 PhoP 突变子的 2.3 倍和 6.4 倍。通过 RT-qPCR 技术分析了 surfactin 合成酶基因 *srfAA* 在不同菌株中的表达情况，结果表明在低磷环境下，PhoR 和 PhoP 突变子中 *srfAA* 基因的表达量比 NCD-2 野生型菌株均降低了 2.1 倍。构建 PhoR 和 PhoP 突变子的互补菌株，证明互补菌株均能使突变子的性状恢复到野生型水平。以上结果证明，PhoR/PhoP 双组分影响 surfactin 的合成。

关键词：枯草芽孢杆菌；双组分；脂肽类抗生素；溶血性；生物膜

---

* 基金项目：公益性行业科研专项（201503109）；国家自然科学基金项目（31272085，31572051）；国家现代棉花产业技术体系（CARS-18-15）；河北省财政专项（F17C10007）

** 第一作者：董丽红，博士研究生，专业方向为植物病害生物防治与分子植物病理学；E-mail：xingzhe56@126.com

*** 通信作者：马平，博士，研究员，主要从事植物病害生物防治研究；E-mail：pingma88@126.com

Identification of the Antagonistic Strain JZB130180 and Its Antimicrobial Activities*

Wang Hongli, Liu Weicheng, Zhang Dianpeng,
Zhao Juan, Liu Dewen, Lu Caige**

(*Institute of Plant and Environment Protection, Beijing Academy of Agriculture and Forestry Sciences, Beijing* 100097, *China*)

Abstract: The antagonistic actinomycete strain JZB130180, originally isolated from the soil samples collected in the Leigongshan areaof Guizhou Province. The antimicrobial spectrum tests indicated that the antimicrobial substance had a strong inhibitory activity against *Botrytis cinerea* and *Xanthomonas vesicatoria* and other plant pathogenic fungi and bacteria. Based on the results of the morphological, physiological and biochemical characteristics, 16S rDNA, *atpD*, *recA* and *rpoB* gene sequence analysis, strain JZB130180 was identified as *Streptomyces blastmyceticus*, and has been stored in the Chinese Academy of Sciences Microbial Culture Collection (Accession No. CGMCC13270); Extracellular enzyme related biological characteristics detection tests showed that the strain had the strong capability of producing chitinase, siderophore, protease and cellulose, but not phosphatase. The virulence of fermentation filtrate of strain JZB130180 against *Monilinia fructicola* was tested by means of mycelium growth rate, the results showed that the EC_{50} was 40.5μg/mL. *In vitro*, the traits of biocontrol activities of *Streptomyces blastmyceticus* strain JZB130180 were characterized and effectivein controlling grey mold on tomato fruits and *Monilinia fructicola* on peach fruits were examined by treating them with the fermentation broth, the control effects were up to 90% and 60%, respectively. In greenhouse experiment, the fermentation broth of strain JZB130180 exhibited significant growth promotion effect on plants such as tomato and pepper by soaking root or seed. Moreover, in this study, we isolated, purified and assayed the physicochemical properties of the antifungal substance from the fermentation broth of strain JZB130180, and got light yellow crystalline powder, there are three inherent characteristic absorption peaks of 340 nm, 357 nm and 378 nm by UV wavelength scanning, and a typical absorption peak at $T_R = 16.439$ min under 357 nm and 70% methanol detection conditions. The active substances arestablein acid and alkaline solution, but less stable under UV and temperature. The research results provide an important basic data for application of the strain in the field of biological control of plant diseases and clarification of its mechanism.

* Funding: Beijing Municipal Science and Technology Plan Projects (No. D151100003915002 and D151100003915003) and Special Project for Innovation Ability Construction of BAAFS (No. KJCX20170415 and KJCX20170107)

** Corresponding author: Lu Caige; E-mail: lcgf88@ sina. com

草莓炭疽病菌鉴定及对双苯菌胺敏感基线建立*

张桂军**，刘　肖，郭　巍，张志勇，毕　扬***
（北京农学院植物科学技术学院植物保护系，北京　102206）

摘　要：草莓炭疽病是全世界草莓种植业的重要病害之一，它是多种炭疽菌复合侵染引起的。本研究中对采集自我国北京、浙江、湖南、山东、辽宁等省市的草莓炭疽病病样进行了致病菌的分离、纯化、回接与鉴定，鉴定结果显示优势致病菌为草莓炭疽菌（*Colletotrichum fragariae*）与胶孢炭疽菌（*C. gloeosporioides*）；此外，在本研究中首次分离获得能引起草莓炭疽病的两种炭疽病菌：平头炭疽菌（*C. truncatum*）和 *C. boninense*。

化学防治仍然是目前防治草莓炭疽病的主要手段，然而，由于一些杀菌剂的长期和大量使用，草莓炭疽病菌对某些杀菌剂产生了一定的抗药性，因此，亟需开发一些具有全新作用机制且与市场上现有杀菌剂无交互抗药性的新型杀菌剂用于防治草莓炭疽病。双苯菌胺（SYP-14288）是沈阳化工研究院自主研发的新型杀菌剂，具有低毒、广谱的特点，对炭疽病菌高效。本研究采用菌丝生长速率法测定了采自全国6个省市共86株草莓炭疽菌（*C. fragariae*）对双苯菌胺的敏感性。结果表明：这些菌株对双苯菌胺的敏感性频率分布呈连续性的单峰曲线，没有出现敏感性下降的抗药亚群体，其 EC_{50} 值介于0.002 6~0.006 4μg/mL，平均 EC_{50} 为（0.004 0±0.000 2）μg/mL，因此可将该单峰曲线作为田间检测草莓炭疽菌（*C. fragariae*）对双苯菌胺抗药性的敏感性基线。

草莓炭疽病可由多种炭疽菌引起，且不同炭疽菌对同种杀菌剂的敏感性存在差异，上述研究对于指导草莓生产上合理用药，提高病害防控手段具有重要的理论意义。

关键词：草莓炭疽病；炭疽菌；双苯菌胺；敏感性基线

---

* 基金项目：北京市教委科研计划项目（KM201610020007）；北京市粮经作物产业创新团队（BAIC09-2017）

** 第一作者：张桂军，硕士研究生；E-mail：1769483624@ qq. com

*** 通信作者：毕扬，讲师，主要从事植物病害化学防治与病原菌抗药性研究；E-mail：biyang0620@ 126. com

二甲基三硫醚对杧果胶孢炭疽菌的抑制作用及机理研究*

唐利华**，郭堂勋，黄穗萍，李其利***，莫贱友***
（广西农业科学院植物保护研究所/广西作物病虫害生物学重点实验室，南宁　530007）

摘　要：杧果胶孢炭疽菌（*Colletotrichum gloeosporioides*）是引起杧果炭疽病的主要病原之一，可为害杧果的花、果、叶、枝条，造成严重的经济损失，影响杧果产业的发展。二甲基三硫醚（Dimethyl trisulfide）是生防菌链霉菌菌株 JK-1 产生的主要活性物质之一，研究发现其对杧果胶孢炭疽菌具有非常好的抑菌活性。毒力测定实验结果表明：二甲基三硫醚浓度在 4.58μL/L 时对胶孢炭疽菌的抑菌率达到 50%，浓度为 12.69μL/L 时抑菌率达到 95%。利用不同浓度的二甲基三硫醚对胶孢炭疽菌分生孢子分别熏蒸处理 15h，结果表明浓度在 2μL/L 时，可完全抑制杧果炭疽菌孢子的萌发。利用荧光染料 FDA（荧光素二乙酸）和 PI（碘化丙啶）染色在荧光显微镜下观察孢子的存活情况并统计孢子的存活率，结果发现在 100μL/L 的二甲基三硫醚熏蒸处理下，随着处理时间的延长，孢子的存活率显著降低，2h 分生孢子的存活率为 90.15%，4h 降低到 8.7%，到 10h 孢子存活率降至 0。通过光学显微镜和透射电镜观察二甲基三硫醚对胶孢炭疽菌菌丝、分生孢子形态和超微结构的影响，结果表明处理的菌丝和分生孢子出现变形坏死，细胞壁膜受损轮廓模糊，内部细胞器结构被严重破坏，原生质发生外漏。另外，为了找到二甲基三硫醚的作用靶标，已通过转录组测序获得了 1 767个上调基因，11 655个下调基因，下一步将对转录组测序结果进行详细分析，找到与二甲基三硫醚互作的主要靶标基因进行深入研究，为揭示其作用机理奠定基础。

* 基金项目：国家现代农业产业技术体系广西杧果创新团队建设专项，国家自然科学基金（31560526，31600029），广西留学回国重点基金（2016GXNSFCB380004），广西农业科学院科技发展基金（桂农科 2016JZ24，2017JZ01）

** 第一作者：唐利华，助理研究员，主要研究方向为热带、亚热带果树病害及其防治；E-mail：gxtanglihua@ gxaas. net
*** 通信作者：李其利，副研究员，主要研究方向为植物真菌病害及植物病害生物防治；E-mail：liqili@ gxaas. net
莫贱友，研究员，主要研究方向为热带、亚热带果树病害及其防治；E-mail：mojianyou@ gxaas. net

Antimicrobial Peptaibols, TK6 could Inhibit the Phytopathogen Fungi *Botrytis cinerea* through Killing the Cell or Inducing the Cell Apoptosis*

Ren Aizhi**, Dong Ping, Zhao Peibao***

(*College of Agriculture*, *Liaocheng University*, *Liaocheng Shandong* 252059)

Abstract: Biocontrols getting more and more important in our mordern times, and *Trichoderma* is an important biocontrol factor and have good effect to many fungi and bacteria.

Trichokonins is an Antimicrobial peptaibols, which extracted from the strains of *Trichoderma pseudokoningii* SMF2. The control mechanism of trichokonins on *Phalaenopsis* gray mold mainly includes two aspects: *fungi* inhibitting effect, induced resistance. Trichokonins could inhibit the mycelial growth, spores and the sclerotia germination of *B. cinerea*, and make the permeability of cell membranes of pathogens increased, the soluble protein contents of pathogens reduced.

During the experiment we also found that after treatment with trichokonins *Phalaenopsis*'s growth was obviously stronger than the control, Aboveground and belowground biomass have increased, the indicators of roots also strengthened.

Especially, TKs could also induce the fungus cell apoptosis of *B. cinerea*.

Key words: *Trichoderma pseudokoningii*; Trichokonins; *Phalaenopsis*; Gray mold; Control mechanism; Induced cell apoptosis.

---

* 基金项目：山东省自然科学基金 ZR2013CM006；山东省科技发展计划 2014GNC110020

** 第一作者：任爱芝，女，主要从事植物病害生物防治相关研究

*** 通信作者：赵培宝，副教授；E-mail：zhaopeibao@163.com

菜心炭疽病菌对咪鲜胺抗药性分子机制初探*

于　琳**，何自福***，蓝国兵，佘小漫，汤亚飞，邓铭光
（广东省农业科学院植物保护研究所，广东省植物保护新技术重点实验室，广州　510640）

摘　要：希金斯刺盘孢（*Colletotrichum higginsianum* Sacc.）引起的炭疽病（anthracnose disease）是我国华南地区菜心等十字花科蔬菜上的最常见和最主要病害之一。菜心（*Brassica rapa* var. *parachinensis*（Baily）Hanelt）是华南地区的主要与特色蔬菜，华南地区高温高湿的气候条件极易造成菜心炭疽病的大面积发生和流行。由于目前尚无优良抗病品种，化学防治是控制该病害的主要措施，咪鲜胺是该病害的主要防控药剂之一。笔者团队长期监测发现广东省田间菜心炭疽病菌已经对咪鲜胺产生抗药性。比较菜心炭疽病菌咪鲜胺抗性代表菌株和敏感代表菌株的抗性相关基因 *ChCYP51A*（甾醇 14-α 脱甲基酶基因）和 *ChCYP51B*（齿孔醇 14-α 脱甲基酶基因）及其编码蛋白质的序列，发现抗性代表菌株 Ch14BL9 的 *ChCYP51A* 基因上游插入了一个长度为 1 879 bp 的新转座元件，命名为 ChTE1，ChCYP51A 蛋白还存在 D161E 和 F508L 突变；抗性代表菌株 Ch14BL1 和 Ch13MX4 的 ChCYP51A 蛋白均存在 D161E 突变，ChCYP51B 蛋白还存在 F508I 突变。比较咪鲜胺抗性菌株和敏感菌株中 *ChCYP51A* 和 *ChCYP51B* 基因的表达量，发现咪鲜胺抗性代表菌株 Ch14BL9、Ch13MX4 和 Ch14BL1 中 *ChCYP51A* 基因的表达量分别为咪鲜胺敏感代表菌株 Ch14KP2 中 *ChCYP51A* 基因表达量的 5.7 倍、7.4 倍和 17 倍；上述三个抗性菌株中 *ChCYP51B* 基因的表达量分别为敏感菌株 Ch14KP2 中 *ChCYP51B* 基因表达量的 5.6 倍、1.8 倍和 10.3 倍。综上所述，笔者推测菜心炭疽病菌对咪鲜胺产生抗性与 *ChCYP51A* 和 *ChCYP51B* 基因的突变和/或上调表达有关。

关键词：菜心炭疽病菌；咪鲜胺；抗药性；分子机制

---

* 基金项目：广州市科技计划项目（2014J4100187）

** 第一作者：于琳，博士，助理研究员；E-mail：yulin@gdaas.cn

*** 通信作者：何自福，博士，研究员；E-mail：hezf@gdppri.com

放线菌 JY-22 发酵液对辣椒疫霉菌的抑制作用和机理*

张鸿铭[1**]，邓永杰[1]，薛　杨[2]，董　鹏[3]，孙现超[1***]
（1. 西南大学植物保护学院，重庆　400715；2. 西南大学柑桔研究所，重庆　400712；
3. 重庆市农业技术推广总站，重庆　400020）

摘　要：辣椒疫病在辣椒全生育期均有感染，但是在开花期最严重。发病时，幼苗期和成株期的症状有所不同。前者茎部会出现水浸状软腐，并有倒伏产生。为了寻找辣椒疫病的生物防治菌株，明确 JY-22 放线菌发酵液对辣椒疫霉的抑菌活性及机理。笔者通过培养皿中培养的方法，测定 JY-22 放线菌对辣椒疫霉病菌萌发和生长的抑制作用。通过电导率法测定 JY-22 放线菌对其细胞膜渗透性、脂质氧化程度、麦角甾醇含量、可溶性蛋白含量的作用。研究结果表明，JY-22 放线菌对辣椒疫霉病菌菌丝生长和孢子萌发均有强烈的抑制作用，且其能导致菌丝体内电解质泄露，细胞膜上麦角甾醇含量下降，脂质过氧化产物即丙二醛含量上升，且能抑制菌丝体内蛋白质的含量。由上述结果得出 JY-22 放线菌对辣椒疫霉病菌有明显的抑制作用。它主要是通过细胞膜脂质的过氧化来抑制细胞膜上麦角甾醇的合成，进而使细胞膜受损和通透性增加，最终阻碍菌丝生长。

关键词：放线菌；辣椒疫霉；抑制作用；抑制机理

---

* 基金项目：重庆市社会事业与民生保障创新专项（cstc2015shms-ztzx80011，cstc2015shms-ztzx80012）；中央高校基本科研业务费专项资金资助项目（XDJK2016A009，XDJK2017C015）

** 第一作者：张永至，男，硕士研究生，从事植物病理学研究

*** 通信作者：孙现超，研究员，博士生导师，主要从事植物病毒学及植物病害生物防治研究；E-mail：sunxianchao@163.com

解淀粉芽孢杆菌 W10 的抗真菌多肽的分离和鉴定*

张　迎**，何玲玲，单海焕，纪兆林，童蕴慧，张清霞***
（扬州大学园艺与植物保护学院，扬州　225009）

摘　要：从番茄根际分离得到的解淀粉芽孢杆菌 W10 抑菌谱广，尤其对番茄灰霉病菌和桃褐腐病菌具有较强的颉颃作用。前期研究结果表明菌株 W10 可产生 40 ku 左右的抗菌蛋白抑制病原真菌的生长。本研究从 W10 无菌滤液中用硫酸铵和 Superdex200 凝胶过滤层析分离到另一种抗真菌的小肽 5240，纯化后的 5240 抑菌谱广。经测定 5240 耐高温，在 100℃下 20min 仍然具有抑菌能力，并且在较宽的 pH 范围（3~10）内仍保持活性，且颉颃活性不受蛋白酶 K 和胰蛋白酶的影响。MALDI-TOF/TOF 飞行质谱显示，5240 主要为伊枯草菌素 A（C14~C16）异构体。以上研究结果表明，iturinA 也是解淀粉杆菌 W10 产生的主要抗菌物质。

关键词：解淀粉芽孢杆菌；灰霉病菌；脂肽；伊枯草菌素 A

---

* 基金项目：江苏省农业科学自主创新基金（CX（15）1037）；江苏省重点研究发展计划（No. 202035354）

** 第一作者：张迎，硕士研究生，植物病理学；E-mail：1369008159@ qq. com

*** 通信作者：张清霞，副教授，从事植物病害生物防治研究；E-mail：zqx817@ 126. com

微生态制剂对甜菜根际细菌群落的影响及甜菜根腐病的防治效果研究*

王震铄**，庄路博，蔚　越，聂　力，王　琦***
（中国农业大学植物保护学院，北京　100193）

摘　要：内蒙古凉城地区甜菜根腐病（Beet root rot）主要为镰刀菌根腐病，由病原菌茄病镰刀菌（*Fusarium solani*）引起，严重威胁甜菜生长发育，造成甜菜大面积减产。本研究中，将基于有机肥和颉颃微生物（优势种群为假单胞菌）发酵而成的生物有机肥（BIO）施用到甜菜单一栽培田块。结果表明：BIO 的应用能够有效防治甜菜根腐病，促进甜菜生长，具体表现为甜菜根腐病发病率与病情指数显著降低、增产 46.9%。为了阐明 BIO 抑病及促生长机制，在内蒙古凉城县的甜菜根腐病的重病田进行试验，利用 Illumina HiSeq 测序平台分别检测施（处理）或不施（对照）BIO 情况下，收获期甜菜根际土壤微生物群落组成。试验结果表明：与对照土壤样品相比，BIO 可以显著改变甜菜根际微生物群落组成，导致变形菌门（Proteobacteria），厚壁菌门（Firmicutes），放线菌门（Actinobacteria）等细菌群落显著变化。这些结果表明：BIO 的应用通过增加对甜菜根腐病病原菌具有颉颃作用的相关细菌群落的数量，显著抑制根腐病的发生，增加甜菜块根的总产量。

关键词：甜菜根腐病；BIO；Illumina 测序；根际微生物群落

* 基金项目：新型高效生物杀菌剂研发（SQ2017ZY060083）
** 第一作者：王震铄，男，博士研究生，研究方向：微生物防治；E-mail：zhenswang@ 163. com
*** 通信作者：王琦，男，教授，植物病害防治与生物与微生态实验室；E-mail：wangqi@ cau. edu. cn

柑橘溃疡病防控的新型杀菌增效剂研究

邓嘉茹，廖金星，仇善旭，常长青
（华南农业大学群体微生物基础理论与前沿技术创新团队，广州 510642）

摘 要：由黄单胞杆菌 *Xcc* 引起的柑橘溃疡病是柑橘生产上重要的细菌性病害，目前柑橘溃疡病防治以化学农药为主，大量的农药使用不仅使病原菌产生明显的耐药性，同时造成日益严重的环境问题。细菌形成生物膜的特性是其本身的一种自我保护机制，笔者实验室的前期研究发现柑橘溃疡病菌 *Xcc* 产生的群体感应信号分子能驱散由病原菌在逆境环境中形成的生物膜。因此本研究探索在市售主流杀菌剂如农用硫酸链霉素、中生菌素和喹啉铜中加入一定浓度的群体感应信号分子是否能够增加杀菌剂的杀菌效果开展了一系列实验。笔者在 3 种杀菌剂中加入 0.1～100.0μmol/L 的不同浓度的信号分子对室内培养基上的 *Xcc* 进行杀菌效果测试，结果发现 5.0～50.0μmol/L 浓度区间的信号分子与 3 种化学农药混配能够明显增加其杀菌效果。进一步的实验在室外的盆栽四季橘上进行了测试，结果也验证了其增效作用。此外，还检测了群体感应信号分子使用后四季橘主要生长指标和有关酶活性，结果表明该浓度区间的信号分子使用并没有对四季橘的生长和抗性造成不良影响。笔者的研究结果表明群体感应信号分子可以作为一种不同于传统的新型的生物杀菌增效剂，能够对农用硫酸链霉素、中生菌素以及铜制剂喹啉铜具有明显的杀菌增效作用，有良好的开发应用前景，对柑橘产业绿色安全健康发展，实现"减药减施"的社会发展目标具有重要意义。

关键词：柑橘溃疡病；群体感应；信号分子；杀菌剂；增效剂

柑橘潜叶蛾化学农药减量替代防治试验初报*

王　超**，田　伟，席运官***，刘明庆，李　刚，杨育文
（环境保护部南京环境科学研究所，南京　210042）

摘　要：柑橘潜叶蛾（*Phyllocnistis citrella* Stainton）属鳞翅目细蛾总科叶潜蛾科，是为害柑橘夏秋梢的重要害虫之一，常以幼虫潜入叶表皮下为害，并使受害植株并发溃疡病，严重时影响柑橘产量。柑橘潜叶蛾的防治主要依赖化学农药，造成害虫抗药性增强，天敌难以形成有效种群，果品农药残留超标，化学农药有机污染物污染环境。本试验评定了化学农药减量替代防治模式对柑橘潜叶蛾的防治效果，以期为柑橘种植生产降低化学农药使用量提供参考。

试验果园地处常州市武进区雪堰镇雅浦村，排灌方便，果树为7年龄挂果树，树冠直径1.5~2m，栽植密度为950株/hm^2，施药时柑橘处于果实膨大末期。药前一个月之内未施用过任何农药，施药时潜叶蛾幼虫处于初孵期，果树叶片有明显潜叶蛾幼虫潜入的虫道，药前虫口基数较大。选取长势和新梢抽发较为一致的柑橘树，设置对照组和3个试验处理组，每组3个重复，每个重复2株柑橘树；对照组施用常规柑橘化学杀虫剂（25g/L溴氰菊酯和522.5 g/L氯氰·毒死蜱），稀释1 000倍喷雾；处理组施用稀释2 000倍的化学杀虫剂，同时施用一种植物源农药（1.5%除虫菊素、0.6%苦参碱或0.5%藜芦碱），植物源农药施用浓度为推荐用有效浓度；采用背负式喷雾器进行整株均匀喷雾处理，以叶片滴水为度。在每棵树的东、西、南、北、中5个方位系标牌标记嫩梢，作为防效定点调查叶片，每个方位调查10片叶。施药前调查虫口基数，药后3h、1d、2d、8d分别用手持放大镜调查叶片上的活潜叶蛾幼虫数量，采用Excel计算各处理组虫口减退率，并用DPS软件进行统计分析，比较各处理组间虫口减退率的差异显著性。结果表明：对照及3个试验处理对柑橘潜叶蛾都有一定的防治效果，对照组平均虫口减退率在药后3h、1d、2d、8d分别为55.83%、77.56%、88.32%、89.60%；试验处理速效性差于对照，试验处理组和对照组的虫口减退率在药后1d、2d、8d均不存在显著差异（$P<0.05$）；3个试验处理组中，化学杀虫剂2 000倍液+1.5%除虫菊素对柑橘潜叶蛾的防治效果最好，平均虫口减退率在药后3h、1d、2d、8d分别达49.37%、72.77%、85.10%、88.15%；对照及试验处理对柑橘植株均安全，未见有害影响，试验处理对非靶标生物影响低于对照。

本试验初步表明以植物源农药减量替代化学杀虫剂能有效防治柑橘潜叶蛾，且减量替代模式对非靶标生物影响较小，是柑橘生产中降低化学农药面源污染、保证柑橘产量、保护生物多样性的防治新手段。

关键词：柑橘潜叶蛾；农药减量；虫口减退率

* 基金项目：国家重大科技专项（2014ZX07206001）

** 第一作者：王超，助理研究员，研究方向为植物病虫害生物防治；E-mail：wcofrcc@126.com

*** 通信作者：席运官，研究员，研究方向为有机生态农业；E-mail：xygofrcc@126.com

橡胶树胶孢炭疽菌生防菌株 QY-3 的鉴定及抗菌活性评价*

辜柳霜**，吴曼莉，李晓宇，柳志强***
（海南大学热带农林学院，海口　570228）

摘　要：橡胶炭疽病是危害我国橡胶树的一种重要的叶部病害，是当前橡胶生产中的四大病害之一，该病可为害橡胶小苗、大田幼树直至成龄开割树，严重时会引起橡胶树的反复落叶和嫩梢回枯，推迟开割时间，产胶量显著下降。胶孢炭疽菌（*Colletotrichum gloeosporioides*）是引起橡胶炭疽病的一种重要病原菌。本研究以胶孢炭疽菌为靶标菌，从土壤中分离到一株对胶孢炭疽菌具有较高颉颃活性的放线菌 QY-3，皿内抑制率达 88%。通过形态特征、生理生化结合 16S rDNA 序列分析，鉴定菌株 QY-3 为 *Streptomyces deccanensis*。对菌株 QY-3 产抗菌物质的发酵培养基进行了筛选，结果发现小米培养基最适于抗菌活性物质的产生，发酵液对胶孢炭疽菌（*C. gloeosporioides*）、杧果蒂腐病菌（*Botryodiplodia theobromae*）、火龙果溃疡病菌（*Neoscytalidium dimidiatum*）、香蕉炭疽病菌（*C. musae*）、香蕉枯萎病菌（*Fusarium oxysporum* f. sp. *cubense*）的抑制率分别为 87%，54%，45.2%，88.9%，54.3%。进一步对发酵液进行乙酸乙酯萃取和浓缩，1L 发酵液可获得 113mg 粗提物，利用生长速率法测得粗提物对胶孢炭疽菌的 EC_{50} 为 26.5μg/mL，且粗提物具有较好的热稳定。由此可见，菌株 QY-3 具有开发为橡胶炭疽病生防菌株的潜力。

关键词：橡胶炭疽病；胶孢炭疽菌；*Streptomyces deccanensis*；生物防治

* 基金项目：海南省重点研发计划（ZDYF2016155）；国家自然科学基金（31560045）

** 第一作者：辜柳霜，硕士研究生；E-mail：1258227086@qq.com

*** 通信作者：柳志强，博士，副教授，主要从事植物病害生物防治研究；E-mail：liuzhiqiang80@126.com

广西香蕉枯萎病根际颉颃生防菌的筛选*

黄穗萍[1,2]**，莫贱友[1]，郭堂勋[1]***，李其利[1]，唐利华[1]，陈　军[3]，吴玉东[3]，韦继光[2]
（1. 广西农业科学院植物保护研究所/广西作物病虫害生物学重点实验室，南宁　530007；
2. 广西大学，南宁　530004；3. 钦州市植保植检站，钦州　535000）

摘　要：香蕉枯萎病是由尖孢镰刀菌古巴专化型（*Fusarium oxysporum* f. sp. *cubense*）侵染引起的全株性土传病害。香蕉枯萎病极难防治，目前缺少园艺性状良好的抗性品种。生物防治是控制其发生危害的重要途径之一。本研究采集不同种类的香蕉根际土壤41份，采用梯度稀释法分离土壤中的微生物，获得真菌476株，细菌1 238株，放线菌752株。经过平板对峙实验，217株真菌，40株细菌和76株放线菌对香蕉枯萎病菌4号生理小种有颉颃活性。通过盆栽防治试验，筛选到2株真菌、2株细菌和3株放线菌有较好的防治效果。其中一株生防真菌的防效为52.45%，一株生防放线菌的防效为59.37%，有较好的应用前景。相关的作用机理和防治应用试验正在进行中。

关键词：香蕉枯萎病菌；颉颃生防菌

* 基金项目：广西农业科学院基本科研业务专项（2015YM01，2015YT39）；广西农业科学院科技发展基金（2015JZ40）；钦州市科学技术局科技创新能力与条件建设计划（2015270204）

** 第一作者：黄穗萍，助理研究员，主要从事植物真菌病害研究工作；E-mail：361566787@qq.com

*** 通信作者：郭堂勋，副研究员，主要从事植物病害防治研究工作；E-mail：guotangxun@gxaas.net

马铃薯种薯疮痂病成因及防治措施

郝智勇*，白艳菊**

（黑龙江省农业科学院克山分院，克山　161606）

摘　要：马铃薯疮痂病属于真菌病害，是马铃薯生产中的四大病害之一，马铃薯生产田发生此病害严重影响其产量及外观，若种薯生产中感染此病害，对其后代影响严重。本文主要对疮痂病的形成因素、危害及防治措施进行了归纳总结。

（1）形成因素。土壤中引起马铃薯疮痂病的致病菌主要有三类：*Streptomyces scabies*、*S. acidiscabies*、*S. turgidcabiesis*，除此之外已报道的其他致病性链霉菌已多达十几种。温度、土壤酸碱性和湿度是影响病害发生最重要的 3 个环境条件，发病最适宜土壤温度为 23~25℃，最适宜 pH 值 6.0~7.5。

（2）对种薯的为害。疮痂病在马铃薯上发生的程度可以分为 5 个等级。对于商品薯来讲，当疮痂病发病指数在 2 级以下时，对其销售影响不大，当达到 3 级时已经影响其品质。

（3）防治措施。严格控制种薯来源；建立良好的轮作机制，及时深翻整地；选育抗病品种，进行种薯消毒；马铃薯小薯（原种）基质的消毒与更换；多施微肥壮苗；发病时及时进行药物防治。

关键词：马铃薯；疮痂病；危害；防治措施

* 第一作者：郝智勇，从事马铃薯病毒及类病毒检测；E-mail：haohzyhhhh@126.com

** 通信作者：白艳菊，女，黑龙江省嫩江县人，博士，研究员，从事植物分子病理学研究；E-mail：yanjubai@163.com

颉颃细菌 JY1-5 的种类鉴定及其防病机制研究*

何玲玲**，张　迎，孔祥伟，张清霞***，童蕴慧

（扬州大学园艺与植物保护学院，扬州　225009）

摘　要：番茄灰霉病是番茄上的重要病害之一，对番茄的生产造成了很大的危害。本实验前期从江苏江阴萝卜根系土壤中分离筛选到一株颉颃细菌 JY1-5，能够抑制番茄灰霉病菌的生长，同时对稻瘟病菌、瓜果腐霉病菌、辣椒疫霉病菌、水稻纹枯病菌、黄瓜枯萎病菌、桃褐腐病菌、油菜菌核病菌等都有抑菌活性，具有抑菌广谱性。连续两年在番茄温室大棚进行防病接种实验，结果表明 JY1-5 对番茄灰霉病防效均在 32%以上，且防病效果显著高于药剂腐霉利。经 16S rRNA 合成基因序列分析及相关生理生化特征检测，菌株 JY1-5 初步鉴定为多粘类芽孢杆菌（*Paenibacillus polymyxa*）。平板对峙培养实验结果显示，菌株 JY1-5 发酵滤液没有明显抑菌活性，将菌体经甲醇萃取后对番茄灰霉病菌具有较好的抑菌活性。实验证明菌株 JY1-5 甲醇萃取物对番茄灰霉孢子萌发和菌丝生长都有明显抑制效果。利用 MALDI-TOF-MS 分析甲醇提取物得出 m/z = 869. 50、883. 54、897. 54、901. 54、905. 50、911. 54、915. 55、919. 52、929. 58、943. 59、947. 53、965. 53. 979. 55、993. 55 峰图，其分子量与 *P. polymyxa* 产生的抗生素 fusaricidins 的分子量相同。以上结果表明：菌株 JY1-5 对番茄灰霉有显著的抑制作用，在生物防治方面有广泛的应用前景。

关键词：番茄灰霉病；种类鉴定；多粘类芽孢杆菌；甲醇萃取物；fusaricidins

---

* 基金项目：江苏省重点研发计划（现代农业）重点项目（BE2015354）

** 第一作者：何玲玲，硕士研究生，从事植物病害生物防治研究；E-mail：2545817765@ qq. com

*** 通信作者：张清霞，副教授，从事植物病害生物防治研究；E-mail：zqx817@ 126. com

小檗碱对水稻白叶枯病菌及细菌性条斑病菌生物活性的影响*

黎芳靖**，覃巧玲，龙娟娟，黎起秦，林　纬，袁高庆***

（广西大学农学院，南宁　530005）

摘　要：水稻白叶枯病和细菌性条斑病是严重威胁水稻生产的两种细菌性病害，药剂防治仍然是当前重要的防治手段之一，但目前可用于防治这两种病害的药剂品种偏少，需要尽快研发出高效安全环保的药剂。笔者项目组在前期研究工作中发现，从狭叶十大功劳中分离出的小檗碱具有抑菌谱广、抑菌活性强的特点，并对水稻白叶枯病和细菌性条斑病表现出较好的防效。本文采用比浊法测定小檗碱对水稻白叶枯病菌和细菌性条斑病菌的抑制活性，并通过含药平板转接培养病原菌后观察小檗碱对病菌致病力的影响。研究结果表明：小檗碱对水稻白叶枯病菌的最小抑菌浓度（MIC）为 10μg/mL，最小杀菌浓度（MBC）为 40μg/mL，抑制中浓度（EC_{50}）为 1.65μg/mL；对水稻细菌性条斑病菌的 MIC、MBC 和 EC_{50} 分别为 15μg/mL、30μg/mL 和 1.72μg/mL。分别测定了小檗碱 EC_{50} 和 MIC 两种浓度对两种病菌生长曲线的影响。当菌液初始浓度为 1×10^7CFU/mL 时，EC_{50} 处理组中病菌进入对数期的时间与对照相比推迟 12~16h，但两者进入稳定期的时间相差无几；当药剂浓度为 MIC 时，2 种病原菌浓度均一直保持初始菌量。而当菌液初始浓度增加为 1×10^8CFU/mL 时，EC_{50} 处理和对照之间的生长情况均无明显差别，但药剂浓度达到 MIC 时，2 种病菌菌体生长缓慢，相比对照，推迟 28h 进入对数期，且进入稳定期的 OD 值比对照减少了一倍左右。说明小檗碱对水稻白叶枯病菌和细菌性条斑病菌生长有较强的抑制作用。在致病力影响试验中，以小檗碱 MIC 浓度对两种病原菌进行连续作用，分别用转接培养第 1 代、第 10 代、第 20 代以及溶剂作用第 20 代的病菌和未经任何处理的野生菌株接种感病水稻品种，不论是对水稻白叶枯病还是细菌性条斑病，所有处理中均是野生菌株的病指最高，但与溶剂处理和药剂处理第 1 代及第 10 代的病指相比未达差异显著水平，而与药剂连续处理 20 代的病指相比达到差异显著水平。由此说明，小檗碱对水稻白叶枯病菌和细菌性条斑病菌的致病力均有一定影响，但并未能使其丧失致病性。本文结果为深入研究小檗碱对两种病菌的抑菌机制打下基础。

关键词：小檗碱；水稻白叶枯病菌；水稻细菌性条斑病菌；抑菌活性；致病性

---

* 基金项目：国家自然科学基金（31560523）

** 第一作者：黎芳靖，硕士研究生，研究方向为植物病害防治；E-mail：1789430758@qq.com

*** 通信作者：袁高庆，副教授，研究方向为植物病害防治；E-mail：ygqtdc@sina.com

L1-9 菌剂对甜瓜的促生防病效果研究*

李　欢**，曹雪梅，陈　茹，贾　杰，马桂珍，暴增海***
（1. 淮海工学院海洋学院，连云港　222005；2. 江苏耕耘化学有限公司，连云港　222005）

摘　要：为了验证 L1-9 菌剂对甜瓜的促生防病效果，采用田间试验，随机区组试验设计见下表。定期调查甜瓜幼苗的株高、径粗、净重及叶片宽度、甜瓜枯萎病的发病情况及甜瓜产量。

表　L_1-9 菌剂甜瓜试验处理

处理组	育苗期	移栽期
处理 1	菌液与基质混匀	菌液灌根
处理 2	菌粉与基质混匀	菌液灌根
处理 3	菌液与基质混匀+菌液浸种	菌液灌根
处理 4	菌粉与基质混匀+菌液浸种	菌液灌根
处理 5	菌液浸种	菌液灌根
处理 6	未用菌剂处理	菌液灌根
CK	未用菌剂处理的空白对照	未用菌液灌根的空白对照

结果表明：L1-9 菌剂对甜瓜幼苗具有明显的促生作用，各处理组间甜瓜生长无显著差别，各处理组甜瓜幼苗株高、径粗、净重及叶片宽度明显高于对照组，移栽时处理组甜瓜苗平均株高为 11. 93cm，对照组平均株高为 8. 87cm，定植时处理组叶片平均宽度 12. 5cm，对照组叶片平均宽度为 11. 25cm。该菌剂对甜瓜枯萎病具有较好的防病效果，各处理组苗期和大田期均无枯萎病的发生，而对照组在花期出现了枯萎病株，发病率为 0. 8%。测产结果表明：处理组单株瓜重 1. 56kg，每公顷产量为 46 890kg，对照组大田单株瓜重 1. 53kg，每公顷产量为 45 900kg，增产 2. 16%，表明该菌剂对甜瓜产量具有一定的增产效果。

关键词：L1-9 菌剂；甜瓜枯萎病；促生防病作用；大田试验

---

* 基金项目：连云港市科技计划项目（现代农业）（NYYQ1619）
** 第一作者：李欢，硕士生，研究方向为食品加工与安全；E-mail：747265424@ qq. com
*** 通信作者：暴增海，博士，教授，研究方向为抗菌微生物及植物病害生物防治；E-mail：baozenghai@ sohu. com

两种涂干剂对苹果枝干轮纹病的防治作用研究*

张林尧**，王午可，胡同乐***，曹克强***
（河北农业大学植物保护学院，保定 071001）

摘 要：苹果轮纹病为害枝干和果实，造成树势衰弱，已经成为我国苹果生产中的重要病害之一，而且有愈发严重的趋势。本实验以自主研发的保护性生物菌剂轮纹终结者 1 号和腐轮 4 号为材料，测定了其田间防治效果、附着期、持效期以及最佳施药时期。结果表明：在田间防治试验中，对已发生轮纹病的树体，轮纹终结者 1 号和腐轮 4 号均表现良好的防治效果，均能有效抑制苹果树体病瘤数的增加。轮纹终结者 1 号和腐轮 4 号在田间涂干 190d 以后，药斑仍然具有抑制苹果轮纹病菌分生孢子萌发的效果。在田间涂刷苹果树主干和 1~6 年生枝条 6 个月后，轮纹终结者 1 号和腐轮 4 号均表现出良好的附着性，腐轮 4 号附着率较轮纹终结者 1 号更高；在 1 年生、5 年生、6 年生枝条涂刷轮纹终结者 1 号和腐轮 4 号 3 个月后，两种涂干剂均表现较高的附着率；2 年生、3 年生和 4 年生枝条上轮纹终结者 1 号的附着率较腐轮 4 号稍差。春季田间不同月份采用轮纹终结者 1 号涂刷树体的试验表明，3 月、4 月和 5 月涂刷轮纹终结者 1 号均对苹果枝干轮纹病有较好的防治效果。结合持效期和农事操作的便利性，在生产上建议 4 月进行涂刷。

关键词：苹果枝干轮纹病；涂干剂；防治效果；持效期

---

* 基金项目：国家苹果产业技术体系（CARS-28）；河北省自然科学基金（c2016204140）

** 第一作者：张林尧，在读硕士研究生，研究方向为植物病害流行与综合防治；E-mail：272881055@qq.com

*** 通信作者：胡同乐，博士，教授，从事植物病害流行与综合防治研究；E-mail：tonglemail@126.com
曹克强，博士，教授，从事植物病害流行与综合防治研究；E-mail：ckq@hebau.edu.cn

苹果轮纹病菌对氟醚菌酰胺的敏感性检测*

张彬彬**，张凯伦，丁　杰，杜若诗，赵晓雨，夏晓明***
（山东农业大学植物保护学院，泰安　271018）

摘　要：检测我国主要苹果产区田间苹果轮纹病菌对新型酰胺类杀菌剂氟醚菌酰胺的敏感性现状。采用菌丝生长速率法，明确我国6个不同省份的106株苹果轮纹病菌对氟醚菌酰胺的敏感性，比较不同省份苹果轮纹病菌对氟醚菌酰胺的差异性。氟醚菌酰胺对供试的106个苹果轮纹病菌菌株的EC_{50}值差别不大，在0.991 3~26.902 2mg/L，平均EC_{50}为6.123 3±0.347 3mg/L，呈不连续单峰曲线。采自山东蒙阴的菌株MY8对氟醚菌酰胺的敏感性最低（EC_{50}为26.902 2mg/L），是最敏感菌株EC_{50}值的27.14倍，所有菌株平均EC_{50}值4.39倍。剔除菌株MY8后，剩余105个菌株对氟醚菌酰胺的敏感性频率分布呈连续偏正态分布，EC_{50}平均值为5.925 4±0.288 1 mg/L，可建立田间苹果轮纹病菌对氟醚菌酰胺的敏感基线。不同省份苹果轮纹病菌的EC_{50}平均值存在明显差异（$P<0.05$）。其中，陕西省的苹果轮纹病菌对氟吡菌酰胺的敏感程度最高，其他依次为河南省、山东省、山西省和河北省，辽宁省的敏感性最低，辽宁省的EC_{50}平均值是陕西省2.61倍。田间苹果轮纹病菌对氟吡菌酰胺的敏感性水平相对较高，并未发现敏感性下降的抗药性群体，可建立田间苹果轮纹病菌对氟吡菌酰胺的敏感基线。

关键词：苹果轮纹病菌；氟吡菌酰胺；敏感性；抗药性

The sensitivity of *Botryosphaeria dothidea* from China to LH-2010A

Zhang Binbin, Zhang Kailun, Ding Jie,
Du Ruoshi, Zhao Xiaoyu, Xia Xiaoming
(*College of Plant Protection*, *Shandong Agricultural University*, *Tai'an* 271018, *Shandong*, *China*)

Abstract: The aim of this study was to determine the sensitivity level and baseline of *B. dothidea* to LH-2010A using 106 field isolates collected from Shandong, Hebei, Henan, Liaoning, Shanxi and Shaanxi respectively. The sensitivities of 106 isolates from 6 different provinces to LH-2010A were determined by using mycelial growth assay on PDA containing serial concentrations of LH-2010A, then compared the differences of isolates from different provinces. The EC_{50} values range of 106 isolates from different provincesin China were 0.991 3 to 26.902 2 mg/L, and the mean EC_{50} value was 6.123 3±

* 基金项目：国家自然基金（31301695）；山东省自然科学基金（ZR2013CQ040）；泰安市科技攻关项目（201540699）；山东省农业科学院青年科研基金（2016YQN32）；山东省果树研究所所长基金

** 第一作者：张彬彬，男，在读研究生，主要从事农药毒理有害生物抗药性研究；E-mail：502011549@qq.com

*** 通信作者：夏晓明，主要从事农药毒理有害生物抗药性研究；E-mail：xxm@sdau.edu.cn

0. 347 3mg/L, which seemed to be discontinuous unimodal curve. The ratio of the highest to the lowest EC_{50} values was 27. 14. The sensitivity of strain from Shandong Mengyin county to fluorine was the lowest (26. 902 2mg/L), which was 4. 39 times of the average EC_{50} value of all strains, so the strain MY8 should be eliminated, and the mean EC_{50} value of the left isolates were 5. 925 4±0. 288 1mg/L, the sensitivity frequency distribution of EC_{50} values of 105 isolates was continuous, unimodal and positively skewed, thus the mean EC_{50} values could be used as the sensitivity baselines. There were obvious differences ($P < 0.05$) in the mean EC_{50} values of *B. dothidea*from different provinces. The sensitivity of isolates collected from Shaanxi was the highest based on the mean EC_{50} values, and the sensitivity of isolates collected from Liaoning was the lowest, the EC_{50} values from high to low was Henan, Shandong, Shanxi, Hebei and Liaoning respectively. The ratio of the Liaoning province to the Shanxi province was 2. 61by mean EC_{50} values. However, there were no differences in sensitivity to LH-2010A in the tested field isolates collected from different provinces using the SNK test at $P = 0.01$ level. The resistance isolates of *B. dothidea* have not been detected, and all the tested field isolates of *B. dothidea*were still sensitive to LH-2010A. The mean EC_{50}values could be used as the sensitivity baselines of LH-2010A to *B. dothidea*.

Key words: *Botryosphaeria dothidea*; LH-2010A; Sensitivity; Resistance

苹果轮纹病（*Botryosphaeria dothidea*）是造成苹果烂果的主要病害之一[1-3]。苹果轮纹病是我国苹果生产中不仅在生长期发病，同时会为害果实和枝干，还可在果实储藏期发病，造成烂果，给苹果产业造成重大损失，严重时可对果园造成毁灭性破坏[4-5]。随着富士等敏感品种在我国的大量推广种植，该病害已成为我国山东、辽宁、河南、河北、山西和陕西等苹果产区苹果生产中的主要病害，且逐年加重，并造成巨大的经济损失[6-7]。

目前，苹果轮纹病主要防治措施以化学药剂防治为主，常用药剂主要包括戊唑醇、多菌灵、甲基硫菌灵等常用杀菌剂，由于常年使用，存在一定抗性风险[8]。王英姿研究也发现，尽管大部分的山东苹果轮纹病菌对戊唑醇仍表现敏感，但部分菌株已产生 2~5 倍的低水平抗性[9]。通过苹果轮纹病菌对杀菌剂的敏感性研究发现，不同地区苹果轮纹病菌对多菌灵产生了不同的抗性，而甲基硫菌灵通过转化为多菌灵对病菌起作用，两种药剂间存在着交互抗性。其中，两种低浓度的药剂存在着对菌丝促进生长的作用，避免低浓度施药[10-12]。刘保友研究表明，山东省不同地区苹果轮纹病菌对苯醚甲环唑和氟硅唑敏感性，存在着敏感性下降的群体，两种药剂存在较大的抗性风险[13-14]。

氟醚菌酰胺（LH-2010A）是山东农业大学与山东联合农药工业有限公司合作，于 2010 年创新合成的一种新型含氟苯甲酰胺类杀菌剂，氟醚菌酰胺是一种高效广谱杀菌剂，具有保护和治疗的作用，其作用机理主要表现为干扰病原菌的细胞膜通透性和破坏细胞壁，还影响菌体的三羧酸循环，作用真菌线粒体呼吸链，抑制琥珀酸脱氢酶活性，阻断电子传递，抑制真菌孢子萌发[15-17]。

为此，本研究采用菌丝生长速率法，检测了我国主要苹果产区的 106 个苹果轮纹病菌对氟醚菌酰胺的敏感性现状，并比较了不同省份苹果轮纹病菌对杀菌剂的敏感性差异，以期为苹果轮纹病防治的药剂选择和合理使用提供依据，并为田间苹果轮纹病菌的抗药性检测和治理提供基础。

1　材料和方法

1.1　材料

1.1.1　供试菌株

供试的106个苹果轮纹病菌株均由山东省果树研究所分离，纯化，鉴定，保存，提供。其中包括陕西省4株、辽宁省11株、河南省7株、山西省5株、河北省20株和山东省59株。

1.1.2　供试药剂

95.8%氟醚菌酰胺原药（LH-2010A），由山东省联合农药工业有限公司生产。准确称取95.8%氟醚菌酰胺原药1.030 9g，并溶于丙酮中，配成10 000mg/L的母液，置4℃冰箱中备用。

1.2　方法

1.2.1　含药PDA培养基的制备

在预试验的基础上，将氟吡菌酰胺按有效成分含量分别设64mg/L、32mg/L、16mg/L、8.0mg/L、4.0mg/L和2.0mg/L系列质量浓度处理。并用含有0.1%的吐温80的水溶液将氟吡菌酰胺母液稀释至上述设计浓度的10倍浓度，备用。

在无菌操作条件下，将预先融化、灭菌的90mL马铃薯葡萄糖琼脂（PDA）培养基加入无菌锥形瓶中，冷却至45℃，然后用移液器从低浓度到高浓度依次定量吸取配好的浓度为设计浓度10倍的各处理药液10mL，分别加入上述锥形瓶中，并充分摇匀，然后等量倒入直径为9cm的培养皿中，制成最终浓度为相应设计浓度（64mg/L、32mg/L、16mg/L、8.0mg/L、4.0mg/L和2.0mg/L）的含药PDA平板。以不含药剂的0.1%的吐温80水溶液处理作空白对照。

1.2.2　苹果轮纹病菌对氟吡菌酰胺的敏感性测定

在超净工作台内，将在PDA培养基上转接活化4 d后的苹果轮纹病菌，使用酒精灯高温灭菌后，内径为7.0mm的灭菌打孔器在菌落边缘部位制作菌碟，用接种针将菌碟接种于含药PDA平板中央，菌丝面朝下，盖上皿盖，放入26℃恒温培养箱内，连续黑暗培养4d。每处理4个重复，每个重复1皿。采用十字交叉法测量各处理菌落直径，减去7.0mm菌碟直径，即为菌落生长直径。统计不同处理菌落生长直径，求出其平均值，并计算各浓度处理对病原菌的菌丝生长抑制率。菌丝生长抑制率（%）＝［（对照菌落生长直径-处理菌落生长直径）/对照菌落生长直径］×100。

1.2.3　数据统计与分析

采用DPS和SPSS16.0统计分析软件，用各浓度梯度值的对数值及其相应的菌丝生长抑制率几率值进行回归分析，计算氟吡菌酰胺对不同苹果轮纹病菌菌株的EC_{50}、相关系数R、回归方程以及95%置信区间。采用SNK检验法比较不同省份苹果轮纹病菌对氟吡菌酰胺的敏感性差异。以EC_{50}值为横坐标，菌株分布频数为纵坐标，绘制频数分布柱状图，EC_{50}值的分类间隔值（柱形图中单根柱的宽度）按Scott（1979）的公式：$H=3.49SN^{-1/3}$，其中H：分类间隔值（bin-width）；S：EC_{50}值的标准偏差（standard deviation）；N：测定的菌株数量。求取分类间隔值后，依据EC_{50}值的分布范围确定柱形图的柱体数量[18]。

2　结果与分析

2.1　苹果轮纹病菌对氟吡菌酰胺的敏感性

采用菌丝生长速率法测定了田间106个苹果轮纹病菌株对氟吡菌酰胺的敏感性，结果显示（表1），氟醚菌酰胺对供试苹果轮纹病菌的EC_{50}值分布在0.991 3~26.902 2mg/L，EC_{50}平均值为6.123 3±0.347 3mg/L。其中，山东泰安的菌株TS7对氟醚菌酰胺最敏感（EC_{50}为0.991 3mg/L），山东蒙阴的菌株MY8对氟醚菌酰胺的敏感性最低（EC_{50}为 26.902 2mg/L），两者相差27.14倍。

Table 1 The resistance of *Botryosphaeria dothidea* to LH-2010A in different region of China

Strain class	Attribution	Regression equation	Coefficient coefficient	EC_{50} (mg/L)	EC_{50} (95%CL) (mg/L)
BD2	Baoding, Hebei	$y=4.9781+1.1679x$	0.9627	1.044	0.8099~1.3459
BD1	Baoding, Hebei	$y=4.9455+1.2395x$	0.9769	1.1046	0.9130~1.3408
BD15	Baoding, Hebei	$y=4.9485+1.2292x$	0.9685	1.1012	0.8779~1.3813
BD3	Baoding, Hebei	$y=4.1080+1.3470x$	0.9648	4.5942	3.6320~5.8112
CL12	Changli, Hebei	$y=3.5414+1.8515x$	0.9930	6.1348	5.7439~6.5523
CL14	Changli, Hebei	$y=3.5469+1.4722x$	0.9910	9.7214	8.9676~10.5387
CL16	Changli, Hebei	$y=3.3549+1.5128x$	0.9894	12.232	11.2776~13.2672
CL17	Changli, Hebei	$y=3.6488+1.9294x$	0.9954	5.0115	4.7344~5.3133
CL18	Changli, Hebei	$y=3.5725+1.6623x$	0.9954	7.2240	6.8617~7.6053
CL19	Changli, Hebei	$y=3.2602+1.6968x$	0.9970	10.6003	10.1380~11.0837
FN2	Funing, Hebei	$y=3.5332+1.5236x$	0.9815	9.1774	8.2624~10.1939
FN3	Funing, Hebei	$y=4.3637+1.4358x$	0.9919	2.7745	2.5097~3.0673
FN4	Funing, Hebei	$y=3.6481+1.8511x$	0.9944	5.3743	5.0562~5.7124
FN8	Funing, Hebei	$y=4.6026+1.2234x$	0.9855	2.0422	1.7445~2.3907
QL19	Qinglong, Hebei	$y=3.7481+1.7228x$	0.9963	5.3295	5.0710~5.6011
QL20	Qinglong, Hebei	$y=3.2958+1.9050x$	0.9965	7.8448	7.5034~8.2017
QL21	Qinglong, Hebei	$y=3.9444+1.4997x$	0.9949	5.0572	4.7653~5.3670
QL24	Qinglong, Hebei	$y=3.8809+1.6179x$	0.9895	4.917	4.5082~5.3629
QL28	Qinglong, Hebei	$y=4.2564+1.4269x$	0.9825	3.3197	2.8987~3.8017
QL32	Qinglong, Hebei	$y=3.8151+1.6565x$	0.9945	5.1915	4.6796~5.7594
MC	Mianchi, Henan	$y=3.3192+1.5768x$	0.9974	11.6390	11.1806~12.1163
MC10	Mianchi, Henan	$y=3.8741+1.5775x$	0.9911	5.1724	4.7827~5.5939
MC6	Mianchi, Henan	$y=3.9137+1.5797x$	0.8647	4.8719	3.4436~6.8927
103XY7	Xinyang, Henan	$y=3.7873+1.6154x$	0.9955	5.6325	5.3372~5.9442
XY3	Xinyang, Henan	$y=3.7612+1.6878x$	0.9962	5.4201	5.1559~5.6979
XY5	Xinyang, Henan	$y=3.7169+1.6588x$	0.9991	5.7713	5.6334~7.9126
XY7	Xinyang, Henan	$y=3.5697+1.5016x$	0.9905	8.9635	8.2268~9.7661
HLD 10	Huludao, Liaoning	$y=3.4485+1.8953x$	0.9929	6.5860	6.1714~7.0285
HLD 11	Huludao, Liaoning	$y=4.4046+1.6522x$	0.9974	9.2389	8.8394~9.6565
HLD 17	Huludao, Liaoning	$y=3.4142+1.8623x$	0.9941	7.1044	6.7010~7.5322
HLD 18	Huludao, Liaoning	$y=3.7820+1.6754x$	0.9837	5.3333	4.7992~5.9267
HLD14	Huludao, Liaoning	$y=3.2509+1.4219x$	0.9900	16.9868	15.7400~18.3324
SZ3	Suizhong, Liaoning	$y=3.8464+1.5881x$	0.9892	5.3258	4.8902~5.8002
SZ5	Suizhong, Liaoning	$y=3.6869+1.8178x$	0.9776	5.2732	5.0622~5.4886

（续表）

Strain class	Attribution	Regression equation	Coefficient coefficient	EC_{50} (mg/L)	EC_{50} (95%CL) (mg/L)
XC21	Xingcheng, Liaoning	y=4.457 8+1.552 5x	0.991 6	9.84 8	9.114 1~10.641 0
XC22	Xingcheng, Liaoning	y=4.581 0+1.257 5x	0.970 4	2.130 4	1.705 0~2.662 0
XC27	Xingcheng, Liaoning	y=3.349 7+1.522 2x	0.995 9	12.138 8	11.546 8~12.761 2
XC49	Xingcheng, Liaoning	y=3.008 0+1.994 1x	0.995 7	9.975 9	9.495 5~10.480 7
1DE2	DongE, Shandong	y=3.818 5+1.308 9x	0.985 4	7.992 2	7.289 1~8.763 1
1DE6	DongE, Shandong	y=3.869 6+1.590 3x	0.992 7	5.138 0	4.786 3~5.515 5
2DE1	DongE, Shandong	y=3.921 3+1.579 1x	0.997 5	4.820 6	0.282 7~0.309 1
DE	DongE, Shandong	y=3.919 6+1.535 0x	0.992 1	5.056 5	4.693 2~5.447 9
DE2	DongE, Shandong	y=3.873 6+1.515 5x	0.997 0	5.536 7	5.299 0~5.785 0
DE21	DongE, Shandong	y=4.004 3+1.394 3x	0.988 8	5.177 8	4.740 7~5.655 1
DE3	DongE, Shandong	y=3.766 6+1.561 0x	0.994 6	6.167 6	5.822 7~6.532 9
DE6	DongE, Shandong	y=3.959 1+1.482 8x	0.995 9	5.034 9	4.772 7~5.311 4
HM 1	Huimin, Shandong	y=3.931 8+1.431 4x	0.995 0	5.575 4	5.266 9~5.902 0
HM 18	Huimin, Shandong	y=7.852 2+1.617 8x	0.993 1	5.122 4	4.781 2~5.488 0
HM 2	Huimin, Shandong	y=4.045 2+1.479 6x	0.927 6	4.419 2	3.438 4~5.679 7
HM 22	Huimin, Shandong	y=3.595 4+1.554 2x	0.985 5	8.011 9	7.314 1~8.776 3
HM 24	Huimin, Shandong	y=3.733 8+1.732 4x	0.996 4	5.352 6	5.097 1~5.621 0
HM 25	Huimin, Shandong	y=4.022 5+1.443 7x	0.988 8	4.754 7	4.340 4~5.208 2
HM 26	Huimin, Shandong	y=4.357 9+1.406 7x	0.972 4	2.860 5	2.376 6~3.443 1
HM 27	Huimin, Shandong	y=3.851 7+1.598 3x	0.997 2	5.228 9	5.005 3~5.462 6
HM 3	Huimin, Shandong	y=3.540 5+1.629 0x	0.984 5	7.869 2	7.156 7~8.652 6
HM 4	Huimin, Shandong	y=3.752 1+1.607 5x	0.998 7	6.210 1	6.038 4~6.386 8
HM 5	Huimin, Shandong	y=4.045 1+1.482 4x	0.994 1	4.407 2	4.117 4~4.717 3
HM 6	Huimin, Shandong	y=3.756 3+1.596 6x	0.998 9	6.011 7	5.856 5~6.171 1
HM 7	Huimin, Shandong	y=3.225 7+2.092 2x	0.984 5	7.047 4	6.404 9~7.754 4
HM8	Huimin, Shandong	y=4.232 8+1.761 5x	0.997 2	2.726 1	2.610 8~2.846 5
HM9	Huimin, Shandong	y=4.253 2+1.881 4x	0.994 9	2.494 3	2.348 4~2.649 2
LK1	Longkou, Shandong	y=3.802 4+1.633 0x	0.994 6	5.412 3	5.098 3~5.745 7
LK2	Longkou, Shandong	y=3.874 7+1.501 2x	0.996 0	5.618 5	5.340 45.911 0
LK3	Longkou, Shandong	y=3.884 3+1.481 1x	0.996 5	5.665 6	5.402 6~5.941 5
LK4	Longkou, Shandong	y=3.782 1+1.684 1x	0.997 7	5.286 3	5.083 6~5.497 0
LK41	Longkou, Shandong	y=3.656 0+1.834 5x	0.995 6	5.402 9	5.118 6~5.702 9
LK42	Longkou, Shandong	y=3.745 6+1.701 7x	0.995 2	5.459 5	5.159 6~5.776 9
LK5	Longkou, Shandong	y=2.778 5+2.360 6x	0.978 2	8.731 2	7.792 8~9.782 5

（续表）

Strain class	Attribution	Regression equation	Coefficient coefficient	EC_{50} (mg/L)	EC_{50} (95%CL) (mg/L)
LK51	Longkou, Shandong	y=3.668 4+1.761 1x	0.996 2	5.703 4	5.429 3~5.991 3
LK52	Longkou, Shandong	y=3.770 5+1.594 2x	0.998 2	5.905 0	5.712 4~6.104 1
LK6	Longkou, Shandong	y=3.708 5+1.723 7x	0.997 9	5.613 5	5.413 0~5.821 5
LK61	Longkou, Shandong	y=3.704 0+1.700 3x	0.999 0	5.784 3	5.641 6~5.930 5
LK7	Longkou, Shandong	y=3.419 7+506 81x	0.994 7	11.189 6	10.554 2~11.863 2
LK9	Longkou, Shandong	y=3.706 9+1.774 6x	0.995 6	5.334 7	5.052 7~5.632 5
LK10	Longkou, Shandong	y=2.960 7+1.865 3x	0.983 6	12.390 6	11.303 9~13.593 6
LK11	Longkou, Shandong	y=3.708 7+1.819 8x	0.993 5	5.123 6	4.792 4~5.477 7
LK12	Longkou, Shandong	y=3.729 0+1.650 8x	0.098 5	5.887 4	5.711 3~6.068 9
LK13	Longkou, Shandong	y=3.460 0+1.730 4	0.996 2	7.761 8	7.406 7~8.133 8
LK14	Longkou, Shandong	y=3.805 7+1.469 2x	0.983 3	6.499 8	5.685 6~7.430 6
LK15	Longkou, Shandong	y=3.318 1+1.641 3x	0.993 0	10.586 7	9.884 4~11.338 8
MY5	Mengyin, Shandong	y=3.811 8+1.719 0x	0.994 4	4.911 6	4.605 2~5.238 4
MY7	Mengyin, Shandong	y=3.956 7+1.494 9x	0.994 2	4.988 2	4.678 1~5.318 8
MY8	Mengyin, Shandong	y=3.005 0+1.395 3x	0.964 5	26.902 2	22.806 0~31.734 0
MY9	Mengyin, Shandong	y=3.797 2+1.672 5x	0.993 9	5.237 4	4.909 8~5.586 9
PL 3	Penglai, Shandong	y=3.991 8+1.383 8x	0.989 5	5.353 2	4.921 0~5.823 4
PL 4	Penglai, Shandong	y=3.942 5+1.510 4x	0.989 1	5.013 4	4.591 5~5.470 4
PL 5	Penglai, Shandong	y=4.583 3+1.278 6x	0.965 2	2.177 9	1.660 4~2.710 4
210PY1	Pingyin, Shandong	y=3.867 4+1.621 9x	0.995 2	4.992 9	4.712 3~5.290 3
210PY2	Pingyin, Shandong	y=3.844 3+1.591 7x	0.991 5	5.322 1	4.933 3~5.741 5
212PY5	Pingyin, Shandong	y=7.722 4+1.738 4x	0.995 6	5.431 5	5.142 7~5.736 5
PY1	Pingyin, Shandong	y=3.869 7+1.493 6x	0.997 0	5.718 8	5.468 4~5.965 9
QX 13	Qixia, Shandong	y=3.690 6+1.623 9x	0.996 9	6.402 1	6.132 6~6.683 5
QX11	Qixia, Shandong	y=3.226 8+1.603 6x	0.997 3	12.757 6	12.257 3~13.278 4
QX12	Qixia, Shandong	y=3.929 1+1.553 3x	0.996 3	4.891 8	4.647 8~5.148 7
TS15	Taishan, Shandong	y=4.484 4+1.408 0x	0.992 0	2.323 6	2.150 3~2.511 0
TS16	Taishan, Shandong	y=4.895 6+1.326 5x	0.984 7	1.198 6	1.032 3~1.391 7
TS7ck	Taishan, Shandong	y=5.004 5+1.177 2x	0.988 8	0.991 3	0.861 7~1.410 4
SXDL18	Daliang, Shanxi	y=4.306 7+1.561 2x	0.996 6	2.780 2	2.652 6~2.913 9
SXDL4	Daliang, Shanxi	y=4.006 5+1.487 1x	0.989 4	4.656 4	4.257 6~5.092 5
SXWRRT15	Wanrong, Shanxi	y=4.467 9+1.417 8x	0.992 5	2.373 0	2.203 1~2.555 9
SXWRRT25	Wanrong, Shanxi	y=2.645 0+1.970 6x	0.995 6	15.669 5	14.903 0~16.475 3
SXWRRT31	Wanrong, Shanxi	y=4.411 5+1.708 8x	0.988 9	2.209 9	2.012 6~2.426 5

（续表）

Strain class	Attribution	Regression equation	Coefficient coefficient	EC_{50} (mg/L)	EC_{50}（95%CL）(mg/L)
SSXDL25	Dali，Shaanxi	$y=3.7592+1.5718x$	0.996 8	6.150 8	5.889 4~6.438 8
SSXDL4	Dali，Shaanxi	$y=3.8761+1.5262x$	0.993 8	5.450 2	5.111 2~5.811 7
YN1	Yannan，Shaanxi	$y=4.3776+1.3850x$	0.995 6	2.814 4	2.667 6~2.969 3
YN32	Yannan，Shaanxi	$y=4.3324+1.5074x$	0.995 4	2.772 3	2.624 1~2.928 9

正态分布检验结果表明：所有供试菌株对氟醚菌酰胺的敏感性频率分布呈不连续单峰曲线。将敏感性最低的菌株 MY8 剔除后，剩余 105 株苹果轮纹病菌对氟醚菌酰胺的敏感性频率分布呈连续单峰曲线（图 1），EC_{50}平均值为 5.925 4±0.288 1mg/L。进一步的正态性检验结果表明：当显著水平为 95% 时，偏度 skew = 1.178，峰度 kurt = 2.227，Shapiro-wilk 检验得，$P=0.00$，$P<0.05$，说明 105 个苹果轮纹病菌菌株菌丝生长对氟醚菌酰胺敏感性的频率分布符合偏正态分布，因此，可以采用 EC_{50}平均值 5.925 4±0.288 1mg/L 作为田间苹果轮纹病菌菌丝生长对氟醚菌酰胺的敏感基线。

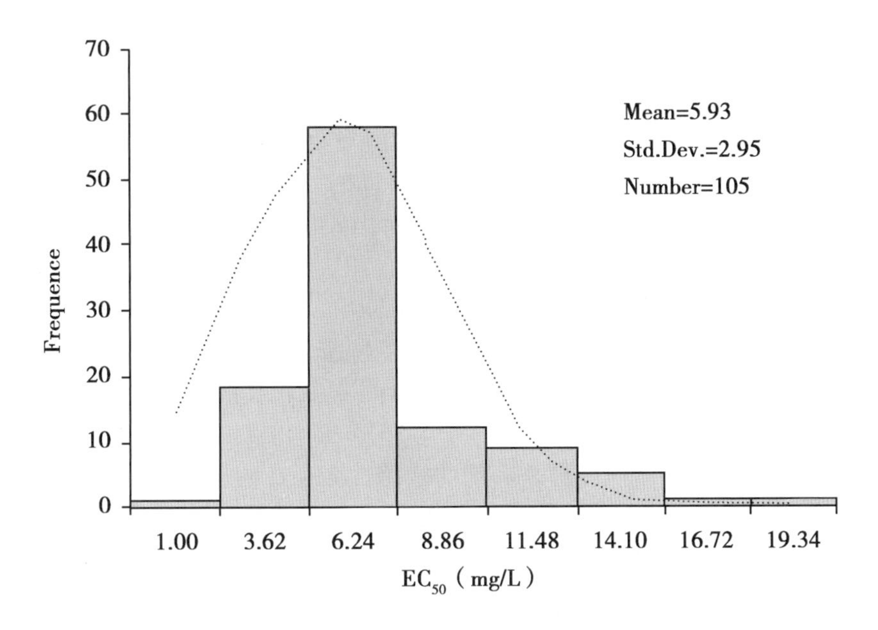

Fig. 1 The frequency histograms of sensitivity for 106 isolates of *Botryosphaeria dothidea* to LH-2010el by EC_{50} values

(The bin width of 2.18was calculated by Scott's rule)

2.2 不同地理来源苹果轮纹病菌对氟吡菌酰胺的敏感性水平

采用 SNK 检验法比较不同省份苹果轮纹病菌对氟吡菌酰胺的敏感性差异，统计分析结果表明（表 2），陕西省的苹果轮纹病菌对氟吡菌酰胺的敏感程度最高，EC_{50}平均值 4.296 9±0.879 8 mg/L，辽宁省的苹果轮纹病菌对氟吡菌酰胺的敏感程度最低，EC_{50}平均值分别为 8.176 5±1.226 0mg/L，陕西省的 EC_{50}平均值与辽宁和河南两省的 EC_{50}平均值存在显著差异（$P<0.05$），但辽宁和河南两省之间差异不明显（$P>0.05$），其中山东省的 EC_{50}平均值是陕西省的 1.90 倍）。山东省、河北省和山西省的 EC_{50}平均值介于中间，分别为 6.050 6±0.461 2mg/L、5.489 8±0.718 1mg/L 和 5.537 8±2.570 2mg/L，且三省之间差异不明显（$P>0.05$）。

Table 2 The sensitivity of *Botryosphaeria dothidea* to LH-2010A in different provinces of China

Province	Isolate numbers	Range of EC_{50} (mg/L)	(Mean±SE)	(CV)	SD
Shandong	58	0. 991 3~12. 757 6	5. 691 1±0. 294 0 bc	0. 585 5	2. 238 7
Hebei	20	1. 044 0~12. 232 0	5. 489 8±0. 718 1 bc	0. 584 9	3. 211 4
Liaoning	11	2. 130 4~16. 986 8	8. 176 5±1. 226 0 a	0. 497 3	4. 066 4
Henan	7	4. 871 9~11. 639 0	6. 781 55±0. 961 0ab	0. 363 8	2. 542 5
Shanxi	5	2. 209 9~15. 669 5	5. 537 8±2. 570 2 bc	1. 037 8	5. 747 2
Shaanxi	4	2. 772 3~6. 150 8	4. 296 9±0. 879 8c	0. 409 5	1. 759 7
Total	105	0. 991 3~16. 986 8	5. 925 4±0. 288 1	0. 409 5	0. 498 2

Note: The small letters within a row indicate a difference at at $P=0.05$ level using the Student Newman Keuls (SNK) test.

3 讨论

氟醚菌酰胺是一种新型含氟苯甲酰胺类杀菌剂，结构式中含有 7 个氟原子。氟原子形成的 C-F 键的键能远大于 C-H 键，增加了有机氟化合物的生理活性和稳定性。含氟有机化合物具有较高的脂溶性和疏水性，能够促进其在生物体内吸收与传递，增强与生物体的结合能力，使生物体的生理作用发生变化。含氟农药同样有这样的功效，对病原菌或害虫的抑制或毒杀作用也大大提高[19-21]。本研究检测了氟醚菌酰胺对 106 株田间苹果轮纹病菌的敏感性，检测到了 1 个敏感性下降的菌株 MY7（采自山东蒙阴，EC_{50}为 26. 902 2mg/L）。剔除该菌株后，剩余 105 个菌株的 EC_{50}值介于 1. 672 2~27. 009 5mg/L，平均为 6. 086 6±0. 384 6mg/L，与张伟测得的 EC_{50}较为接近。且这 105 个菌株对氟醚菌酰胺敏感性的频率分布符合偏正态分布，因此平均值 6. 086 6±0. 384 6mg/L 可以用来作为田间苹果轮纹病菌菌丝生长对氟醚菌酰胺的敏感基线。

不同地理来源的病原菌会对同一种杀菌剂的敏感性会存在明显差异[22]。本研究发现，同样的 6 个省份苹果轮纹病菌对氟醚菌酰胺的平均 EC_{50}值存在明显差异（$P<0.05$）。事实上，田间植物病原菌对药剂的敏感性水平差异，除了与病原菌的采集时间和地理来源分布、寄主作物不同种植年限、病害发生的轻重程度有一定关系外，当地农民所采用的药剂种类，用药历史、习惯和水平均会导致杀菌剂敏感水平差异的产生。这也进一步说明，氟醚菌酰胺很可能在我国不同省份被广泛应用于苹果轮纹病的防治。

田间苹果轮纹病菌对氟醚菌酰胺的敏感性水平相对较高，并建立了田间苹果轮纹病菌菌丝生长对氟醚菌酰胺的敏感基线；不同省份田间苹果轮纹病菌对同一种杀菌剂的敏感性水平存在明显差异。

参考文献

[1] Kang L, Hao H M, Yang Z Y, , *et al.* The advances in the research of apple ring rot (in chinese) [J]. Chinese Agricultural Science Bulletin (中国农学通报), 2009, 25 (9): 188-191.

[2] Feng B, Liu L, Wu H, *et al.* The intraspecific genetic diversity of pathogenic fungi of apple ring rot (in chinese) [J]. Scientia Agricultura Sinica (中国农业科学), 2011, 44 (6): 1125-1135.

[3] Yan Z L, Zhang Q J, Zhang S N, *et al.* Identification of apple cultivars for their resistance to ring rot disease

(in chinese) [J]. Journal of Fruit Science (果树学报), 2005, 22 (6): 654-657.

[4] Li G X, SHEN Y B, GAO Y M, *et al*. Study on the infection mechanism of apple ring rot disease (in chinese) [J]. Journal of Fruit Science (果树学报), 2007, 24 (1): 16-20.

[5] Zhang G L, Li B H, Wang C Z, *et al*. Curative effects of six systemic fungicides on tumor development on apple branches caused by *botryosphaeria dothidea* (in Chinese) [J]. Journal of Fruit Science (果树学报), 2010, 27 (6): 1029-1031.

[6] Wang Y Z, Zhang W, Liu B Y, *et al*. Research on resistance and geographical distribution of Botryosphaeria dothidea from apple to tebuconazole in Shandong province (in Chinese) [J]. Journal of Fruit Science (果树学报), 2010, 27 (6): 961-964

[7] Liu P, Zhou Z Q, Guo L Y. Sensitivity of Botryosphaeria berengeriana f. sp. piricola to carbendazim, imibenconazole and propiconazole fungicides (in Chinese) [J]. Journal of Fruit Science (果树学报), 2009, 26 (6): 907-911.

[8] Fan K, Qu JL, LI L G, *et al*. Study on baseline-sensitivity of Botryosphaeria dothidea to tebuconazole and the biological characteristics of tebuconazole-resistant mutants (in Chinese) [J]. Journal of Fruit Science (果树学报), 2013, 30 (4): 650-656.

[9] Ma Z Q, Li H X, Yuan Z H, *et al*. Preliminary research on detection of resistance of apple ring rot (Macrophoma kawatsukai) to carbendazim (in Chinese) [J]. Chinese Journal of Pesticide Science (农药学学报), 2000, 2 (3): 94-96.

[10] Li X J, Fan K, Qu J L, Zhang Y, *et al*. Determination of the sensitivity of Botryosphaeria berengriana to carbendazim fungicide (in Chinese) [J]. Journal of Fruit Science (果树学报), 2009, 26 (4): 516-519.

[11] Zhang C L, Zhang T T, Lu X T, *et al*. Toxicity test of ten fungicides to Macrophoma kawatsukai and Colletotrichum gloeosporioides in laboratory (in Chinese) [J]. Chinese Agricultural Science Bulletin (中国农学通报), 2012, 28 (27): 236-240.

[12] Yang W H, Liu K Q. Resistance detection of *Botryosphaeria berengeriana* f. sp. piricola tocarbendazim and thiophanate-meth (in chinese) [J]. Journal of Plant Protection (植物保护学报), 2002 (02): 191-192.

[13] Liu B Y, Zhang W, Luan B H, *et al*. Sensitivity of *Botryosphaeria dothidea* to difenoconazole and flusilazole and cross-resistance of different fungicides (in chinese) [J]. Acta Phytopathologica Sinica (植物病理学报), 2013, 43 (5): 541-548.

[14] Liu B Y, Wang P S, Zhang W, *et al*. Pathogenicity of Botryosphaeria dothidea derived from defferent locations of circum-Bohai Bay region on current-year shoots (in Chinese) [J]. Journal of Shandong Agricultural University (Natural Science) (山东农业大学学报（自然科学版）), 2010, 41 (4): 508-512.

[15] Yang J C, Wu Q, Liu Y Y, *et al*. Recent advance of flourine-containing pesticides (in Chinese) [J]. Agrochemicals (农药), 2011, 50 (4): 290-295.

[16] TOQUIN V, BARJA F, SIRVEN C, *et al*. A new mode of action for fluopicolide: modification of the cellular localization of a soectrin-like protein [J]. Pflanzenschutz-nachrichten bayer, 2006, 59: 2-3, 171-184.

[17] Jiang L L, Wu Y G. Laboratory Toxicity Test of 9 Fungicides to Phytophthora infestans (in chinese) [J]. World Pesticides, 2015 (02): 59-61.

[18] David W. Scott. On optimal and data-based histograms [J]. Biometrika, 1979, 66 (3): 605-610.

[19] Meng R J, Wang W Q, Lu F, *et al*. Synergis Interaction and Field Control Effect of Mixtures of *LH*-2010*A* with Dimethomorph against *Phytophthora infestans* (in Chinese) [J]. Agrochemicals (农药), 2017 (02): 141-144.

[20] Zhang H S, Zhai M T, Wang K Y, *et al*. Study on fungicidal activity and mode of action of a novel fungicidal agent, LH-2010A against Rhoizoctonia solani (in Chinese) [J]. Chinese Journal of Pesticide

Science（农药学学报），2013，15（4）：405-411.

[21] Keinath A P，Fillippeli E，Hausbeck M K，*et al.* Baseline sensitivity to fluopicolide in phytophthora capsici isolates from the eastern United States [J]. Phytopathology，2010，100：S201.

[22] Yuan S K，Liu X L，Si N G，*et al.* Sensitivity of Phytophthora infestans to flumorph：In vitro determination of baseline sensitivity and the risk of resistance [J]. Plant Pathology，2006，55（2）：258-263.

生防菌处理后微型薯蛭石基质中可培养微生物种类变化分析*

关欢欢**，杨德洁，邱雪迎，赵伟全***，刘大群***
（河北农业大学植物保护学院，河北省农作物病虫害
生物防治工程技术研究中心，保定　071000）

摘　要：以蛭石为基质生产微型薯过程中马铃薯疮痂病发生较为普遍，为从微生物的动态变化角度探索病害的发生发展规律，本研究对生防菌防治微型薯疮痂病后蛭石中可培养微生物的动态变化进行了初步分析，旨在了解微型薯疮痂病防治前后基质中微生物的种群变化规律，为进一步研究疮痂病的发生及防病机理提供依据。本研究利用生防菌 ZWQ-1+农用链霉素、生防菌 ZWQ-1+柠檬酸分别对重复使用的旧蛭石苗床进行处理，利用玫瑰黄链霉菌 Men-myco-93-63 发酵液对新旧混合蛭石的苗床进行处理。采用平板稀释法对不同处理后以及对照的蛭石中的主要可培养真菌、细菌和放线菌进行了分析。测试结果表明：生防菌 ZWQ-1+农用链霉素处理后蛭石中的微生物总量约为 3.52×10^7 CFU/g，其中细菌类群有 3 个属：金黄杆菌属、假单胞菌属、芽孢杆菌属，约占 78.05%；放线菌类群有 4 个属：微杆菌属、考克氏菌属、链霉菌属、红球菌属，约占 21.85%；真菌类群有 2 个属：青霉属、镰孢菌属，约占 0.09%。生防菌 ZWQ-1+柠檬酸处理后的微生物总量约为 8.15×10^7 CFU/g，其中细菌类群有 4 个属：金黄杆菌属、假单胞菌属、根瘤菌属、无色杆菌属，约占 80.53%；放线菌类群有 4 个属：链霉菌属、考克氏菌属、节杆菌属、微杆菌属，约占 19.40%；真菌类群有 2 个属：青霉属、镰孢菌属，约占 0.08%。玫瑰黄链霉菌 Men-myco-93-63 发酵液+新旧混合蛭石的微生物总量约为 7.02×10^7CFU/g，其中细菌类群有 3 个属：金黄杆菌属、假单胞菌属、根瘤菌属，约占 91.13%；放线菌类群有 4 个属：链霉菌属、节杆菌属、考克氏菌属、红球菌属，约占 8.83%；真菌类群 2 个属：青霉属和镰孢菌属，约占 0.04%。对照中微生物总量约为 9.19×10^7 CFU/g，其中细菌类群有 6 个属：假单胞菌属、根瘤菌属、变形杆菌属、动性球菌属、无色杆菌属、寡养单胞菌属，约占 74.13%。放线菌类群有 4 个属：链霉菌属、微杆菌属、节杆菌属、红球菌属，约占 25.25%。真菌类群有 6 个属：轮枝孢属、枝顶孢属、青霉属、枝孢属、镰孢菌属、曲霉属，占 0.62%。本研究对不同处理蛭石中微生物的数量和种类进行了初步分析后发现，利用生防菌防治疮痂病后蛭石中真菌和放线菌含量减少而细菌含量变化不显著，说明蛭石中微生物数量变化与疮痂病的消长密切相关。

关键词：马铃薯疮痂病；蛭石；平板稀释法；可培养微生物

---

* 基金项目：河北省自然科学基金（C2014204109）；河北农业大学科研发展基金计划项目
** 第一作者：关欢欢，硕士研究生；E-mail：guanhhuan@foxmail.com
*** 通信作者：赵伟全，教授，博士生导师；E-mail：zhaowquan@126.com
刘大群，教授，博士生导师；E-mail：ldq@hebau.edu.cn

生防菌株JY-1对几种植物病原真菌和细菌的抑制作用*

欧阳慧**，王　新，王园秀，贺　哲，周　英，鄢明峰，蒋军喜***
（江西农业大学农学院，南昌　330045）

摘　要：西姆芽孢杆菌（*Bacillus siamensis*）JY-1是本实验室前期筛选到的对猕猴桃果腐病菌具有显著颉颃作用的生防菌株。为了进一步探究其抗菌谱范围，本研究主要采用平板对峙法和滤纸片法，对常见的9种植物病原真菌和4种植物病原细菌进行抑菌效果测定。结果表明：JY-1菌株对9种植物病原真菌和4种植物病原细菌菌落生长有抑制效果。在病原真菌方面，该菌株对水稻稻瘟病菌的抑菌效果最好，抑制率高达93.37%；对其余8种病原真菌菌丝生长的抑制率均高于59%。通过光学显微镜观察，发现受抑制的菌丝表现出不同程度的变形、膨大、扭曲，甚至导致菌丝消亡现象。在病原细菌方面，该菌株对水稻白叶枯病菌产生的抑菌效果最佳，抑菌圈平均直径高达18.58mm。本文首次在国内报道西姆芽孢杆菌对植物病原细菌具有抑制作用。

关键词：西姆芽孢杆菌；平板对峙法；植物病原真菌；植物病原细菌

* 国家自然科学基金（31460452）；江西省科技计划项目（20141BBF60019）
** 第一作者：欧阳慧，硕士研究生，研究方向为分子植物病理学；E-mail：jxndoyh2014@126.com
*** 通信作者：蒋军喜，教授，博士，主要从事植物病害综合治理研究；E-mail：jxau2011@126.com

生防芽孢杆菌2-1的定殖及其对土壤微生态的影响*

闫 杨**，刘月静，陈 芳***

（聊城大学药学院，聊城 252059）

摘 要：枯草芽孢杆菌2-1是分离自中药材天南星根际土壤的颉颃细菌，对多种植物根腐病有明显的抑制作用，并且能够有效促进植株生长。以生长周期相对较短的黄瓜作为试验对象，温室盆栽试验的结果表明2-1对黄瓜根腐病存在有效抑制作用，并能使土壤中植物抗病的相关酶活性明显升高，如脲酶、碱性磷酸酶、几丁质酶和过氧化氢酶。此外，生防菌株在土壤中的定殖情况也能明显影响其生防效果。在生物防治过程中将大量生防菌施入植物根部土壤中并使其有效定殖，虽在植物土传病害防治中起到积极作用，但也应对生防菌对土壤生态平衡可能存在的影响作出深入了解。

本研究通过对生防菌2-1进行绿色荧光蛋白基因标记，构建出具有四环素抗性的2-1-GFP工程菌株，用于研究2-1在土壤中定殖后的动态过程。工程菌株2-1-GFP的遗传稳定性实验表明质粒稳定性达到95%，故可推断出该工程菌株可以满足黄瓜特定生育期内研究的需要。温室盆栽试验表明2-1可以稳定定殖在黄瓜根围土壤中，2-1在黄瓜整个生育周期内根围的活菌数可以稳定在10^6~10^7cfu/g，可以满足作物整个生育期的需要。

在温室试验过程中，每隔4周从黄瓜根际土壤取样，在黄瓜生育期取样4次。使用相应试剂盒检测与土壤中植物抗病相关的脲酶、碱性磷酸酶、几丁质酶和过氧化氢酶酶活性，检测结果表明4种酶在黄瓜根际土壤中的酶活性呈现动态变化且与植株长势相关。利用变性梯度凝胶电泳（DGGE）法研究土壤微生态，操作方便快捷，而且可以避免传统的平板培养法不能培养土壤中全部微生物所带来的弊端。用土壤DNA提取试剂盒（北京天根）提取高纯度的土壤总DNA，用16S rDNA特异性引物27f和1492r进行PCR扩增，用DGGE引物357f-GC和518r进行巢式PCR扩增，将PCR产物进行变性梯度凝胶电泳分析。DGGE电泳图谱结果反映出，在黄瓜苗期不同处理之间的微生物与播种前土壤中的微生物类群相比无明显变化。播种一个月后，土壤中微生物类群随着植株生长而变化，电泳条带亮度及数量均有明显增加。两个月后，与植物长势相对应，微生物菌群增长达到峰值并趋于稳定。3个月后，植株进入生长衰亡期且土壤中菌群数量减少，不同处理间的差异减小。根据DGGE图谱可见，植物根际微环境影响植株根部土壤中的微生物类群增殖，且其变化与植物生长状态紧密相关。该研究将为今后2-1菌株在中药材及蔬菜作物根腐病的大田应用奠定理论基础。

关键词：生防芽孢杆菌；定殖；酶活性；微生态

* 项目资助：国家自然科学基金项目（31401799）

** 第一作者：闫杨，硕士，研究方向为植物病害生物防治；E-mail：yanyang0906@126.com

*** 通信作者：陈芳，博士，副教授，主要从事植物病害生物防治研究；E-mail：chenfang20045@163.com

甜瓜尾孢菌叶斑病病原鉴定及药剂防治研究*

叶云峰[1]，杜婵娟[2]，付 岗[2**]，柳唐镜[1]，覃斯华[1]，洪日新[1]，李桂芬[1]
(1. 广西壮族自治区农业科学院园艺研究所，南宁 530007；
2. 广西壮族自治区农业科学院微生物研究所，南宁 530007)

摘 要：广西的厚皮甜瓜栽培以保护地大棚种植为主，近两年在广西厚皮甜瓜各主产区发现一种叶部新病害，该病害侵染北甜一号、好运 11 号、新金凤凰等厚皮甜瓜品种，其症状与以往发生的叶部病害症状不同。该病在叶片上形成中央灰白色、边缘褐色、外围有黄色晕圈的近圆形病斑，还可形成带黄色晕圈的不规则形大斑，或沿着叶缘的叶脉向叶片内部扩展呈倒“V”字形的褐色大斑，病斑大小差异较大，有时多个病斑融合成片，致叶片大面积干枯坏死，造成植株早衰。该病在春茬的发病率为 30%～50%，而秋茬发生更为严重，多数大棚发病率达 80%以上，严重的大棚发病率达 100%，成为 2015—2016 年广西大棚甜瓜最严重的叶部病害之一。

本研究对该病害的病原菌进行分离和致病性测定，并通过形态学特征和 ITS rDNA 基因序列分析将该病原菌鉴定为瓜类尾孢菌（*Cercospora citrullina*）。初步开展了甜瓜叶斑病药剂防治试验，试验结果表明：在测试的 4 种药剂中，70%甲基硫菌灵 600 倍液的防治效果最好，达 80.28%；其次为 30%苯醚甲环唑 3 000倍液，防治效果为 70.27%。试验结果可为该病害的有效防控提供参考依据。

关键词：厚皮甜瓜；叶斑病；瓜类尾孢菌；病原鉴定；药剂防治

* 基金项目：广西农业科学院科技发展基金资助项目（桂农科 2017JM17）；广西农业科学院基本科研业务专项（2015YT49）；国家星火计划项目（2014GA790006）

** 通信作者：付岗，博士，副研究员，主要从事植物病害及其生物防治研究；E-mail：fug110@gxaas.net

五种三唑类杀菌剂对禾谷镰刀菌 *CYP51A* 基因敲除菌株的敏感性研究*

钱恒伟[1**]，迟梦宇[1]，赵　颖[1]，黄金光[1,2***]
（1. 青岛农业大学农学与植物保护学院，青岛　266109；
2. 山东省植物病虫害综合防控重点实验室，青岛　266109）

摘　要：小麦赤霉病主要是由禾谷镰刀菌（*Fusarium graminearum*）引起的一种麦类病害，在全球范围内均有发生，目前生产上对于小麦赤霉病的防治主要依靠化学药剂防治。三唑类药剂是防治小麦赤霉病三大类杀菌剂之一，防治效果较好。该类杀菌剂主要作用于甾醇生物合成中的14α-脱甲基酶（CYP51），使真菌麦角甾醇的合成受阻，从而破坏细胞膜功能，起到杀菌的功能。禾谷镰刀菌 *FgCYP51A* 基因编码 14α-脱甲基酶，与病原菌对三唑类杀菌剂的敏感性密切相关。为了探究 *CYP51A* 基因在三唑类杀菌剂对禾谷镰刀菌的药效中的作用。作者采用 PEG 介导的原生质体转化 split-marker PCR 片段法将扩增获得的带有潮霉素标签的混合片段转化到 *F. graminearum* PH-1 制备的原生质中，获得了 *FgCYP51A* 的敲除转化子，并进行了验证，筛选出了 Δ*FgCYP51A* 菌株。随后采用菌丝生长速率法测定了禾谷镰刀菌 Δ*FgCYP51A* 菌株对五种三唑类杀菌剂的 EC_{50}值，Δ*FgCYP51A* 菌株对丙环唑、戊唑醇、烯唑醇、三唑醇和三唑酮的 EC_{50}值分别为0. 024mg/L，0. 012mg/L，0. 148mg/L，0. 154mg/L 和0. 474mg/L。根据 *FgCYP51A* 基因敲除转化子对于不同三唑类杀菌剂敏感性的差异变化，将五种三唑类杀菌剂分为两类。第一类包括戊唑醇、三唑醇、三唑酮和丙环唑，*FgCYP51A* 基因敲除体敏感性显著上升；第二类为烯唑醇，敲出体对其敏感性无明显变化。本研究证明了 *FgCYP51A* 基因与禾谷镰刀菌对三唑类杀菌剂的敏感性相关，为生产上合理使用杀菌剂防治小麦赤霉病以及减少抗药性现象发生提供了理论依据。

关键词：禾谷镰刀菌；CYP51A；基因敲除；三唑类杀菌剂；敏感性

---

* 基金项目：国家自然基金项目（31471735）；青岛市现代农业产业技术体系（6622316106）
** 第一作者：钱恒伟，硕士研究生，植物病理专业
*** 通信作者：黄金光，教授，研究方向：蛋白质结构生物学，分子植物病理学；E-mail：jghuang@ qau. edu. cn

多功能生防菌 *Bacillus amyloliquefaciens* W1 研究初报

李兴玉[1,2,3]，何鹏飞[2,3]，吴毅歆[2,3]，何月秋[2,3]
（1. 云南农业大学基础与信息工程学院，昆明 650201；2. 云南农业大学/农业生物多样性应用技术国家工程中心，昆明 650201；3. 云南省微生物发酵工程研究中心有限公司，昆明 650217）

摘　要：基于筛选多功能生防菌株资源，拓展生防菌的应用空间，本文筛选到一株多功能生防菌 *Bacillus amyloliquefaciens* W1（CGMCC No. 11949）。菌株最优发酵条件为温度 35℃，pH 值 8.0，装液量 0.15/0.5（*v/v*），在培养 4~24h 期间转速为 170r/min，24~48h 转速变为 160r/min。该菌株对三七根腐菌，玄参炭疽病菌，火龙果果腐病菌，小麦雪腐病菌，苹果斑点落叶病菌和甘蓝黑腐病菌都有颉颃作用；对二斑叶螨有明显致死作用（CN105543151A），在温室试验中，对玉米二斑叶螨的防治效果可达到 82.5%；对番茄生长也具有显著的促进作用，经菌液处理的番茄长势明显好于对照，株高超出对照 36.8%，且叶色更为浓绿。该菌株能在玉米、番茄、蚕豆等作物体内的定殖和转运，喷洒于玉米叶片的标记菌可以传导至茎、叶组织，具有开发成多功能微生物叶面肥的潜力。

关键词：解淀粉芽孢杆菌；定殖；二斑叶螨；抑菌作用

铁皮石斛内生真菌分离及对灰葡萄孢菌颉颃作用的筛选*

徐艳芳[1**]，张佳星[1**]，李 玲[1]，王华弟[2]，张传清[1***]

（1. 浙江农林大学农学与食品科学学院，临安 311300；
2. 浙江省农药检定管理所，杭州 310004）

摘 要：铁皮石斛（*Dendrobium officinale* Kimura et Migo）含有多种石斛多糖具有抗氧化、抗炎、抗肿瘤等多种功效，是中国传统中药材。其植物体内内生真菌由于与药用植物的协同进化，可能具有与宿主植物相似的活性成分，在生物防治中具有较大的潜力。本研究从浙江省临安市采集圣兰8号、晶品1号和临安野生种3个品种的健康铁皮石斛，利用组织分离法进行了内生真菌的分离纯化，并利用平板对峙法，初步筛选了对铁皮石斛灰霉病病原菌灰葡萄孢（*Botrytis cinerea*）具有抑制活性的内生菌菌株，并对其发酵液的初提取物进行了抑菌活性测定。结果表明：从3个品种的铁皮石斛健康组织中共分离得到122株内生真菌，其中从叶片分离得到20株，茎秆57株，根部45株。筛选获得内生真菌DA-1对灰葡萄孢菌有明显抑制作用，在PDA培养基上对峙培养时产生明显的抑菌圈；且发酵液粗提取物对灰葡萄孢菌具有明显的抑制效果。

关键词：铁皮石斛；内生真菌；分离纯化；抑菌活性；灰葡萄孢菌

---

* 基金项目：浙江省重点研发项目（2015C02G1320008）；浙江省公益项目（2016C32002）

** 第一作者：徐艳芳，硕士研究生，主要从事植物病理学研究；E-mail：1005020125@ qq. com
张佳星，硕士研究生，主要从事植物病理学研究；E-mail：275893077@ qq. com

*** 通信作者：张传清，博士，教授，主要从事植物病害治理及杀菌剂药理学与抗药性研究；E-mail：cqzhang@ zafu. edu. cn

一株耐高温长枝木霉 TW20741 的分离鉴定及其功能评价*

赵晓燕，周红姿，吴晓青，周方园，张新建

（山东省科学院生态研究所，山东省应用微生物重点实验室，济南　250014）

摘　要：采用稀释平板法从新疆巴彦淖尔市临河湿地土壤分离到一株耐 50℃ 高温的木霉 TW20741。为了明确木霉菌株 TW20741 的分类地位，并评价其生物学功能，通过形态学特征结合 ITS、TEF 序列分析对其进行了种类鉴定，利用菌丝生长速率法测定了菌株的抑菌能力，利用透明圈大小定性检测了菌株的水解酶活性。结果表明：菌株 TW2074150℃ 高温条件下 72h 生长速率为 3.49%。菌株 TW20741 的形态学特征与 *Trichoderma longibrachiatum* 一致，其 ITS、TEF 序列与 *Trichoderma longibrachiatum* 的同源性均达到 99%以上。菌株 TW20741 在 PDA 平板上对 8 种植物病原真菌均有抑制作用，对终极腐霉的抑制率最高，达 98.50%。菌株 TW20741 具有蛋白酶、几丁质酶、β-1，3-葡聚糖酶及纤维素酶活性，其中蛋白酶活性和纤维素酶活性较高，透明圈宽度分别为 1.5cm 和 0.6cm。综合形态学特征和序列分析，将菌株 TW20741 鉴定为长枝木霉。菌株 TW20741 不但对植物病原真菌具有广谱抗性，而且具有多种水解酶活性，是一株非常有潜力的生物防治菌株。

关键词：长枝木霉；分离鉴定；功能评价

* 基金项目：国家自然基金-哈茨木霉 LTR-2 与立枯丝核菌、畸雌腐霉重寄生互作的转录组学解析（31572044）；山东省优秀中青年科学家科研奖励基金-木霉消除草酸作用的关键基因筛选及功能识别（No. BS2015SW029）；省攻关-杀虫防病工程内生细菌的构建及其生防菌剂研制（2010GSF10206）

解淀粉芽孢杆菌D2WM的抑菌物质初步研究

陈嘉敏[1,2]，张　伟[1,2]，路　露[1]，傅本重[1]，魏　蜜[1]
（1. 湖北工程学院生命科学技术学院/特色果蔬质量安全控制湖北省重点实验室，
孝感　432000；2. 湖北大学生命科学学院，武汉　430062）

摘　要：解淀粉芽孢杆菌D2WM（*B. amyloliquefaciens*）是一株对十字花科蔬菜软腐病病原菌欧文氏菌有显著颉颃作用的生防菌株，本文通过发酵获得该颉颃菌的发酵液，再通过硫酸铵沉淀法、盐酸法以及有机溶剂萃取法提取该颉颃菌发酵液中起颉颃作用的活性物质；通过抑菌圈实验初步确定该解淀粉芽孢杆菌D2WM的颉颃物质主要为非蛋白的有机化合物。并通过4种不同极性有机溶剂萃取后抑菌效果分析，发酵液中抑菌物质的最佳提取剂为正丁醇，将正丁醇萃取的粗提物经液质联用仪分析，结果表明抑菌物质主要有8种，其分子量介于166~927u。此外，该颉颃菌D2对炭疽病菌、溃疡病菌等8种病原真菌也具有很好的颉颃作用，是一株能同时颉颃欧文氏菌和常见病原真菌的广谱颉颃菌。本课题为进一步阐明欧文氏菌颉颃菌的抑菌物质奠定基础，为十字花科软腐病的生物防治提供参考依据。

关键词：欧文氏菌；软腐病；解淀粉芽孢杆菌；抑菌物质

胶孢炭疽菌生防放线菌 gz-8 的分离、鉴定及抗菌活性评价*

张　凯**，吴曼莉，辜柳霜，柳志强***，李晓宇
（海南大学热带农林学院，海口　570228）

摘　要：胶孢炭疽菌（*Colletotrichum gloeosporioides*）是热带、亚热带经济作物的重要病原菌之一，其寄主范围广且种内变异多，可以侵染多种作物并引起炭疽病，已报道的主要寄主植物有橡胶、可可、杧果、木瓜、荔枝、番石榴等经济作物。目前，生产上对胶孢炭疽菌的防治主要依靠化学防治，然而化学防治带来的环境污染、抗药性等问题已逐渐显现，生物防治作为一种无污染、高效的防治措施已经引起了广泛的关注。

从山东临沂地区分离到一株放线菌 gz-8，对胶孢炭疽菌具有较强的颉颃作用，皿内抑制率达到了 72.5%。通过形态特征和生理生化测定，结合 16S rDNA 序列分析，鉴定该菌株为栗褐链霉菌（*Streptomyces badius*）。利用 PDB 培养基对菌株 gz-8 进行了发酵培养，分别提取了发酵液及菌丝的粗提物，抑菌谱测定表明菌株 gz-8 粗体物对橡胶炭疽病菌（*C. gloeosporioides*）、香蕉炭疽病菌（*C. musae*）、杧果蒂腐病菌（*Botryodiplodia theobromae*）、香蕉枯萎病菌（*Fusarium oxysporum* f. sp. *cubense*）、火龙果溃疡病菌（*Neoscytalidium dimidiatum*）、油菜菌核病菌（*Sclerotinia sclerotiorum*）均具有较好的抑制活性。进一步对菌株 gz-8 粗提物的活性进行了评价，发酵液粗提物可以明显抑制胶孢炭疽菌菌丝生长（EC_{50} = 3.18μg/mL）和分生孢子萌发（IC_{50} = 1.92μg/mL），且活性明显高于菌体粗提物。发酵液粗提物具有良好的热稳定性，1mg/mL 发酵液粗提物对感炭疽病杧果果实的防效达到 35.94%，与同浓度下百菌清的防效无显著差异。

关键词：炭疽病；胶孢炭疽菌；栗褐链霉菌；鉴定；生物防治

---

* 基金项目：海南省重点研发计划（ZDYF2016155）；国家自然科学基金（31560045）
** 作者介绍：张凯，男，硕士研究生；E-mail：157942451@qq.com
*** 通信作者：柳志强，博士，副教授，主要从事植物病害生物防治研究；E-mail：liuzhiqiang80@126.com

没食子酸在水稻叶片上对水稻细菌性条斑病防效的持效期*

张锡娇**，魏昌英，汪锴豪，袁高庆，林　纬，黎起秦***

（广西大学农学院，南宁　530004）

摘　要：没食子酸（Gallic acid，GA），学名3，4，5-三羟基苯甲酸，分子式$C_7H_6O_5$，为植物中的天然酚类化合物，具多种生物活性，广泛应用于食品、生物、医药和化工等领域。本研究的前期研究发现，GA对水稻细菌性条斑病菌（*Xanthomonas oryzae* pv. *oryzicola*（Fang *et al.*）Swings *et al.*）具有较强的抑制作用。本研究利用高效液相色谱法（HPLC）检测GA在水稻叶片上的作用方式和残留期，并在检测其在植株叶片上残留量的同时接种病原菌，接种后8d调查水稻细菌性条斑病的发病情况，以确定GA在水稻叶片上对水稻细菌性条斑病防效的持效期。研究结果表明：GA不具有内渗作用，只能残留在水稻叶片的表面。不同浓度GA的处理，其在水稻叶片上的残留期不同，在浓度为100mg/L下，GA在水稻叶片上的残留期为16d，浓度为200mg/L下，GA在水稻叶片上的残留期为24d，浓度为400mg/L下，GA在水稻叶片上的残留期为28d。叶片用200μg/mL的GA处理后，在0~4d时GA的残留量为120.5μg/g，能完全抑制水稻细菌性条斑病的发生；在8~16d时GA的残留量在98.3~49.8μg/g，能部分抑制水稻细菌性条斑病的病情扩展，病指在7.53~29.77；在施药第16天以后，GA的残留量等于或低于27.5μg/g，病指达43.76，与CK的病指（46.25）差异未达显著水平，说明此时的残留量完全不能抑制水稻细菌性条斑病的病情扩展，所以GA在温室条件下的持效期为16d。本研究未在田间稻叶上测定GA对水稻细菌性条斑病抑制作用的持效期，由于田间气候因素多变，田间持效期可能低于温室的持效期，因此，建议用GA防治水稻细菌性条斑病时，施药间隔期以7~10d为宜，且田间用药量应不低于200mg/L。

关键词：水稻细菌性条斑病；没食子酸；作用方式；持效期

---

* 基金项目：广西自然科学基金（2014GXNSFAA118073）

** 第一作者：张锡娇，女，硕士研究生，研究方向为植物细菌性病害及其防治；E-mail：550598404@qq.com

*** 通信作者：黎起秦，教授，研究方向为植物病害及其防治；E-mail：qqli5806@gxu.edu.cn

枯草芽孢杆菌 BAB-1 挥发性抑菌物质对番茄灰霉菌的抑菌作用*

张晓云**，鹿秀云，郭庆港，李社增，王培培，赵卫松，马　平***

（河北省农林科学院植物保护研究所/河北省农业有害生物综合防治工程技术研究中心/农业部华北北部作物有害生物综合治理重点实验室，保定　071000）

摘　要：枯草芽孢杆菌（*Bacillus subtilis*）菌株 BAB-1 是本实验室筛选出的一株生防细菌，对番茄灰霉菌（*Botrytis cinerea*）具有较强的抑制作用。菌株 BAB-1 产生的挥发性抑菌物质（Volatile organic compounds，VOCs）是其主要的抑菌物质之一，前期研究表明该菌株至少能够产生 17 种挥发性物质。为评价挥发性物质的抑菌作用，本实验通过二分格培养皿熏蒸法测定了其中 12 种挥发性化合物对番茄灰霉菌的抑制作用，结果表明二甲胺的抑菌作用最强，EC_{50} 为 0.126μL/mL，其次为甲酸、丙酸、二甲基二硫及丙胺，EC_{50} 分别为 0.135μL/mL、0.137μL/mL、0.141μL/mL 和 0.171μL/mL；采用离体叶片法和离体果实法评价了这 5 种挥发性化合物对番茄灰霉病的生防效果，结果表明二甲胺及甲酸对番茄叶片灰霉病的防效达到 100%，160μL/皿的二甲胺、甲酸及丙酸处理亦能够完全防止番茄果实灰霉病的发生；通过观察 5 种挥发性化合物对番茄灰霉菌菌丝生长及孢子萌发的影响发现，5 种挥发性化合物均能够造成菌丝发生消解、断裂以致不能正常生长，对孢子萌发的抑制率均在 80% 以上，其中二甲胺与甲酸的孢子萌发抑制率达到 100%；研究了 5 种挥发性化合物对番茄灰霉菌的抑菌作用方式，结果表明二甲胺、甲酸、丙酸及丙胺处理均能够杀死番茄灰霉菌，二甲基二硫处理 5d 与 7d 也能够杀死番茄灰霉菌，而处理 1d 与 3d 则可以暂时的抑制灰霉菌的生长，将处理过的番茄灰霉菌菌盘转接至 PDA 平板上能够恢复正常生长。

关键词：枯草芽孢杆菌；挥发性物质；番茄灰霉菌；抑制作用

---

* 基金项目：国家自然科学基金（31501697）；公益性行业（农业）科研专项（201303025）

** 第一作者：张晓云，硕士，助理研究员，主要从事植物病害生物防治研究；E-mail：zxy_ zxl@163.com

*** 通信作者：马平，博士，研究员，主要从事植物病害生物防治研究；E-mail：pingma88@126.com

防治烟草青枯病的药剂筛选*

张耀文[1**]，卢燕回[2]，霍　行[1]，袁高庆[1]，林　纬[1]，黎起秦[1***]
（1. 广西大学农学院，南宁　530005；2. 中国烟草总公司
广西壮族自治区公司，南宁　530022）

摘　要：由青枯雷尔氏菌 *Ralstonia solanacearum*（Smith）Yabuuchi *et al.* 引起的烟草青枯病是一种广泛分布于烟区的毁灭性土传细菌性病害，对烟草生产威胁很大。目前生产上种植的烟草品种多为感病品种，化学药剂防治仍然是防控烟草青枯病的重要措施之一。但是青枯雷尔氏菌基因型多样，生理及致病力分化明显，单一化学药剂的使用不仅会恶化土壤环境，防效也不稳定。本研究在筛选有效化学药剂种类和复配比例的基础上，结合生防菌剂及诱抗剂的使用，探索其对烟草青枯病的防治效果。采用比浊法测定了乙蒜素、农用硫酸链霉素、代森锰锌、氢氧化铜等16种药剂对烟草青枯病菌的最低抑菌浓度（MIC）。结果表明：乙蒜素和农用硫酸链霉素对病原菌的抑制作用最强，其MIC值均为10μg/mL；其次是代森锰锌、苯醚甲环唑、氢氧化铜、氯溴异腈尿酸、恶霉灵、春雷霉素和没食子酸甲酯等，MIC值均为100μg/mL，叶枯唑的为200μg/mL，其他药剂的MIC值大于500μg/mL；因国家已经停止受理和批准农用硫酸链霉素的农药登记，本研究选择广谱性杀菌剂乙蒜素和代森锰锌分别以质量比5∶1、2∶1、1∶1、1∶2和1∶5进行复配，复配剂对烟草青枯病菌的抑制中浓度（EC_{50}）分别为5.854 9μg/mL、6.526 5μg/mL、7.733 2μg/mL、5.245 6μg/mL和4.406 8μg/mL，增效系数（SR）值分别为1.004 5μg/mL、0.870 9μg/mL、0.459 2μg/mL、1.015 4μg/mL和1.172 1μg/mL。将乙蒜素·代森锰锌复配剂（2∶1，m/m，有效浓度0.18mg/mL）与枯草芽孢杆菌（中保冠蓝，有效浓度2×10^{8}cfu/g）以及植物免疫诱抗剂（阿泰灵，有效浓度0.06mg/mL）单独或不同组合施用，测定其对烟草青枯病的室内防治效果。研究发现，前期灌施枯草芽孢杆菌和喷施诱抗剂后再结合灌施复配剂的防效为83.43%，单独灌施复配剂的防效为75.65%，枯草芽孢杆菌灌施与诱抗剂喷施相结合的处理对烟草青枯病的防效是62.44%，而其他单独施用乙蒜素或代森锰锌单剂（有效浓度均为0.18mg/mL）、生防菌剂或者诱抗剂的防效为55%左右。

关键词：烟草青枯病；乙蒜素；代森锰锌；生防菌剂；诱抗剂

---

* 基金项目：中国烟草总公司广西壮族自治区公司科技创新项目（2014-07）

** 第一作者：张耀文，本科生，研究方向为植物病害防治；E-mail：2297887061@qq.com

*** 通信作者：黎起秦，教授，研究方向为植物病害防治；E-mail：qqli5806@gxu.edu.cn

香蕉炭疽病生防放线菌 CY-14 的鉴定及发酵液活性评价*

张月凤**，吴曼莉，柳志强，李晓宇***

（海南大学热带农林学院，海口 570228）

摘 要：香蕉炭疽病是香蕉重要的采后病害，由香蕉炭疽菌（*Colletotrichum musae*）侵染引起，在全世界产蕉区普遍发生，并造成严重的经济损失。本研究以香蕉炭疽菌为靶标菌，筛选到一株颉颃放线菌 CY-14，皿内抑制率达 51.4%。通过形态特征，生理生化及 16S rDNA 序列分析，初步将菌株 CY-14 鉴定为链霉菌属。对菌株 CY-14 产抗菌物质的发酵培养基进行了筛选，结果发现利用小米培养基培养 7d，发酵液对香蕉炭疽菌的抑制率达 92.8%，发酵液 50 倍稀释液的抑制率为 71.1%。此外，菌株 CY-14 小米发酵液对香蕉枯萎病菌（*Fusarium oxysporum* f. sp. *cubense*）、杧果蒂腐病菌（*Botryodiplodia theobromae*）、橡胶炭疽病菌（*C. gloeosporioides*）的抑制率均在 60%，且发酵液的抗菌活性具有一定的热稳定性。孢子萌发试验发现菌株 CY-14 发酵液 1 000倍稀释液可以完全抑制香蕉炭疽菌分生孢子，且多数孢子出现明显变长的现象。综上所述，菌株 CY-14 具有开发为香蕉炭疽病生防菌株的潜力，具有一定的应用前景。

关键词：香蕉炭疽菌；链霉菌；鉴定；生物防治

---

* 基金项目：海南省重点研发计划（ZDYF2016155）；国家自然科学基金（31560045）

** 作者介绍：张月凤，硕士研究生；E-mail：1092758978@ qq. com

*** 通信作者：李晓宇，博士，副教授，主要从事植物病原真菌学研究；E-mail：hdlixiaoyu@ 126. com

几种新型杀菌剂对工艺葫芦炭疽病菌的室内毒力测定*

程 娇**，刘梦铭，任爱芝，赵培宝***
（聊城大学农学院植物保护系，聊城 252000）

摘 要：山东聊城堂邑镇种植工艺葫芦 1 217hm^2，被称为“江北葫芦之乡”，葫芦寓意“福禄”，成品有激光雕刻、烙铁刻绘、彩绘、粘贴画等工艺产品。葫芦炭疽病是为害果实影响其商品性、观赏性及收藏价值的重要病害，因多年连作病菌积累及抗药性的产生，生产上葫芦炭疽病难以控制，严重影响老百姓经济效益。从土壤和病组织中分离病菌鉴定为黑刺盘孢菌（*Colletotrichum nigrum*）和瓜类炭疽菌（*Colletotrichum orbiculare*），笔者选择了 4 种不同作用类型的杀菌剂，测定了它们对两种病原菌的抑制作用，旨在明确该病原菌对不同杀菌剂敏感性差异，并筛选有效药剂，为有效控制葫芦炭疽病提供科学依据。

供施药剂：96%苯醚甲环唑、98.4%嘧菌环胺、98.1%咪鲜胺、96%噻呋酰胺；对照药剂：99%多菌灵。应用生长速率测定法，测定 4 种药剂对葫芦炭疽病菌黑刺盘孢（*C. nigrum*）和瓜类炭疽菌（*C. orbiculare*）的抑菌效果，对照药剂 99%多菌灵，试验结果如下：4 种杀菌剂和多菌灵对黑刺盘孢菌（*Colletotrichum nigrum*）的 EC_{50}分别为：98.1%咪鲜胺 0.171 2μg/mL、96%苯醚甲环唑 0.601 2μg/mL、98.4%嘧菌环胺 5.801 1μg/mL、96%噻呋酰胺 8.797 0μg/mL、多菌灵 22.36μg/mL，大小顺序为咪鲜胺<苯醚甲环唑<嘧菌环胺<噻呋酰胺<多菌灵；对瓜类炭疽菌（*Colletotrichum orbiculare*）的 EC_{50} 值分别为：98.1%咪鲜胺 0.147 3μg/mL、96%苯醚甲环唑 0.701 2μg/mL、98.4%嘧菌环胺 5.001 1μg/mL、96%噻呋酰胺 7.903 0μg/mL、多菌灵 23.012 0μg/mL，大小顺序为咪鲜胺<苯醚甲环唑<嘧菌环胺<噻呋酰胺<多菌灵。由此可得，4 种新型杀菌剂对两种为害葫芦的炭疽菌的抑菌效果一致，均明显好于传统药剂多菌灵，其中抑制效果最好的为咪鲜胺，其次毒力效果较好的为苯醚甲环唑，在生产中可推荐使用。

关键词：葫芦炭疽病；杀菌剂；毒力测定；EC_{50}

* 基金项目：聊城市科技计划项目（2014GJH03）
** 第一作者：程娇，研究生
*** 通信作者：赵培宝，博士，副教授，从事植物与病原物分子互作研究；E-mail：zhaopeibao@163.com

海洋细菌 GM1-1 菌株抗菌活性物质的分离

曹雪梅[1**]，李　欢[1]，贾　杰[2]，陈　茹[1]，
王　琦[1]，马桂珍[1***]，暴增海[1]
（1. 淮海工学院海洋学院，连云港　222005；2. 江苏耕耘化学有限公司，连云港　222005）

摘　要：GM-1-1 菌株是从连云港海域分离到的一株对多种病原真菌具有较好抑制作用的海洋解淀粉芽孢杆菌（*Bacillus amyloliquefaciens*），该菌株对核盘菌（*Sclerotinia sclerotiorum*）、葡萄白腐病菌（*Coniothyrium diplodiella*）、小麦根腐病菌（*Bipolaris sorokiniana*）、番茄早疫病菌（*Alternaria solania*）、禾谷镰刀菌（*Fusarium graminearum*）、斑点落叶病菌（*Alternaria alternate*）、丝核菌（*Rhizoctonia cerealis*）、尖镰孢菌（*Fusarium oxysporum* f. sp. *cucumarimum*）等多种病原真菌具有强烈的抑制作用，且抗菌作用具有热稳定性和紫外线稳定性。为了获得该发酵液中的抗菌活性物质，采用硅胶柱层析、制备型薄层层析以及 HPLC 等技术对海洋解淀粉芽孢杆菌 GM-1-1 抗菌活性组分进行分离纯化，以白色念珠菌（*Candida albicain*）为指示菌，采用牛津杯法测定不同组分的抑菌活性。本试验通过硅胶柱层析一共收集到 6 组组分，分别测定不同组分的抑菌活性，得到两组活性组分，命名为 C1、C2，通过制备型薄层层析法和高效液相色谱法来进一步检测活性组分的纯度。

本研究通过分离纯化 GM-1-1 菌株发酵液中的抑菌活性物质，旨在为该菌株生物防治剂的大规模开发和应用提供相应的理论基础。

关键词：解淀粉芽孢杆菌；抑菌作用；分离纯化

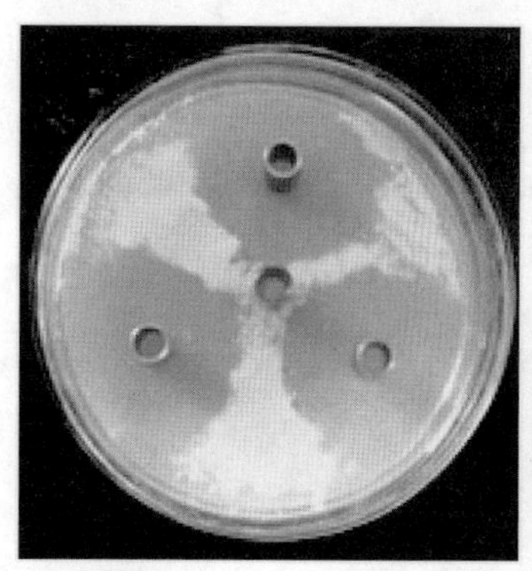

GM-1-1 菌株发酵液对白色念珠菌的抑制作用

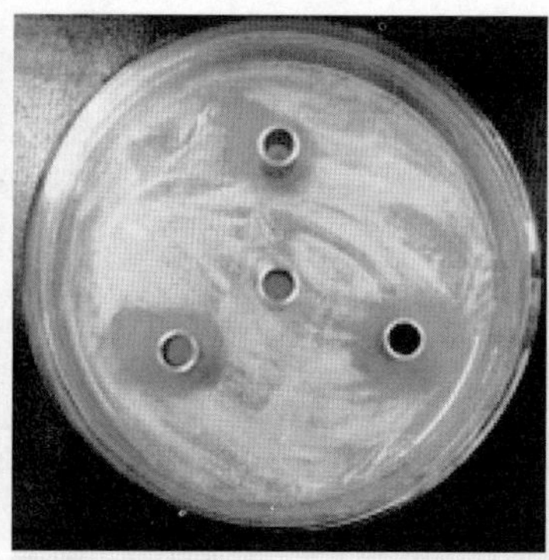

C1 组分对白色念珠菌的抑制作用

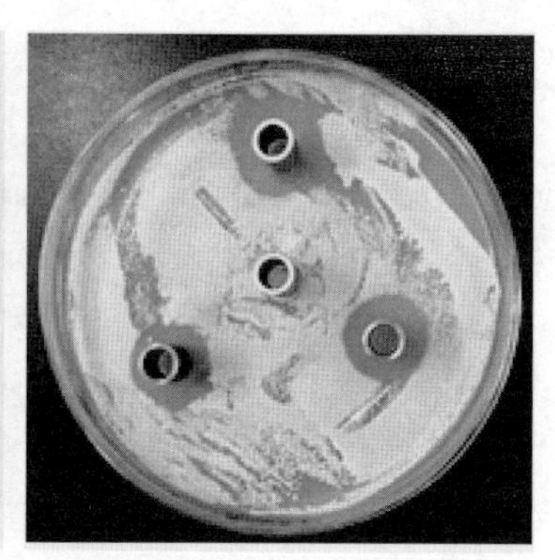

C2 组分对白色念珠菌的抑制作用

* 基金项目：江苏省科技厅（现代农业）重点研发计划（BE2016335）；连云港市科技局产业前瞻与共性关键技术项目（CG1513）；国家级大学生实践创新训练计划项目（G201411641105002）；江苏省“十二五”高等学校水产类重点专业项目资助

** 第一作者：曹雪梅，硕士生，研究方向：生物化工；E-mail：2224256423@qq.com

*** 通信作者：马桂珍，教授，研究方向：抗菌微生物及植物病害生物防治；E-mail：415400420@qq.com

西瓜根系分泌物对西瓜枯萎病菌及其生防菌的化感作用*

李　丹**，秦伟英，刘正坪，魏艳敏，任争光，赵晓燕***
（北京农学院植物科学技术学院，农业应用新技术北京市重点实验室，北京　102206）

摘　要：西瓜枯萎病俗称“死秧病”，是西瓜上危害最严重的病害之一，西瓜枯萎病在世界各西瓜产区均造成较大损失，不仅影响生产而且对于农民的经济效益也有很大的影响。该病属土传病害，病菌在土壤中可长期存活，防治十分困难。植物的根系分泌物不仅为植物根际土壤微生物提供营养，同时也影响着根际土壤微生物的群落多样性。对西瓜根系分泌物与西瓜枯萎病菌以及其生防菌之间的作用的研究可以为明确西瓜枯萎病在田间的发生规律以及为生防菌的应用技术研究打下基础。本实验主要研究了西瓜根系分泌物中的 3 种有机酸（苹果酸、琥珀酸、柠檬酸）和 6 种氨基酸（精氨酸、丙氨酸、谷氨酸、酪氨酸、甲硫氨酸和天冬氨酸）对西瓜枯萎病菌的产孢量的影响以及对生防菌 TR2 的趋化作用。

结果表明：3 种有机酸和 6 种氨基酸均能促进西瓜枯萎病菌 Xg-3 的孢子形成。其中谷氨酸和琥珀酸在浓度为 400mg/L 时促进效果最明显，分别达到 2.60×10^6 CFU/mL 和 2.20×10^6 CFU/mL，天冬氨酸在浓度为 200mg/L 时促进效果最明显，达到 2.56×10^6 CFU/mL，为对照的 4~5 倍；其次是苹果酸、柠檬酸、丙氨酸和酪氨酸，分别在 800mg/L、400mg/L、100mg/L 和 100mg/L 时促进效果最好，达到 1.56×10^6 CFU/mL、1.57×10^6 CFU/mL、1.28×10^6 CFU/mL、和 1.69×10^6 CFU/mL，为对照的 3 倍左右；而精氨酸和甲硫氨酸相对较低，但其促进效果也为对照的 2 倍左右。说明西瓜根系分泌物能促进西瓜枯萎病菌的生长和繁殖，导致病菌在根系周围的积累，加重病害的发生。

解淀粉芽孢杆菌 TR2 是对西瓜枯萎病菌 Xg-3 具有较好颉颃活性的生防细菌，试验结果表明：生防菌 TR2 对供试的 3 种有机酸和 6 种氨基酸表现出不同的趋化作用。其中对甲硫氨酸、酪氨酸和天冬氨酸的趋化效果最为明显，达到对照的 2.5 倍左右；其次是谷氨酸，达到对照的 2 倍左右；而对苹果酸、琥珀酸、柠檬酸、丙氨酸和精氨酸趋化作用相对不是很明显。说明生防芽孢杆菌对西瓜根系分泌物中的不同成分具有不同的趋化作用。

综上所述，西瓜枯萎病菌和生防芽孢杆菌对西瓜根系分泌物的不同成分的响应具有明显的差异，而且不同植物种类或不同西瓜品种的根系分泌物的成分也有明显不同，本实验的研究结果可以为研究西瓜枯萎病菌和生防芽孢杆菌对不同植物种类或不同西瓜品种的根系分泌物的成分响应打下基础，从而进一步指导西瓜不同品种或西瓜与其他植物的轮作，以及生防菌的应用奠定基础。

关键词：西瓜枯萎病；西瓜根系分泌物；产孢；趋化作用

---

* 基金项目：两株生防芽孢杆菌对西瓜枯萎病的生防作用机理研究；北京市农委“菜篮子”新型生产经营主体科技能力提升项目“北京八达岭小浮沱蔬菜专业合作社对接技术服务”（20150203-04）

** 第一作者：李丹，女，在读硕士研究生；E-mail：289773003@qq.com

*** 通信作者：赵晓燕，女，博士，主要从事植物病理学研究；E-mail：zhaoxy777@163.com

水稻白叶枯病菌颉颃菌筛选及其抗菌代谢产物分离鉴定

杨丹丹，周　莲，何亚文*

（上海交通大学生命科学技术学院，微生物代谢国家重点实验室，上海　200240）

摘　要：水稻白叶枯病是亚洲稻区的重要水稻病害之一，病原为稻黄单胞菌，包括白叶枯菌（*Xanthomonas oryzae* pv. *oryzae*，*Xoo*）和稻细菌性条斑病菌（*Xanthomonas oryzae* pv. *oryzicola*，*Xoc*）两个致病变种。目前由于病原菌的抗药性的增强，化学防治白叶枯病的常用杀菌剂，如叶枯唑、农用链霉素等，防治效果逐年下降；生物农药以其高效、低残留的优势特点，在水稻白叶枯病防治中将发挥重要作用。笔者利用纸片琼脂扩散法筛选得到对稻黄单胞菌 PXO99A 有颉颃作用的一株细菌，命名为 RX1101 菌株；通过 16SrRNA 基因序列分析，RX1101 被鉴定为链霉菌（*Streptomyces zagrosensis* sp. nov.）。RX1101 发酵液对 PXO99A 有很强的颉颃作用。为了分离和鉴定发酵液中活性抑菌物质，笔者首先通过正交试验优化发酵条件，并以此为基础进行大规模发酵。离心收集发酵液，用漏斗过滤除去固体残留物，用等体积乙酸乙酯萃取，弃去水相，有机相用旋转蒸发仪蒸干得到粗体物。利用制备型液相色谱对粗产物进行制备，色谱柱为 C18（20. 0mm×250mm，5μm），流动相 A 相为水（经 Millipore 公司 Milli-Q 纯水器处理并抽滤）和 0. 1%甲酸，B 相为乙腈，梯度洗脱得到 20 种组分。利用纸片琼脂扩散法筛选对稻黄单胞菌 PXO99A 有颉颃作用的组分，结果显示组分 16 和 17 保留对 PXO99A 强抑菌活性。通过 LC-MS 和核磁共振 NMR 确定 16 和 17 组分为同一种物质，解析了其结构，将该物质命名为 SPRXO。SPRXO 不仅对稻白叶枯病菌 PXO99A 有强抑制作用，对稻细菌性条斑病菌 RS105、野油菜黄单胞菌 XC1 和 8004 等也具有强抑制作用；但对其他细菌，如铜绿假单胞菌、金黄色葡萄球菌、洋葱伯克霍尔德菌以及致病真菌白色念珠菌、水稻纹枯菌等，无明显抑菌效果，显示该代谢产物对黄单胞菌属病原菌具有特异的抑菌活性。通过优化发酵条件，RX1101 菌株抑菌代谢产物 SPRXO 的发酵效价可以达到 2000 mg/L，有望实现产业化开发和应用。

关键词：水稻白叶枯病；链霉菌；筛选；鉴定

* 通信作者：何亚文；E-mail：yawenhe@ sjtu. edu. cn

黑龙江省水稻恶苗病菌对氰烯菌酯敏感性及其对咪鲜胺抗性风险评估*

Muhammad Waqas Younas**，彭　钦，刘　莹，刘西莉***
（中国农业大学植物病理学系，北京　100193）

摘　要：水稻恶苗病是黑龙江省水稻产区的常见毁灭性病害，引起该病害的主要病原菌为 *Fusarium fujikuroi*。2013—2015 年，该病害在黑龙江省尤为严重，对水稻生产造成了极大威胁。本研究测定了分离自黑龙江地区水稻样本上的恶苗病菌对新型杀菌剂氰烯菌酯的敏感性，评估了水稻恶苗病菌对咪鲜胺的抗性风险，为田间生产中进行抗药性监测和寒冷地区水稻恶苗病的防治提供科学参考。

首先采用菌丝生长速率法，测定了采自黑龙江地区的 95 株水稻恶苗病菌对氰烯菌酯的敏感性。结果表明：供试水稻恶苗病菌对氰烯菌酯的 EC_{50}分布于 0. 031 7～1. 717μg/mL，平均 EC_{50}值为 0. 274±0. 210μg/mL。获得的水稻恶苗病菌对氰烯菌酯的敏感性频率分布呈单峰曲线，未发现敏感性下降的亚群体。因此，可将其作为水稻恶苗病菌对氰烯菌酯的敏感性基线，该结果为田间抗药性监测提供了参考和依据。

通过鉴别浓度和 EC_{50}测定方法，检测了 95 株水稻恶苗病菌对咪鲜胺的敏感性，结果表明：6 株水稻恶苗病菌对咪鲜胺表现为敏感，其余 89 株病原菌均对咪鲜胺表现为抗性，抗性频率达到 93. 6%。生物学性状研究表明，部分抗性菌株的菌丝生长速率、产孢能力和分生孢子萌发率发生了显著变化，抗性菌株的分生孢子产量稍低于敏感菌株，但抗性菌株菌的致病力显著高于敏感菌株。交互抗药性研究结果表明：咪鲜胺与氰烯菌酯、种菌唑、戊唑醇、多菌灵和氟吡菌酰胺之间没有交互抗药性。结合抗性菌株的分离频率和生存适合度研究结果，综合评估黑龙江省水稻恶苗病菌对咪鲜胺具有中-高等抗性风险。

关键词：水稻恶苗病菌；敏感性；抗性；风险评估

---

* 基金项目：公益性行业（农业）科研专项（编号：201303023）
** 第一作者：Muhammad Waqas Younas，在读硕士研究生；E-mail：waqasyounas591@ cau. edu. cn
*** 通信作者：刘西莉，女，教授；主要从事植物病原菌与杀菌剂互作研究；E-mail：seedling@ cau. edu. cn

上海地区番茄灰霉病菌对几种重要杀菌剂的抗药性检测*

武文帅**，胡志宏，刘鹏飞，刘西莉***
（中国农业大学植物保护学院植物病理系，北京　100193）

摘　要：由灰葡萄孢菌（*Botrytis cinerea*）引起的番茄灰霉病，是番茄生产中最具毁灭性的真菌病害之一。在生产实际中，主要通过化学杀菌剂来防控该病害的发生和发展。但由于病原菌寄主范围广、遗传变异性强，对其防治药剂极易产生抗性，从而导致田间防治效果的下降。因此，明确田间番茄灰霉病菌对几种重要杀菌剂的抗药性现状，对于指导药剂的科学使用，优化灰霉病菌抗药性管理方案具有重要意义。

本研究于 2016 年 2—4 月，从上海市宝山、崇明、奉贤、金山、浦东新区、青浦和松江等地区采集、分离和纯化获得 699 株灰霉病菌，选择了具有地理代表性的菌株，分别采用鉴别浓度法（MIC）和菌丝生长速率法测定了其对多菌灵、啶酰菌胺、氟吡菌酰胺、嘧菌酯、嘧霉胺、咯菌腈和苯醚甲环唑 7 种供试杀菌剂的抗性。根据敏感性差异，将供试菌株分为敏感、中抗和高抗 3 个敏感性水平。研究发现，对多菌灵表现为高抗的菌株占供试菌株的 29%，对啶酰菌胺表现为中、高抗的菌株分别占 27%和 14%，20%供试菌株对氟吡菌酰胺表现为高抗；30%供试菌株对嘧菌酯高抗；对嘧霉胺表现为中、高抗的菌株分别为 80%和 3%；对咯菌腈表现为中、高抗菌株分别为 15%和 19%；20%供试菌株对苯醚甲环唑表现为高抗。上海市不同地区的番茄灰霉病菌普遍产生了抗药性，其中，有 71 株灰霉病菌分别对上述某一种杀菌剂表现为抗性，54 株灰霉病菌同时对两种杀菌剂产生抗性，另外 54 株灰霉病菌同时对 3 种杀菌剂产生抗药性，35 株灰霉病菌同时对 4 种杀菌剂产生抗性，6 株灰霉病菌同时对 5 种杀菌剂产生抗性。抗性菌株的生物学性状研究结果表明：不同菌株的菌丝生长速率、温度敏感性、产孢能力、孢子萌发能力和致病力等指标与其对杀菌剂的敏感性之间没有明显的相关性，与敏感菌株相比，抗性菌株在田间具有一定的生存适合度。

关键词：番茄灰霉病菌；杀菌剂；抗药性

* 基金项目：公益性行业（农业）科研专项（编号：201303023）
** 第一作者：Muhammad Waqas Younas，在读硕士研究生；E-mail：waqasyounas591@ cau. edu. cn
*** 通信作者：刘西莉，女，教授，主要从事植物病原菌与杀菌剂互作研究；E-mail：seedling@ cau. edu. cn

褪黑素对苹果采后灰霉病的防效及防病机制初探*

于春蕾**，曹晶晶，于子超，李保华，王彩霞***

（青岛农业大学植物保护学院，山东省植物病虫害综合防控重点实验室，青岛　266109）

摘　要：由灰葡萄孢菌（*Botrytis cinerea*）引起的灰霉病是苹果采后贮藏期的主要病害之一，该病害在苹果采后运输、贮藏等各个环节均可发生，造成严重经济损失，有效控制灰霉病对于提高苹果采后贮藏保鲜具有重要意义。褪黑素广泛存在于高等植物体内，可在植物组织内自由移动，能够清除自由基且具有抗氧化作用，且外源褪黑素可有效缓解病原菌、干旱、重金属等逆境胁迫对植物造成的损伤。本研究以“富士”果实为材料，采用刺伤接种法，测定了外源褪黑素对苹果采后灰霉病的防效，并对其防病机制进行了初步探究。结果表明：外源褪黑素处理可有效降低灰霉病的发病率和病斑面积，0.2mmol/L 褪黑素处理后间隔 72 h 以上接种病原菌的处理，防效高达 83%以上，且该浓度褪黑素对灰霉病菌孢子萌发和菌丝生长均无明显的抑制作用，但褪黑素可显著提高果实内过氧化物酶（POD）、过氧化氢酶（CAT）、超氧化物歧化酶（SOD）、多酚氧化酶（PPO）和苯丙氨酸解氨酶（PAL）活性，其峰值是对照的 1.86~8.73 倍。表明外源褪黑素通过持续提高果实内防御酶活性诱导苹果对灰霉病的抗性。

关键词：褪黑素；苹果；灰霉病；防御酶

---

* 基金项目：现代农业产业技术体系建设专项资金（CARS-28）；国家级大学生创新训练项目（200610435005）

** 第一作者：于春蕾，女，硕士研究生，研究方向果树病理学；E-mail：ycl10190818@163.com

*** 通信作者：王彩霞，女，教授，主要从事果树病害研究；E-mail：cxwang@qau.edu.cn

Preliminary Study on the Antifungal Activity of the Volatile Substances Produced by *Bacillus subtilis* Czk1[*]

He Chunping[1**], Tang Wen[2], Li Rui[1], Liang Yanqiong[1], Wu Weihuai[2], Zheng Jinlong[1], Xi Jingen[1], Huang Xin[1], Yi Kexian[1***]

(1. *Environment and Plant Protection Institute*, *CATAS*, *Haikou*, 571101, *China*; 2. *Institute of Tropical Agriculture and Forestry*, *Hainan University*, *Haikou* 570228, *China*)

Abstract: We obtained a strain of *Bacillus subtilis*, which wenamed Czk1, from the aerial roots of rubber trees. This bacterial isolate its volatiles substances exhitits strong antipathogentic activities against *Ganoderma pseudoferreum*, *Phellinus noxius*, *Helicobasidium compactum*, *Rigidoporus lignosus*, *Sphaerostilbe repens*, and *Colletotrichum gloeosporioides*. Our research has shown that LB Broth medium as the optimal medium of strain Czk1. When the bacterial concentration is 10^5CFU/mL, the inhibition ratio reaches up to more than 50%. When the bacterial concentration is 10^8CFU/mL, the inhibition ratio reaches up to more than 90%. The volatiles can promote and inhibit the pathogen pigment, and cause the growth malformed of fungal mycelium, such as *H. compactum*, *P. noxius* and *C. gloeosporioides*. The inhibition rate of spore germination was 71.30% of *C. gloeosporioides bcause of* volatiles substances of strain Czk1, nevertheless, they have no influence on the spore morphology. Meanwhile, the volatiles which the strains Czk1generate in sterilized soil still have a good inhibitory effect to pathogens growth. The inhibition rate of the strain Czk1volatiles was 63.37% and 54.57% for *G. pseudoferreum* and *P. noxius* espectively.

Key words: *Bacillus subtilis*; Rubber treediseases; Volatiles; Antifugal activity

* 基金项目：现代农业产业技术体系建设专项资金（CARS-34-BC1）资助

** 第一作者：贺春萍，女，研究员，研究方向：植物病理学；E-mail：hechunppp@163.com

*** 通信作者：易克贤，博士，研究员；E-mail：yikexian@126.com

都市大叶黄杨白粉病的防治药剂筛选

郑晓露，刘　悦，黄志岳，毕　扬，陈　艳*
（北京农学院植物保护系，北京　102206）

摘　要：大叶黄杨白粉病在北京地区每年都有发生，但市面上针对大叶黄杨白粉病防治的特效药并不多见，并且随着喷施某一类药剂次数的增多，频率的加快，病原菌的抗药性问题也日益突出。居住区内大叶黄杨由于缺少药剂防治，发病率及严重度更是明显高于公共绿地。大叶黄杨白粉病的严重发生，不仅会严重影响大叶黄杨园林绿化作用的发挥；其白粉随气流扩散通过呼吸道进入人体内，更会严重影响人体健康。

为筛选出防治效果良好的药剂，本文选取北京一居民小区内的大叶黄杨，对其白粉病的发生过程经过连续3年的病害检测，从杀菌剂中筛选出可能防治大叶黄杨白粉病的15种药剂进一步进行田间药效防治的筛选，包括，露娜森、25%腈菌唑乳油、40%氟硅唑乳油、40%嘧菌·乙嘧酚、50%醚菌酯、4%农抗120、0.5%大黄素甲醚、25%乙嘧酚、60%嘧菌酯、25%丙环唑、20%三唑酮、10%苯醚甲环唑、寡雄腐霉、70%代森锰锌、75%百菌清。连续2年在发生严重的大叶黄杨上进行防治，每年喷施2次，观察4次的喷施效果，结果证明，露娜森的防效最好，兼具保护和治疗的作用；25%腈菌唑、40%氟硅唑、25%丙环唑、20%三唑酮的药效次之，对大叶黄杨白粉病的进一步发展具有显著的抑制作用；60%嘧菌酯、50%醚菌酯、40%嘧菌·乙嘧酚、0.5%大黄素甲醚水剂亦有一些作用，后期病害发展严重，喷施2次药剂亦不能控制病害的发展。未经药剂处理的对照和寡雄腐霉、70%代森锰锌、75%百菌清的病害症状发展严重，新叶、老叶均布满白粉，后期出现枯萎的现象。药剂的防效从高到低依次为：露娜森> 25%腈菌唑乳油、40%氟硅唑>丙环唑、20%三唑酮 > 苯醚甲环唑>40%嘧菌·乙嘧酚、50%醚菌酯> 农抗120水剂>0.5%大黄素甲醚、25%乙嘧酚> 60%嘧菌酯。本实验的结果对都市大叶黄杨白粉病的控制具有指导意义。

关键词：大叶黄杨；白粉病；防治；药剂筛选

---

* 通信作者：陈艳，副教授，博士，硕士生导师，主要从事杀菌剂开发与利用研究；E-mail：cheny@bua.edu.cn

嘧肽霉素与氨基寡糖素混配增效及抗 TMV 机制初步研究*

董蕴琦**，吴元华**，安梦楠***

（沈阳农业大学植物保护学院，沈阳 110866）

摘 要：嘧肽霉素是本实验室研发的一种低毒、低残留胞嘧啶核苷类农用抗生素，是由不吸水链霉菌（*Streptomyces achygruscopicus*）产生，具有抑制病毒复制、降低病毒粒体浓度的作用。氨基寡糖素是农用壳寡糖，通过激发植物自身的免疫反应，使植物获得系统性抗性从而起到抗病和增产作用。本文将具有治疗效果的嘧肽霉素和诱抗作用的氨基寡糖素按照一定比例混配，以希获得一种新型的诱导、预防和治疗多重作用的广谱抗病毒剂。结果表明：得到了最适二者混配的比例，在枯斑寄主心叶烟上对 TMV 的枯斑抑制率高于嘧肽霉素 10.2%，高于氨基寡糖素 15.4%；在系统花叶寄主烤烟 K326 上，对 TMV 防效分别提高 12.5%和 17.6%。

BY-2 原生质体是由烟草悬浮细胞（Bright Yellow 2）进行酶解细胞壁得到。为了明确混配制剂在单细胞内对烟草花叶病毒是否产生抑制作用，在 BY-2 原生质体中接种 TMV 后，采用混配药剂的不同浓度处理并培育 20h，分别提取 BY-2 原生质体的蛋白和 RNA，Western blot 检测结果表明：原生质体内 TMV 外壳蛋白（CP）的积累量受到显著抑制；通过 RT-qPCR 对六对抗性基因的表达量进行分析，发现植物热激同源蛋白 HSP70 表达量明显上调；不同浓度混配剂处理的 BY-2 细胞，活性氧的含量在 3h 内显著增加，并且细胞死亡率不受药剂影响。

关键词：嘧肽霉素；氨基寡糖素；烟草花叶病毒（*Tobacco mosaic virus*，TMV）

---

* 基金项目：国家烟草重大专项［中烟办（2016）259 号］

** 第一作者：董蕴琦，女，在读硕士，研究方向：植物病毒；E-mail：846942932@qq.com

吴元华，男，教授，博士生导师，研究方向：植物病毒学和生物农药；E-mail：wuyh09@vip.sina.com

*** 通信作者：安梦楠，男，讲师，研究方向：植物病毒；E-mail：anmengnan1984@163.com

氟吡菌酰胺与阿维菌素复配拌种对小麦孢囊线虫病的田间防治效果*

迟元凯[1**]，汪　涛[1]，赵　伟[1]，彭德良[2]，黄文坤[2]，戚仁德[1***]

(1. 安徽省农业科学院植物保护与农产品质量安全研究所，合肥　230031；

2. 中国农业科学院植物保护研究所，北京　100193)

摘　要：小麦孢囊线虫（cereal cyst nematode，CCN）是严重为害小麦、大麦等麦类作物的一类植物病原线虫，目前在我国16个省市区均有发生，在河南、河北、安徽等小麦主产区为害尤为严重，减产达20%~70%，筛选安全高效的小麦孢囊线虫防治药剂迫在眉睫。作者在前期试验中发现氟吡菌酰胺和阿维菌素拌种对小麦孢囊线虫有较好防效，为进一步优化拌种配方，于2015—2016年在安徽省宿州市埇桥区进行了氟吡菌酰胺+阿维菌素、氟吡菌酰胺、阿维菌素等药剂拌种防治小麦孢囊线虫试验，通过调查苗期小麦出苗率、株高、根长、分蘖数评价拌种剂安全性；通过调查小麦返青期根系中侵入的小麦孢囊线虫数和扬花期产生的白雌虫数，结合收获期测产，综合评价拌种对小麦孢囊线虫的防控效果。结果发现：用供试药剂拌种处理后，小麦出苗率、株高、根长、分蘖数与对照相比均无显著差异；阿维菌素、氟吡菌酰胺+阿维菌素拌种对返青期小麦孢囊线虫侵染有较好抑制效果，其防效分别为57.7%和65.3%；氟吡菌酰胺、氟吡菌酰胺+阿维菌素显著减少扬花期小麦根系白雌虫数量，防效分别为70.4%和78.2%；氟吡菌酰胺、氟吡菌酰胺+阿维菌素处理区小麦产量分别为535.3kg/亩和541.2kg/亩，相比对照区（493.6kg/亩）分别增产8.4%和9.6%。以上结果表明：用氟吡菌酰胺、阿维菌素等药剂拌种对苗期小麦生长安全，其中氟吡菌酰胺+阿维菌素拌种后既可抑制小麦孢囊线虫侵入，减少线虫侵入对根系造成的损伤，有效减少根系孢囊的产生，显著增加小麦产量，其防效和增产效果优于氟吡菌酰胺和阿维菌素单剂，适合在生产上用于小麦孢囊线虫的防治。

关键词：小麦孢囊线虫；氟吡菌酰胺；阿维菌素；拌种；防效

---

* 基金项目：公益性行业（农业）科研专项（201503114）

** 第一作者：迟元凯，助理研究员，从事植物线虫学研究；E-mail：chi2005112@163.com

*** 通信作者：戚仁德，研究员，从事植物土传病害综合防控技术研究；E-mail：rende7@126.com

苹果白绢病的发病流行条件与防治方法*

高常燕**，李保华***

（青岛农业大学农学与植物保护学院，山东省植物病虫害综合防控重点实验室，青岛 266109）

摘　要：苹果白绢病（Southern Blight of apple）是苹果根茎部的重要病害，是由齐整小核菌（*Sclerotium rolfsii* Sacc）侵染所致。苹果白绢病在高温高湿条件下容易发病，若防治不及时，病菌很快环绕茎基部，导致树体死亡。近年来，矮砧密植果园受害严重，个别果园一年的死树率高达 10%。本试验对苹果白绢病的发病流行条件与防治技术进行了研究，结果如下。

白绢病菌在 10～35℃的温度条件下都能生长，其中 25～35℃下菌丝生长较快，当温度低于 15℃时，病菌生长很慢。当土壤含水量高于 33%时病菌才能生长；当土壤含水量在 33%～88%之间时，随土壤湿度增加，菌丝生长扩展速度加快；饱和的土壤湿度不利于病菌菌丝生长，菌丝在湿度饱和的土壤中生长速度较慢。白绢病菌的菌丝主要在土壤表面或表层生长扩展，最适条件下每天的生长扩展距离达 2.9cm；白绢病菌在土壤深处生长扩展很慢，而且容易形成菌核。侵染苹果根茎部的白绢病菌主要来源于表层土壤，并从土壤表面扩展到树体根茎部。

用吡唑醚菌酯、甲基立枯磷、异菌脲、噁霉灵、咯菌腈、乙烯菌核利、福美双、多菌灵、硫酸铜 9 种药剂，按厂家推荐浓度处理含有菌核的土壤，研究杀菌剂对菌核的杀菌和抑菌效果。结果表明：所有药剂都不能够杀死菌核中的病菌。多菌灵和乙烯菌核利抑菌效果最差，药剂处理 1 周后，对菌核萌发的抑制率为 0%；甲基立枯磷和吡唑醚菌酯的抑菌效果最好，药剂处理 1 周后的抑菌率分别为 73.3%和 33.3%。

用 95%甲基立枯磷、50%福美双、生石灰、及生石灰与硫酸铜配制药土，撒于菌核外围。结果表明：生石灰、以及石灰与硫酸铜混合物拌成的药土能够有效抑制白绢病菌菌丝的生长，萌发后的菌核只能产生新菌核，但菌丝生长扩展距离很短；其他药剂都不能明显的抑制菌丝的生长扩展。因此，在树体茎基部外围撒施生石灰，或生石灰与硫酸铜的混合物能有效抑制病菌接近或接触寄主，有效防治茎基部白绢病的发生。

关键词：苹果白绢病；发生与流行；防治方法

---

* 基金项目：国家苹果产业技术体系（CARS-28）；山东省农业科学院农业科技创新（CXGC2016B11）-主要农作物病虫害绿色防控关键技术

** 第一作者：高常燕，女，硕士研究生；E-mail：changyangao@126.com

*** 通信作者：李保华，教授，主要研究方向为植物病害流行和果树病害研究；E-mail：baohuali@qau.edu.cn

8种杀菌剂对苹果树白绢病菌的室内毒力测定*

李　栋**，董向丽，李平亮，李保华***
（青岛农业大学农学与植物保护学院，山东省植物病虫害综合防控重点实验室，青岛　266109）

摘　要：苹果白绢病（*Selerotium rolfsii* Sacc）主要侵染苹果树的根茎部，造成根部皮层腐烂和树体死亡，是苹果苗期和幼树期的重要病害。近年来，随着矮砧密植栽培技术的发展，白娟病呈上升趋势，病重园死树率高达10%，苗圃的死苗率也高达5%，成为制约苹果发展的主要病害。对于白绢病，目前生产上缺乏有效的防治技术，实际生产中主要以多菌灵灌根防治白绢病，但效果差。

为了筛选有效的防治药剂，测定了95%甲基立枯磷原粉，250g/L吡唑醚菌酯乳油，80%多菌灵可湿性粉剂，50%乙烯菌核利，25g/L咯菌腈悬浮种衣剂，99%噁霉灵粉剂，50%福美双可湿性粉剂和45%异菌脲悬浮剂8种杀菌剂对苹果白绢病菌的抑制中浓度。其中，吡唑醚菌酯对苹果白绢病菌菌丝的生长抑制效果最好，其抑制中浓度 EC_{50} 为0.086μg/mL；其次为咯菌腈和甲基立枯磷，两种药剂抑制中浓度 EC_{50} 值分别为0.359μg/mL和1.486μg/mL；乙烯菌核利对白绢病菌丝生物的抑制效果最差，其抑制中浓度 EC_{50} 为754.646μg/mL；多菌灵的抑制中浓度为141.188μg/mL。研究结果表明：多菌灵对苹果白绢病菌生长扩展的抑制效果较差，不能作为防治白绢病的有效药剂，建议将吡唑醚菌酯、咯菌腈和甲基立枯磷作为防治苹果白绢病的主推药剂，并以三种药剂为主要有效成分，开发适合茎基部使用的剂型，研究其使用技术。

关键词：苹果；白绢病菌；杀菌剂；毒力测定

* 基金项目：国家苹果产业技术体系（CARS-28）；山东省农业科学院农业科技创新工程（CXGC2016B11）-主要农作物病虫害绿色防控关键技术

** 第一作者：李栋，硕士研究生，研究方向为植物保护；E-mail：ld1826765930@163.com

*** 通信作者：李保华，教授，主要从事植物病害流行和果树病害研究；E-mail：baohuali@qau.edu.cn

套袋苹果黑点病发病条件研究及防治药剂筛选*

栾　梦**，李保华***
（青岛农业大学农学与植物保护学院，山东省植物病虫害综合防控重点实验室，青岛　266109）

摘　要：套袋果实黑点病是自苹果大面积推广套袋栽培后出现的新病害，主要由粉红单端孢 *Trictothecum roseum* 侵染所致，每年可以造成 3%～30%的产量损失。目前，由于对该病害发生规律和流行特点了解甚少，生产上缺乏有效的防治措施。本研究通过人工控制条件诱导果实发病，研究温度、湿度对黑点病发病的影响，测试了不同杀菌剂对黑点病形成的抑制效果。结果如下。

6 月和 7 月，自未喷施过杀菌剂的苹果园内采集苹果幼果，在人工控温、控湿条件下诱导果实发病。结果表明：在高湿条件下培养 3d，幼果果面上就可以产生典型的黑点病斑，再继续培养 1～2 周，病斑上会形成稀疏的粉红色霉层，镜检为粉红单端孢的分生孢子和分生孢子梗。在 6 个处理温度（10℃、15℃、20℃、25℃、30℃、35℃）中，20℃下诱发病斑数量最多，每个果实平均病斑数为 4.97 个；35℃下诱发的病斑最少，每果平均 0.08 个病斑。在 4 个湿度（100%、95%、85%、75%）处理中，100%的相对湿度诱导的病斑数最多，每果平均 6.72 个，75%相对湿度下诱发的病斑数量最少，平均病斑数为 0.33 个。在所有处理中，20℃下 100%的相对湿度诱发的黑点病斑最多，平均每个果实 18.9 个病斑。然而，在室内用粉红单端孢接种的苹果幼嫩果实，没有诱导出黑点病斑，其原因有待于深入研究。

组织学观察结果表明：黑点病病菌在果肉组织内生长扩展时，能诱导果肉外围细胞木栓化，而木栓化的细胞阻止了病菌的进一步扩展。因此，套袋果实上的黑点病斑小，直径多为 1～3mm，且后期不再扩展。

用吡唑嘧菌酯、戊唑醇、苯醚甲环唑、代森锰锌、丙森锌、甲基硫菌灵、百菌清、多抗霉素、克菌丹 9 种杀菌剂，以厂家推荐的浓度对幼果进行喷雾处理，然后将处理后的果实在 20℃ 100%相对湿度条件下诱发病斑。结果表明：吡唑醚菌酯能有效的抑制幼果上黑点病斑的形成，抑制率为 91.9%，其次是多抗霉素，抑制率为 85.2%，甲基硫菌灵对幼果上黑点病斑的形成抑制效果最差，为 65.5%。

关键词：套袋苹果黑点病；温湿度；药剂筛选

---

* 基金项目：国家苹果产业技术体系（CARS-28）；山东省农业科学院农业科技创新工程（CXGC2016B11）-主要农作物病虫害绿色防控关键技术

** 第一作者：栾梦，硕士研究生，研究方向为植物保护；E-mail：luanmeng0626@126.com

*** 通信作者：李保华，教授，主要从事植物病害流行和果树病害研究；E-mail：baohuali@qau.edu.cn

北京地区番茄灰霉病菌的多重抗药性检测*

乔广行**，李兴红，黄金宝，刘　梅，周　莹，张　玮
（北京市农林科学院植物保护环境保护研究所，北京　100097）

摘　要：2015年10月至2016年5月，在北京9个郊区县采集番茄灰霉病病样本180个，室内分离纯化得到166个灰葡萄孢（*Botrytis cinerea*）单孢菌株，采用最低抑制浓度法（Minimum Inhibition Concentration，MIC）测定了其对苯并咪唑类（多菌灵）、氨基甲酸酯类（乙霉威）和二甲酰亚胺类（腐霉利与异菌脲）四种杀菌剂的抗药性。结果表明：北京地区番茄灰霉病菌对多菌灵、乙霉威、腐霉利和异菌脲产生抗性菌株的频率分别为98.80%、97.59%、91.57%和92.77%；供试菌株对4种杀菌剂的抗性类型有$Car^R Die^R Pro^R Ipr^R$、$Car^R Die^R Pro^S Ipr^S$、$Car^R Die^S Pro^S Ipr^R$、$Car^S Die^S Pro^R Ipr^R$、$Car^R Die^S Pro^S Ipr^S$、$Car^R Die^R Pro^R Ipr^S$和$Car^S Die^R Pro^S Ipr^S$共7种，所占比例分别是90.36%、5.42%、1.81%、0.6%、0.6%、0.6%和0.6%，表明北京地区番茄灰霉病菌对苯并咪唑类（多菌灵）、氨基甲酸酯类（乙霉威）、二甲酰亚胺类（腐霉利和异菌脲）三类杀菌剂的抗药性存在抗药性风险，抗药性水平与严重程度有待于进一步研究。在生产中需慎用，一方面注意轮换用药以及施药适期，另一方面生产上注意选择替代新型杀菌剂和生物农药。

关键词：番茄灰霉病菌；多菌灵；乙霉威；腐霉利；异菌脲；多重抗药性

---

* 基金项目：北京市农林科学院创新能力建设专项（储备项目：KJCX20150406）

** 第一作者：乔广行，主要从事果蔬病害诊断与综合治理研究；E-mail：qghang98@126.com

一株生防芽孢杆菌的筛选、鉴定及颉颃活性测定*

李晶晶**，任　莉，孙新林，王　京，乔　康***
（山东农业大学植物保护学院，泰安　271018）

摘　要：从多年连作的大棚番茄根际土样中分离、筛选得到一株对番茄灰霉病原菌 *Botrytis cinerea* 具有较强颉颃活性的菌株 TA-1。通过菌落形态观察和生理生化试验初步将其鉴定为芽孢杆菌（*Bacillus* spp.）菌群；通过进一步的 16S rDNA 及 gyrB 基因序列与模式菌株序列构建系统发育树分析，结果表明该菌株与甲基营养型芽孢杆菌 *Bacillus methylotrophicus* 具有更近亲缘关系；综合几种鉴定结果最终将菌株 TA-1 鉴定为甲基营养型芽孢杆菌 *B. methylotrophicus*。室内毒力试验结果表明：菌株 TA-1 对番茄灰霉病菌具有显著颉颃活性；温室盆栽条件下，该菌株对番茄灰霉病菌的侵染具有较好的防效。

关键词：生防芽孢杆菌；筛选鉴定；甲基营养型芽孢杆菌；颉颃活性测定

---

* 基金项目：国家自然科学基金项目（31601661）；公益性行业（农业）科研专项（201503130）；国家级大学生创新创业训练计划（201610434107）

** 第一作者：李晶晶，女，本科生，农药学专业；E-mail：17863800979@ 163. com

*** 通信作者：乔康，男，副教授，博士，主要从事土壤消毒技术和农药毒理学研究；E-mail：qiaokang11-11@ 163. com；qiaokang@ sdau. edu. cn

玉米茎基腐病颉颃菌的筛选及其防治效果*

程星凯[1**]，王红艳[1]，王　东[2]，乔　康[1***]，王开运[1]
（1. 山东农业大学植物保护学院，泰安　271018；2. 山东农业大学农学院，泰安　271018）

摘　要： 芽孢杆菌由于具有生防作用成为近几年的研究热点，为了开发玉米茎基腐病的生防资源，本研究通过稀释平板涂布法、平板对峙法从玉米植株内筛选出一株对玉米茎基腐病菌有较强抑制作用的内生菌株 TA-1，并进行了温室盆栽试验。结合形态观察、生理生化特征以及 16S rDNA 基因序列同源性比对分析，该菌鉴定为甲基营养型芽孢杆菌 *Bacillus methylotrophicus*。平板对峙结果表明该颉颃菌对玉米茎基腐病菌有显著的抑制效果，抑菌率达 72.16%；温室盆栽试验结果显示，TA-1 发酵液、菌体悬液和无菌滤液对玉米茎基腐病原菌都有一定的颉颃效果，防效在 65%左右，其中发酵液>菌体悬液>无菌滤液。

关键词： 玉米茎基腐病菌；甲基营养性芽孢杆菌；颉颃细菌；温室盆栽试验

---

* 基金项目：公益性行业（农业）科研专项（201503130）

** 第一作者：程星凯，男，在读硕士，主要从事农药毒理与有害生物抗药性研究；E-mail：17863800979@ 163. com

*** 通信作者：乔康，男，副教授，博士，主要从事土壤消毒技术和农药毒理学研究；E-mail：qiaokang11 - 11@ 163. com；qiaokang@ sdau. edu. cn

生姜根系内生真菌的分离鉴定及生防菌株筛选*

刘增亮[1,2]，汪　茜[1,2]，龙艳艳[1,2]，龙　游[1,2]，杨矩静[1,2]，陈廷速[1,2]**

（1. 广西农业科学院微生物研究所，南宁　530007；2. 广西物宝农业微生物应用联合实验室，南宁　530007）

摘　要：植物根系内生真菌广泛地存在于各种植物根系中，其具有促进植物生长，增强植物抗病性和耐旱等功能，近年来越来越被重视。植物根系内生真菌抑制土传病害的研究更成为其中关注的热点。生姜是一种重要的经济作物，但由于连作所导致的土传病害已严重影响生姜产业的可持续发展。因此，开展生姜根系内生真菌研究对利用土壤微生物改善土壤生态和抑制土传病害具有重要的意义。

本研究以广西柳州、桂林、百色地区采集到的健康生姜为供试材料，采用组织分离法分离生姜内生真菌，并加以纯化和保存。所获得的 22 株内生真菌，通过形态学鉴定，并结合 ITS 序列比对，明确了这些真菌分离物的分类属性。它们覆盖 11 个属 15 个种。在这些分离物中，青霉属（*Penicillium*）和毛克菌属（*Chaetomium*）是优势菌群，分别占总菌株数的 32%和 18%。通过平板对峙培养的方法筛选对姜瘟病菌（*Ralstonia solanacearum*）和生姜茎腐病菌（*Pythium myriotylum* Drechsler）有颉颃效果的菌株，结果表明：有 4 株菌株对姜瘟病菌有较强的颉颃效果，有 3 株菌株对生姜茎腐病菌有较强的颉颃效果，这些菌株值得进一步深入研究。

关键词：生姜；根系内生真菌；生防作用

---

* 项目资助：广西科技攻关项目（桂科 AD16380054）

** 通信作者：陈廷速；E-mail：chen20409@hatmail.com

Screening, Identification and Inhibition effect of Antagonistic Bacteria against *Fusarium oxysporum* f. sp. *nevium*[*]

Li Ping[1**], Liu Dong[1,2], Gao Zhimou[2***], Bi Zhangyou[1]

(1. *Department of Horticulture and Landscape*, *Anqing Vocational and Technical College*, *Anqing* 246003, *China*; 2. *School of Plant Protection*, *Anhui Agricultural University*, *Hefei* 230036, *China*)

Abstract: An antagonistic bacterial strain BTF5 against *Fusarium oxysporum* f. sp. *niveum* was isolated from rhizosphere soil collected of continual watermelon mono-cropping by dual-culture, and identified as *Burkholderia vietnamiensis* according to morphology, biochemical assay and homology analysis of 16S rDNA sequence. Inhibition rate of the strain BTF5 against *F. oxysporum* f. sp. *niveum* was up to 63.09% in the antagonistic experiment. Less change in the inhibition effect of antagonizing against growth of *F. oxysporum* f. sp. *niveum* was observed during 8 generations of transfers on artificial media, indicating its potential biocontrol efficacy and application prospect.

Key words: *Fusarium oxysporum* f. sp. *niveum*; Antagonistic bacteria; Screening

---

* 基金项目：安徽高校省级自然科学研究项目（KJ2015A368）；安徽省高校优秀青年人才支持计划重点项目（gxyqZD2016516）；国家自然科学基金（31671977）

** 第一作者：李萍，博士，副教授，主要从事植物真菌病害防治研究；E-mail：liping05515156@163.com

*** 通信作者：高智谋，博士，教授，主要从事真菌及真菌病害研究；E-mail：gaozhimou@126.com

微生态制剂对烟草叶内生菌群落结构的影响及烟草赤星病防治效果的研究*

聂　力**，王震铄，蔚　越，庄路博，王　琦***

（中国农业大学植物保护学院，北京　100193）

摘　要：烟草是商业性非食物叶用经济作物，是我国重要的财政收入。福建三明主栽品种云烟 87 的主要叶部病害是烟草赤星病，赤星病（Tobacco brown spot）由链格孢（*Alternaria alternata*）引起的真菌性病害，在烟叶近成熟时开始发生，采收期湿度大会严重影响烟叶的产量和品质，造成巨大经济损失。

目前主要的防治方法是化学防治，化学药剂防治会引起病原菌产生抗性，并且农残会影响烟叶的安全性。随着生物防治技术的进展，使用 BCAs（生物控制剂）抑制病害的发生已有报道。本研究使用植物微生态制剂，对烟草赤星病进行绿色防控。研究微生态制剂与其他常规防治方法防治烟草赤星病的差别。在烟草团颗期、旺长期、成熟期、采收期采取烟叶，通过提取叶内生菌基因组 DNA、高通量测序，分析叶内生菌群落结构，研究微生态制剂对叶内生菌群落结构的调控作用。根据微生物群落的分析结果，构建烟叶—微生态环境—菌群平衡体系。研究表明，微生态制剂的施用会对烟叶内生菌群落结构产生一定影响，并对烟草赤星病有一定的防治效果。

关键词：微生态制剂；烟叶内生菌；烟草赤星病；群落结构

* 基金项目：新型高效生物杀菌剂研发（SQ2017ZY060083）

** 第一作者：聂力，女，硕士研究生，研究方向：微生物防治；E-mail：451089475@ qq. cpm

*** 通信作者：王琦，男，教授，植物病害防治；E-mail：wangqi@ cau. edu. cn

九种杀菌剂对荔枝霜疫霉病菌的室内毒力测定*

王荣波**，常红洋，刘裴清，李本金，翁启勇，陈庆河***
（福建省作物有害生物监测与治理重点实验室，福建省农业
科学院植物保护研究所，福州　350003）

摘　要：荔枝霜疫霉病是目前荔枝生产上最重要的病害之一，由荔枝霜疫霉菌（*Peronophythora litchii*）侵染引起的，该病严重影响荔枝品质、产量以及鲜果的贮运和外销。在离体条件下比较了八大类9种杀菌剂10%烯肟菌胺、10%苯醚菌酯、45%烯肟菌酯、100g/L氰霜唑、250g/L双炔酰菌胺、70%代森锰锌、75%百菌清、60%百泰及62.5%霜霉威对荔枝霜疫霉菌丝生长的影响。结果表明：供试药剂对荔枝霜疫霉的菌丝扩展表现出不同程度的抑制作用，其中10%苯醚菌酯和62.5%霜霉威的抑制效果最好，其EC_{50}值分别为0.727 7mg/L、0.85mg/L，而10%烯肟菌胺抑制效果较差，其EC_{50}值达到1 029mg/L，推测该菌株可能对10%烯肟菌胺产生了一定的抗药性，其具体作用机制还需进一步研究。本研究结果为有效控制荔枝霜霉病提供了实验依据。

关键词：荔枝霜疫霉；杀菌剂；毒力测定；EC_{50}值

* 基金项目：福建省农业科学院博士基金项目（DC2017-9）；公益性行业（农业）科研专项（201303018）；福建省属公益类科研专项（2016R1023-1）

** 第一作者：王荣波，山东泰安人，博士，助理研究员
*** 通信作者：陈庆河，研究员；E-mail：chenqh@faas.cn

七种杀菌剂对菜豆灰霉病菌的室内毒力测定*

董　玥，蒋　娜，李健强，罗来鑫**

（中国农业大学植物病理学系，种子病害检验与防控北京市重点实验室，北京　100193）

摘　要：油豆角是我国东北地区特有的一种菜豆品种，外观亮丽，营养丰富，纤维少，口感好。由灰葡萄孢（*Botrytis cinerea*）侵染引起的灰霉病是该地区油豆角生产中常见病害之一，花、果、叶、茎均可发病，病斑初为淡褐色至褐色，后造成果实软腐并在发病果实表面着生灰色霉层，严重制约了油豆角的生产。

由于缺少高抗灰霉病的品种，生产中灰霉病的防治以化学药剂为主。因为连续使用同一杀菌剂易产生抗药性，筛选不同作用方式的防控油豆角灰霉病的安全高效杀菌剂具有重要意义。本研究以抑制病菌信号转导、呼吸作用、氨基酸及蛋白质合成、细胞壁水解酶的分泌四类不同作用机制的 7 种杀菌剂原药为供试药剂，采用菌丝生长速率法进行室内毒力测定。结果表明：咯菌腈、异菌脲、腐霉利、啶菌噁唑、啶酰菌胺、嘧菌环胺和嘧霉胺 7 种杀菌剂对油豆角灰霉病菌表现出较高抑菌活性，其中咯菌腈、啶菌噁唑、嘧菌环胺、嘧霉胺、异菌脲对供试的灰葡萄孢（*B. cinerea*）室内抑制效果最好，其 EC_{50} 分别为 0.001 2μg/mL、0.010 2μg/mL、0.017 6μg/mL、0.203 7μg/mL、0.573 5μg/mL；啶酰菌胺、腐霉利次之，EC_{50} 分别为 1.283 2μg/mL、1.669 2μg/mL。生产中，以咯菌腈为有效成分的杀菌剂常用于种子包衣处理，分析认为，后续可以选用以啶菌噁唑、嘧菌环胺、嘧霉胺为有效成分的杀菌剂制剂实施对油豆角灰霉病的田间防控试验。

关键词：油豆角灰霉病；灰葡萄孢；杀菌剂；室内毒力测定

* 第一作者：董玥，硕士研究生；E-mail：648366398@qq.com

** 通信作者：罗来鑫，副教授，博士生导师，主要从事种子病理学研究；E-mail：luolaixin@cau.edu.cn

三株芽孢杆菌防治韭菜灰霉病研究*

岳鑫璐[1,2]**，高淑梅[1,3]，况卫刚[4]，罗来鑫[1]，李志强[2]，李健强[1]***

（1. 中国农业大学植物病理学系，北京　100193；2. 深圳市农业科技促进中心，深圳　518055；
3. 北京社会管理职业学院，燕郊　101601；4. 江西农业大学农学院，南昌　330054）

摘　要：韭菜灰霉病是由葱鳞葡萄孢菌（*Botrytis squamosa*）引起的真菌病害，造成韭菜产量和品质损失严重，设计开发环境友好和高效的技术产品是韭菜灰霉病防控和安全生产的需要。本研究从病害生物防治的角度出发，探究了三株芽孢杆菌 SWS、HL29、JM-3 的室内抑菌率和田间防病效果，并对其抑菌机制进行了初步解析。结果表明：①芽孢杆菌 SWS、HL29 和 JM-3 对韭菜灰霉病菌菌丝生长平均抑制率分别为 87.5%，62.3%，65.0%。②2015 年、2016 年两次韭菜田间试验防效：SWS 灌根处理的防效分别为 80.53%、62.70%；HL29 叶面喷施防效为 64.67%、63.91%；JM-3 叶面喷施防效为 80.13%、64.18%。③芽孢杆菌抑菌防病机制：SWS、HL29、JM-3 菌株产生的挥发性物质对韭菜灰霉病菌菌丝生长平均抑制率分别为 31.17%、35.82%、37.08%；SWS、HL29、JM-3 菌株的蛋白粗提物对病菌菌丝生长没有明显抑制作用，脂肽粗提物对韭菜灰霉病菌菌丝生长平均抑制率分别为 98.67%、16.33%、42.81%；其中 SWS 脂肽粗提物的抑菌效果最显著，且对韭菜灰霉病菌孢子萌发的抑制率达 93.22%。分子检测结果显示，菌株 SWS 基因组中含有编码抗菌物质 Fengycin、Surfactin 的基因，HPLC 分析显示，SWS 代谢产物中的抗菌物质 Fengycin 浓度为 23.9mg/mL。综合评价室内抑菌、田间药效和抑菌机制研究结果，本研究认为 SWS 作为生物防治因子，值得加速其在韭菜灰霉病防控实践中的应用。

关键词：韭菜灰霉病；生物防治；芽孢杆菌；代谢产物；抗菌肽

---

* 基金项目：北京市科委项目子课“韭菜灰霉病绿色防控技术研究”（Z151100001215009）资助

** 第一作者：岳鑫璐，硕士，主要从事种子病理学和作物病害生物防治研究；E-mail：yxl20071029@126.com

*** 通信作者：李健强，教授，博士生导师，主要从事种子病理及杀菌剂药理学研究；E-mail：lijq231@cau.edu.cn

Biocontrol of Grey Mould and Promotion of Tomato Growth by *Aspergillus* sp. Isolated From the Qinghai-Tibet Plateau in China*

Zhao Juan[1**], Liu Weicheng[1***], Liu Dewen[1], Lu Caige[1], Zhang Dianpeng[1], Wu Huiling[1], Dong Dan[1], Meng Lingling[2]

(1. *Institute of Plant and Environment Protection*, *Beijing Academy of Agriculture and Forestry Sciences*, *Beijing* 100097; 2. *Department of Plant Pathology*, *China Agricultural University*, *Beijing* 100193)

Abstract: Grey mould caused by *Botrytis cinerea* is one of the primary fungal diseases occurred on tomato plants worldwide including China. Since public concerns regarding the use of chemical fungicides have increased due to the impacts on environment and human health, biological control is a promising alternative strategy to restrict the spread of the disease. This study deals with the species identification, grey mould biocontrol and plant growth promotion effects of a fungal strain QF05, which was isolated from the sandy soil collected at the Qinghai-Tibet plateau in China. Morphological observation and multi-gene phylogeny based on the combination of the rDNA-ITS, β-tubulin and calmodulin genes classified the strain QF05 as *Aspergillus tubingensis*. In vitro dual culture test indicated an obvious inhibitory activity of the strain QF05 against *Botrytis cinera*. In the in vivo biocontrol assays, the strain QF05 showed favorable grey mould reduction effect on tomato fruits and plants. Treatments with the culture filtrate of strain QF05 and its 10 fold dilution gave disease biocontrol efficiency of 34. 5 and 31. 0% on tomato plants, respectively. The strain QF05 also displayed definite promotion effect on seed germination and plant growth of tomato. The plant length, total fresh and dry weights of tomato plants were all significantly increased by 17. 9%, 26. 1% and 64. 1%, after treated with the 10 fold dilution of the culture broth of strain QF05. The strain QF05 also presented the positive biological features such as indoleacetic acid and siderophore production, phosphate solubilization, as well as rhizosphere colonization. All these results indicated the good performance of *A. tubingensis* QF05 for the biocontrol of tomato grey mould and the growth promotion of tomato plants.

Key words: *Aspergillus tubingensis*; Species identification; Grey mould; *Botrytis cinera*; Biocontrol potential; Growth promotion

---

* 基金项目：北京市农林科学院所级科技创新团队（JNKST201607）；北京市科技计划（No. D151100003915003）；北京市优秀人才（No. 2015000020060G140）；北京市农林科学院科技创新能力建设专项（KJCX20170107）

** 第一作者：赵娟，博士，助理研究员，研究方向：微生物资源利用及植物病害生物防治；E-mail：zhaojuan119882@163. com

*** 通信作者：刘伟成，博士，研究员，研究方向：资源微生物及其发酵工艺；E-mail：liuwich@ 163. com

微生态制剂对大蒜根腐病防治效果及根际微生物群落结构影响的研究*

庄路博**，蔚　越，王震铄，聂　力，王　琦***
（中国农业大学植物保护学院，北京　100193）

摘　要：山东省金乡县大蒜种植面积较大，种植历史悠久，连年重茬种植导致大蒜根腐病发生严重，蒜头较小，商品率低，严重影响蒜农的经济收入。大蒜根腐病是由腐霉菌引起的土传病害。有研究证明根际微生物的多样性和动态变化很大程度上决定着土传病害的发生情况。本实验室已有研究表明，将抗重茬微生态制剂施用到田间，可影响根际微生物群落结构，从而达到防治土传病害，促进植物生长的效果。

本研究旨在初步探明抗重茬微生态制剂的应用对根际微生物群落结构的影响，以及其抑制大蒜根腐病的作用机制，并评估抗重茬微生态制剂维持大蒜根际土壤良好微环境的能力。在本实验中，将抗重茬微生态制剂施用在山东省金乡县的大蒜根部，收获前采集土壤和植物样品，提取样品总 DNA，利用高通量测序等技术分析相关样品的微生物群落结构。现有结果表明抗重茬微生态制剂不仅对根际微生物的群落结构产生一定影响，而且对大蒜植株有良好的防病促生效果。本研究对阐明抗重茬微生态制剂防治土传病害的机制具有重要的意义。

关键词：微生态制剂；大蒜根腐病；根际；微生物群落结构

* 基金项目：新型高效生物杀菌剂研发（SQ2017ZY060083）
** 第一作者：庄路博，男，硕士研究生，研究方向：微生物防治；E-mail：zhuanglubocau@ 163. com
*** 通信作者：王琦，男，教授，从事植物病害防治研究；E-mail：wangqi@ cau. edu. cn

柑橘黄龙病脱除关键技术研究进展*

姚林建**，易　龙***，李双花，谢昌平，黄爱军
（赣南师范大学国家脐橙工程技术研究中心/生命与环境科学学院，赣州　341000）

摘　要：柑橘黄龙病是全世界柑橘产业最具破坏性的病害。柑橘苗木无病化对于可持续柑橘产业发展至关重要。本文简述了柑橘黄龙病脱除方法、茎尖嫁接脱毒的关键技术及要点等方面的内容。

关键词：柑橘黄龙病；无病化；茎尖嫁接

柑橘黄龙病是靠柑橘木虱传播的柑橘产业上的一种细菌性病害，也是柑橘产区特别关注的传播蔓延速度极快的一种毁灭性病害。大多数柑橘品种都易感病，表现为生长缓慢、枯枝、衰落或死亡等，被称为“梢枯病”[1]。1919 年在广东省潮汕地区发现柑橘黄龙病之后，相继在巴西、美国、墨西哥等地被报道。柑橘黄龙病被认为是一种新出现和重新出现的细菌性病害，严重威胁全世界的柑橘产业[2]。我国柑橘黄龙病主要分布于广东、广西、福建、江西、云南、贵州、台湾等地[3]。为防止柑橘黄龙病的发生与扩散，极力寻找新的方法和抗病材料的选育之外，无病苗木的繁育对恢复柑橘产业尤为重要[4]。

1　柑橘黄龙病脱除方法研究进展

1.1　珠心胚培养脱毒

柑橘类种子多为多胚种子，除具有合子胚外，还有多个珠心胚[5]。珠心胚由珠心组织细胞即体细胞分化形成，具有和母体相同的遗传特性。病毒一般不通过种子传播，因而通过珠心胚培养获得的再生植株是无毒的，同时保留了母株的遗传特性。郭惠珊等[6]选用四会贡柑珠心胚进行离体培养，得到的珠心苗与生长两周的江西红橘实生苗在无菌条件下，在试管内进行微型嫁接，成活率达 90%以上。虽然成活率很高，且能够脱除柑橘黄龙病，但由珠心胚栽培的柑橘需要较长的时间才能具备商业价值。

1.2　热处理脱毒

热处理脱毒法是应用较早的一种脱毒技术[7]。柑橘黄龙病热处理技术的主要原理是利用外界的传热介质，如光照、湿热空气、热水或蒸汽等，向感染黄龙病柑橘植株传递足够热量，在不影响其正常生长的前提下，使黄龙病病菌钝化甚至死亡，黄龙病症状消失。林孔湘等[8]选取已经感染柑橘黄龙病的 2 年生椪柑嫁接苗和 3 年生蕉柑树，前者经湿热空气 48℃、49℃及 50℃处理 31min、35min 及 40min 后定植，大部分可恢复健康，但有一小部分起初生长正常，但 8 个月

* 基金项目：江西省高等学校科技落地项目（KJLD13079）；江西省重点研发项目（20161BBF60070）
** 第一作者：姚林建，男，在读硕士研究生，主要从事柑橘病害防治工程研究；E-mail：854144600@qq.com
*** 通信作者：易龙，男，教授，博士，主要从事柑橘病害防控研究；E-mail：yilongswu@163.com

后仍发病。后者经湿热空气48℃、49℃、50℃分别处理45min、50min、55min、60min、65min和51℃处理35min、40min、45min、50min、55min后重植，全部恢复健康，且28个月后仍未见一株发病。1978—1980年间骆学海[9]选取355条感染黄龙病的接穗，采用47~50℃的热汤处理6~12min，在嫁接后经46个月全部没有发病，对照在8~11个月后发病率达70%~100%。

1.3 茎尖嫁接脱毒法

黄龙病在柑橘树体内分布不均匀，易通过维管系统而转移，茎尖（约0.1~0.5mm）缺少维管系统，故病菌含量极少甚至不含病菌。1972年Murashinge等[10]研究人员在试管中获得黄化砧木苗，用老系柑橘树的茎尖作接穗，在无菌条件下进行茎尖嫁接，得到的少数成活柑橘苗经检测证明其中部分植株已脱除裂皮病而且能保留老系品种的优良栽培性状。1975年Navarro[11]通过对茎尖嫁接成活率及脱毒率的研究，建立了常规的茎尖嫁接操作方法。经广大研究者的长期研究，发现通过茎尖嫁接来培育柑橘苗木，可脱除黄龙病外的多种病害。我国在20世纪70年代后期开始开展柑橘茎尖嫁接的工作，并已取得显著的效果。Song Ruilin等[12]利用0.18~0.20mm长的茎尖和13~18d的砧木进行茎尖嫁接，并在移植前分别滴入一滴激动素和玉米素在砧木切口处，其成活率可达40%~60%，且不同品种的茎尖存在显著差异，对柑橘黄龙病的脱毒率可达100%。

1.3.1 倒“T”嫁接法

在无菌条件下在试管中取出培养了14~18d的且根颈直径大于2mm的砧木苗。留取根端约4~6cm、茎约1.5~2cm，其余用嫁接刀切除。在靠近茎端处横切一刀，竖切两刀，深度到木质部，去掉切除的部分。切口如图1-C所示，将茎尖（含2~3个叶原基）放入切口中，茎尖切面与砧木切面贴合。用镊子将嫁接苗放入液体培养基内，在光照1 600lx、(28±2)℃，培养16h，暗培养8h，待嫁接芽萌发2~3片叶后，进行移栽。

1.3.2 改进的“⊥”压法

在离茎上端约5cm处切出“⊥”切口，并利用刀尖在“⊥”的交叉处向形成层轻压出一个空隙，切口如图1-A，B所示。将茎尖放入交叉处，切口对齐即可。改进的方法较传统的倒“T”法操作更为简便、快速，且克服了因倒“T”法伤口大、易干枯等缺点[13]。

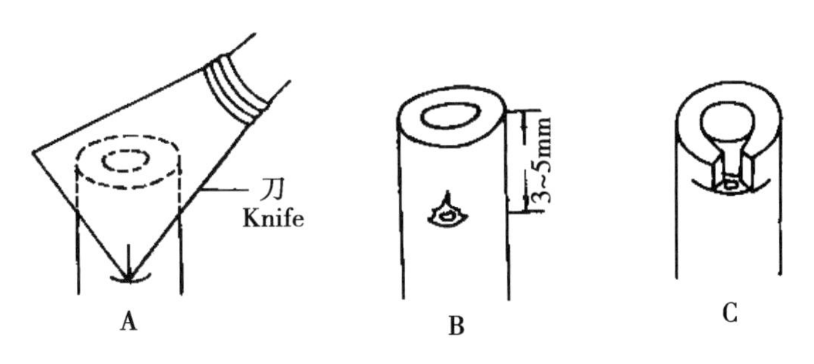

图1 两种嫁接方法示意图（引至姜玲[14]）

1.4 柑橘黄龙病的检测方法

柑橘茎尖嫁接脱毒是实现柑橘苗木无病化的重要手段，而柑橘黄龙病检测技术则是柑橘无病化的重要保证。血清学检测法、电镜观察法、PCR检测法等，是柑橘黄龙病检测中应用较为广泛的技术方法。目前，PCR检测法是检测柑橘黄龙病的主要手段，该法能够对大量的样品进行检测，且准确率高，是建立柑橘良种无病苗木繁育体系可靠的技术保障。

3 展望

柑橘黄龙病严重影响柑橘的生长和结果，使柑橘果实失去经济价值。已对我国多数地区的柑橘产业造成了巨大的危害，病树被大面积清除，而对于产业的恢复急需大量的无病毒苗木，所以柑橘无病化处理和无病苗木的快繁在我国柑橘业也越来越重视。目前，柑橘黄龙病脱除技术已逐步完善，茎尖嫁接结合热处理可以有效地脱除黄龙病。因此，当务之急是通过借鉴国外经验，并结合我国实际情况，加大柑橘良种无病毒苗木繁育体系，从而保障我国柑橘产业的健康可持续。

参考文献

[1] Asana R D. The citrus dieback problem in relation to cultivation of citrus fruits in India [J]. Indian Journal of Horticulture, 1958, 15: 283-286.

[2] Janse, J. D. Bacterial diseases that may or do emerge, with (possible) economic damage for Europe and the mediterranean basin: notes on epidemiology, risks, prevention and management on first occurrence [J]. Journal of Plant Pathology, 2012, 94 (4): S4. 5-S4. 29.

[3] Catara, V D *et al.* HLB (huanglongbing) nuova minaccia per gli agrumi [J]. LInformatore agrario, 2015, 2: 52-55.

[4] 郭文武，邓秀新．柑橘黄龙病及其抗性育种研究 [J]．农业生物技术学报，1998，16 (1)：37-41.

[5] 许智宏．植物生物技术 [M]．上海：上海科学技术出版社，1998.

[6] 郭惠珊，邹韵霞，许丽萍．贡柑珠心胚培养与微型嫁接研究 [J]．中山大学学报论丛，1989 (4)：118-122.

[7] Desvignes J C. Virus diseases of fruit trees [M]. Ctifl, tremprim, 1999.

[8] 林孔湘，骆学海．柑桔黄梢（黄龙）病热治疗的初步研究 [J]．植物保护学报，1965，4 (2)：169-175.

[9] 骆学海．柑桔黄龙病热水间歇消毒试验 [J]．华南农业大学学报：自然科学版，1983，4 (1)：97-103.

[10] Murashinge T, WP Bitters, TS Rangan, *et al.* A technique of shoot-apex grafting and its utilization towards recovering virus-free citrus clones [J]. Hortscience, 1972, 7: 118-119.

[11] Navarro L. The citrus variety improvement program in Spain [J]. Ivia Es, 1976: 198-202.

[12] Song Rui-lin, Wu Ru-jian, Chen Qing-ying etc. Increasing the survival rate of shoot-tip grafting (STG) and its application in production of disease-free citrus seedlings. [J]. Fujian Journal of Agricultural Sciences, 1990, V (1): 20-26.

[13] 马风桐．柑桔属茎尖嫁接脱毒研究 [J]．植物学报，1989，31 (7)：565-568.

[14] 姜玲，万蜀渊，王映红等．柑桔茎尖嫁接操作方法的改进及研究 [J]．华中农业大学学报，1995 (4)：381-385.

拟康氏木霉菌种子包衣技术探究*

盛宴生**，刘梦明，任爱芝，赵培宝***

（聊城大学农学院植物保护系，聊城　252059）

摘　要：为了采用拟康式木霉菌防治种传病害和土传病害，笔者探究了该菌种子包衣技术和防病效果。通过对小麦种子、葫芦种子包衣防病研究，找到了木霉菌包衣的添加材料，建立了拟康氏木霉菌种子包衣技术体系，并掌握了木霉菌包衣能够极大的提高种子的发芽率，当木霉菌的浓度在10^8次方时对种子发芽的促进效果最好，低浓度也有一定的效果，浓度过高对种子的发芽具有一定的抑制作用；测试了种子包衣后防病效果，表明包衣后镰刀菌侵染率相比对照显著降低。

关键词：木霉菌；包衣研究；发芽率；抗病性

---

* 基金项目：山东省自然科学基金 ZR2013CM006；山东省科技发展计划 2014GNC110020

** 第一作者：盛宴生，男，在读硕士研究生，主要从事植物病害生物防治相关研究

*** 通信作者：赵培宝，副教授；E-mail：zhaopeibao@163. com

微生态制剂对马铃薯疮痂病的防治效果及内生菌群落结构影响的研究*

蔚　越**，王震铄，庄路博，聂　力，王　琦***
（中国农业大学植物保护学院，北京　100193）

摘　要：马铃薯疮痂病是我国马铃薯主要病害之一。但目前仍缺乏有效手段防治马铃薯疮痂病。随着生物防治技术的发展，利用生物有机肥、有机改良剂、生防菌株发酵液等抑制病害发生的事例已有报道。大量研究证明，应用微生物制剂可以防治作物病害并提高作物产量。先前实验结果表明：微生态制剂对于马铃薯疮痂病有很好的防治效果。本实验旨在从微生态水平上揭示其防病机制。

本研究比较微生态制剂防治马铃薯疮痂病与其他常见方法防治马铃薯疮痂病之间的差别。在收获前采集马铃薯块茎样品，提取样品总 DNA，分析比较不同处理之间的马铃薯块茎内生菌群落结构之间的差异。从群落结构的角度解释微生态制剂对马铃薯疮痂病的防病机制，明确微生态制剂对马铃薯块茎内生菌群落的调控作用，并比较不同的处理对疮痂病的防治效果以及对马铃薯产量的影响。研究表明，微生态制剂的施用会对马铃薯块茎内生菌群落结构产生影响，并对马铃薯疮痂病有一定的防治效果。

关键词：微生态制剂；马铃薯疮痂病；块茎；内生菌；群落结构

---

* 基金项目：新型高效生物杀菌剂研发（SQ2017ZY060083）

** 第一作者：蔚越，男，硕士研究生，研究方向：微生物防治；E-mail：njauyykzd@ 163. com
*** 通信作者：王琦，男，教授，从事植物病害防治研究；E-mail：wangqi@ cau. edu. cn

颉颃菌株 Q2 的鉴定及其对苦瓜枯萎病生防潜力的初步研究*

田叶韩**，王永阳，李雪玲，高克祥***，何邦令***
（山东农业大学植物保护学院，山东省蔬菜病虫生物学重点实验室，泰安　271018）

摘　要：苦瓜枯萎病是由尖孢镰刀菌苦瓜专化型（*Fusarium oxysporum* f. sp. *momordicae*）引起的土传病害，广泛分布在世界各苦瓜产区。该病害在苦瓜整个生育期都可发生，病害流行时可使瓜田出现大量死藤，严重影响苦瓜的产量和品质。

本研究从健康的苦瓜植株根际土壤中新分离得到 1 株对苦瓜枯萎病菌具有较强抑菌活性的颉颃真菌菌株 Q2，采用形态学特征、生物学特征和 rDNA-ITS、Beta-tubulin 蛋白基因及 *CaM* 钙调蛋白基因序列信息分析比对等方法对菌株 Q2 进行了鉴定，将其定名为产紫篮状菌 *Talaromyces purpurogenus*，无性型为产紫青霉 *Penicillium purpurogenum*。平皿对峙试验结果表明：产紫篮状菌菌株 Q2 能抑制苦瓜枯萎病菌的菌丝生长，抑制率为 60.95%，并使病原菌的菌丝出现扭曲、变粗、断裂及原生质浓缩等畸变现象；3 次温室盆栽防病试验结果证实：在苦瓜苗移栽时，用 1×10^7 个孢子/mL 的分生孢子悬浮液灌根处理，对苦瓜苗期枯萎病的防治效果达到 52.41%~68.19%。

本研究还对产紫篮状菌菌株 Q2 的防病机理进行了初步研究，研究结果表明：经分生孢子悬浮液处理后，可提高苦瓜苗的根系活力和苯丙氨酸解氨酶（PAL）、过氧化物酶（POD）、多酚氧化酶（PPO）等植物防御酶的活性；对苦瓜植株的生长具有一定的促生作用，可以明显促进植株毛细根的生长发育。

产紫篮状菌菌株 Q2 能在 9 种不同培养基上生长和产孢，对不同碳源、氮源、pH 值和温度适应性强，可在中高温条件下较快生长和产孢。因此，产紫篮状菌菌株 Q2 是一株具有开发潜力的生防菌株，有希望将其开发成为具有防病促生功能的新型生防微生物菌剂。

关键词：苦瓜枯萎病；产紫篮状菌；颉颃真菌；生防潜力

---

* 基金项目：公益性行业（农业）科研专项经费项目（201503110-12）

** 第一作者：田叶韩，硕士研究生，研究方向：植物病害生物防治；E-mail：tianyehan@ 163. com

*** 通信作者：高克祥，博士，教授，博士生导师，研究方向：植物病害生物防治；E-mail：kxgao63@ 163. com
何邦令，副教授，硕士生导师，研究方向：园林、花卉、果树等植物病虫害防治新技术的研究与推广应用；E-mail：hebangling@ 126. com

Comparative research of effective utilization and efficacy duration of phenamacrilon rice bakanae disease control by different seed treatment methods Department of Plant Pathology, China Agricultural University

Liu Panqing, Wang Mingqi, Liu Pengfei

(*Department of Plant Pathology*, *China Agricultural University*, *Beijing* 100193)

Phenamacril is a domestically created new fungicide, which showed a good control effecton rice bakanae disease. To clarify the effective utilization and efficacy duration of phenamacril on rice bakanae disease control, the flowable concentrate for seed coating (FSC) and flowable concentrate for seed treatment (FS) of phenamacril were prepared and used by seed treatments. The residue of phenamacril on seeds after soaking was studied by means of QuEChERS—High performance liquid chromatography (HPLC). The sample (3g) was extracted with 1% acetic acid acetonitrile (3mL), purified by graphite carbon black (GCB) (7.5mg/mL). The solution was detected by HPLC with UV detection at 290 nm. The linear regression formulations of phenamacril in seed and plant were $y=41.1060x-2.7224$ ($R^2=0.9995$) and $y=45.739x-26.764$ ($R^2=0.9994$) respectively. The recoveries were ranged from 88% to 115%. The results showed that the method of quantitative determination was of a high accuracy and repeatability. Results showed that there was no significant difference in pesticide utilization between seed soaking with water after coated with FSC and seed soaking with FS. Fungicides degradation dynamics study of phenamacril in rice plants used by seed coating was in conformity with the first-order dynamic equation. The decline coefficient was K = 0.0855, T1/2 = 8.11d and a residue of 2.78mg/kg could be detected after 25 days of planting. The residue is quite more than the EC_{90} of phenamacril on *Fusarium fujikuroi* in vitrowhich shows the efficacy duration on rice bakanae disease can be more than 25d in field. The study may offer an effective reference for the research and development of phenamacril by seed coating.

Key words: Phenamacril; Seed treatment; QuEChERS; HPLC

山东主要棉区铃病发生及其综合防控*

赵 鸣**，王红艳，薛 超，隋 洁

（山东棉花研究中心，济南 250100）

摘 要：棉花铃病是棉花生产中铃期棉铃病害的统称，生产中棉铃可受多种病原菌侵染引发烂铃，为摸清山东省棉花铃病的发生现状，2015—2016 年对山东省主要植棉地区的棉花铃病发生情况进行了多点调查，调查发现各植棉区棉铃疫病、红腐病、红粉病、炭疽病、黑果病、曲霉病和软腐病均有发生，疫病、红腐病和红粉病的发病比例最高；利用 SPSS 软件对 46 个调查点平均单株烂铃数和烂铃株率进行了聚类分析，按照发病程度划分成 3 类，结果表明相同调查点不同年份发病轻重并不相同，2015 年和 2016 年两年发病都归为相同类别的调查点仅有 13 个，占调查点总数的 28.26%，而两年发病归为不同类别的调查点有 33 个，占调查点总数的 71.74%；从种植区域来看，不同地区、不同种植模式条件下铃病发生种类没有特别明显的区域划分，从病情发展来看，铃期不同病害发病所占比例不同，铃病发病前期，疫病所占比例最高，随着时间推移，疫病比例下降，红粉和红腐所占比例增高。明确山东主要棉区铃病发生情况，对下一步药剂筛选和防治策略制定有重要的指导意义。

关键词：山东；棉花；铃病

* 基金项目：山东省自然科学基金项目（ZR2015YL072）；山东省农业科学院青年科研基金项目（2015YQN17）

** 第一作者：赵鸣，男，主要从事植棉花植保相关研究；E-mail：scrczhm@163.com

不同杀菌剂对两种玉米穗粒腐病菌的室内毒力测定*

赵清爽**，席靖豪，袁虹霞***，李洪连

（河南农业大学植物保护学院，郑州 450002）

摘　要：穗粒腐病是玉米上一种常见的籽粒病害。随着玉米品种的更换、种植密度增加及秸秆还田措施的普及推广，玉米穗粒腐病的发生日益严重，已成为我国各玉米产区的主要病害之一。玉米穗粒腐病不仅会造成果穗腐烂而导致直接减产，更重要的是一些病原菌能产生真菌毒素，严重威胁人类和动物的健康和安全。据本课题组调查，在我国玉米主产区河南省，玉米穗粒腐病的发生与危害逐渐加重，一般年份穗腐发生率为 10%~20%，严重年份发生率为 30%~40%，感病品种发生率可高达 50%以上。过去报道的能够引起玉米穗粒腐病的真菌有 20 多种，在我国大部分玉米产区，以拟轮枝镰刀菌（*F. verticillioides*）和禾谷镰刀菌（*F. graminearum*）为穗腐病病原菌的优势种。防治玉米穗粒腐病的方法包括抗穗粒腐病品种的选育和利用、农业防治以化学药剂。但目前抗穗粒腐病的玉米品种较少，农业防治难以实施，为探索利用化学药剂防治玉米穗粒腐病的技术，本研究采用含毒介质法分别测定玉米穗粒腐病禾谷镰刀菌和拟轮枝镰刀菌对 5 种不同杀菌剂的敏感性。

测定结果表明：咯菌腈、醚菌脂、戊唑醇、百菌清、代森锰锌 5 种杀菌剂对两种病原镰刀菌具有不同的抑制效果。对于禾谷镰刀菌，戊唑醇和咯菌腈抑制作用最强，其 EC_{50} 值分别为 3.242 8μg/mL 和 4.782 0μg/mL，百菌清和代森锰锌次之，醚菌脂效果最差；对于拟轮枝镰刀菌，戊唑醇毒力抑制作用最强，EC_{50} 值为 0.005 1μg/mL，而咯菌腈对拟轮枝镰刀菌毒力抑制效果相对较差，EC_{50} 值达到 90.493μg/mL，百菌清基本上没有抑制效果，EC_{50} 值达到 1 324.193μg/mL。研究结果初步发现，禾谷镰刀菌对三唑类杀菌剂戊唑醇的敏感性显著低于拟轮枝镰刀菌，应引起关注。

关键词：杀菌剂；玉米穗粒腐病；毒力测定

---

* 基金项目：河南省玉米产业体系植保岗位科学家科研专项（S2015-02-G05）

** 第一作者：赵清爽，女，硕士研究生

*** 通信作者：袁虹霞，教授，主要从事玉米病害研究；E-mail：yhx2156@163.com

马铃薯疮痂病菌颉颃菌室内筛选与鉴定*

邱雪迎**，关欢欢，杨德洁，赵伟全***，刘大群***

（河北农业大学植物保护学院，河北省农作物病虫害生物防治工程技术研究中心，保定　071000）

摘　要：马铃薯疮痂病是目前马铃薯生产中的重要土传病害，当前尚无切实有效防治该病的化学药剂，因此对该病的生物防治资源进行挖掘是探索防治方法的有益途径。本研究从生物防治角度入手，对采集自不同地区的马铃薯栽培土壤和微型薯生产蛭石进行了微生物分离和疮痂病菌颉颃菌株筛选。以我国存在的不同马铃薯疮痂病原菌 *Streptomyces scabies*（典型菌株 CPS-1）、*S. galilaeus*（典型菌株 CPS-2）和 *S. acidiscabies*（典型菌株 CPS-3）、*S. turgidiscabies*（典型菌株 St-2）的典型菌株为指示菌，将采集的土壤样品自然风干后，采用筛土法皿内筛选对疮痂病原菌具有抑制效果的颉颃菌株，挑取并纯化后，将所得颉颃菌株用滤纸片法、平板对峙培养法进行活性验证并测量抑菌带，最终获得有明显抑菌效果的菌株 6 株。其中对 CPS-1 有抑制作用的 4 株，分别为 H2-T7、H2-T8、H2-T9、H2-T10，抑菌带分别为 2.3mm、5.0mm、2.0mm 和 3.3mm；对 CPS-2 有抑制作用的 2 株 Sx-1-X2 和 Sx-1-T8，抑菌带分别为 3.0mm 和 4.0mm。对 6 个菌株进行培养后，形态学特征表明其中含有放线菌 2 株、真菌 2 株、细菌 2 株。为进一步确定所得颉颃菌株的种类，将获得的细菌采用细菌基因组提取试剂盒，真菌和放线菌采用 CTAB 法，提取菌株的基因组。放线菌选用 16S rRNA 通用引物 Pu/Pd，细菌选用 16S rRNA 通用引物 27F/1492R，真菌选用 ITS 引物 ITS4/ITS5 进行 PCR 扩增。所得目的条带用琼脂糖凝胶 DNA 回收试剂盒回收后，交由上海生工进行测序。经过在 NCBI GenBank 中进行 Blast 比对，发现颉颃菌株 H2-T8 为玫瑰浅紫链霉菌，H2-T9 为淡紫灰链霉菌，H2-T10 和 SX-1-T8 均为巴西株青霉菌，H2-T7 为嗜烟碱节杆菌，Sx-1-X2 为产酶溶杆菌。这些疮痂病菌颉颃菌株的获得为马铃薯疮痂病的生物防治拓展了资源，接下来将对颉颃菌株的防病效果进行温室盆栽试验，进一步筛选可用的生防菌株。

关键词：马铃薯疮痂病；颉颃菌；生物防治

---

* 基金项目：河北省自然科学基金（C2014204109）；河北农业大学科研发展基金计划项目

** 第一作者：邱雪迎，硕士研究生；E-mail：qiuxying@foxmail.com

*** 通信作者：赵伟全，教授，博士生导师；E-mail：zhaowquan@126.com

刘大群，教授，博士生导师；E-mail：ldq@hebau.edu.cn

第八部分　其　他

榆瘿蚜取食侵染榆树叶片形成虫瘿过程中转录组学分析*

李　轩**，黄智鸿***
（河北北方学院，张家口　075000）

摘　要：榆瘿蚜取食侵染榆树叶片形成了榆树虫瘿，为了明确榆瘿蚜侵染下的榆树叶片的抗性机理，本研究以榆瘿蚜自然取食刺激的榆树叶片为研究材料，利用 Illumina Hi Seq™ 2000 技术测序平台进行转录组测序和功能注释分析。共获得 102 017条 Unigenes，通过 NR 与 BLAST 等数据库比对其中有 37 899条（37.15%）Unigense 被注释。利用 COG、GO、KEGG 等数据库对榆树虫瘿叶片进行基因功能预测，COG 按功能将匹配的 Unigenes 基因划分 25 大类；GO 注释将信息归纳为基因的 3 大主类，57 个亚类；KEGG 注释将 Unigene 分为 110 个不同的代谢通路，还发现与氧化应激防御、信号转导等代谢相关的 Unigenes。本结果为今后研究榆瘿蚜侵染下的榆树虫瘿叶片的抗性机理提供了理论基础，对进一步揭示榆树抗榆瘿蚜的分子机制，挖掘我国抗榆瘿蚜种质资源和抗虫育种具有重要的指导作用。

关键词：转录组测序；榆树虫瘿；榆瘿蚜

Analysis of transcriptomics in the process of larvae entering the leaves of *Tetraneura akinire* Sasaki

Li Xuan，Huang Zhihong
（*Hebei North University*，*Zhangjiakou*，*Hebei* 075000，*China*）

Abstract：*Tetraneura akinire* Sasaki feeding infiltration of elm leaves formed the elm insects gall，in order to clarify the resistance mechanism of elm leaves under the infection of *Tetraneura akinire* Sasaki，the study of Elm leaf with natural feeding stimulation of *Tetraneura akinire* Sasaki，sequencing and functional annotation analysis were performed using the Illumina Hiseq ™ 2000 technology sequencing platform. Received a total of 102 017 Unigenes，A database of 37 899（37.15%）Unigense is commented on the NR and BLAST. Using the database of COG，GO and KEGG to predict the gene function of the elm insects，COG by function will match the Unigenes gene divided into 25 categories；GO comments summarize the information into three major categories of genes，57 subclasses；The KEGG annotation divides Unigene into 110 different metabolic pathways，also found that with the oxidative stress defense，signal transduction

* 基金项目：河北省自然科学基金项目（C2013405094）
** 第一作者：李轩，硕士研究生，研究方向：植物保护；E-mail：1639352729@qq.com
*** 通信作者：黄智鸿，教授，研究方向：植物有害生物生态学和综合治理；E-mail：hbnuhzh@163.com

and other metabolic related Unigenes. The results provide a theoretical basis for the study of the resistance mechanism of *Tetraneura akinire* Sasaki blades infected by aphids in the future. To further reveal the molecular mechanism of Elm against *Tetraneura akinire* Sasaki, it has important guiding function to excavate the germplasm resources and insect resistance breeding of *Tetraneura akinire* Sasaki in China.

Key words: Transcription group sequencing; Elm insects gall; *Tetraneura akinire* Sasaki

转录组（transcriptome）是研究生物体在特定生理条件下所有基因的整体表达情况，而转录组学[1]是揭示生物体之间基因的序列组成和功能，以及发掘病虫害侵染、抗病和衰老及非生物胁迫等特定的生理过程和代谢途径中相关的功能基因，是对基因进行精确定量分析的理想途径。转录组测序（RNA-Sep）技术作为一种新转录组研究手段，能够注释校正基因、基因表达、差异基因的鉴定和转录起始位点鉴定等内容[2,3]，彭振[4]等人利用 RNA-Sep 鉴定出 26 个与耐盐相关的转录因子家族 124 个转录因子基因，并采用荧光定量验证转录数据的准确性。黄娟[5]等人利用 Illumina Hi Seq™ 2000 进行了荞麦根的转录组学分析及黄酮合成基因的鉴定，37546 条基因得到了有效注释，有 21 个基因注释到黄酮合成途径中的关键酶。本研究针对榆瘿蚜取食诱导榆树叶片形成虫瘿过程中的差异基因的表达和新基因的挖掘，应用转录组学研究可以从分子水平上揭示榆瘿蚜取食后叶片对逆境胁迫的应答机制[6]。

本研究以榆瘿蚜为害榆树叶片形成虫瘿的 3 个发育阶段初始形成期、成长分化期、开裂期的叶片和未被取食的叶片为材料，通过 RNA-Seq 测序平台对榆瘿蚜为害前后叶片转录组进行分析，旨在探索榆树叶片在侵染过程中可能存在的基因的功能类别和作用以及代谢途径等，为后续高通量筛选抗病相关抗性基因的克隆奠定基础。

1 材料与方法

1.1 试验材料与取样方法

材料取自河北省张家口市郊区，根据榆瘿蚜的生活周期，在虫瘿发育的初始形成期、成长分化期和开裂期选取榆瘿蚜侵染的样株，在树冠的不同部位枝条随机采集形成虫瘿的叶片 3 个；同时在未被榆瘿蚜侵染的样株相应部位的枝条上采集正常叶片 3 个。取样时多次取样以备用，每个样株为一次重复，试验设 3 次重复，戴一次性手套摘取叶片，叶片采集后立即置于液氮中，放于冰箱-80℃中冻存备用。

1.2 试验方法

1.2.1 RNA 的提取

榆树叶片总 RNA 的提取参照 Invitrogen 公司的 TRIzol 试剂盒说明书。采用 Agilent 2100 Bioanalyzer 检测总 RNA 质量，检测 RNA 完整性可用于后续实验。

1.2.2 转录组测序流程

样品合格后进行文库构建，具体流程为：

（1）用 DNaseI 处理提取 RNA 样品，通过带有 oligo（dT）磁性珠富集真核生物 mRNA；

（2）加入 Fragmentation Buffer 将 mRNA 碎成 200bp 的片段；

（3）以 mRNA 为模板用引物随机六聚体合成 cDNA 的第一链，再添加缓冲液、d NTPs、dNTPs、RNase H 和 DNA 聚合酶 I 合成第二链；

（4）利用试剂盒（AMPure XP beads）进行纯化 cDNA；

（5）纯化 cDNA 再进行粘性末端修复，在 cDNA 的 3′末端加上“A”碱基连接头，然后 AMPure XP beads 进行大小片段的选择，通过 PCR 富集扩增；

（6）用安捷伦 2100 生物分析仪和 ABI Step One Plus 实时荧光定量 PCR 进行检验构建好的文库，合格后用（Illumina Hi Seq™ 2000）进行测序。

1.2.3 转录组生物信息学分析

（1）测序产出统计：在测序过程中由于使用的化学试剂不断消耗，因此要对测序得到的原始数据进行评估过滤，获得干净的 reads。对测序数据进行处理过程：带接头（Adaptor）的 reads 应被去除带掉；碱基信息（>5%）的 reads 无法被确定的去除；去除低质量 reads（Q≤10 的碱基数占整个 read 的 20%以上 reads）；

（2）用 Trinity 软件将测序 reads 按指定的 K-mer 打断来构建库，去除包含错误的 K-mer 并对过滤得到的 Clean 进行从头组装，并分别对组装出来（Contigs 和 Unigenes）做长度分布统计；

（3）进行 Unigene 功能注释：分别将 NR、Swiss-Prot、KEGG、COG、GO 数据库与转录组组装的 Unigenes 进行比对，对 Unigenes 成功匹配到每个库的结果进行注释统计并绘制 COG 分类图；

（4）进行 Unigenes GO 的分类：使用 Blast[161]（版本号 2.5.0）进行比对，对比对成功的 Unigene 根据 NR 注释信息进行 GO 的功能注释并统计分类，对 Unigene 做 GO 功能分类统计图；

（5）代谢通路分析：根据 KEGG 数据库上的信息将获得的 Unigenes 进行注释，明确 Unigenes 在生物体生命代谢活动中参与活动的代谢路径和行使的基因功能。

2 结果与分析

2.1 RNA 提取及质量检测分析

使用安捷伦 2100 分析仪对提取的榆树叶片总 RNA 质量进行质检与评估，结果表明：各组织叶片 RNA 浓度均在 300ng/μL 以上，$OD_{260}/OD_{280} \geq 1.8$，各叶片组织完整性较高（图 1），质量符合 cDNA 建库可用于上机测序要求。

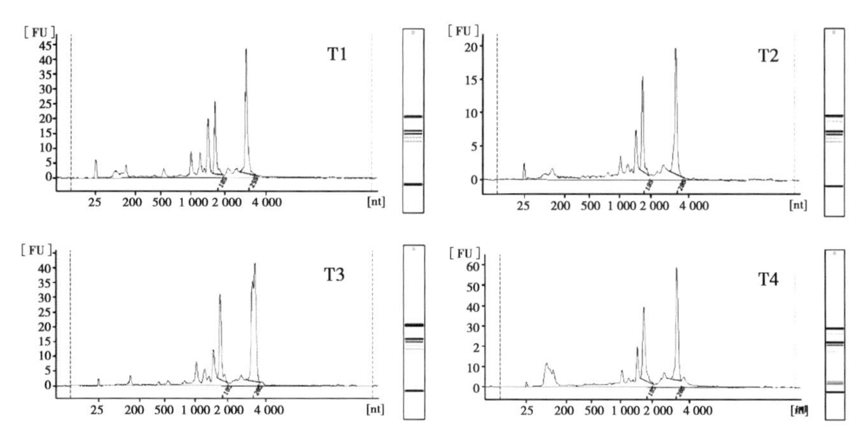

图 1 榆树叶片 RNA 提取质量检测

Fig. 1 Quality of Total RNA of the *Ulmus pumila L* leave

注：T1，T2，T3，T4 分别代表未被榆瘿蚜取食的榆树叶片、榆瘿蚜为害榆树叶片形成虫瘿的初始形成期、成长分化期、开裂期的总 RNA 质量检测图。

2.2 测序产出统计

对榆树叶片经过测序质量控制，共得到 23.19Gb Clean Data，各样品 Q30 碱基百分比均不小于 90.30%。结果如表 1 可知，4 个文库的测序结果中，数据质量可靠，由此可以看出，测序结果质量较好，可以用于后续的数据组装及处理要求和生物信息学分析。

表 1　叶片测序数据评估统计表

Table 1　statistics for sequencing the *Ulmus pumila L* leave

Samples	BMK-ID	Read Number	Base Number	GC Content	%≥Q30
正常	T1	23 520 545	5 920 658 801	46.50%	90.30%
初始形成期	T2	23 838 189	6 000 727 673	46.43%	90.60%
成长分化期	T3	22 055 062	5 551 005 583	46.02%	90.87%
开裂期	T4	22 725 685	5 718 762 474	46.24%	90.34%

注：Samples：不同样品测序名称；BMK-ID：百迈客对样品的统一编号；Read Number：Clean Data 中 pair-end Reads 总数；Base Number：Clean Data 总碱基数；GC Content：Clean Data GC 含量，即 Clean Data 中 G 和 C 两种碱基占总碱基的百分比；%≥Q30：Clean Data 质量值大于或等于 30 的碱基所占的百分比。

2.3 组装结果统计

榆树叶片组装共得到 102 017条 Unigene 中，Unigene 的 N50 为 1 255nt，组装完整性较高，具体的统计信息见图 2。

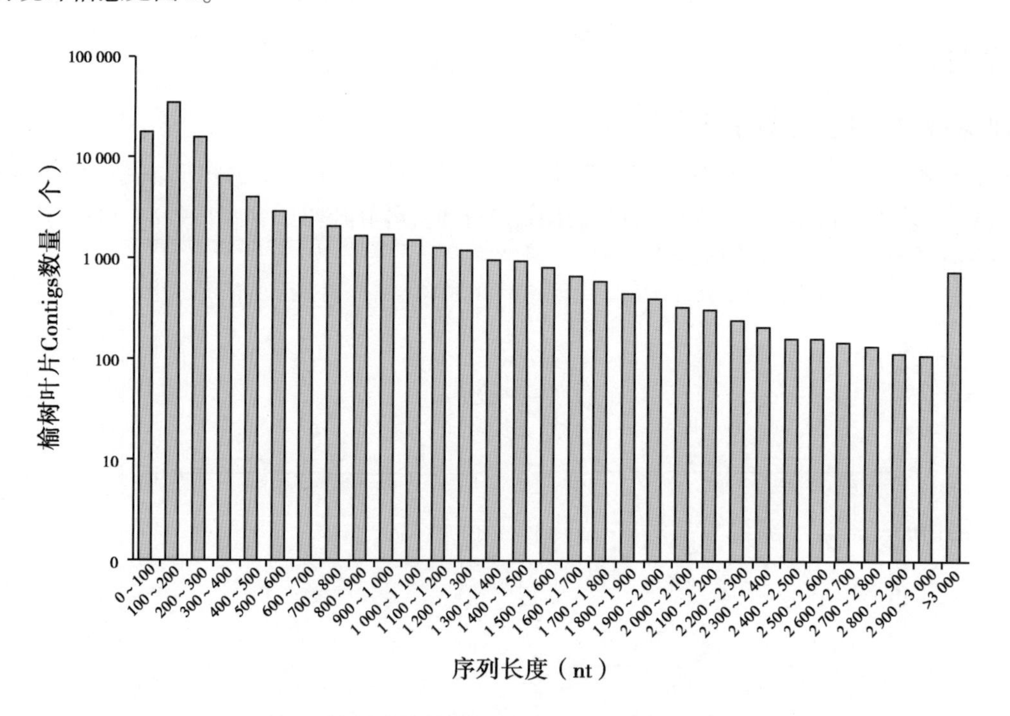

图 2　榆树叶片转录组的 Contigs 长度分布统计

Fig. 2　Length distribution of assembly contigs of the *Ulmus pumila L* leave

2.4 榆树叶片转录组 Unigene 功能注释、分类和代谢途径分析

2.4.1　基因注释结果统计汇总如表 2 所示，在 102 017个 Unigene 基因中，使用 BLAST 软件设定

参数 E-value 不大于 10^{-5}，最终获得 66 619个有注释信息的 Unigene，其中将 Unigene 序列与 NR、Swiss-Prot 数据库比对，NR 数据库比对到 66 125个 Unigene（64. 82%），Swiss-Prot 数据库比对到 37 899个 Unigene（37. 15%）；通过对 102 071个 Unigene 和 KEGG 数据库比对，有 17 430个 Unigene 参与了榆瘿蚜为害榆树叶片形成虫瘿过程中的不同代谢通路；在 COG 数据库中有 26 257个 Unigene 可以被预测出功能；有 47 003个 Unigene 能映射到 GO 的不同功能节点上。

有 35 398个 Unigenes 没有 NR、NT 和 Swiss-Prot 数据库比对到，可能是新发现的转录本，也可能是由于在取样过程中榆瘿蚜没有剥离干净，或者是榆树叶片区别于其他物种而特有的基因，这有待于进一步的研究来验证。

表 2　注释结果统计

Table 2　Statistics of annotation results

Annotated databases	Unigene	≥300nt	≥1 000nt
COG	26 257	17 475	8 760
GO	47 003	30 979	15 529
KEGG	17 430	11 500	4 976
Swiss-Prot	37 899	27 953	14 078
nr	66 125	41 972	18 767
All	66 619	42 108	18 771

注：Annotated databases：表示各功能数据库；Unigene：表示注释到该数据库的 Unigene 数；≥300nt：表示注释到该数据库的长度大于 300 个碱基的 Unigene 数；≥1 000nt：表示注释到该数据为的长度大于 1 000个碱基的 Unigene 数。

2. 4. 2　榆树叶片转录组 Unigene COG 功能分类

所得的基因产物通过 COG（Clusters of Orthologous Groups）数据库进行同源分类，将比对得到的所有榆树叶片 Unigenes 基因在 COG 库上进行 Blast 比对，并对 Unigenes 进行功能分类同时预测其可能的功能，从宏观上把握榆树叶片的基因功能的特征分布。

榆树叶片转录组中有 33 923个 Unigenes 通过 Blast 比对可与 COG 数据库中基因匹配而建立对应关系，可将 Unigenes 基因按功能划分 25 类（用 A-Z）如图 3 所示。其中，R 类 General function prediction only（一般功能预测）最多，有 6 298个基因（占 18. 57%）；J 类 Translation，ribosomal structure and biogenesis（翻译、核糖体结构和生物起源）次之，相关基因有 3 398个，（占 10. 01%）；W 类 Extracellular structures（保外基质结构）最少，基因数量仅 1 个，其他类的基因表达丰富度都各不相同。

2. 4. 3　榆树叶片转录组 Unigene GO 功能分类

经 BLAST 软件进行 GO 分析，有 47 003个相应的 GO 注释能在 66 619个注释序列中被发现。用 WEGO 软件处理统计并分类，进一步将这些 GO 信息归纳为基因的 3 大主类，57 个亚类，具体如下：分子功能分类中主要聚集在催化活性（21 857个）和结合作用（20 118个）；在主细胞组分中主要聚集在细胞成分（16 254个）、细胞（17 253个）和细胞器（12 123个），在生物学过程分类中主要聚集于单个有机体过程（12 329个）、细胞过程（18 505个）和代谢过程（17 142个），其

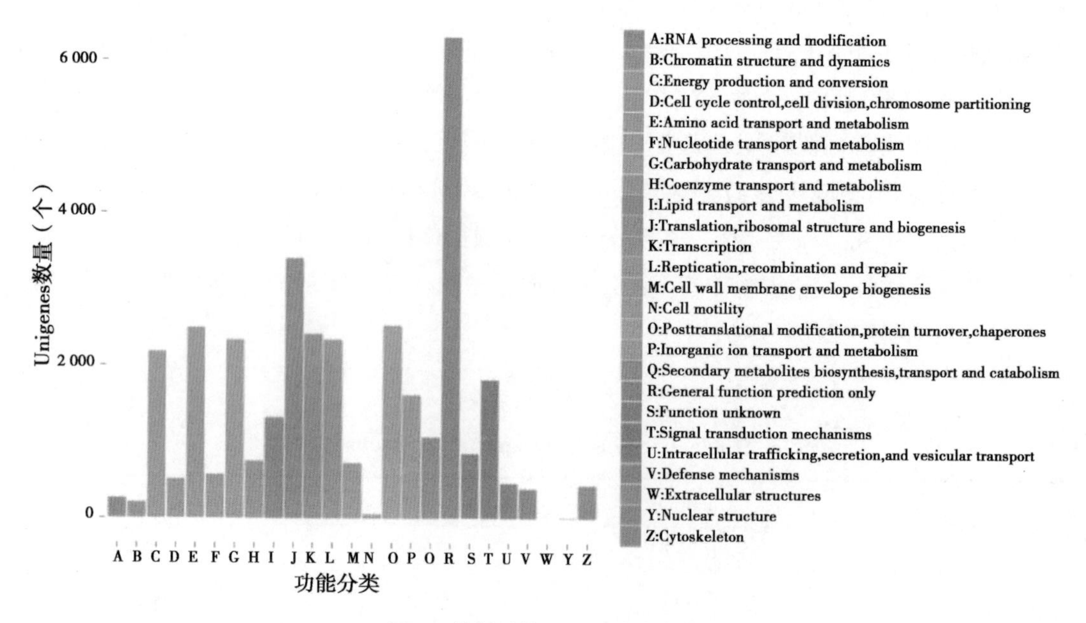

图 3 榆树叶片 COG 功能注释

Fig. 3 COG function classification of unigene in all unigenes

注：COG 功能分类编号的全称注释见图右边的批注；横坐标为分类编号，纵坐标是每个对应 COG 功能注释的 unigenes 数量。

他涉及细胞增殖的有 2 354个，信号的有 2 442个，生物调节的有 5 949个，免疫反应的有 1 872个，应激反应的有 8 585个，这些功能基因在榆瘿蚜为害榆树叶片形成虫瘿过程中起着重要作用。进行 GO 分类基于 NR 的注释具体见图 4。

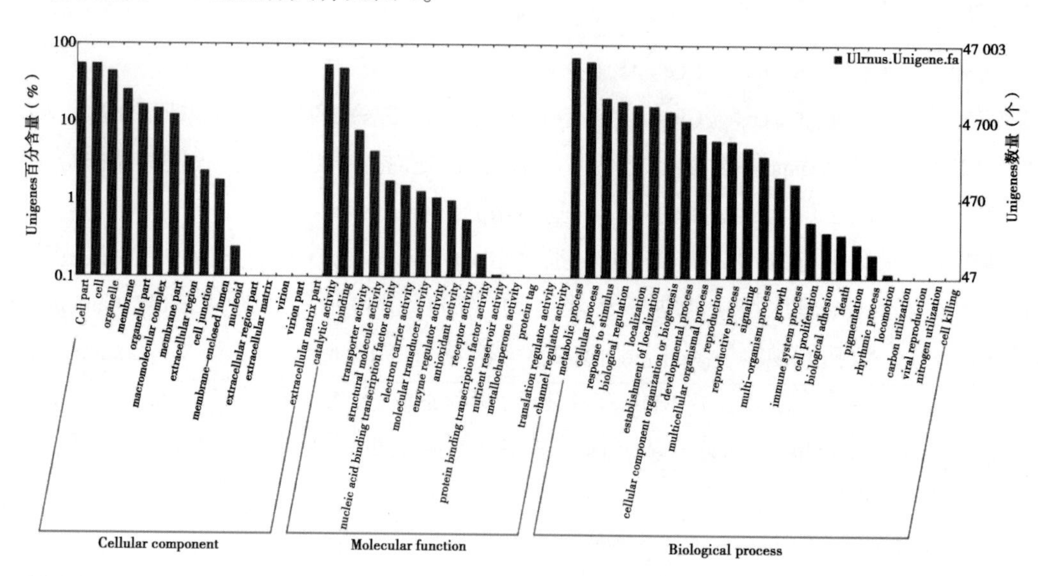

图 4 榆树叶片 Unigenes 的 GO 分类

Fig. 4 Gene ontology（GO）term assignments of the *Ulmus pumila L* leave unigenes

注：横轴表示为 GO 的功能种类，左边的纵轴是 Unigenes 占总数量的百分比，右边的纵轴是被注释到的对应 GO 功能的 Unigenes 数量。

2.4.4　榆树叶片转录组 Unigenes 代谢通路分析

KEGG（Kyoto Encyclopedia of Genes and Genomes）数据库是系统分析记录了细胞内生物在特定生物体中的变化以及细胞内分子间相互作用的网络，通过 KEGG Pathway 的分析可以进一步了解基因产物在细胞中的生物功能和基因产物间的相互作用。为全面系统分析榆瘿蚜为害榆树叶片的 Unigenes 在细胞中的代谢途径以及功能，使 BLAST 软件将 KEGG 数据库和本次转录组中所得的测序 46 619个 Unigenes 进行比对，发现有 12 640个（37.4%）基因被成功注释，共有 110 个不同的代谢通路被映射到。其中包括糖酵解、三羧酸循环、磷酸戊糖代谢通路，苯丙氨酸生物合成，次生化合物生物合成，植物激素信号转导，黄酮类物质合成、植物-病原互作，核苷酸剪切修复，过氧化物酶体合成、嘌呤代谢，玉米素的生物合成、脂类代谢，萜类生物合成等，具体列于表 3。

榆树转录组中涉及的 Unigenes 在不同代谢通路中数量差异比较大，其中核糖体（Ribosome）的 Unigenes 数量最多，为 1 380个（11.92%），其次是代谢途径（Metabolic pathways），相关的 Unigenes 有 913 个（10.9%），植物-病原物互作（Plant-pathogen interaction）相关的 Unigenes 有 741 个（5.78%），过氧化物酶体（Peroxisome）相关基因 438 个（4.76%）植物激素信号传导（Plant hormone signal transduction）相关的 Unigenes 有 296 个（2.15%）。

表 3　榆树叶片转录组的 Unigenes KEGG 代谢途径分类（前 10）

Table 3　KEGG classification of *Ulmus pumila L* leave unigenes in digital transcriptome（first 10）

	通 路	注释到的基因	通路地址
1	Ribosome	1380（11.92%）	ko03010
2	Oxidative phosphorylation	879（8.95%）	ko00190
3	Metabolic pathways	913（10.9%）	ko01100
4	Plant-pathogen interaction	705（7.62%）	ko04626
5	Phagosome	326（2.58%）	ko04145
6	Citrate cycle（TCA cycle）	417（3.3%）	ko00020
7	RNA transport	650（5.14%）	ko03013
8	Plant hormone signal transduction	296（2.15%）	ko04075
9	Proteasome	330（2.61%）	ko03050
10	Pentose phosphate pathway	245（1.94%）	ko00030

2.5　虫瘿叶片基因的挖掘

在榆瘿蚜取食榆树叶片的转录组中，和本研究相关的 Unigenes 有氧化应激防御、信号转导等代谢需要进一步分析，如图所示：图 5、图 6 与氧化应激防御相关的 Unigenes 有 438 个，与信号转导相关的 Unigenes 有 764 个，对这些 Unigenes 功能进一步挖掘，将为榆瘿蚜为害榆树形成虫瘿的分子机制奠定基础。

3　讨论与结论

转录组测序是近年来发展的一种对 mRNA 高通量的测序技术，它能够对样品任意条件下进

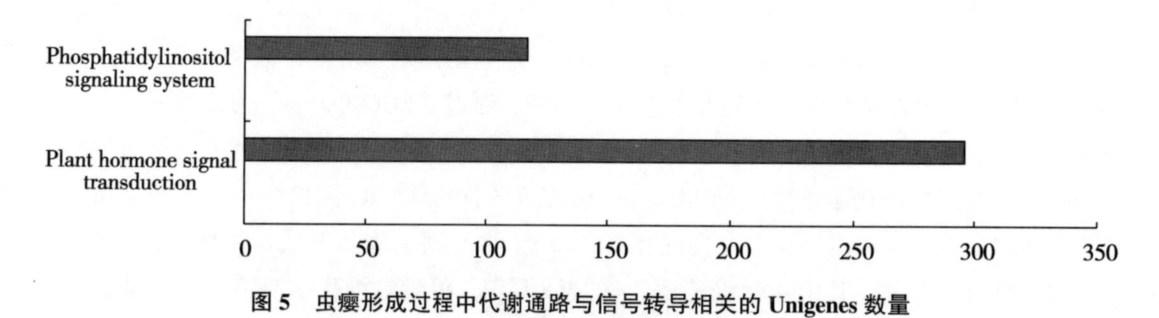

图 5　虫瘿形成过程中代谢通路与信号转导相关的 Unigenes 数量

Fig. 5　Unigenes related to signal transduction in metabolic pathways in the galls formation

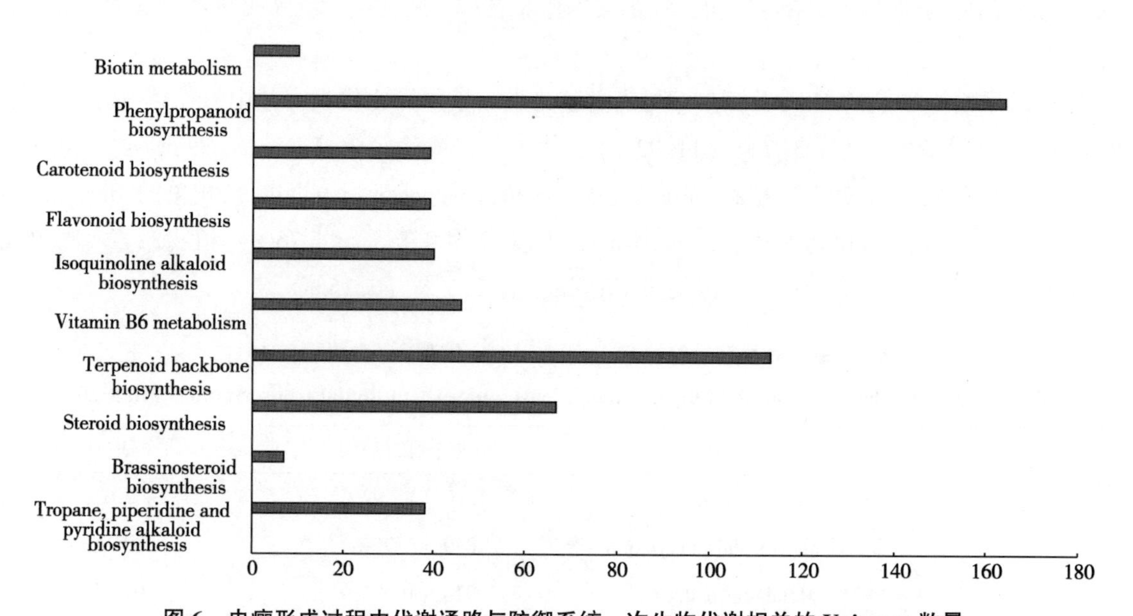

图 6　虫瘿形成过程中代谢通路与防御系统、次生物代谢相关的 Unigenes 数量

Fig. 6　Unigenes related to defense system and secondary metabolism in metabolic pathways in the galls formation

行测序，对基因组序列依赖性不强，因此，对于无标准基因组的生物体也可以进行转录组测序，通过对 cDNA 片段进行从头组装得到转录本来构建 Unigene 库。本研究用 Illumina 平台测序构建榆瘿蚜为害榆树叶片形成虫瘿过程的转录组文库，获取了 102 017个 Unigenes，用 Blast 软件进行比对，结果有 66 619个 Unigenes 比对到 NR 数据库中（75%），已知黄瓜已经发表基因组序列匹配达 85.6%，匹配率达最高，而哈密瓜响应青霉菌侵染转录组基因序列匹配率达 71.28%。发现有 35 398个 Unigenes（约 35%）没有与 NT、NR 和 Swiss-Prot 等数据库比对成功，它们可能是新的转录本，或者是由于榆树没有标准的基因组区别于其他物种而特有的基因，这需要进行进一步研究。

基因本体论（gene ontology，GO）是国际标准化语言中的一种基因功能分类数据库，提供了一套三个动态水平的系统定义包括生物学过程（Biological Process）、细胞组成（Cellular Component）和分子功能（Molecular Function），分别描述转录组测序中产物基因的特性与功能，而且物种没有相关性。本研究中发现有 47 003个相应的 GO 注释能在 66 619个注释序列中发现，有许多与氧化应激防御、信号转导等相关的基因在榆瘿蚜为害榆树叶片形成虫瘿过程中起着重要作用。

利用 BLAST 软件将 KEGG 数据库和本次转录组中所得的测序 46 619个 Unigenes 进行比对，发现有 12 640个（37.4%）基因被成功注释，共有 110（37.5%）个不同的代谢通路被映射到。本研究重点关注免疫系统和植物激素等代谢通路，功能基因在榆瘿蚜为害榆树叶片形成虫瘿过程中起着重要作用，对没有比对的 Unigenes 的功能需进一步挖掘，为榆瘿蚜为害榆树形成虫瘿的分子机制奠定基础。随着榆树抗虫基因资源的不断积累，转录组测序所得到的基因序列等数据有助于榆树抗虫性状在分子遗传育种方向进行改良，也为榆树相近的物种在基因组学研究中提供了有价值的参考基因。

参考文献

[1] 房学爽，徐刚标．表达序列标签技术及其应用［J］．经济林研究，2008，26（2）：127-130.

[2] Mafra V，Martins P K，Francisco C S，*et al*. Candidatus Liberibacter americanus induces significant reprogramming of the transcriptome of the susceptible citrus genotype［J］. BMC genomics，2013，14（1）：247-261.

[3] Allardyce J A，Rookes J E，Hussain H I，*et al*. Transcriptional profiling of *Zea mays* roots reveals roles for jasmonic acid and terpenoids in resistance against *Phytophthora cinnamomi*［J］. Functional & Integrative Genomics，2013，13（2）：217-228.

[4] 彭振，何守朴，龚文芳．陆地棉幼苗 NaCl 胁迫下转录因子的转录组学分析［J］. 作物学报，2017，43（3）：354-370.

[5] 黄娟，邓娇，陈庆富．荞麦根的转录组学分析及黄酮合成基因的鉴定［J］. 中国农业科技导报，2017，19（2）：9-19.

[6] 陈全助．福建桉树焦枯病菌鉴定及其诱导下桉树转录组和蛋白组学研究［D］. 福州：福建农林大学．

基于环介导等温扩增技术的植物病原物检测技术的应用研究*

彭丹丹**，张源明，舒灿伟，周而勋***
（华南农业大学农学院植物病理学系，广州 510642）

摘　要：植物病害病原物是能侵染和寄生于植物体并导致侵染性病害发生的生物，多为异养型的非专性寄生物。对植物病原物的快速检测是进行植物检疫、监测、预报及病害防治不可或缺的基础工作。近年来，以环介导等温扩增技术（Loop-mediated isothermal amplification，LAMP）为基础的分子检测技术发展迅速，使得植物病原物的检测更简单、快速、灵敏和可靠。

LAMP 技术是一种利用特异识别靶序列上 6 个位点的 2 对特殊设计的引物和一种具有链置换活性的 DNA 聚合酶，在恒温条件下特异、高效地扩增 DNA 的新型恒温核酸扩增技术，具有操作简单、灵敏度高、特异性强及检测结果迅速等优点。其扩增产物为一系列反向重复的靶序列构成的茎环结构和多环花椰菜样结构的 DNA 片段混合物，电泳图谱呈阶梯式带状分布，反应伴有白色的副产物焦磷酸镁沉淀产生，可通过肉眼定性判断反应结果。

近年来，LAMP 技术已检测植物病原物领域得到了广泛的应用。本文通过总结以 LAMP 技术为基础的检测技术快速检测和诊断植物病原物导致的植物病害的案例，对近年来基于 LAMP 技术的植物病原物检测方法进行了介绍与评述，这些方法包括多重 LAMP、原位荧光 LAMP、实时荧光定量 LAMP 及 LAMP-LFD 等检测技术。预测了 LAMP 技术的应用前景，具有替代 PCR 方法的可能性，从而为其研究和应用提供一个平台。

关键词：植物病原物；分子检测；LAMP 技术

---

* 基金项目：国家公益性行业（农业）科研专项资助项目（201403075）

** 第一作者：彭丹丹，硕士生，研究方向：植物病原真菌及真菌病害；E-mail：1129961515@ qq. com

*** 通信作者：周而勋，教授，博导，研究方向：植物病原真菌及真菌病害；E-mail：exzhou@ scau. edu. cn

Bioactive Constituents from the Peels of *Clausena lansium* *

Deng Huidong[1,2,3**], Luo Zhiwen[1,2], Guo Lijun[1,2,3], Fan Hongyan[1,2,3], Hu Fuchu[1,2,3], Yu Naitong[4], Liu Zhixin[4], Hua Min[1,2***]

(1. *Institute of Tropical Fruit Trees*, *Hainan Academy of Agricultural Science*, *Haikou* 571100, *China*; 2. *Key Laboratory of Tropical Fruit Tree Biology of Hainan Province*, *Haikou* 571100, *China*; 3. *Haikou Scientific Observation and Experiment Station of Tropical Fruits*, *Ministry of Agriculture*, *Haikou* 571100, *China*; 4. *Institute of Tropical Bioscience and Biotechnology*, *Chinese Academy of Tropical Agricultural Sciences*, *Haikou*, 571101)

Abstract: *Clausena lansium* Skeels (Wampee) is a tropical species of the Rutaceae family, whichis widely distributed in southern China. The fruits of wampee have high medicinal and edible value, which can be used for digestive disorders, abdominal pain, colic, phlegm, cough, and asthma. In order to discover the bioactive constituents with medicinal value in peels of *C. lansium* in Hainan, one new monoterpenoid, nerol oxide-8-carboxylic acid, and one new flavonoid glycoside, claulansoside A, together with six known compounds, clausenamide, quercetin, isorhamnetin, dihydromyric, 2″, 3″-dihydroxyanisolactone and (*E*, *E*) -8- (7-hydroxy-3, 7-dimethylocta-2, 5-dienyloxy) psoralen, have been isolated from the peels of *C. Lansium*by column chromatography. Their structures were determined using a combination of 1D, and 2D NMR (HMQC, HMBC, COSY and NOESY) techniques, and HR-ESI-MS analyses. The antimicrobial activities against *Staphylococcus aureus* of compounds 1-8 were measured using paper disc diffusion method, among which compounds 1 and 7 exhibited antibacterial activity against *Staphylococcus aureus* with the diameter of inhibition zones of 11.5 mm and 14.2 mm. The hypoglycemic activities of compounds 1-8 were evaluated in vitro by the inhibition of α-glucosidase method and compounds 3 and 6 showed α-glucosidase inhibitory activity *in vitro*.

Key words: *Clausena lansium*; Monoterpenoid; Flavonoid; Antibacterial activity; Hypoglycemic activity

---

* Funding: Natural Science Foundation of Hainan Province (20153101); Agricultural science and technology innovation project of Hainan Academy of Agricultural Science (CXZX201502)

** First author: Deng Huidong, Male, Research Intern, Arboriculture and Natural Products Research of Fruit; E-mail: denghuidong2010@ aliyun. com

*** Corresponding author: Hua Min, researcher; E-mail: huamin2528@ 139. com

杧果果实动态发育及 CPPU 施用方案初步筛选*

郭利军[1**]，范鸿雁[1]，邓会栋[1]，罗志文[1]，何舒[2]，胡福初[1]，
王祥和[1]，何　凡[1]，余乃通[3]，刘志昕[3]，华　敏[1***]
（1. 海南省农业科学院热带果树研究所/海南省热带果树生物学重点实验室/农业部海口热带果树科学观测试验站，海口　571100；2. 海南省农业科学院植物保护研究所，海口　571100；3. 中国热带农业科学院热带生物技术研究所，海口　571101）

摘　要：本文以台农 1 号杧果为研究对象，于谢花后 7d 至谢花后 119d 每间隔 2 周测定果实纵径、横径和侧径值，绘制杧果果实动态发育曲线，并设置 CPPU 喷施时期（谢花后 7d 和谢花后 21d）、喷施次数（1 次和 2 次）和 CPPU 浓度（0mg/L、4mg/L、8mg/L、16mg/L、32mg/L、64mg/L、128mg/L、256mg/L）三因素的混合正交试验。结果表明：谢花后随着时间增长，台农 1 号杧果的纵径、横径和侧径度值呈典型的“S”形曲线增长。谢花后 35d 前，果实纵横径及侧径度值缓慢增长，花后 35d 至花后 77d，各指标值快速增长，谢花后 77d 后，各指标值逐渐达到高峰值而趋于稳定；初步适宜的 CPPU 喷施方法为，采用低于 64mg/L CPPU 应于谢花后 7d 和谢花后 21d 各喷施 1 次处理，采用高于 128mg/L CPPU 应于谢花后 7d 喷施 1 次。

关键词：杧果；CPPU；喷施方法

---

* 基金项目：海南省自然科学基金（314161）和公益性行业（农业）科研专项经费项目（201203092）

** 第一作者：郭利军，男，硕士，助理研究员，从事热带果树栽培及生理研究；E-mail：guolijunjunliguo@ 163. com

*** 通信作者：华敏，研究员，从事热带果树栽培研究；E-mail：huamin2528@ 139. com

生防链霉菌 *S. exfoliatus* FT05W 与 *S. cyaneus* ZEA17I 固体培养基的优化

李继业[1]，陈孝玉龙[2*]

（1. 贵州大学农学院，贵阳　550002；2. 贵州大学烟草学院，贵阳　550002）

摘　要：链霉菌（*Streptomyces*）通常能够产生多种有益次生代谢产物及酶。另外，其孢子的产生及菌丝的生长在其植物病害生物防治和促生功能中扮演者重要的角色。*S. exfoliatus* FT05W 与 *S. cyaneus* ZEA17I 是两株对核盘菌（*Sclerotinia sclerotiorum*）和寄生疫霉菌烟草致病型（*Phytophtora parasitica* var. *nicotianae* Tucker）有良好颉颃作用的生防菌株，并且对不同的园艺植物有特异性的促生效果。在之前的研究中发现，*S. exfoliatus* FT05W 和 *S. cyaneus* ZEA17I 在察氏酵母琼脂培养基（Czapek agar medium）上生长良好，且产孢能力强。高氏 1 号合成培养基（Gause′s No. 1 synthetic agar medium）是放线菌的常用培养基，在相关实验操作及工业发酵中应用较为广泛。为了优化 *S. exfoliatus* FT05W 与 *S. cyaneus* ZEA17I 在固体培养基上的生长能力及初步摸清其发酵条件，笔者测试了其在高氏 1 号合成培养基上的生长情况。结果显示，两菌株都不能在高氏 1 号合成培养基上正常生长，但往该培养基中加入 2g/L 的酵母提取物（yeast extract）后，两株生防链霉菌都能够正常生长，且长势良好。随后，笔者比较了二者在察式酵母琼脂培养基和和高氏 1 号合成培养基（酵母提取物作为辅料）上的产孢能力。研究结果表明：*S. exfoliatus* FT05W（1×10^8 CFU/mL）和 *S. cyaneus* ZEA17I（1×10^8CFU/mL）分别在以上两种培养基上的产孢能力没有明显差异，所以用高氏 1 号合成培养基（酵母提取物作为辅料）来培养以上生防链霉菌是一种可以替代察氏酵母琼脂培养基的新选择。

关键词：生防链霉菌；培养基；生长

* 通信作者：陈孝玉龙；E-mail：chenxiaoyulong@ sina. cn

六盘水市猕猴桃周年主要病害调查及病原鉴定*

潘　慧[1**]，李　黎[1]，胡秋舲[2]，张胜菊[1]，祖　达[2]，钟彩虹[1***]

（1. 植物种质创新与特色农业重点实验室，中国科学院武汉植物园，武汉　430074；

2. 六盘水市农委植保植检站，六盘水　553001）

摘　要：自 2007 年以来，贵州省六盘水市猕猴桃产业蓬勃发展。但 2014 年以来，该市猕猴桃病害问题日益严重。中国科学院武汉植物园联合六盘水市农委分别于 2015 年 10 月，2016 年 4 月及 8 月对六盘水市共 12 个乡镇 27 个代表性栽培园区进行了猕猴桃周年病害调查。该调查结果将为六盘水市后期开展有针对性的检疫、预测预报及病害综合防治提供理论依据。

基于感病症状采集大量病害样本，运用生物学特征分析及分子鉴定对分离得到的病原菌进行分析，研究结果表明水城县猴场镇、米箩镇、顺场乡，六枝特区郎岱镇和盘县普古乡共 6 个栽培园区检测到细菌性溃疡病病原菌（丁香假单胞杆菌猕猴桃致病变种，*Pseudomonas syringae* pv. *actinidiae*，简写 Psa）。各园区鉴定到真菌性果实软腐病病原菌种类较多，包括拟盘多毛孢菌（*Pestalotiopsis* sp.），拟茎点霉菌（*Phomopsis* sp.），间座壳菌（*Diaporthe* sp.，拟茎点霉菌 *Phomopsis* sp. 的有性态），链格孢菌（*Alternaria alternata*），葡萄座腔菌（*Botryosphaeria dothidea*）。同时，三次病害调查均检测到链格孢菌（*A. alternata*），该菌会引起猕猴桃的褐斑病及果实黑腐病。秋冬季和夏季的部分园区检测到拟盘多毛孢菌（*Pestalotiopsis* sp.）和胶孢炭疽菌（*Colletotrichumgloeosporioides*），证实分别会引起猕猴桃的灰斑病和炭疽病。

综上，六盘水市猕猴桃夏季和秋冬季主要病害为细菌性溃疡病和真菌性灰斑病、褐斑病、炭疽病、果实软腐病和黑腐病，春季病害相对轻微，部分园区感染了细菌性溃疡病，个别园区存在的病原菌可能会导致褐斑病，果实软腐病和黑腐病。

关键词：猕猴桃；病害；生物学特征；ITS 分子鉴定；Psa 鉴定

---

* 基金项目：中国科学院科技服务网络计划研究项目（KFJ-EW-STS－076）；农业部作物种质资源保护与利用项目（2015NWB027）；国家公派访问学者项目（Grant NO. 201504910013）；湖北省技术创新专项（重大项目）（2016ABA109）

** 第一作者：潘慧，硕士，研究助理，主要从事猕猴桃病害的分离及鉴定等研究；E-mail：panhui@ wbgcas. cn

*** 通信作者：钟彩虹，研究员，主要从事猕猴桃资源挖掘、育种、病害等相关研究；E-mail：zhongch1969@ 163. com

解淀粉芽孢杆菌（*Bacillus amyloliquefaciens*）TWC2 摇瓶发酵条件优化*

梁艳琼[1**]，黄　兴[1]，吴伟怀[1,2]，李　锐[1]，郑金龙[1]，
习金根[1]，贺春萍[1***]，易克贤[1]

（1. 中国热带农业科学院环境与植物保护研究所/农业部热带农林有害生物入侵监测与控制重点开放实验室/海南省热带农业有害生物检测监控重点实验室，海口　571101；
2. 农业部橡胶树生物学与遗传资源利用重点实验室/省部共建国家重点实验室培育基地—海南省热带作物栽培生理学重点实验室/农业部儋州热带作物科学观测实验站，儋州　571737）

摘　要：从台湾草叶片上筛选到 1 株对甘蔗赤腐病菌有较强颉颃作用的解淀粉芽孢杆菌（*Bacillus amyloliquefaciens*）TWC2，抗菌谱测定其对多种植物病原真菌具有较好的抑制效果。为了探讨 TWC2 菌株液体发酵条件，提高其活性次级代谢产物的产量，以菌体生物量和发酵液颉颃甘蔗赤腐病菌 *Colletotrichum falcatum* Went. 活性为指标，采用单因素和正交试验对 TWC2 菌株的发酵培养基营养成分和发酵条件进行了优化。结果表明：TWC2 菌株最佳发酵培养基成分为 0.5%麦芽糖，1% 果糖，1%蛋白胨，1%酵母浸出粉，0.1% $CaCl_2$。最适发酵条件为培养温度 34℃，摇床转速 200r/min，发酵起始 pH 值=6.0~6.5，接种量 7%，250mL 三角瓶装液量 40mL，发酵时间为 48h。在最佳发酵培养基和培养条件下，发酵液抑菌能力较优化前提高了 39.91%。

关键词：解淀粉芽孢杆菌；发酵条件优化；单因子试验；正交试验；*Colletotrichum falcatum*

---

* 资金项目：国家天然橡胶产业技术体系病虫害防控专家岗位项目（No. CARS-34-GW8）；中央级公益性科研院所基本科研业务费专项资助项目（NO. 2014hzs1J013；NO. 2015hzs1J014）；中国热带农业科学院橡胶研究所省部重点实验室/科学观测实验站开放课题（RRI-KLOF201506）

** 第一作者：梁艳琼，女，助理研究员，研究方向：植物病理学

*** 通信作者：贺春萍；E-mail：hechunppp@163.com，易克贤；E-mail：yikexian@126.com

西瓜蔓枯病菌群体的遗传多样性与交叉抗药性

李昊熙

（贵州大学烟草学院，贵阳 550025）

摘　要：植物病害的病原、宿主和环境 3 个决定性因素都应该是复杂多变的动态体系，而每一种植物病害的病原物更是在多个层面呈现着遗传多样性，影响着病害随时间与空间的发生和发展。本研究用 16 个微卫星标记鉴定了蔓枯病 9 个种群、528 个病原菌个体的基因型，并用几种数理模型测量了病原菌种、区域群体结构与微观群体结构，并选取其中已知基因型的 132 个病原菌个体检测了对 3 种作用机制的 4 种杀菌剂的抗药性与交叉抗药性，以及抗药性与基因型的对应关系。对病原菌遗传多样性的深入理解和精确监控，有助于掌握蔓枯病的流行规律，更是有针对性地设计综合防治措施的重要依据。

关键词：西瓜蔓枯病；遗传多样性；交叉抗药性

枣疯病植原体 LAMP 检测技术及其在抗病种质资源筛选中的应用*

王圣洁[1**]，王胜坤[2]，张文鑫[1]，林彩丽[1]，于少帅[1]，
汪来发[1]，朴春根[1]，郭民伟[1]，田国忠[1]
（1. 中国林业科学研究院森林生态环境与保护研究所国家林业局森林保护学重点实验室，
北京　100091；2. 中国林业科学研究院热带林业研究所，广州　510520）

摘　要：枣疯病是枣树上一种毁灭性的病害，由于病原传播快，无有效的防治措施，造成了严重的经济损失。高效、精确的植原体分子诊断与检测技术，可以帮助预测病害发生情况、筛选优良的抗性种质资源、为枣树产业的健康发展提供保障，本研究旨在建立一种能够特异性检测和鉴定 16SrV-B 亚组植原体的环介导恒温扩增技术，实现 16SrV-B 组植原体的快速检测。并借助此快速检测技术，对以不同品系，不同传病方式以及不同表现症状的种质资源进行的抗病性测定，以期获得具有稳定的抗枣疯病植原体的枣树种质品系。以植原体 *tuf* 基因作为靶标，通过序列比对，设计了具有特异性的 LAMP 引物组。建立适用于 16SrV-B 亚组植原体的特异性 LAMP 检测体系。以 16SrII 组、16SrV 组、16SrXIX 组植原体样品按照此检测体系进行反应，来验证其特异性；同时将稀释后的感病组织样品同步进行 PCR 和 LAMP 检测，以确定其检测灵敏度。以河北唐县军城、北京玉泉山苗圃和昌平流村 3 个枣树抗性资源保存与测定圃中共约 80 余个品系为研究对象，对通过嫁接传病或自然感染的显症株（枝）、疑似株、感染康复株、表观无症株及未感染枣树进行采样，采用 PCR、巢式 PCR 和 LAMP 技术对表现丛枝、黄叶、卷叶、花叶、花变叶等症状及未显症株的共 160 份枝叶样品进行了植原体带菌情况检测。比较不同品系在病原菌入侵后所产生的症状表现，从中筛选出对枣疯病植原体具有稳定的抗性的种质资源。同时对在资源圃中通过嫁接传病第 4 年恢复开花和结果能力，但表现为畸形萎缩的病果（青果）的枣果组织进行诊断，确定植原体在果肉细胞中是否存在，并以此病果的种仁进行组培，确定枣疯病是否存在通过种子进行病原传播的可能。16SrVb-LAMP 在恒温 63℃条件下在 30 min 内完成扩增，通过在反应液中加入钙黄绿素肉眼判断为绿色为阳性结果，褐色为阴性结果，并且与用荧光定量设备进行判读扩增曲线的结果一致。在特异性方面，16SrVb–LAMP 检测体系能检测出 3 种分别引起枣疯病、槐树丛枝、樱桃致死黄化病的 16SrV-B 亚组植原体，显示为阳性结果，健康对照和其他组包括 16SrI 组、16SrII 组和 16SrXIX 组植原体，甚至于 16SrV-H 亚组重阳木丛枝也不能产生阳性结果，具有较强的特异性。相比于 PCR 检测，16SrVb-LAMP 检测的灵敏度提高了约 100 倍；对于抗病资源圃的田间样品的检测结果显示，普通 PCR、巢式 PCR 和 16SrVb-LAMP 检测技术的植原体总体检出率分别为 38. 13%、61. 25%和 45%；其中 3 种方法皆检测到植原体的样品 61 份（包含丛枝小叶 47 份、黄叶 7 份、卷叶 11 份、花叶 1 份、无症 4 份），3 种方法皆未检测到植原体的样品

* 基金项目：国家自然科学基金项目（31370644）

** 第一作者：王圣洁，男，博士研究生，主要方向：分子植物病理；E-mail：wsjguoyang@ 126. com

62 份（包含无症 34 份、花叶 8 份、卷叶 17 份、黄叶 9 份）。检测结果与症状统计分析表明丛枝小叶症状与病原定殖、枣树感病性非常相关，单纯黄叶可能是寄主的抗病反应，而花叶则非植原体侵染所致；验证了不同品种枣树抗病性与其体内固有或诱导抗性对病菌生长和繁殖的抑制作用有关；揭示了在河北唐县军城开展的用感染植原体的病枝与病皮接种抗病品系 JL24 和壶瓶枣诱发了枣树长效的抗病免疫反应，并进一步肯定了在昌平抗性测定圃中从北京市古枣树单株筛选的抗病品系的持续抗病稳定性。对于病果（青果）进行的病原检测结果显示，在其果肉组织中确定了植原体的存在，但是以这些病枣树产果的种子（胚）进行的组织培养苗中均未检测出植原体。本研究首次以 *tuf* 基因作为靶标，建立了能同时检测 3 种 16SrV-B 亚组植原体的环介导恒温扩增技术，具有高效、特异、操作简便、检测时间短及成本较低等优点。借助以此检测技术，对抗病种质资源进行筛选，确定了不同抗性品系在植原体入侵后的抗性反应，同时确定了畸形萎缩病果的果肉中植原体的存在，但不能确定枣疯病能通过种子进行传病。

关键词：植原体；环介导恒温扩增；*tuf* 基因；快速检测；抗病性鉴定；种质资源

拟康氏木霉多肽抗菌素基因簇 TPX2 启动子片段的克隆*

盛宴生**，余 薇，任爱芝，赵培宝***
（聊城大学农学院植物保护系，聊城 252059）

摘 要：拟康式木霉菌（*Trichoderma* sp.）能够通过非核糖体途径产生肽类抗菌素 TK6，在该菌生防过程中发挥重要作用。为了进一步开发应用多肽抗菌素笔者计划探究其转录调控机制，以通过转录调控提高其产量，为此笔者拟通过克隆其基因簇 TPX2 的启动子，了解其生物活性，为进一步应用提供帮助。本实验采用 CTAB/NaCl 方法提取拟康氏木霉（*Trichoderma pesudokoninii*）的 DNA，采用巢氏 PCR（*polymerase chain reaction*）方法对拟康氏木霉基因组 DNA 扩增，以得到拟康氏木霉多肽抗菌素基因簇 TPX2 的启动子编码序列。经过两轮 PCR，扩增到了预期长度的条带，回收测序表明笔者得到了基因簇 TPX2 的启动子序列，目前正在通过融合 GFP 来探讨其转录活性。

关键词：拟康氏木霉菌；非核糖体肽；启动子

---

* 基金项目：山东省自然科学基金 ZR2013CM006；山东省科技发展计划 2014GNC110020

** 第一作者：盛宴生，男，在读硕士研究生，主要从事植物病害生物防治相关研究

*** 通信作者：赵培宝，副教授；E-mail：zhaopeibao@ 163. com

烟草赤星病菌 ISSR-PCR 最佳反应体系的建立与优化*

李六英[1**]，窦彦霞[1]，高　敏[1]，马冠华[1***]，陈国康[1]，张　勇[2]

（1. 西南大学植物保护学院，重庆　400716；2. 中国烟草总公司
重庆市公司酉阳分公司，重庆　409600）

摘　要：采用正交试验设计，对影响重庆烟草赤星病菌 ISSR-PCR 反应体系的各主要因子（Mg^{2+}，dNTP，*Taq*DNA 聚合酶和引物）进行了优化试验。在此基础上，通过梯度 PCR 试验确定最适退火温度和循环次数。结果表明：25μL 总反应体系中：含 1. 00mmol/L Mg^{2+}，0. 15 mmol/L dNTP，0. 30μmol/L 引物，0. 75U *Taq* DNA 聚合酶，1×PCR Buffer 和 30. 00ng DAN 模板。扩增程序为：94℃预变性 4min，94℃变性 1min，50℃退火 1min，72℃延伸 90s，共 36 个循环，最后 72℃延伸 10min。并利用 24 份重庆烟草赤星病菌基因组 DNA 和引物 UBC841 对最佳反应体系进行 PCR 扩增，证实了该反应体系稳定可靠。本研究为重庆烟草赤星病菌的遗传多样性分析等研究提供了一种稳定的技术。

关键词：烟草赤星病菌；ISSR-PCR；正交设计；体系优化

---

* 基金项目：重庆烟草有害生物综合治理专项（NY20140401070002）

** 第一作者：李六英，女，硕士研究生，主要从事植物病理学研究；E-mail：2356562502@ qq. com

*** 通信作者：马冠华，副教授，主要从事植物病原学研究；E-mail：nikemgh@ swu. edu. cn

棉花黄萎病菌 LAMP 检测方法的建立与应用*

廖富荣[1**]，阿热艾·拉孜木别克[1,2]，方志鹏[1]，陈　青[1]，
陈红运[1]，黄蓬英[1]，林振基[1]，林石明[1]
（1. 厦门出入境检验检疫局检验检疫技术中心，厦门　361026；
2. 福建农林大学植物保护学院，福州　350002）

摘　要：棉花黄萎病菌（*Verticillium dahliae*）是一种我国禁止进境的检疫性有害生物，具有十分广泛的寄主范围。根据特异基因设计引物，经条件优化后，建立了棉花黄萎病菌的环介导等温扩增（Loop-mediated isothermal amplification，LAMP）检测方法。所建立的方法具有良好的灵敏度，比普通 PCR 方法高 10 倍。利用该方法对菠菜种子进行检测，10 批种子的 20 个检测样品中有 17 个检测样品显示阳性，该结果与普通 PCR 方法相一致。因此，所建立的 LAMP 检测方法，具有简单、灵敏、快速等特点，适合棉花黄萎病菌的快速检测。

关键词：棉花黄萎病菌；环介导等温扩增；菠菜种子

---

* 项目资助：福建省自然科学基金项目（2015J01148）；国家质量监督检验检疫总局科技计划项目（2015IK190）；厦门市科技计划项目（3502Z20154080）

** 第一作者：廖富荣，高级农艺师，主要从事植物病原鉴定及检测方法研究；E-mail：lfr005@163.com

营养物质对枯草芽孢杆菌 NCD-2 菌体生长和芽孢形成的调控*

付一帆**，郭庆港，鹿秀云，董丽红，赵卫松，张晓云，李社增，马　平***
（河北省农林科学院植物保护研究所，河北省农业有害生物综合防治工程技术研究中心，农业部华北北部作物有害生物综合治理重点实验室，保定　071000）

摘　要：枯草芽孢杆菌（*Bacillus subtilis*）是一种应用广泛的生防细菌，对多种植物病害表现良好的生防效果。枯草芽孢杆菌由于能形成抗逆、耐热的芽孢而有利于产品货架期的延长和产品开发，因此，成为开发微生物源农药的重要资源。由我实验室筛选得到的枯草芽孢杆菌NCD-2 菌株对多种植物土传病害表现较好的防治效果，以该菌株为主要活性成分的微生物农药已获得正式农药登记。目前已建立了 NCD-2 菌株的规模化发酵工艺，但其发酵水平还有提高的空间。为了能进一步提高发酵水平，筛选能提高芽孢总量、芽孢形成率，缩短发酵时间的碳源、氮源等营养物质变得势在必行。利用微生物表型芯片技术（Phenotype Microassays）发现了 5 种对 NCD-2 菌株生长有明显影响的营养物质，碳源物质有：木糖、阿拉伯糖、核糖；氮源物质有：谷氨酰胺、半胱氨酸。以 M9 为基础培养基，通过单因素摇瓶发酵实验发现：含有阿拉伯糖或者谷氨酰胺的培养基，芽孢形成的时间早，以阿拉伯糖为碳源的改良培养基在 24h 时就开始形成芽孢，此时以葡萄糖为碳源的 M9 基础培养基中没有发现芽孢形成。更值得注意的是，以谷氨酰胺为氮源的改良培养基，在 36h 时芽孢形成率就达到了 78.8%，菌体总量达到 6.1×10^7 CFU/mL，而此时以氯化铵为氮源的 M9 基础培养基中菌体总量达到 6.8×10^7 CFU/mL，但是还没有芽孢的形成。这一结果可以为实际生产过程中的菌体发酵和产品加工提供思路。

关键词：枯草芽孢杆菌；芽孢；营养物质

---

* 基金项目：国家自然科学基金项目（31572051）
** 第一作者：付一帆，硕士研究生；E-mail：15230233219@163.com
*** 通信作者：马平，博士，研究员，主要从事植物病害生物防治研究；E-mail：pingma88@126.com

基于拟南芥-微生物共培养系统筛选植物促生菌的研究*

王梅菊，刘　晨，杨　龙，张　静，吴明德，李国庆**
（华中农业大学植物科技学院，武汉　430070）

摘　要：植物促生微生物是指能够促进植物生长，增加田间产量，并能够降低植物周围生物和非生物压力的一类有益微生物。关于植物促生菌株的筛选，传统的方法是利用微生物浸种以及直接处理植株的方法进行筛选。这种方法的优点是筛选环境与植物正常的环境接近，缺点是工作量大、效率低。本研究旨在建立一种新的植物促生菌的筛选模式，筛选植物促生微生物菌株。其原理是利用模式植物拟南芥-微生物共培养系统，在室内快速筛选促进拟南芥生长的微生物菌株。

采用稀释平板分离法从不同植物表面得到 83 个菌株，其中细菌 29 株，酵母 54 株。经过拟南芥-微生物共培养筛选，69 个菌株处理的拟南芥幼苗与对照处理拟南芥（没有接种微生物）幼苗生长状况没有明显差别，10 个菌株对拟南芥幼苗生长有不同程度的抑制作用（幼苗死亡或长势差），4 个菌株（CanL-30、jh1、mgh1 和 qcg）对拟南芥幼苗生长有促进作用。

采用油菜种子萌发实验对 CanL-30、jh1、mgh1 和 qcg 等 4 个菌株进行复筛。分别用 10^7 CFU/mL 的菌悬液对油菜种子进行 1h 浸种处理，以水浸种处理为对照。将不同处理种子分别置于垫有潮湿滤纸的培养皿中，放在 20℃、16h 光照/8h 黑暗光照培养箱中培养，一周之后测量油菜种子萌发率以及油菜苗根长和芽长。其中对照组种子萌发率为 73.3%。菌株 CanL-30、jh1、mgh1 和 qcg 处理组种子萌发率分别为 78.9%、72.2%、88.9%和 75.6%。菌株 CanL-30、mgh1 和 qcg 处理的油菜苗根明显伸长，而菌株 jh1 却抑制了油菜苗根伸长。各处理油菜苗芽长与对照比没有显著差异（$P>0.05$）。

对菌株 CanL-30、mgh1 和 qcg 进行了分子鉴定。通过对 16S rDNA 序列分析显示，菌株 CanL-30 与 *Bacillus subtilis* 的同源性为 99%，菌株 qcg 与 *Serratia liquefaciens* 的同源性为 99%。通过 ITS 序列分析结果显示，菌株 mgh1 与 *Cryptococcus magnus* 的同源性为 100%。

分别对 CanL-30、mgh1 和 qcg 进行促生机制探究，发现菌株 CanL-30 和 qcg 能产生挥发性气体促进植物生长。GC-MS 分析 CanL-30 产生的挥发性物质主要为酮类物质，包括 2-Undecanone、2-Dodecanone 和 2-Tetradecanone。qcg 产生的挥发性气体主要是烷类物质和醇类物质。此外，菌株 mgh1 和 qcg 能够产生 IAA，其产量分别为 17.53μg/mL 和 65.22μg/mL。同时还发现菌株 qcg 具有溶磷能力。

上述结果说明：采用拟南芥-微生物共培养系统筛选植物促生菌是可行的。

---

* 基金项目：公益性行业（农业）科研专项（项目编号：201303025）

** 通信作者：李国庆，教授；E-mail：guoqingli@mail.hzau.edu.cn

金黄杆菌 PYR2 对小麦和棉花种子萌发及幼苗生长的促进作用

刘 缨，王梦雨*，罗来鑫，李健强**

（中国农业大学植物病理学系/种子病害检验与防控北京市重点实验室，北京 100193）

摘 要：金黄杆菌属（*Chryseobacterium*）是一类广泛存在于环境中的革兰氏阴性细菌。已有研究表明，该属菌株 PYR2 可以有效降解 DDT，能够产生植物生长素吲哚-3-乙酸（IAA）。据此，本试验开展了 PYR2 菌株对小麦和棉花促生效果的研究，采取发芽纸和盆栽育苗的方法，测定其菌悬液和发酵上清液对供试作物种子发芽及幼苗生长的影响。发芽纸试验中，使用 1×10^9 CFU/mL、2×10^8 CFU/mL、2×10^7 CFU/mL 三种浓度的菌悬液和相应的发酵上清液浸种处理小麦（品种：新乡 58）和棉花（品种：KV3）种子，以清水处理作为对照。结果显示，菌悬液和发酵上清液处理的种子发芽势、发芽率、畸变率、苗高、根长、侧根数和鲜重指标在不同程度上均优于对照组。其中，三种不同浓度菌悬液处理后的小麦和棉花种子，发芽后小麦苗高和棉花根长分别比对照组增加 12.3%、11.1%、11.1%和 48%、26%、26%；不同浓度发酵上清液处理后的小麦和棉花种子，发芽后小麦苗高和棉花根长分别比对照组提高 18.5%、11.1%、9.9%和 20.0%、32.0%、56.0%。盆栽试验中，使用三种浓度的菌悬液和相应的发酵上清液处理小麦和棉花幼苗，结果显示，菌悬液和发酵上清液灌根处理后，小麦、棉花幼苗的株高、根长、鲜重和干重等指标均在不同程度上优于对照组。本研究首次验证了 *Chryseobacterium* sp. PYR2 菌株的菌悬液和发酵上清液对小麦和棉花种子萌发及幼苗生长具有一定的促生作用。该结果对后续开发 DDT 降解及植物促生双效微生物菌剂有重要的参考价值。

关键词：*Chryseobacterium* sp. PYR2；小麦；棉花；促生作用

* 第一作者：王梦雨，硕士研究生，主要从事种传病害研究；E-mail：wangmengyu@cau.edu.cn

** 通信作者：李健强，教授，主要从事种子病理学和杀菌剂药理学研究；E-mail：lijq231@cau.edu.cn

不同激素浓度和配比对番茄遗传转化的影响*

张银山，杜 鹃，逯麒森，王 珂，李 宇**，李洪连
（河南农业大学植物保护学院，郑州 450002）

摘 要：番茄（*Lycopersicon esculentum* Mill）是系茄科番茄属一年生或多年生草本双子叶植物，同时也是植物遗传转化研究过程中的模式植物，基因组较小，遗传学基础雄厚。其具有富含多种营养成分，培育时间较短，基因组信息清楚，遗传转化较容易等优点。植物遗传转化技术，是应用DNA重组技术有目的地将外源基因或DNA片段通过生物、物理或化学等手段导入受体植物基因组中，以获得外源基因在受体植物中稳定遗传和表达的一门技术。番茄作为模式植物，对其遗传性状、外源基因的合成和载体的构建、基因的整合和表达和植株再生以及优良株系的选育等，已经形成一套较为成熟的技术流程，在植物介导的RNAi、植物寄生线虫功能基因的验证以及抗病虫害等方面取得了巨大的成果，具有广泛的应用前景。

本研究以一种常见的番茄品种（粉都80）为材料，通过对子叶的离体培养，测定了不同浓度激素对番茄外植体再生的影响，为下一步农杆菌介导的植物遗传转化奠定基础。研究结果表明：16个不同浓度下玉米素（ZT）和吲哚乙酸（IAA）组合培养对诱导番茄外植体形成愈伤组织和芽的分化具有明显的差异，其中以MS+ZT（1.0mg/L）+IAA（0.1mg/L）的诱导效果最佳，该处理组外植体的出愈率为100%，出芽率为95.56%；所有添加激素处理组的出愈率和出芽率均大于不添加任何激素的对照处理组，这说明添加激素能明显提高番茄外植体的诱愈率和诱芽率。11个不同浓度的IAA和NAA对番茄芽梢生根的结果表明：添加微量的生长素IAA或NAA能极大的提高番茄外植体的生根率，但不同处理之间差异显著，其中添加0.15mg/L IAA的生根率最高，20d后的生根率达到100%，平均出根数为45.73，绝大多数可以发育成正常植株；而未添加生长素的对照组，少数芽梢能生根，但生根缓慢，很难长出完整的根系；多数芽梢基部膨大并无根系生成，并逐渐黄化萎蔫死亡。因此，在1/2 MS培养基中添加0.15mg/L的IAA能获得根系发达、生长健壮的再生番茄植株。

关键词：激素；浓度；遗传转化

---

* 基金项目：国家自然科学基金（31601619）

** 通信作者：李宇，博士，主要从事植物线虫学研究；E-mail：limiao04@163.com

水杨酸诱导水稻叶片的磷酸化蛋白质组学分析*

孙冉冉[1]，聂燕芳[2]，张 健[1]，王振中[1]，李云锋[1]**

（1. 华南农业大学农学院；2. 华南农业大学材料与能源学院，广州 510642）

摘 要：水杨酸（Salicylic acid，SA）是一种重要的信号分子，能够诱导水稻的抗病性。蛋白质磷酸化是一种重要的蛋白质翻译后修饰，在植物抗性信号转导途径中起着重要的作用。开展SA诱导的水稻磷酸化蛋白质组变化，对于全面了解SA诱导水稻的抗病性机制具有重要的意义。

以抗稻瘟病近等基因系水稻CO39（不含已知抗稻瘟病基因）及C101LAC（含*Pi*-1抗稻瘟病基因）为材料，用SA喷雾接种水稻，于接种后的12 h和24 h取样。经叶片总蛋白质的提取、磷酸化蛋白质的富集、双向电泳（2-DE）和凝胶染色，获得了不同时间段的磷酸化蛋白质Pro-Q Diamond特异性染色2-DE图谱和硝酸银染色2-DE图谱。用PDQuest 8.0软件进行图像分析，共获得了47个差异表达的磷酸化蛋白质。

采用MALDI-TOF/TOF质谱技术对差异表达的磷酸化蛋白质进行了分析，成功鉴定了其中的40个磷酸化蛋白质点；主要参与光合作用、防卫反应、抗氧化作用、蛋白质合成与降解、氨基酸代谢和能量代谢等功能。

关键词：水杨酸；水稻；磷酸化蛋白质组；双向电泳

---

* 基金项目：国家自然科学基金（31671968）；广东省自然科学基金（2015A030313406）；广东省科技计划项目（2016A020210099）

** 通信作者：李云锋；E-mail：yunfengli@ scau. edu. cn

适合于 2-DE 分析的水稻质膜磷酸化蛋白质富集方法的建立*

聂燕芳[1]，邹小桃[2]，王振中[2]，李云锋[2**]

（1. 华南农业大学材料与能源学院，广州 510642；2. 华南农业大学农学院，广州 510642）

摘 要：细胞质膜（plasma membrane，PM）作为植物细胞与外界的屏障，在细胞壁的合成、离子转运和细胞信号转导过程等方面起着重要的作用。质膜蛋白质的磷酸化也被证实广泛参与了病原菌的信号识别等反应。开展植物细胞质膜蛋白质磷酸化变化的研究，对于了解植物和病原菌的相互识别和信号跨膜转导等具有重要意义。

采用双水相分配法和离心法对水稻叶片的细胞质膜进行了纯化，结果表明由 6.3/6.3%（*w/w*）Dextran T 500/PEG 3350 组成的双水相体系可以获得高纯度的质膜微囊，其质膜标志酶 VO_{43-}-ATPase 的相对活性达 93%以上。采用 Al（OH）$_3$-MOAC 法，通过质膜蛋白质与 Al（OH）$_3$ 的孵育、非磷酸化蛋白质的清洗、磷酸化蛋白质的洗脱等步骤，对水稻质膜磷酸化蛋白质进行了富集；结果表明 9mg 的质膜蛋白质经过富集可以得到约 300～350μg 质膜磷酸化蛋白质，即质膜磷酸化蛋白质的富集得率为 3.33%～3.89%。

为了检验 Al（OH）$_3$-MOAC 法对水稻叶片质膜磷酸化蛋白质富集的特异性，进一步对质膜磷酸化蛋白质进行了 2-DE 分析。结果表明：同一凝胶的 Pro-Q Diamond 染色和硝酸银染色的 2-DE 图谱背景清晰，蛋白质点分布均匀，没有明显的纵向拖尾和水平横纹现象；其中 Pro-Q Diamond 染色图谱中的蛋白质点数为 296 个，硝酸银染色图谱中的蛋白质点数为 316 个；表明该技术体系可以获得高分辨率和高重复性的 2-DE 图谱。采用 PDQuest 8.0 图像分析软件对 2 种染色图谱进行叠加分析，结果表明两种图谱中蛋白质点的吻合度高。

上述结果表明：应用双水相分配法和 Al（OH）$_3$-MOAC 法，并结合 2-DE 技术和 Pro-Q Diamond 磷酸化蛋白质特异性染色技术，建立了高效的水稻叶片质膜磷酸化蛋白质的富集和 2-DE 分析技术体系。

关键词：水稻；质膜磷酸化蛋白质；富集；双向电泳

---

* 基金项目：国家自然科学基金（31671968）；广东省自然科学基金（2015A030313406）；广东省科技计划项目（2016A020210099）

** 通信作者：李云锋；E-mail：yunfengli@ scau. edu. cn

本生烟瞬时表达系统快速检测 RNAi 载体沉默效果

武亚丹，张秀春，冼淑丽，余乃通，王健华，刘志昕
（中国热带农业科学院热带生物技术研究所，农业部热带作物
生物学与遗传资源利用重点实验室，海口　571101）

摘　要：RNA 干扰（RNAi）技术是研究基因功能的一种常用方法。为快速检测 RNAi 载体的沉默效果，通过构建木薯 *eIF4E3* 基因的过量表达载体 pAI-Ca4E3 以及靶标木薯 *eIF4E3* 基因的 RNAi 表达载体 p1300-Ca4E3，利用农杆菌注射法将表达载体在本生烟叶片中进行瞬时表达。半定量 RT-PCR 检测结果显示单独注射过量表达载体 pAI-Ca4E3 或与表达载体 pCAMBIA1300 共同注射的样品，农杆菌注射后第 4 天、5 天和 6 天都能检测到木薯 *eIF4E3* 基因的高效表达，并且二者木薯 *eIF4E3* 基因表达量基本相同，说明木薯 *eIF4E3* 基因能在本生烟中高效瞬时表达，但是过量表达载体 pAI-Ca4E3 与干扰载体 p1300-Ca4E3 共同注射的样品，农杆菌注射后第 4 天、第 5 天还是第 6 天后都几乎检测不到木薯 *eIF4E3* 基因的表达，说明过量表达载体表达的木薯 *eIF4E3* 基因已被干扰载体有效沉默。研究结果表明：已建立同时表达靶标基因和干扰靶标基因的干扰载体的本生烟瞬时表达系统，并结合半定量 RT-PCR 进行 RNAi 载体沉默效果检测的方法。该方法在农杆菌注射 4 天后就能有效的检测 RNAi 载体的沉默效果，具有快捷、操作简便等特点，为 RNAi 技术应用于基因功能研究提供简单、快捷、有效的证据。

关键词：本生烟；瞬时表达；干扰载体；半定量 RT-PCR；沉默效果检测